# Springer Series in Materials Science

Volume 199



For further volumes:
http://www.springer.com/series/856

The Springer Series in Materials Science covers the complete spectrum of materials physics, including fundamental principles, physical properties, materials theory and design. Recognizing the increasing importance of materials science in future device technologies, the book titles in this series reflect the state-of-the-art in understanding and controlling the structure and properties of all important classes of materials.

George S. Nolas
Editor

# The Physics and Chemistry of Inorganic Clathrates

Springer

*Editor*
George S. Nolas
Department of Physics
University of South Florida
Tampa, FL
USA

ISSN 0933-033X ISSN 2196-2812 (electronic)
ISBN 978-94-017-9126-7 ISBN 978-94-017-9127-4 (eBook)
DOI 10.1007/978-94-017-9127-4
Springer Dordrecht Heidelberg New York London

Library of Congress Control Number: 2014942455



Printed on acid-free paper

Springer is part of Springer Science+Business Media (www.springer.com)

# Foreword by C. Cros

Clathrates are atypical crystalline compounds whose structures are built up from a three-dimensional framework of cage-like polyhedrons of one chemical species, the host lattice that encloses atoms or molecules of a second species, the guest lattice. Their name comes from the Latin word *clatratus* meaning enclosed or protected by crossed bars of a grating. Clathrates are also called host-guest complexes and are sometimes indexed as inclusion compounds. Several families of clathrates are known, the host lattices of which are of organic or inorganic nature. The most important of them by the number of its members and the variety of encountered structures are obviously the so-called clathrate hydrates, in which face-sharing cages of solid water enclose gaseous or liquid- or both-guest species. Whereas the first of these hydrates have been known for two centuries, a new series of isostructural compounds has been more recently evidenced, in which elements of group 14, Si, Ge and Sn form the host lattice and other ones such as alkali-metals, alkaline-earth or less-common metals the guest species. The main reason for the at first sight surprising similarities between hydrates and group 14 clathrates comes from the fact that the atmospheric pressure structure of solid $H_2O$, as well as those of C, Si, Ge and α-Sn, are tetrahedrally bonded ones, and this surrounding is in fact preserved in clathrate structures. This new series of clathrates are now called inorganic clathrates and are the subject matter of the present book.

A very large majority of the characteristic polyhedral cages of clathrate structures exhibit 12 pentagonal faces combined (or not) with a lower number of hexagonal ones, according to Euler's rule about polyhedrons. The smallest and most frequent of them is the pentagonal dodecahedron, with only 12 pentagonal faces, 20 vertices and 30 edges, more simply referred to as $5^{12}$. The next ones by increasing size include the addition of 2 ($5^{12}6^{2}$), 3 ($5^{12}6^{3}$), 4 ($5^{12}6^{4}$) and 8 ($5^{12}6^{8}$) hexagonal faces. Most of the 10 indexed clathrate-type structures are composed of a combination of two or three of those cages, filling in the space and forming large and highly symmetrical unit-cells, involving for example 46, 136, 172 or even more atoms of the host lattice. Whatever the arrangement of polyhedral cages, the host species are tetrahedrally bonded with almost equivalent inter-atomic distances and bond angles close to 108° (pentagonal angle) and 120° (hexagonal angle). Clathrate host lattices can be thus considered as open forms of tetrahedrally bonded structures, and it is not so surprising that clathrate analogues called clathrasils also exist with silica as the host lattice, another well-known tetrahedrally bonded structure.

Unexpected, however, were the relations of most of the clathrate structures with those of classical metal alloys such as the A15 ($Cr_3Si$) or C15 ($MgCu_2$) ones, in which the metallic atoms are located in the same sites as the guest alkali-metals and where many tetrahedral voids exist corresponding to the positions occupied by the host species in the corresponding clathrate. Thus, such alloys can be considered as templates for clathrate structures.

Another very interesting feature of clathrates is that their characteristic cages made of edge-sharing pentagonal and hexagonal faces are also related to the fullerene forms of carbon, the first element of group 14, the structures of which are based on the same geometrical rules. The $C_{60}$ buckyball is built up from 12 pentagonal faces sharing edges with 20 hexagonal ones: $5^{12}6^{20}$. Clathrate cages correspond in fact to the smallest possible fullerene clusters and form 3-D frameworks in order to neutralize the strains due to dangling bonds, whereas all the valence electrons are engaged in single or double bonds in a $C_{60}$ buckyball, enabling it to form isolated clusters. According to recent calculations, clathrate forms of carbon could exist, but have not, as yet, been found.

Due to their special cage-like architecture and given the relations they have with hydrates, clathrasils, inter-metallic alloys and fullerenes, the structures of inorganic clathrates are really fascinating.

The two first inorganic clathrates, $Na_8Si_{46}$ and $Na_xSi_{136}$ ($3 \leq x \leq 11$), were discovered in 1965 and their structures identified by comparison with those of the gas and liquid hydrates. Then, clathrates with other alkali-metal guests and group 14 host lattices were studied, and the first physical properties investigated. At the same time, the ability of clathrate host lattices to be slightly non-stoichiometric or modified by substitution of one part of the group 14 elements by neighbouring ones (group 13 or 15) was demonstrated, involving either positively or negatively charged host lattices compensated by ionised guest species and giving rise to a large variety of semi-conducting clathrates. The results of these studies were quite interesting, but on the whole, relatively little attention was paid to the clathrate compounds by the scientific community, perhaps because there was no obvious potential application at that time.

In the mid-80s of the last century, the discovery of the fullerene forms of carbon and the high $T_C$ superconducting properties of Ba-La cuprates focused the attention of a great number of scientists on these two new and exciting research fields. A few years later, superconductivity was found in alkali-metal fullerides $A_3C_{60}$ (A = K, Rb, Cs) with $T_C$ values ranging from 19 to 32 K. In the search for other kinds of structures able to exhibit similar properties, and given the obvious similarities with fullerides, cage-like inorganic clathrates began to attract increasing attention.

After 30 years of intensive research, a great amount of work has now been done. Interesting results have been obtained and promising applications as thermoelectric materials are in progress, making these inorganic clathrates one of the major recent research fields. Given the present stage of development, the time has come to publish a thorough and up to date book on inorganic clathrates, including the contribution of the most famous experts on the subject. To the best of my knowledge, this has not been done so far.

Professor George S. Nolas has now undertaken the management of this task as Editor, and has entrusted Springer with its publication. No doubt this book will be successful and useful to the scientific community and hopefully will be available in the libraries of universities and research institutes and laboratories all over the world.

Bordeaux, France, July 2013 Christian Cros

# Foreword by G. A. Slack

The use of clathrate semiconductors as thermoelectric materials has developed over the last 40 years. The starting point was a study of high thermal conductivity nonmetallic crystals [1] which showed that adamantine or diamond-like, tetrahedrally bonded crystals had the highest phonon-dominated thermal conductivity near room temperature of any known solids. The prime example being diamond itself, which has a Debye temperature for the phonons of 2240 K. In 1980 it was shown [2] that the common form of water-ice ($H_2O$) is also a typical adamantine compounds with a wurtzite-type (ZnS) structure. It has analogues in bromellite (BeO), zincite (ZnO), greenockite (CdS), and iodyrite (AgI), among others. The oxygen in the $H_2O$ molecule replaces both the metallic and nonmetallic atoms in the wurtzite structure. In comparison to diamond the Debye temperature of ice is 226 K, a factor of 10 less than diamond. Its room temperature thermal conductivity is 1,000 times smaller.

The connecting feature of dihydrogen oxide and C, Si, Ge, Sn, etc., is that they all possess four non-bonded, outer-shell electrons to use for covalent bonding. Hence they can all form various types of clathrate structures. The molecular diameter of $H_2O$, 2.76 Å, places it between Ge and Sn in size. The curious feature is that these "high thermal conductivity" crystals with tetrahedral bonding can form open-structure crystals with very low thermal conductivity. Once it was understood [3] that for each particular crystal structure there is a minimum possible phonon thermal conductivity where each phonon only travels a mean-free-path equal to one wavelength, then it became possible to calculate this minimum value for any crystal. These minimum values showed [4] that many known thermoelectric materials such as PbTe or InSb could be greatly improved by further reduction in the phonon mean free paths. The clathrate structures of Si, Ge and ice have reduced thermal conductivities due to the increase in the number of atoms per primitive unit cell as the diamond lattice with two atoms per cell expands into the multi-atom clathrate. The second possibility with clathrates is that guest atoms or molecules in the cages can be tailored to preferentially interact with the energy-transporting acoustic phonons and substantially reduce their mean-free paths. This reduction shows up nicely [4] in the ice clathrates of methane ($CH_4$), tetrahydrofuran ($C_4H_8O$), and xenon (Xe). These solids have thermal conductivity values even lower than that of amorphous ice, i.e. water. The calculated phonon spectrum of the xenon-ice clathrate shows [5] that the Xe vibrations, whose density of states peaks at 23, 35, and 46 K, interact strongly with the propagating acoustic phonons in ice.

The fact that one adamantine crystal could be produced in which the thermal conductivity was reduced to values near the theoretical minimum by an uncharged, neutral atom like xenon suggested that the phonon heat transport could be lowered in other semiconducting crystals without changing the electron concentration or their mobilities. This is the phonon glass-electron crystal, or PGEC, concept [4].

At this point I remembered that Christian Cros [6, 7] and a group at the University of Bordeaux had made Si and Ge clathrate compounds, and that my friend and coworker John Kasper [6] had noticed their structure was the same as some of the rare gas hydrates. Maybe, these could be turned into useful thermoelectrics? Some of them have been.

The $Sr_8Ga_{16}Ge_{30}$ crystals grown by Nolas et al. [8] come fairly close to being a good PGEC thermoelectric material. It is important to use single crystals in studying both the electron and phonon transport at temperatures between 4 and 300 K. In polycrystalline samples the grain boundaries scatter both kinds of carriers and can lead to many erroneous conclusions about both the fundamental properties and ultimate utility of the material.

The field of open-structure solids, sometimes called “holey crystals”, is expanding rapidly at present, and promises to yield a large number of practical applications. The present book will greatly aid researchers in their endeavour to determine what these are.

Scotia, New York, July 2013 Glen A. Slack

## References


1. G.A. Slack, J. Phys. Chem. Solids. **34**, 321 (1973)
2. G.A. Slack, Phys. Rev. B **22**, 3065 (1980)
3. G.A. Slack, Solid State Phys. **34**, 1 (1979)
4. G.A. Slack, in *Handbook of Thermoelectrics* ed. by D.M. Rowe, (CRC Press, Boca Raton, FL, 1995), pp. 407–440
5. J. Baumert, C. Gun, V.P. Shpakov, J.S. Tse, M. Krisch, M. Müller, H. Requardt, D.D. Klug, S. Janssen, and W. Press, Phys. Rev. B **68**, 174301 (2003)
6. J.S. Kasper, P. Hagenmuller, M. Pouchard and C. Cros, Science **150**, 1713 (1965)
7. C. Cros, M. Pouchard, P. Hagenmuller and J.S. Kasper, Bull. Soc. Chim. Fr. **7**, 2737 (1968)
8. G.S. Nolas, G.A. Slack and S.B. Schujman, in *Recent Trends in Thermoelectric Materials Research*, ed. by T.M. Tritt, Vol 1, (Academic Press, NY, 2001), pp. 255–300

# Preface

The interest in inorganic clathrates has grown substantially over the past two decades, due to their unique physical properties as well as their potential for technological applications. New compositions as well as new synthetic approaches for the synthesis and crystal growth of inorganic clathrate compositions have recently been reported. This has led to fundamental insight into the unique physical properties these materials possess, specifically due to their unique crystal structures. The inspiration for this volume came about as a result of several conversations that convinced me of the interest in a book that encompasses these developments in addition to presenting an overview of the history of these materials and some of the different structure types that have been reported. The book is intended as a reference for both those working on specific aspects of research on inorganic clathrates and those who are interested in understanding the basics of what has been developed thus far in order to begin research on these materials.

The volume contains chapters written by prominent experts in the field as well as junior researchers that have been directly involved in clathrate-related research, in some cases for most of their careers. It naturally has an international character as the contributors are from several different countries. The choice of topics is partly due to the particular research interests of the contributors and partly from the research that has been in the spotlight most recently.

I am grateful to Springer and Dr. Maria Bellantone in particular for giving me the opportunity to publish this volume. I am also grateful for the support of the U.S. Department of Energy, Basic Energy Sciences, Division of Materials Science and Engineering, under Award No. DE-FG02-04ER46145, for funding much of the fundamental research on clathrates at the University of South Florida. I am particularly indebted to all my former students, Dr. Matt Beekman and Dr. Stevce Stefanoski in particular, whose efforts on the synthesis and characterization of different clathrate compounds has not only allowed for substantial development in the field but their enthusiasm has, and continues to, increase my enjoyment in research on these interesting materials.

March 2014 George S. Nolas

# Contents

# Chapter 1
# The Early Development of Inorganic Clathrates

**Michel Pouchard and Christian Cros**

**Abstract** In this chapter the authors relate the discovery of the first inorganic clathrates, $Na_8Si_{46}$ and $Na_xSi_{136}$ ($3 \leq x \leq 11$), whose cage-like structures were determined by comparison with those of the two most classical gas and liquid clathrate hydrates. The main characteristics of clathrate compounds are recalled and a brief review of clathrate hydrates is given. The different polyhedral cages and their arrangements in the so-called type I structure ($Na_8Si_{46}$) and type II structure ($Na_xSi_{136}$) are described in details. The synthesis, composition and structure of other inorganic clathrates of silicon, germanium and tin with potassium, rubidium and cesium as guest atoms are reported. The crystal structure (type I or type II) and corresponding composition is closely related to the size of the guest alkali atoms. The formation of the characteristic polyhedral cages with a majority of pentagonal faces is discussed, and results from the arrangement of all the tetrahedrons in eclipsed position. The relation between clathrate structures and those of clathrasils (silica-based clathrates), Frank-Kasper alloys and fullerene forms of carbon is also discussed. The first measurements of the physical properties of inorganic clathrates are reviewed, including electrical conductivity, thermal properties, high pressure behavior, NMR and ESR investigations. The ability for the silicon, germanium and tin host lattices to form non-stoichiometric and mixed frameworks with elements of neighboring groups is briefly described, giving rise to a large variety of new inorganic clathrates with ionic guest-host interactions and semiconducting properties.

---

M. Pouchard · C. Cros (✉)
ICMCB, UPR-CNRS 9048, University of Bordeaux 1, 87, Avenue du Docteur Albert Schweitzer, 33608 Pessac, France
e-mail: christian.cros@numericable.fr

G. S. Nolas (ed.), *The Physics and Chemistry of Inorganic Clathrates*,
Springer Series in Materials Science 199, DOI: 10.1007/978-94-017-9127-4_1,

## 1.1 Introduction

This chapter relates the discovery of the first inorganic clathrates in the early sixties and their development until the late eighties of the last century, i.e., the beginning of the period during which much attention was paid to the many interesting properties of these atypical compounds. Their clathrate-like structures were determined by comparison with those of the gas and liquid hydrates, in which gaseous or liquid molecules are trapped in mostly two kinds of solid water host lattices. The main characteristics of clathrate-type compounds are recalled, as well as a brief history on the very important family of gas and liquid hydrates, which has many common features with the inorganic clathrates considered in this book. The type I and type II structures of $Na_8Si_{46}$ and $Na_xSi_{136}$ ($3 \leq x \leq 11$) are described. The synthesis, the composition and the structure of other inorganic clathrates of silicon, germanium and tin with potassium, rubidium and cesium as guest atoms are reported. The obtained structure (type I or type II) and corresponding composition is interpreted on the basis of the size of the guest alkali atoms. The formation of the characteristic polyhedral cages with a majority of pentagonal faces is discussed, which results from the arrangement of tetrahedrons in eclipsed position. The existence of other clathrate type-structure is recalled. The relation between clathrate structures and those of clathrasils (silica-based clathrates), Frank-Kasper alloys and fullerene forms of carbon is discussed. The first measurements of the physical properties of inorganic clathrates are reported, such as the electrical conductivity, the thermal properties (Seebeck coefficient), the influence of high pressure on the stability of the clathrate type-structures, NMR and ESR investigations. The ability for the host lattices of silicon, germanium and tin to form non-stoichiometric and mixed frameworks with other elements of neighboring groups is described, giving rise to a large variety of new inorganic clathrates with ionic host-guest interactions and semiconducting properties. Finally, a brief introduction to the next chapters of the book is given.

## 1.2 The First Inorganic Clathrates

The first inorganic clathrates were discovered in 1965 when NaSi, a highly reactive Zintl-type phase of composition $[(Na^+)_4(Si_4)^{4-}]$, was submitted to a careful thermal decomposition under argon in the temperature range 320–450 °C during about 10 days [1–5]. In a previous work on the synthesis of alkali-metal silicides and germanides, MSi and MGe (M = Na, K, Rb, Cs), Hohman [6] had reported the formation of intermediate phases of compositions $MSi_8$ and $MGe_4$ for M = K, Rb, Cs, unlike NaSi and NaGe which decomposed directly into the elements. This work had been later revisited by Schäfer and Klemm [7], who found that the intermediate phases corresponded more precisely to the compositions $KSi_6$, $RbSi_6$ and $CsSi_8$ for the silicides and $KGe_4$, $RbGe_4$ and probably $CsGe_4$

for the germanides. All of them were assumed to crystallize in a cubic structure with unit-cell constants of about 1.34 nm for silicides and 1.40 nm for germanides, respectively.

Unlike the results of previous authors, the X-ray diffraction patterns of the residue of decomposition of NaSi showed numerous new lines, revealing the formation of an intermediate phase with lower sodium content. It was also observed that the relative intensity of these new lines was variable as a function of temperature, indicating the presence of two different crystalline phases, one called A, occurring in majority at the lowest temperatures and a second one called B at higher ones. Above 450 °C the characteristic lines of diamond-type silicon began to appear and above 500 °C silicon was the only observable phase. In another series of experiments, NaSi was submitted to decomposition under high vacuum. Surprisingly, the B phase was almost exclusively obtained at temperatures ranging from 320 to 450 °C, with a very minor amount of the A one. Another observation was that the relative intensity of some lines of the B phase changed as a function of temperature, indicating a variable composition in sodium. This tendency was also observed in the B samples obtained at temperatures ranging from 450 to 500 °C, in which increasing amounts of silicon were present, showing that the sodium content became weaker and weaker. On the other hand, no change in the intensity of the diffraction lines of the A phase was observed.

Pure samples of the A and B phases were finally obtained. They consisted in finely divided and bluish grey powders for the A phase and brownish grey for the B ones. The sodium content was determined by chemical analysis (flame spectroscopy) after dissolution in diluted HF. Another method consisted in an ultimate thermal decomposition into silicon above 500 °C and measurement of the weight loss due to sodium evaporation. The obtained Na/Si atomic ratio for the A phase was in any case close to 0.158 and, consequently, it was attributed the formula $NaSi_6$. In three samples of the B phase obtained at 340, 380 and 445 °C, much lower Na/Si atomic ratios were obtained: 0.076, 0.064 and 0.020, respectively. These results indicated that the B phase was a kind of solid solution of a few sodium atoms in a silicon matrix. The $Na_xSi$ ($0.02 \leq x \leq 0.076$) general formula was thus adopted.

The X-ray diffraction patterns of $NaSi_6$ and $Na_xSi$ consisted in many fine and well resolved lines compatible with large and probably highly symmetrical unit-cells. A tentative of indexation of the observed lines of $NaSi_6$ on the basis of a simple cubic unit-cell was successful and a lattice constant $a$ = 1.019 nm was determined. With an experimental density value of 2.27 g/cm$^3$, the number of $NaSi_6$ formula units was 8. The same method of indexation applied to the observed d-spacing of the three samples of $Na_xSi$ led to unit-cell constants very close to $a$ = 1.462 nm. Furthermore, systematic extinctions consistent with an f.c.c. (face-centered-cubic) mode and an Fd-3m space group were observed. Experimental density values ranging from 2.03 to 2.12 g/cm$^3$ for $0.02 \leq x \leq 0.08$ involved approximately 132 $Na_xSi$ formula units per unit-cell. The search for isostructural compounds was unsuccessful and in the absence of single crystals, the two structures remained at first unsolved.

The solution was found by Kasper [3], who made the unexpected relation between the structure of the two above compounds of silicon and those of the two main clathrate type hydrates that he knew quite well, the gas hydrates and the liquid hydrates. The $NaSi_6$ phase proved to correspond more exactly to $NaSi_{5.75}$, i.e., $Na_8Si_{46}$ ($Pm\bar{3}n$ space group), and the $Na_xSi$ one to $Na_xSi_{136}$, in which the x value is multiplied by 136 in comparison with the previous one, i.e., $3 \le x \le 11$ (Fd-3m space group). The latter formulation is used in the following pages. In both structures, silicon atoms form a cage-like framework in which the sodium atoms are inserted.

## 1.3 What is a Clathrate?

The name of clathrate, from the Latin word *clatratus*, i.e., enclosed or protected by cross bars of a grating, was proposed by Powell [8] and adopted thereafter to characterize atypical compounds in which *cage structures of suitable form imprison molecules of a second kind to give compounds of either fixed or variable molecular ratios*. The other characteristics of clathrate compounds are:

(i) The unit-cell is usually large, highly symmetrical and the lattice constants almost independent of the nature of the guest species, which can be an atom or a molecule;
(ii) The host lattice itself can be considered as a possible allotropic form of the major species, which is stabilized by the guest species;
(iii) The guest lattice can be non-stoichiometric and even free-guest clathrates exist;
(iv) At high temperature, they decompose into the starting chemical species, and correspond to a kind of association rather than a combination;
(v) The host-guest interactions affect the properties of clathrates. They are generally weak as in clathrate hydrates, but they can be stronger as in inorganic clathrates.

Several families of clathrates are known, the host lattice of which are of inorganic or organic nature. The most important and representative of them is the series of the so-called gas and liquid hydrates, in which the host lattice is made of corner sharing $H_2O$ tetrahedrons forming large cages of 20 to 36 vertices, in which are enclosed guest molecules of a gas ($Cl_2$, $SO_2$, $CH_4$, $H_2S$, Kr, etc.), or a liquid ($C_6H_6$, $CH_3I$, $CH_3Cl$, $CH_2Cl_2$, $CHCl_3$, $CCl_4$, etc.).

## 1.4 Brief History of Gas and Liquid Hydrates

### 1.4.1 Synthesis and Composition

Before to describe in details the structures of $Na_8Si_{46}$ and $Na_xSi_{136}$ ($3 \leq x \leq 11$), it is interesting to give a brief history of the clathrate hydrates, which served not only as models to solve the structure of the two silicides, but also to the understanding of the role of the size of the guest atoms on the resulting structure and composition.

The first clathrate hydrate was synthesized by Davy in 1811 [9], by cooling a mixture of water and chlorine gas. Its composition, determined by Faraday [10], corresponded to one proportion of chlorine for ten of water. Later, hydrates of other gaseous or liquid chemical species were obtained: sulfur dioxide (1828), bromine (1829), hydrogen sulfide (1840), hydrogen chloride (1876), carbon dioxide, hydrogen phosphide, hydrogen selenide (1882), chloroform (1885). During the end of the nineteen and the beginning of the twentieth century, clathrate hydrates were extensively studied, mainly in France (Berthelot [11], Le Chatelier [12], de Forcrand [13, 14], etc.) and to a less extent in Germany, first, then inversely. In his thesis work, Villard [15] established and revisited the P-T conditions of synthesis of most of the gas hydrates known so far, as well as their decomposition pressure. He also synthesized many hydrocarbon hydrates, such as $CH_4$, $C_2H_6$, $C_3H_8$, $C_2H_4$, $C_2H_2$ [16, 17]. The composition of a large majority of gas hydrates was very close to $G(H_2O)_6$ (G = gas) (Villard's formula) [14, 18]. Their crystals exhibited several habits, but all of them were without effect on polarized light, indicating that they belonged to the cubic system. The property of all these gas hydrates to decompose into their two components led Berthelot [11] to point out that *the two components keep their integrity in the compound and are there in a molecular state as close as possible to that it possesses in its free state*. Villard also studied the conditions of formation of liquid hydrates with molecules such as methyl chloride, ethyl chloride, methylene chloride, chloroform, carbon tetrachloride, carbon disulfide, etc. He observed that these hydrates were stabilized under pressure of a help gas such as nitrogen, oxygen, or air. A great similarity with the gas hydrates was observed and their crystals also belonged to the cubic system. In a thorough review paper on gas and liquid hydrates by Schroeder in 1927, about three hundred references are cited [19].

During the next 20 years, the data on gas and liquid hydrates were improved. The composition of the liquid hydrates were more accurately determined and a $H_2O/L$ (L = liquid) ratio close to 15 was systematically found.

### 1.4.2 Structures of Gas and Liquid Hydrates

The first X-ray diffraction study on clathrate hydrates was carried out by Von Stackelberg [20] on powder and single crystals of the gas hydrate $SO_2(H_2O)_6$ and the liquid hydrate $CHCl_3$–$H_2S(H_2O)_{15}$. In both cases, a cubic unit-cell with a lattice constant close to 1.2 nm was determined. The space group was $Pm\bar{3}n$ ($O_h^3$) for the gas hydrate and the experimental value of the density consistent with 8 formula units, involving a total number of 48 water molecules per unit-cell. For the liquid hydrate, the probable space group was $O^2$ ($P4_232$).

In the structure of the gas hydrate, the 48 oxygen atoms of water were found to occupy the 48(l) positions (x, y, z, with x = 0.1083, y = 0.275, z = 0.1916), forming a total of 8 large voids of slightly different shape and volume, having both 24 vertices, at the center of which were located the 8 $SO_2$ molecules. Two of these were centered on the 2(a) (000) positions and the 6 others on the 6(b) (0½½) ones. Each oxygen atom was tetrahedrally surrounded by four neighbors at a distance close to 0.26 nm, as is the case in solid water. Such structure was considered as a "modified ice", only stable in the presence of guest molecules in the available sites.

Claussen [21] found the correctness of the above structure questionable, because some of the bond angles between water molecules were close to 60°, much smaller than the expected value for a tetrahedral arrangement. By means of molecular models, he came to the conclusion that the most probable kind of cage which might exist in the structure of hydrates would be a regular pentagonal dodecahedron having 12 pentagonal faces, 20 vertices and 30 edges (Fig. 1.1.a). Its size would be large enough to accommodate a gas molecule in its center, all the bond angles close to the expected value of 109.47° and the $H_2O/G$ ratio equal to 5, a value not so far from the experimental ones. However, a long range packing of pentagonal dodecahedrons could not fill in the space, due to its icosahedral symmetry, and it appeared necessary to combine the dodecahedrons with another kind of polyhedron. Claussen found a satisfying combination leading to a cubic system *by a slight compression of a pentagonal dodecahedron, so that the angles around two opposite atoms were exactly tetrahedral and then superimposing these two tetrahedral points on pairs of carbon atoms in a diamond type lattice*. The resulting large cubic lattice included a total of sixteen pentagonal dodecahedrons and in between them, eight large voids were generated, corresponding to a new kind of cage in shape of an hexakaidecahedron (16-faced polyhedron), having 12 pentagonal and 4 hexagonal faces, 28 vertices and 42 edges (Fig. 1.1.d). The total number of water molecules in the proposed unit-cell was 136, with all the bond angles ranging from 108 to 120°. The large hexakaidecahedral voids could accommodate 8 large liquid molecules with a $H_2O/L$ ratio of 17, close to the average experimental value of 15. Sixteen gaseous species could get into the smaller dodecahedral voids ($H_2O/G$ = 8.5), forming a double hydrate.

All the above assumptions were confirmed by Von Stackelberg and Müller [22] in the study of the crystal structure of liquid hydrates with $CHCl_3$, $CH_2Cl_2$ and $C_2H_5Cl$ as guest molecules. The space group was found to be Fd-3m ($O_h^7$) and the unit-cell constant close to 1.72 nm.

Some months after, Claussen [23] proposed another possible structure for the gas hydrates, based on a combination of pentagonal dodecahedrons and a new type of cage, the tetrakaidecahedron (14-faced polyhedron), with 12 pentagonal faces

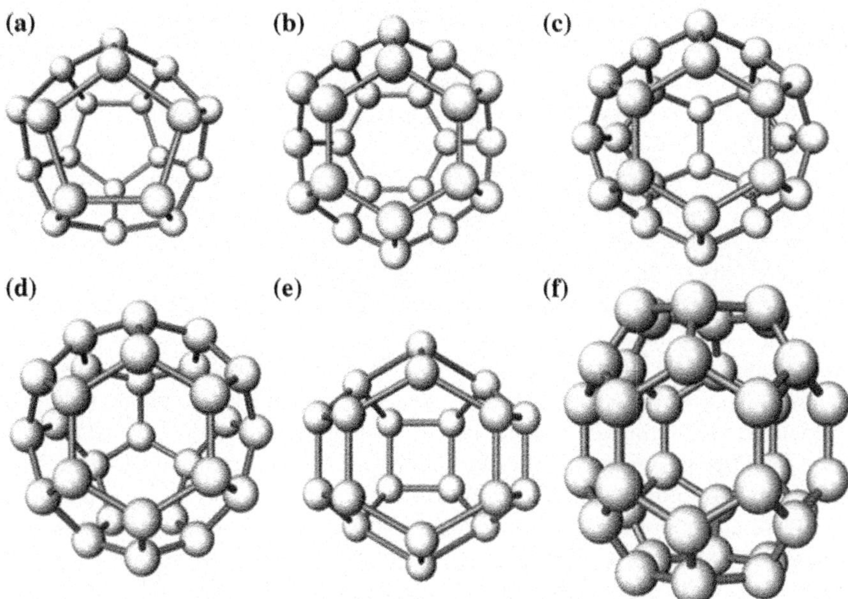


**Fig. 1.1** The characteristic polyhedrons of clathrate-type structures: **a** the pentagonal dodecahedron, **b** the tetrakaidecahedron, **c** the pentakaidecahedron, **d** the hexakaidecahedron, **e** the 12-faced $4^3 5^6 6^3$ polyhedron, **f** the 20-faced $5^{12} 6^8$ polyhedron [5]

and 2 opposite hexagonal ones, 24 vertices and 36 edges (Fig. 1.1b). The structure appeared as a modified body centered cubic lattice, in which pentagonal dodecahedrons occupy the corners and the center. The addition of six extra water molecules in special positions between adjacent dodecahedrons at the corner of the cube led to the formation of hexagonal faces linking together all the pentagonal dodecahedrons. The result was that six tetrakaidecahedrons were generated, exactly filling in the space between dodecahedrons.

The total number of water molecules per unit-cell would be 46 and the number of available voids 8, i.e., 2 small and 6 slightly larger ones, located at the center of the dodecahedral and tetrakaidecahedral cages, respectively. These two kinds of voids could be fully or partially occupied by gas molecules and the corresponding formula would be $G_8(H_2O)_{46}$ or $G_6(H_2O)_{46}$ if the smallest voids remained empty. The calculated $H_2O/G$ ratios of 5.75 and 7.67 were in good agreement with the experimental values.

Almost at the same time, Müller and von Stackelberg [24] reported the structure of Kr, $H_2S$, $Cl_2$ and $SO_2$ gas hydrates, as determined from a X-ray diffraction study on single crystals. In all cases, a cubic unit-cell with a lattice constant close to 1.2 nm and an $O_h^3$ ($Pm\bar{3}n$) space group was obtained. The results of the structural analysis were consistent with the above assumptions of Claussen. For the smallest gas molecules, Kr and $H_2S$, a total occupancy of the available sites of the water host lattice was determined. In the case of the slightly larger $Cl_2$ and $SO_2$

molecules, an overall occupancy rate of 96 % was calculated, in agreement with the experimental density. For larger gas molecules, such as $Br_2$, the two smaller dodecahedral sites were found to remain empty and the resulting formula was $Br_2$,7.67($H_2O$), i.e. $(Br_2)_6(H_2O)_{46}$, in rather good agreement with the analytical data: $Br_2$,7.9($H_2O$).

Also at the same time, the crystal structure of $Cl_2$ hydrate was investigated by Pauling and Marsh [25]. Their results were consistent with those of the two previous authors. However, they clearly showed that the six largest hexakaidecahedral sites were only occupied, leading to the formula $(Cl_2)_6(H_2O)_{46}$.

The structures of the gas hydrates, $G_x(H_2O)_{46}$ (x = 6 or 8) and the liquid hydrates, $L_x(H_2O)_{136}$ ($x \leq 8$), stabilized (or not) by molecules of a help gas, are now referred to as type I and type II clathrates, respectively. The main characteristic of the two host lattices is the presence of face-sharing cages of different size, involving only pentagonal faces for the smallest of them and a combination of a majority of pentagonal and a minority of hexagonal ones for the largest ones. These characteristic polyhedrons are now conveniently identified by the descriptor $n(i)^{m(i)}$, where n(i) is the number of edges in the i type of face and m(i) the number of faces of i type. The pentagonal dodecahedron is identified by $5^{12}$ (12 pentagonal faces), the tetrakaidecahedron by $5^{12}6^2$ (12 pentagonal faces and 2 hexagonal ones) and the hexakaidecahedron by $5^{12}6^4$ (12 pentagonal faces and 4 hexagonal ones).

In the classification proposed by Jeffrey [26], five other clathrate-type structures are indexed, III, IV, V, VI, VII:

(i) Type III (tetragonal, $P4_2/mnm$) is built up from a combination of ten pentagonal dodecahedrons, sixteen tetrakaidecahedrons and four pentakaidecahedrons (15-faced polyhedron (Fig. 1.1.c), i.e., $10(5^{12}) + 16(5^{12}6^2) + 4(5^{12}6^3)$, for a total of 172 $H_2O$ molecules and 30 possible guest species ($H_2O/G = 5.73$);
(ii) Type IV (hexagonal, P6/mmm) corresponds to $6(5^{12}) + 4(5^{12}6^2) + 4(5^{12}6^3)$ for a total of 80 $H_2O$ and 14 possible guest species ($H_2O/G = 5.71$);
(iii) Type V (hexagonal $P6_3/mmc$) is related to type II and is two times smaller in volume. It involves $8(5^{12}) + 4(5^{12}6^4)$ for a total of 68 $H_2O$ and 12 possible guest species ($H_2O/G = 5.67$). It is only hypothetical and has been inferred from less symmetrical complex hydrates;
(iv) Type VI (cubic, I-43d) is made of a combination of unusual and complex polyhedrons with 8 and 17 vertices: $12(4^45^4) + 16(4^35^96^27^3)$. The number of water molecules is 156 including 28 possible guest species ($H_2O/G = 5.57$);
(v) Type VII (cubic, Im−3m) is built up from two semi-regular and space filling truncated octahedrons, $2(4^66^8)$, with no pentagonal face. It is related to a sodalite-type structure and contains 12 $H_2O$ molecules and 2 guest species ($H_2O/G = 6$).

More recently, a new type of clathrate, H, has been added to the above list [27]. The host lattice is built up from of a combination of pentagonal dodecahedrons and two new kinds of cage. The first one is a non-regular dodecahedron with 3 square, 6 pentagonal and 3 hexagonal faces, 20 vertices and 30 edges (Fig. 1.1e), which size is about the same as that of a pentagonal dodecahedron. The other one is a large non regular icosahedron with 12 pentagonal and 8 hexagonal faces, 36 vertices and 46 edges (Fig. 1.1f). The structure corresponds to $3(5^{12}) + 2(4^3 5^6 6^3) + (5^{12} 6^8)$ and involves 34 $H_2O$ molecules and 6 (5 small + 1 large) possible guest species ($H_2O/G = 5.67$).

Some authors use other notations for clathrate-type structures: CS-1, CS-2, CS-3 and CS-4 for the four cubic I, II, VI and VII types, respectively, TS-1 for the tetragonal type III and HS-1, HS-2 and HS-3 for the three hexagonal IV, V and H types. More details about the structure and the corresponding representative clathrate hydrates will be given further (see Sect. 1.10 and Tables 1.3 and 1.4).

## 1.5 Structure of Type I $Na_8Si_{46}$ and Type II $Na_xSi_{136}$

The type I structure of $Na_8Si_{46}$ is depicted in Fig. 1.2. The crystallographic data, as obtained from powder diffraction patterns and assuming a gas hydrate type-structure are reported on Table 1.1 [3]. The 46 silicon atoms of the host lattice are distributed as follows. Forty of them are provided by the two pentagonal dodecahedra located on the corners and the center of the cube and correspond to (Si(2) and Si(3) in 16(i) and 24(k) atomic positions, respectively. The 6 other silicon atoms (Si(1)), forming the hexagonal faces of the 6 tetrakaidecahedrons, occupy the 6(c) atomic positions (¼, 0, ½). Two of the 8 sodium atoms are located at the center of the 2 pentagonal dodecahedrons (Na(1) in 2(a)) and the 6 others at the center of the 6 slightly larger tetrakaidecahedrons (Na(2) in 6(d)). The calculated intensities assuming a total occupancy of both guest sites by sodium were consistent with the experimental X-ray diffraction data, as well as with the measured density of 2.27 g/cm$^3$ ($d_{calc}$ = 2.316 g/cm$^3$).

Each silicon atom forms four bonds of 0.237 nm to other silicon atoms in a distorted tetrahedral arrangement (the interatomic distance in diamond type silicon is 0.235 nm). The Na(1) sodium atoms in the smaller dodecahedral cage is 0.323 nm from 8 of the silicon atoms and 0.337 nm from 12 others. In the largest cage, the Na(2) sodium atom is 0.341 nm from 12 silicon atoms, 0.379 nm from 8 others and 0.360 nm from 4 others. All these interatomic distances appeared two large for a direct Na–Si bonding. In the metallic guest lattice, the Na(1) atoms are surrounded by 12 Na(2) at a distance of 0.570 nm ($a \times (5^{½}/4)$), forming a semi-regular icosahedron (20 vertices and 12 isosceles triangular faces. The Na(2) atoms are surrounded by 4 Na(1) at 0.570 nm and 10 Na(2) at 0.510 nm ($a/2$) for 2 of them and 0.624 nm ($a \times (6^{½}/4)$) for the 8 others, forming an irregular polyhedron having 14 vertices and 24 triangular faces.

The type I clathrate structure can be described in two ways. The first one is to consider that it corresponds to a pseudo body centered cubic arrangement of pentagonal dodecahedrons, in which the remaining space is filled by six tetrakaidecahedrons generated by the addition of six extra silicon atoms (Si(1)) forming hexagonal faces. The second one is to describe it as built up from of a 3-D network of non-intersecting tetrakaidecahedron rows sharing their opposed hexagonal

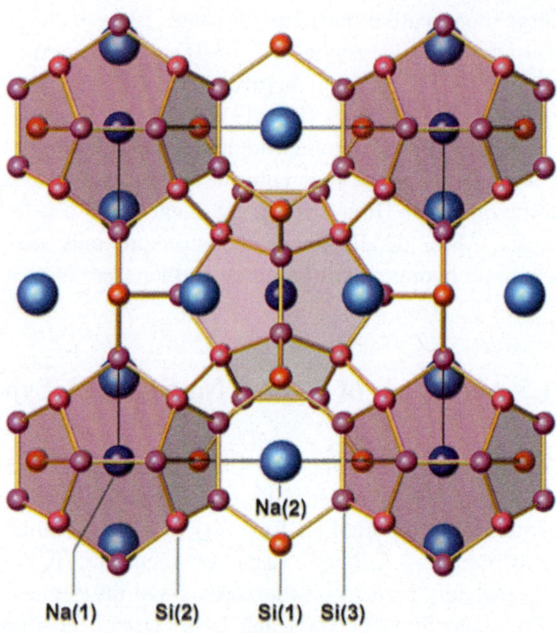


**Fig. 1.2** Structure of $Na_8Si_{46}$ viewed along [100]. For the sake of clarity, the pentagonal dodecahedrons are shaded and the tetrakaidecahedrons partially shown [5]

**Table 1.1** Crystallographic structural data for $Na_8Si_{46}$ and $Na_xSi_{136}$ (x ~ 9.52) [3]

| Clathrate | $Na_8Si_{46}$ | $Na_{9.36}Si_{136}$ |
|---|---|---|
| System | Cubic | Cubic |
| Lattice constant | $a$ = 1.019 ± 0.002 nm | $a$ = 1.462 ± 0.002 nm |
| Space group | $Pm\bar{3}n$ (# 223) | Fd-3m (# 227) |
| Atomic positions | 6 Si(1) in 6(c) ¼/0½ | (000, ½½0, ½0½, 0½½)+ 8 Si(1) in 8(a) 000, ¼¼¼ |
| | 16 Si(2) in 16(i) xxx, x = 0.183 | 32 Si(2) in 32(e) xxx, x = −0.094 |
| | 24 Si(3) in 24(k) 0yz, y = 0.310, z = 0.116 | 96 Si(3) in 96(g) xxz, x = −0.058, z = −0.246 |
| | 2 Na(1) in 2(a) 000 | 16αNa(1) in 16(c) ⅛⅛⅛ |
| | 6 Na(2) in 6(d) ¼½0 | 8βNa(2) in 8(b) ½½½, α = 0.19; β = 0.79 |
| Calculated density (g/$cm^3$) | 2.316 | 2.146 |
| Observed density (g/$cm^3$) | 2.27 | 2.102 |

faces, and developing along the three main axis of the unit-cell, such arrangement giving rise to the formation of voids in shape of dodecahedrons located at the corners and the center of the unit-cell.

The type II structure of $Na_xSi_{136}$ is shown in Fig. 1.3. The refined crystallographic data for the composition x ~ 11 are reported in Table 1.1. They were more difficult to obtain than above because of the non-stoichiometry of the guest species and, furthermore, the non-equivalent occupation rate for the two kinds of

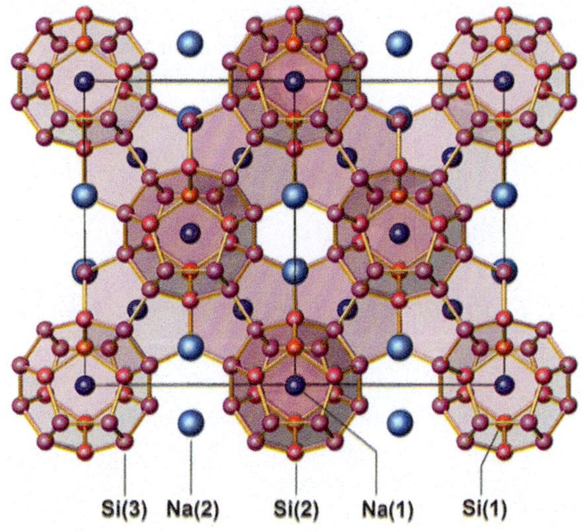


**Fig. 1.3** Structure of $Na_xSi_{136}$ with all Na available sites occupied (x = 24) viewed along [110]. For the sake of clarity, only the pentagonal dodecahedrons are shaded. The voids in between correspond to hexakaidecahedrons [5]

available sites. The 136 silicon atoms of the unit-cell are distributed on the vertices of 16 pentagonal dodecahedrons and 8 larger hexakaidecahedrons. They occupy the three following atomic positions of the Fd-3m ($O_h^7$) space group: 8(a), 32(e) and 96(g). Whatever the composition of the studied samples, x ~ 3 and x ~ 11, the sodium atoms were found to partially occupy the center of both kinds of cages: 16(c) for the smaller and 8(b) for the larger, but preferably the latter. For $Na_{11}Si_{136}$, the best agreement between calculated and experimental diffracted intensities was obtained assuming an occupancy rate of 19 and 79 %, respectively. Given the small size of the sodium atoms, this result was rather surprising, but it was later confirmed by more accurate experimental data and refinement methods [28–30].

The calculated Si–Si interatomic distances are 0.230, 0.234, 0.238 and 0.240 nm, i.e., not significantly different than in diamond type silicon (0.235 nm). In the smallest cage the Na–Si distances are 0.317, 0.325 and 0.332 nm and in the largest one 0.398 and 0.399 nm. These distances are comparable to those already observed in the type I structure. However, the dodecahedral cages are here very slightly smaller. In the metallic guest lattice, the Na(1) atoms are surrounded by 6 Na(1) at a distance of 0.517 nm and 6 Na(2) at 0.606 nm, forming an irregular icosahedron (12 vertices, 20 triangular faces). The Na(2) atoms are surrounded by 12 Na(1) and 4 Na(2) at a distance of 0.606 and 0.633 nm, respectively, forming an irregular polyhedron having 16 vertices and 28 triangular faces.

The structure of $Na_xSi_{136}$ can be also described in two ways. In the first one, the 16 pentagonal dodecahedrons are arranged in groups of four, each of them being centered on the middle of the 16 bonds of a large diamond-type structure. The voids left in between correspond to the 8 hexakaidecahedrons, without the need to add any more atoms. In the second one, the 8 hexakaidecahedrons, sharing their four hexagonal faces, are centered on the positions occupied by silicon atoms in the same diamond-type structure, but shifted by ½½½ in comparison with the previous one. Provided the addition in the unit-cell of 8 extra silicon atoms (8(a) positions), the large voids left in between result in the formation of the 16 pentagonal dodecahedrons filling in the space.

## 1.6 Other Clathrates

### *1.6.1 Silicon and Other Alkali-Metals*

Following the study on NaSi, the other alkali metal silicides, KSi, RbSi and CsSi, were submitted to thermal treatments under argon atmosphere and high vacuum in the temperature range 300–500 °C [4]. The results are summarized in Table 1.2. The data concerning NaSi are added for comparison.

Whatever the experimental conditions, the decomposition product of KSi was a type I clathrate $K_xSi_{46}$, without trace of a type II one. Chemical analysis and ultimate thermal decomposition above 500 °C led to x $\sim$ 7, involving the occupation of almost all the available sites of the structure [31]. The same result was observed in the case of RbSi, but the experimental x value was 5, in agreement with the probable occupancy of only the six largest sites of the structure. With CsSi, only a type II clathrate $Cs_xSi_{136}$ was observed. By ultimate decomposition, an x value close to 7 was determined, indicating the occupancy of almost all the eight largest cages of the structure.

The experimental values of the *a* unit-cell constant for a given type I or type II structure vary very weakly with the size of the guest alkali metal (Table 1.2). The calculated intensities of the diffraction lines of $Rb_xSi_{46}$ and $Cs_xSi_{136}$ were consistent with the above assumptions on the site occupancies in both type structures. However, a broadening of the diffraction lines was observed in the case of the two heaviest alkali-metals, involving more finely divided samples.

By direct synthesis from the elements at 800 °C under argon atmosphere, Gallmeier et al. [32] obtained single crystals of $K_xSi_{46}$ with x = 8 and a lattice constant $a$ = 1.03 nm.

### *1.6.2 Germanium and Alkali-Metals*

The thermal treatment under vacuum in the temperature range 320–400 °C of NaGe led to the formation of a mixture of a type II $Na_xGe_{136}$ clathrate and another phase of composition close to $NaGe_4$, occurring essentially at low temperature (320–360 °C), the structure of which was quite different from that of a type I clathrate and not identified so far. The sodium content in the minority phase $Na_xGe_{136}$ was not possible to determine accurately.

**Table 1.2** Thermal decomposition of alkali-metal silicides and germanides [5]

| Alkali silicide or germanide | Experimental conditions | Struct. type | x | Lattice const. *a* (nm) | Remarks |
|---|---|---|---|---|---|
| NaSi | 420–450 °C/10 days | $Na_xSi_{46}$ | 8 | 1.019 | Frequent presence of $Na_xSi_{136}$ |
| | Dry argon atm. 320–450 °C/60 h under vacuum | $Na_xSi_{136}$ | $3 \le x \le 11$ | 1.462 | Frequent presence of traces of $Na_8Si_{46}$ |
| KSi | 420–480 °C/10 days | $K_xSi_{46}$ | 7 | 1.026 | No trace of a type II clathrate, whatever the experimental conditions |
| | Dry argon atm. 320–480 °C/60 h under vacuum | $K_xSi_{46}$ | 7 | 1.026 | |
| RbSi | 320–470 °C/60 h Under vacuum | $Rb_xSi_{46}$ | 5 | 1.027 | No trace of a type II clathrate |
| CsSi | 320–470 °C/60 h under vacuum | $Cs_xSi_{136}$ | 7 | 1.464 | Broad X-ray diffraction lines. Finely divided powder |
| NaGe | 320–400 °C/60 h under vacuum | $Na_xGe_{136}$ | ? | 1.540 | Presence of another phase of composition close to $NaGe_4$ |
| KGe | 320–380 °C/60 h under vacuum | $K_xGe_{46}$ | 8 | 1.066 | No trace of a type II clathrate |
| RbGe | 320–370 °C/60 h under vacuum | $Rb_xGe_{46}$ | ? | 1.070 | Very broad diffraction lines. |
| CsGe | 320–460 °C/60 h under vacuum | ? | ? | | Very broad diffraction lines. |

The residue of decomposition of KGe between 320 and 380 °C was identified as a type I clathrate, of composition close to $K_8Ge_{46}$ (the structure analysis gave $K_{7.4}Ge_{45}$ [31] ). Similar results were observed for RbGe which led to the formation of a type I $Rb_xGe_{46}$ clathrate, but the x value was not possible to determine, due to the presence of small amounts of a reddish phase identified as germanium hydride, $(GeH)_n$. In the case of CsGe, the X-ray diffraction lines of the residue of decomposition were so broad and diffuse that neither the structure-type, nor the corresponding x value could be determined.

For the crystals obtained by direct synthesis of $K_8Ge_{46}$ between 600 and 700 °C, Gallmeier et al. [33] reported a full occupancy of all the available sites by potassium and a lattice constant of 1.071 nm. This structure was later revisited and it was shown that the germanium host lattice was in fact slightly non-stoichiometric and the exact composition $K_8Ge_{44}\square_2$ ($\square$ = vacancy), involving the presence of 8 $K^+$ ions that compensate the negative charge of the host lattice $(Ge_{44})^{8-}$ [34]. The two vacancies are located on the 6(c) Ge positions which are occupied at 67 %.

### 1.6.3 Tin and Alkali-Metals

$K_8Sn_{46}$, with a unit-cell constant $a$ = 1.203 nm, was the first reported clathrate having a tin host lattice. It was obtained by direct synthesis from the elements in the temperature range 600–700 °C [33]. More recently, the tendency for alkali-metal tin clathrates to be non-stoichiometric has been demonstrated in $K_{1.6}Cs_{6.4}Sn_{44}$, $K_8Sn_{44}$, $Rb_8Sn_{44.6}$ and $Cs_8Sn_{44}$ [35, 36].

## 1.7 Influence of the Size of the Guest Atoms on the Structure and Occupancy Rate of Available Sites

Given the numerous gas and liquid clathrate hydrates and in order to explain the experimental results, Von Stackelberg [20] was the first one to compare in a table the size of the gaseous and liquid guest species with the observed structure type, the corresponding lattice parameter and the occupancy of the available sites. This table was thereafter revisited and differently presented [26, 37]. It showed that the observed structure corresponds to the best fit between the size of the available sites and that of the guest species, the existence of critical size values of the guest which govern the formation of one or the other structure type, as well as the occupancy or the vacancy of the small dodecahedral sites.

A similar diagram was established in the case of the inorganic silicon and germanium clathrates [4, 38]. Figure 1.4 shows the case of silicon, where $r_1$(I) and $r_2$(I) correspond to the *free radii* of the dodecahedral and tetrakaidecahedral cages in the type I structure, respectively, and $r_1$(II) and $r_2$(II) those of the dodecahedral and hexakaidecahedral cages in the type II structure. The free radius of a cage corresponds to the shortest M-Si inter-atomic distances (M = alkali-metal) minus the average radius of silicon, as determined from Si–Si distances. Due to slightly shorter Si–Si interatomic distances (see Table 1.1), the $r_1$(II) free radius is very slightly smaller than $r_1$(I), although they both correspond to dodecahedral voids.

The problem was more complicated to determine the *effective radii* of the guest species, because it depends of how far the electrons extend. The data found in handbooks for alkali-metals gave three series of values: the ionic, the covalent and the metallic radii. As expected from the previous remark that alkali-metals in clathrates were essentially neutral, the ionic and the metallic radii were considered as inappropriate. The covalent radii as determined by Pauling seamed more realistic and, actually, proved to be consistent with experimental results, as shown in Fig. 1.4. The recently revisited values of covalent radii are about the same as the previous ones [39].

Sodium, the smallest of the alkali-metal considered here, with a covalent radius of 0.157 nm, can enter the type I and type II structures and occupy the two kinds of available cages. However, the type I was obtained in milder experimental

Fig. 1.4 Correlation diagram between the host cage free radii and the effective guest alkali-metals radii in the structure of type I and type II silicon clathrates. The *bold arrows* on the left correspond to previous covalent radii and the *thin lines* to the recent ones [5, 35]

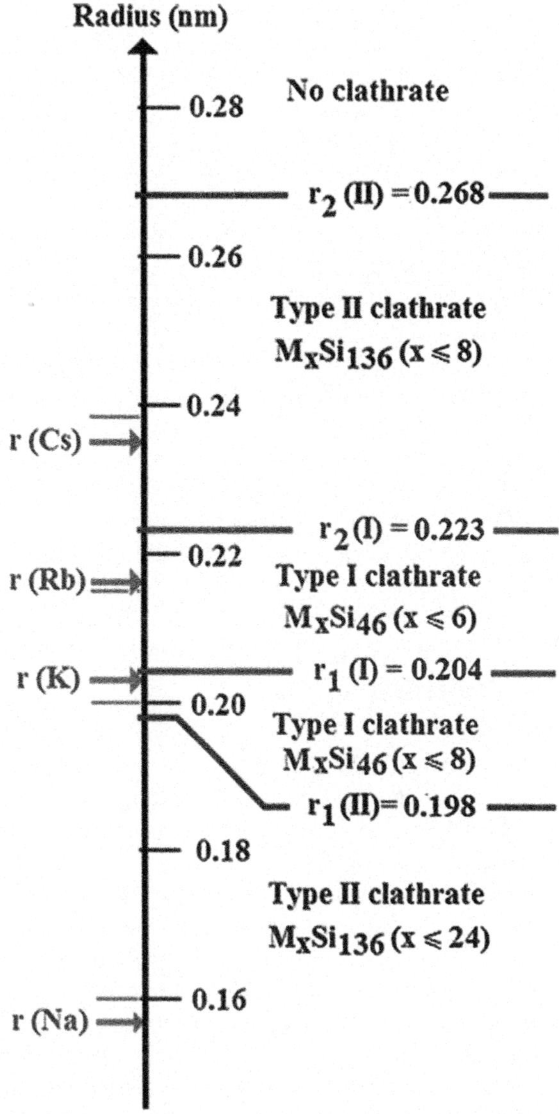


conditions, and appeared less stable than the other one. The effective radius of sodium r(Na) is thus lower than all the free radii of the cages in the two structures.

Potassium, whatever the experimental conditions, gives rise to the formation of only the type I structure and occupy both the small and large available cages. Its effective radius is in between $r_1(II) = 0.200$ and $r_1(I) = 0.204$ nm, very close to the covalent radius (r(K) = 0.203 nm). The same type I structure was obtained with rubidium as guest species, but only the large cages were found to be occupied. Its effective radius is in between $r_1(I) = 0.204$ and $r_2(I) = 0.223$ and consistent with the covalent radius of 0.216 nm.

In the case of cesium, the largest of the alkali-metals, a type II structure was again observed, in which the eight largest cages are only occupied. Its effective radius is therefore higher than $r_2(I) = 0.223$ nm and lower than $r_2(II) = 0.275$, in rather good agreement with a covalent radius of 0.235 nm.

A similar interpretation was given for germanium clathrates, in which the free radii of the available cages were ~4 % larger. With sodium as guest lattice, a minority type II phase was observed, without trace of a type I, which confirmed a higher stability of the type II. Another structure, possibly of clathrate type was in majority obtained, but not identified. With potassium, only a type I structure, $K_xGe_{46}$, (x ~ 8) was obtained. The effective radius of potassium, r(K), is in between the free radii $r_1(II)$ and $r_1(I)$ of the germanium host lattices.

The formation in the series of silicon and germanium inorganic clathrates of a type II structure with the smallest guest species appeared unusual, but logic, because such behavior was not reported for a long time in the series of clathrate hydrates. However, recent studies on the stability of hydrates enclosing small molecules showed that such guest species should form a type II structure [40]. This assumption was confirmed by Davidson et al. [41] in the case of argon and krypton, and Mao et al. [42] in the case of hydrogen.

## 1.8 Clathrates as Open Forms of Tetrahedrally Bonded Structures

Although described as built up from face sharing large polyhedral cages with a majority of pentagonal faces and a minority of hexagonal ones, clathrate-type host lattices correspond, nevertheless, to special arrangements of slightly distorted tetrahedral units. There are two main tetrahedrally bonded type-structures: the cubic diamond type and the hexagonal diamond type (lonsdaleite) having both the same atomic volume. In the cubic diamond type-structure, characteristic of group 14 elements (C, Si, Ge, α-Sn), but also of many compounds such as 1c-$H_2O$ and high-cristobalite $SiO_2$, all adjacent tetrahedrons are in staggered (or trans) position and form intersecting zigzag chains directed along the diagonals of the cube faces, forming bond angles of 109.47°. In the hexagonal diamond structure, which is also adopted by compounds such as 1 h-$H_2O$ and high-tridymite $SiO_2$, adjacent tetrahedrons along the c axis of the unit-cell are alternatively in staggered and eclipsed (or cis) positions, and only staggered in the a,b plane. This gives rise along the c axis to the formation of parallel chains in form of a crankshaft.

In the clathrate-type structures, adjacent tetrahedrons are all in eclipsed position, giving rise to the formation of pentagonal faces after a small decrease of the bond angles from 109.47 to 108° [3, 4, 43]. Figure 1.5 shows how three consecutive tetrahedrons generate the formation of a pentagonal face and then a pentagonal dodecahedron. In order to obtain a complete filling of space, the presence of a minority (about 12 %) of hexagonal faces is needed, which result from the opening of some of the bond angles of the tetrahedrons from 109.47 to 120°.

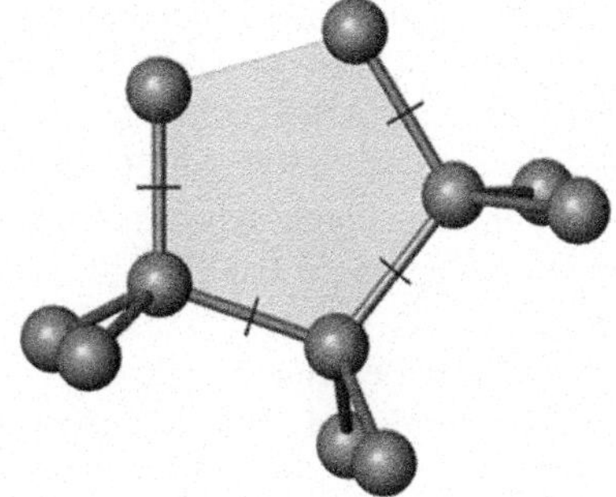

**Fig. 1.5** Arrangement of three adjacent tetrahedrons in eclipsed position showing the formation of pentagonal faces [5]

Due to the presence of large voids, clathrate-type host lattices can be considered as open forms of tetrahedrally bonded structures. Whereas the atomic volume of silicon in its diamond-type structure is 0.020 $nm^3$/atom, its atomic volume in type I and type II host lattices is for both of them 0.023 $nm^3$/atom, i.e., 15 % less compact.

It is interesting to note that the so-called Weaire-Phelan structure of mono-disperse liquid foam, made of connected bubbles with the least area of surface between them, theoretically predicted and experimentally observed, is the same as that of a type I clathrate [44, 45].

## 1.9 Clathrates and Clathrasils

Clathrate-type structures also exist with $SiO_2$ host lattices, another well known tetrahedrally-based compound, which has several analogous structural polymorphs with $H_2O$. These clathrates are called *clathrasils* and classified as a subgroup of porous tectosilicates. In addition to their pure silica frameworks, they differ from zeolites by the presence of almost spherical and medium large voids with small apertures limited to six atoms, instead of wide interconnected channels [46].

A type I clathrate structure was evidenced by Kamb [47] in 1964 in an impure silica mineral, *melanophlogite*, in which the guest species were a mixture of organic and sulfur compounds ($CH_4$, $CO_2$, $SO_2$, etc.) in variable proportions. This clathrate was later synthesized by hydrothermal process [48]. Its formula is $(CH_4, N_2, CO_2)_x(Si_{46}O_{92})$ and the unit-cell pseudo-cubic at room temperature and cubic above 40 °C ($a$ = 1.3436 nm at 200° C). By heating at about 400 °C, the guest species can be removed without destruction of the host lattice, giving a clathrate form of pure silica, stable up to 900 °C [49]. In the classification of tectosilicates, melanophlogite is referred to as MEP [50].

Later on (1984), the corresponding type II clathrasil structure was hydrothermally synthesized by Gies [51], in which the 8 large cages were occupied by molecules such as trimethylamine ($(CH_3)_3N$) and $CO_2$ and the 16 small ones by a mixture of nitrogen, argon and methane. The cubic unit-cell (space group Fd-3m; $a = 1.99$ nm) corresponds to the formula $(CH_3)_3N,CO_2)_y$ $(N_2,Ar,CH_4)_x(Si_{136}O_{272})$. Here also, a guest-free type II clathrate form of silica was obtained. This structure is referred to as dodecasil 3C or NTM.

A third clathrate-type structure of silica, called dodecasil 1H (or DOH) was reported by Gerke and Gies [52]. It is isostructural with the type H hydrates (see Sect. 1.4.2). The lattice constants of the hexagonal unit-cell of dodecasil 1H enclosing piperidine as guest species in the large $5^{12}6^8$ cage are $a = 1.378$ and $c = 1.119$ nm and the space group P6/mmm.

## 1.10 Clathrates and Frank-Kasper Alloy-Structures

In two famous papers, Frank and Kasper [53, 54] showed that a large number of $A_mB_n$ complex alloy structures can be described as the result of different stackings of a very limited number of atomic layers, mainly arranged in the two basic $3^2$, 4, 3, 4 and Kagome 3, 6, 3, 6 nets (Schläfli symbols), giving rise to the formation of four kinds of triangular-shaped polyhedrons with C.N. (coordination number) of 12, 14, 15 and 16, the most abundant of them being the C.N.12 and 20-faced icosahedron. Each $A_mB_n$ alloy structure is built up from the combination of at least two of these Frank-Kasper polyhedrons (abbr. F–K), which include a single A or B atom in their centers. The consequence of such arrangements is that small tetrahedral voids (which remain vacant in the alloys) occur everywhere in the structure. Many of the existing alloy-structures, such as the $AB_2$ Laves phases (C14, C15 and C36) or the $A_3B$ A15 (β-tungsten type) could be explained and other hypothetical ones could be proposed (A15, C14, C15 and C36 are the Strukturbericht notations for elements, alloys and compounds). In a recent computer search, a total of eighty-four of such possible alloy-structures have been reported [55].

Frank and Kasper [54] were the first ones to point out that *it is a fascinating matter that there exists a strict correspondence between the structures of such chemically different structures as the gas hydrates and the inter-metallic compounds*. The type I clathrate structure of gas hydrates proved to be related to that of the A15 alloy (β-tungsten, $Cr_3Si$, etc.) and the type II liquid hydrates ones to that of the cubic Laves phase C15 alloy ($MgCu_2$), in which the oxygen atoms of the water molecules occupy the center of all the empty tetrahedrons of the two alloy-structures. According to Frank and Kasper, *the relationship between clathrates and alloy structures is a logically natural one, since the sphere-packing with all triangulated coordination polyhedrons are ipso-facto those with all tetrahedral interstices, and further have the property that each tetrahedral interstice is tetrahedrally surrounded by four more, providing the right coordination of water molecule*. They suggested that other alloy structures, such as the C14 hexagonal Laves structure ($MgZn_2$-type), the σ-phase ($Cr_{46}Fe_{54}$-type) and the μ-phase ($Mo_6Co_7$-type) could have a related clathrate-type structure. Other possible structures have been more recently proposed by O'Keeffe et al. [56].

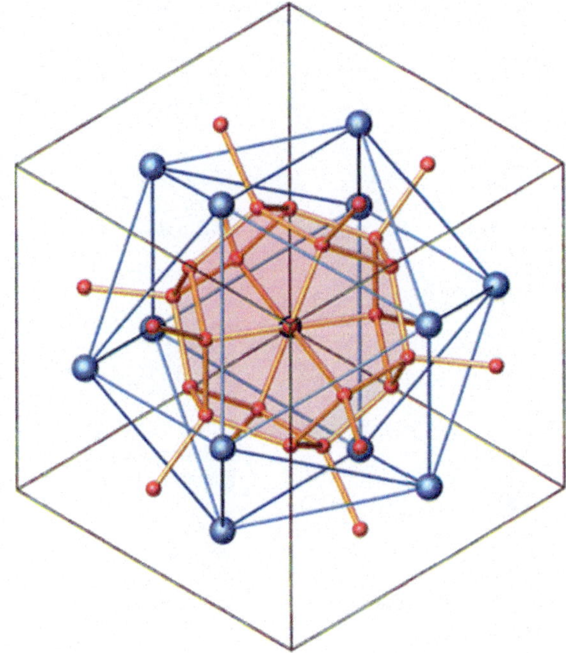

**Fig. 1.6** Partial view along [111] of the structure of type I $Na_8Si_{46}$, showing one of the two C.N.12 Frank-Kasper polyhedron of Na(2) atoms surrounding the central $Si_{20}$ dodecahedron, itself centered on a Na(1) atom

Thus, the F–K structures can be considered as templates for clathrate-type ones. Some authors use the name of dual structures [56, 57]. In fact, the C.N.12, C.N.14, C.N.15 and C.N.16 F–K polyhedrons are the duals of the $5^{12}$ dodecahedral, $5^{12}6^2$ tetrakaidecahedral, $5^{12}6^3$ pentakaidecahedral and $5^{12}6^4$ hexakaidecahedral characteristic cages of the clathrate-type structures (the C.N. of an F–K polyhedron equals the number of faces of the corresponding clathrate cage and the number of triangular faces of an F–K polyhedron is the same as the number of vertices of the related clathrate cage). Figure 1.6 shows a partial view along the [111] direction of the type-I clathrate structure of $Na_8Si_{46}$. The central Na(1) atom is surrounded by a pentagonal dodecahedron of silicon and at longer distance by a C.N.12 F–K icosahedral polyhedron of Na(2). The polyhedron surrounding a Na(2) atom has a C.N. of 14 (10 Na(2) + 4 Na(1)), which is the dual polyhedron of a tetrakaidecahedral cage.

The comparison between the structures of F–K alloys and those of clathrate-type hydrates, corresponding inorganic clathrates and clathrasils are summarized in Tables 1.3 and 1.4. In addition to type I and type II already cited, the F–K σ-phase ($Cr_{46}Fe_{54}$) is the template for type III clathrate hydrates, such as $(Br_2)_{20}\square_{10}(H_2O)_{172}$ [58] as well as that of the inorganic clathrates $M_{30}(Na_{(1\cdot33x-10)}Sn_{172-x})$ (M = Cs or Rb + Cs; x ∼ 9.6) [59] and $Te_y(P_xSi_{172-x})$ (x = 2y; y > 20) [35]. In the same way, the Z phase type-alloy ($Zr_4Al_3$ for example) is the template for the type IV clathrate structure, which is in fact a hypothetical structure inferred from a less symmetrical complex hydrate [26]. This structure-type does not seem to exist, neither in inorganic clathrates, nor in clathrasils. It is also the case for the hexagonal

**Table 1.3** Structural relations between Frank-Kasper alloys and I, II and III clathrate types

| | Type I | Type II | Type III |
|---|---|---|---|
| Template alloy | $Cr_3Si$ (A15) | $MgCu_2$ (C15) Cubic Laves phase | $Cr_{46}Fe_{54}$ σ-phase |
| C.N. of F-K polyhedrons | 2×12 + 6×14 | 16×12 + 8×16 | 10×12 + 16×14 + 4×15 |
| Average C.N. | 13.5 | 13.33 | 13.47 |
| System | Cubic | Cubic | Tetragonal |
| Space group | Pm-3n (#223) | Fd-3m (#227) | $P4_2$/mnm (#136) |
| Clathrate structure | | | |
| Cage Nb. | $2(5^{12})+6(5^{12}6^2)$ | $16(5^{12})+8(5^{12}6^4)$ | $10(5^{12})+16(5^{12}6^2)+4(5^{12}6^3)$ |
| Formula | $G_8H_{46}$ | $G_{24}H_{136}$ | $G_{30}H_{172}$ |
| *H*/*G* ratio[a] | 5.75 or 7.67[b] | 5.67 or 17[b] | 5.73 or 8.6[b] |
| Clathrate hydrates | Gas hydrates | Liquid hydrates, mixed hydrates, small gas-hydrates | Small-sized liquid molecules ($Br_2$, $(CH_3)_2O$) |
| Examples | $(H_2S)_8(H_2O)_{46}$ $\square_2(CO_2)_6(H_2O)_{46}$ | $\square_{16}(C_3H_8)_8(H_2O)_{136}$ $(H_2S)_{16}(CCl_4)_8(H_2O)_{136}$ $(N_2)_{24}(H_2O)_{136}$ | $\square_{10}Br_{20}(H_2O)_{172}$ |
| Inorganic clathrates | Si, Ge, Sn | Si, Ge, Sn | Si+P, Sn |
| Examples | $Na_8Si_{46}$ $\square_2Rb_6Si_{46}$ | $\square_{(24-x)}Na_xSi_{136}$ $0 \leq x \leq 24$ $\square_{16}Rb_8Si_{136}$, $\square_{16}Cs_8Si_{136}$ | $Te_x(P_{2x}Si_{(172-2x)})$ (x>20) $Cs_{30}(Na_{1.33x-10}Sn_{172-x})$ (x~ 9.6) |
| Clathrasils | Melanophlogite MEP | Dodecasil-3C MTN | |
| Examples | $(CH_4, N_2, CO_2)_x$ $(SiO_2)_{46}$ | $((CH_3)_3N, CO_2)_y(N_2, Ar, CH_4)_x$ $(SiO_2)_{136}$ | No example |

[a] *G* and *H* stand for the number of guest and host species, respectively
[b] Calculated when the smallest voids are vacant

type V clathrate, which is related to the type II one and corresponds to the AB instead of ABC stacking of Kagomé-type nets. The corresponding template is the hexagonal C14 Laves-structure of $CaZn_2$. The VI and VII clathrate-type structures of Geffrey are not taken into account, because none of the F–K polyhedrons are concerned in both types and even no pentagonal face exists in type VI. For the new H-type clathrate structure identified by Repmeister et al. [27] in the double-hydrate $(H_2S)_5M(H_2O)_{34}$ (M = methylcyclohexane), the corresponding template is the $CaZn_5$ alloy-structure, which is built up from C.N.12, C.N.12' and C.N.20 polyhedrons. A related clathrasil, dodecasil 1H (DOH) also exists, in which a molecule such as piperidine is located in the large $5^{12}6^8$ cage and $N_2$ in the small $5^{12}$ and $4^35^66^3$ ones. No corresponding clathrate structure to the μ-type alloy ($Fe_7W_6$) cited by Frank and Kasper as a possible template has been identified so far. As indicated in Tables 1.3 and 1.4, the average C.N. in the alloys is close to 13.4 and the *H*/*G* ratio in the fully occupied clathrates in between 5.67 and 5.75.

**Table 1.4** Structural relations between Frank-Kasper alloys and IV, V and *H* clathrate types

| | Type IV | Type V[a] | Type H[b] |
|---|---|---|---|
| Template alloy | $Zr_4Al_3$ Z phase | $CaZn_2$ (C14) Hexag. Laves phase | $CaZn_5$ |
| C.N. of F-K polyhedrons | 6×12 + 4×14 + 4×15 | 8×12 + 4×16 | 3×12 + 2×12'[c] + 1×20 |
| Average C.N. | 13.43 | 13.33 | 13.33 |
| System | Hexagonal | Hexagonal | Hexagonal |
| Space group | P6/mmm (#191) | $P6_3$/mmc (#194) | P6/mmm (#191) |
| Clathrate Structure | | | |
| Cage Nb. | $6(5^{12})+4(5^{12}6^2)+4(5^{12}6^3)$ | $8(5^{12})+4(5^{12}6^4)$ | $3(5^{12})+2(4^35^66^3)+(5^{12}6^8)$ |
| Formula | $G_{14}H_{80}$ | $G_{12}H_{68}$ | $G_6H_{34}$ |
| *H*/*G* ratio[d] | 5.71 or 10[e] | 5.67 or 17[e] | 5.67 or 34[e] |
| Clathrate hydrates | Extrapolated from pseudo-hexag. hydrate | Related to type II | New type |
| Examples | $(i\text{-}C_4H_9)_4NF(H_2O)_{38}$ | No example | $(H_2S)_5(C_7H_{14})(H_2O)_{34}$ ($C_7H_{14}$= methyl-cyclohexane) |
| Inorganic clathrates | | | |
| Examples | No example | No example | No example |
| Clathrasils | | | Dodecasil-1H DOH |
| Examples | No example | No example | $(N_2)_5C_5H_{11}N(SiO_2)_{34}$ ($C_5H_{11}N$ = piperidine) |

[a] Hypothetical type
[b] Does not strictly correspond to a F-K structure: C.N.>16
[c] Dual polyhedron to the $4^35^66^3$ cage (see Fig. 1.1.e)
[d] *G* and *H* stand for the number of guest and host species, respectively
[e] Calculated when the smallest voids are vacant

Two new inorganic clathrate type-structures have been recently discovered, which analogues do not exist in hydrates and clathrasils. The first one, type VIII, identified in the low-temperature forms of $Ba_8(Ga_{16}Ge_{30})$ [60] and $Eu_8(Ga_{16}Ge_{30})$ [61] is cubic (space group I-43m) and built up from only one kind of cage with 23 vertices, corresponding to a distorted and modified pentagonal dodecahedron (20 + 3 vertices). At high temperature, both structures undergo a phase transition to a type I clathrate. The second one, type IX, also cubic (space group $P4_132$), has been evidenced in a series of phases of composition such as $Ba_6In_4Ge_{21}$ [62] or $Ba_6Si_{24}$ [63]. Their chiral-type structure involves the presence of pentagonal dodecahedrons sharing three of their faces. It is related to a type I clathrate by substitution of Ba atoms for pairs of host species in the hexagonal faces of the tetrakaidecahedrons. No corresponding template alloy-structure has been identified for those clathrate types.

## 1.11 Clathrate Structures and Fullerene Forms of Carbon

The buckyball fullerenes, $C_N$, correspond to hollow clusters of carbon, with any even $N \geq 20$, $N \neq 22$, built of edge-sharing pentagonal and hexagonal faces, as is the case of most of the characteristic cages of the clathrates. It is well known from Euler's rule that in a convex polyhedron, the number of faces (F), vertices (V) and edges (E) are related by $F + V = E + 2$. Applying this relation to the case of pentagonal and hexagonal faces leads to the following condition that such polyhedron must contain 12 pentagonal faces and any n number of hexagonal ones so that $0 \leq n \leq \infty$, except 1. In a fullerene cluster, n and N are related by $N = 20 + 2n$, 20 vertices coming from the 12 pentagonal faces and $2 \times n$ more from the n hexagonal ones. Increasing n values generates bigger and bigger clusters until a giant one corresponding to a close graphene sheet with twelve pentagonal faces and an infinite number of hexagonal ones. The smallest possible fullerene is $C_{20}$ ($n = 0$) and corresponds to a pentagonal dodecahedron, and the following ones, $C_{24}$ ($n = 2$), $C_{26}$ ($n = 3$), $C_{28}$ ($n = 4$) and $C_{36}$ ($n = 8$) to the 14-, 15-, 16- and 20-faced cages encountered in the clathrate-type structures. Thus, the characteristic cages of inorganic clathrates are the same as those of the smallest fullerenes clusters of carbon, the first element of group 14.

In the well known archetype $C_{60}$ fullerene, the 12 pentagonal share faces with only hexagonal ones and its high stability has been interpreted as the result of the presence of isolated pentagonal faces [64]. This empirical rule is no more fulfilled in fullerenes with $N < 60$ and below $N = 32$, the fullerene clusters are highly unstable because of strong strains due to the presence of many pentagonal faces. These strains are the strongest in $C_{20}$, which has not been obtained by laser vaporization, but synthesized by dehydration of dodecahedrane, $C_{20}H_{20}$ [65]. Another reason for the high stability of $C_{60}$ is that it corresponds to the smallest fullerene in which all the carbon valence electrons are engaged in simple and double bonds, the latter being localized along the edge of the hexagonal faces. For $N < 60$, dangling bonds are created at the surface of the corresponding clusters, which tend to increase their reactivity and make them less stable. The carbon hybridization in fullerenes is intermediate between $sp^2$ and $sp^3$. According to Haddon [66, 67], it is $sp^{2.28}$ in $C_{60}$ and increases with decreasing N values.

Clathrate-type cages are quite comparable to the smallest fullerene-type clusters. The need to saturate the dangling bonds (one per atom) is there satisfactorily solved by face-sharing of two or three of such small clusters under the form of 3-D host lattices with all bond angles close to 109.47°. In the f.c.c. crystal structure of $C_{60}$, there are strong intra-molecular bonds and weak Van der Waals inter-molecular interactions, whereas all bonds are the same in the clathrate networks. Alkali-metals (Na, K, Rb, Cs) can be inserted in the tetrahedral and octahedral voids of the structure of $C_{60}$, forming fullerides $A_xC_{60}$ (A = alkali metal) with x = 1, 2, 3, 4, 6, 10, all of them being exohedral fullerides. Unlike the inorganic clathrates enclosing alkali metals, which are insensitive to moisture and most of chemicals, the exohedral fullerides are highly reactive. However, endohedral fullerides also exist.

Recent theoretical calculations on hypothetical $Si_N$ fullerene-like clusters with $N > 28$ showed that they would be unstable. However, the addition of Si atoms inside the cluster (stuffed fullerenes) make them more stable, giving rise to some magic numbers of N, such as 33, 36, 39, 45, 50 [68, 69]. According to Li et al. [70], the most probable structure of $Si_{60}$ would be a distorted truncated icosahedron with $T_h$ symmetry instead of $I_h$. On the other hand, the existence of clathrate forms of carbon seems possible [71–73].

## 1.12 First Studies on Physical Properties of Inorganic Clathrates

### *1.12.1 Electrical Conductivity of Inorganic Clathrates*

The thermal evolution of the electrical conductivity of $Na_8Si_{46}$, $K_7Si_{46}$, $K_8Ge_{46}$ and $Na_xSi_{136}$ with x = 3, 7, 11 and 22 was studied on pelletized samples sintered under vacuum at 450 °C only, in order to avoid decomposition. The data obtained in such conditions were inaccurate, due to the influence of grain boundaries. However, a satisfying general tendency was observed, later confirmed by new measurements made on better samples and experimental conditions, as well as by theoretical calculations [4, 38, 74]. The results are reported in Fig. 1.7.

For the $Na_8Si_{46}$ and $K_7Si_{46}$ samples, in which all or almost all the available sites are occupied by alkali-metal atoms, a metallic type behavior was observed in the temperature range 77–600 K. The three $Na_xSi_{136}$ samples (x = 3, 7 and 11) were obtained by thermal decomposition, as reported before, and the last one by treatment of a $Na_xSi_{136}$ sample under sodium vapor in order to increase as much as possible the Na content: $Na_xSi_{136} + (x'-x)Na \rightarrow Na_{x'}Si_{136}$ ($x' > x$) [74]. The observed behavior was consistent with that of a heavily doped semiconductor characterized in the low temperature domain by a activation energy, δE, which increases gradually at higher temperature. The observed δE value increases from 0.001 eV for the almost full and quasi-metallic x = 22 sample to 0.013 eV for x = 11, 0.018 eV for x = 7 and 0.04 eV for x = 3. For the sake of comparison, the thermal behavior of a sample of diamond-type silicon, free of sodium, obtained by thermal decomposition at 500 °C of $Na_xSi_{136}$, was also studied. The extrapolated value of the activation energy at high temperature gave: $\Delta E = 1.2$ eV.

The above results on $Na_xSi_{136}$ were first interpreted by the formation of a narrow impurity band below the conduction band of silicon. With increasing x, this band becomes more and more filled and for the upper x value of 22 almost merges with the conduction band, giving rise to a quasi-metallic behavior, which is effectively observed for $Na_8Si_{46}$ and $K_7Si_{46}$. This interpretation was revisited by Mott [75] who found there an example of semiconductor doping by electrons issued from foreign atoms in interstitial rather than in substitutional position which do not affect the silicon network. The change in conductivity with decreasing x

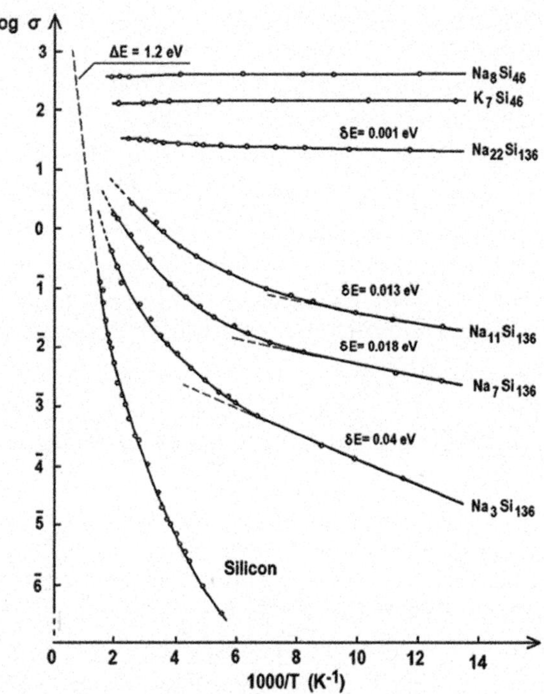


**Fig. 1.7** Electrical conductivity as a function of reciprocal temperature of $Na_8Si_{46}$, $K_7Si_{46}$ and $Na_xSi_{136}$ (x = 3, 7, 11, 22) [74]

value occurs through a classical metal-insulator Mott transition. Numerous recent theoretical calculations have been performed on the band structure of empty and doped inorganic clathrates and the corresponding conductivity properties [76, 77].

A semiconducting behavior was also observed for the germanium clathrate $K_8Ge_{46}$: δE = 0.01 eV. This rather surprising difference in behavior with the corresponding fully occupied silicon analogues was interpreted by Von Schnering [34] as due to a slight non-stoichiometry of the germanium host lattice involving the presence of ionized potassium atoms, $(K^+)_8(Ge_{44}\square_2)^{8-}$.

## *1.12.2 Thermal Properties of Inorganic Clathrates*

Figure 1.8 shows the thermal evolution of the Seebeck coefficient, S, of $K_7Si_{46}$, $Na_xSi_{136}$ with x = 3, 7 and 11 and $K_8Ge_{46}$. The results are consistent with the above conductivity measurements. For metallic $K_7Si_{46}$, a negative and low value of S is observed, linearly varying from −2 at 77 K to −25 μV/K at 600 K. Higher negative S value were obtained for the three compositions of the semi-conducting $Na_xSi_{136}$ (−61 and −143 μV/K at 350 K for x = 11 and x = 7, respectively, and −303 μV/K at 400 K for x = 3), as well as for $K_8Ge_{46}$ (−100 μV/K at 300 K) [4, 38].

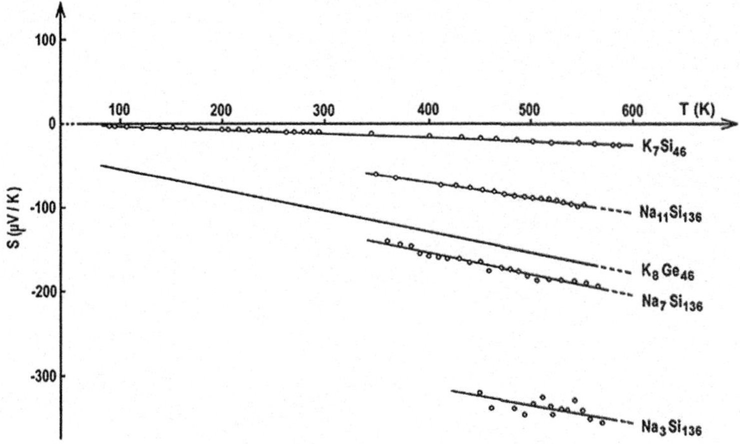


**Fig. 1.8** Thermal evolution of the Seebeck coefficient of $K_7Si_{46}$, $K_8Ge_{46}$ and $Na_xSi_{136}$ for x = 3, 7 and 11 [38]

### 1.12.3 Pressure Behavior of Inorganic Clathrates

The influence of high pressure on the crystal structure of $Na_xSi_{136}$ with x = 3 and 11 was first studied by Bundy and Kasper [78]. A high-compression belt apparatus which was capable of obtaining pressures approaching 20 GPa was used. The pressure evolution in the structure was followed by the change in resistivity of the sample at room temperature at increasing and decreasing pressure. The results are reported in Fig. 1.9. For $Na_3Si_{136}$ and up to about 10 GPa, only a slight change in resistivity was observed, followed by a very sharp decrease by four orders of magnitude to a metallic structure. For $Na_{11}Si_{136}$, a smoother change in resistivity was observed at pressure above 14 GPa. The observed change in conductivity was not reversible on decreasing pressure and the structure of the recovered phases was not identified in the X-ray diffraction patterns.

In spite of their apparent fragility, the two silicon clathrates proved to be remarkably stable under high pressure and their behavior appeared quite comparable to that of diamond-type silicon in the same experimental conditions. Above 8 and up to 15 GPa, a smooth drop of seven degrees of magnitude was observed, giving a metallic tin-like structure. The sharp transition in the conductivity of $Na_3Si_{136}$ at 10 GPa was proposed to be used as a pressure calibration standard.

### 1.12.4 NMR and ESR Characterizations

Static and MAS NMR spectroscopy techniques were used to characterize both the guest and host-lattice atoms in type I $Na_8Si_{46}$ and $(Na,Ba)_xSi_{46}$, as well as in type II $Na_xSi_{136}$ (x ~ 24), $Na_{16}Rb_8Si_{136}$ and $Na_{16}Cs_8Si_{136}$ [79–87]. Large Knight shift (δ)

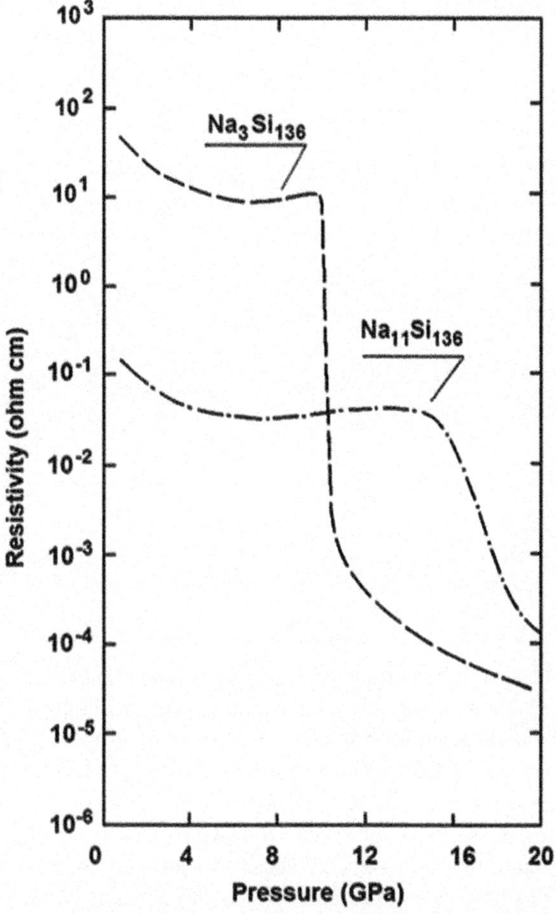


**Fig. 1.9** Pressure evolution at room temperature of the resistivity of $Na_3Si_{136}$ and $Na_{11}Si_{136}$ [78]

values ranging from 1200 to 2100 ppm of the $^{23}$Na resonance line were observed, slightly higher than in metallic sodium itself (δ = 1123 ppm) and much higher than in the Zintl-type ionic compound NaSi (δ = 45 ppm), indicating a partial localization of the 3 s electrons in the vicinity of the Na atoms [79]. Similar Knight shift values were reported by several other authors and after some discrepancies in the results, the observed resonance lines of the two kinds of sodium atoms in the small and large cages of the type I and type II structures were clearly assigned, taking into account the respective number of nuclei and the symmetry of the occupied site [80–87]. For $Na_8Si_{46}$, the Knight shift values are 2019 and 1768 ppm for Na(1) atoms in 2(a) at the center of the dodecahedral cages and Na(2) ones in 6(d) position at the center of the tetrakaidecahedral cages, respectively. For $Na_xSi_{136}$ (x ∼ 24), the δ values are 1601 and 1810 ppm for Na(1) atoms in 16(c) at the center of the dodecahedral cages and Na(2) in 8(b) at the center of the hexakaidecahedral ones, respectively [86]. Calculations have been made to correlate the experimental data to the electronic density of states near the Fermi level [79, 80, 82–86].

The $^{29}Si$ NMR spectra showed the presence of three shifted lines corresponding to the three independent sites for silicon in both type I and type II clathrates, with intensities approximately proportional to the number of Si atoms. For $Na_8Si_{46}$, Knight shift values referred to TMS ($(CH_3)_4Si$) of ~615 ppm for Si(2) in 16 (i), ~650 for Si(1) in 6(c) and ~840 ppm for Si(3) in 24(k) were observed [83, 86]. For $Na_xSi_{136}$ ($x \sim 24$) the resonance lines appeared at ~600 ppm for Si(1) in 8(a) and Si(2) in 32(e) and 713 for Si(3) in 96(g) [86]. In the type II $Na_{16}Rb_8Si_{136}$ and $Na_{16}Cs_8Si_{136}$ clathrates, Knight shift values of 275 and 205 ppm for Si(1) in 8 (a), respectively, ~420 ppm for Si(2) in 32(e) and ~720 ppm for Si(3) in 96(g) were reported [84, 85, 87]. In the almost empty $Na_xSi_{136}$ ($x \rightarrow 0$), much weaker shifts were observed: 0.3 ppm for Si(3) in 96(g), 50.4 for Si(1) in 8(a) and 98 ppm for Si(2) in 32(e), in comparison to −79 ppm for diamond-type silicon [87].

All the above results are consistent with a rather high probability of the presence of 3 s electrons on the sodium atoms, although they contribute to the density of states at the bottom of the conduction band of silicon atoms, for which the NMR resonance lines are also notably shifted. They are in agreement with the conclusions of the qualitative comparison between the free radii of the cages and the effective radius of sodium atoms in clathrates (Fig. 1.4 in Sect. 1.7).

Electron Spin Resonance (ESR) spectroscopy was used to characterize sodium atoms in the highly non-stoichiometric $Na_xSi_{136}$ clathrates [28, 88, 89]. In addition to an intense and broad line centered at $g = 2.0021$ due to the presence of Na clusters, four sharp lines of a hyperfine quadruplet with an isotropic hf splitting of 13.3 mT was observed, corresponding to isolated Na atoms ($I = 3/2$). From the observed hf splitting value, the spin density in the 3 s orbitals was estimated to be ~45 % of one unit spin density [88]. With decreasing x values, the intensity of the cluster line decreases, whereas those of the hyperfine quadruplet lines increased. For the lowest x content, the contribution of the cluster line to the total intensity was very low and a quantitative determination of the residual x content was possible by comparison of the observed intensities with those of standards (diluted solutions of $Cr^{3+}$ doped alum). The ultimate ε sodium content in $Na_\varepsilon Si_{136}$ was found to be close to 0.006, i.e. ~35 ppm [89].

## 1.13 Non-stoichiometry and Mixed Host Lattices in Inorganic Clathrates

At the end of the 1960 decade, the ability of type I structure of inorganic clathrates to form non-stoichiometric and mixed host lattices was demonstrated. In crystal growth experiments on the germanium-phosphor system using iodine as transport gaseous species, von Schnering and Menke [90, 91] and Menke and von Schnering [92] obtained a cubic phase of composition $I_8Ge_{38}P_8$, which was found to be isostructural with a type I clathrate. Replacing phosphorus by other elements of the same group 15, arsenic or antimony, and iodine by chlorine or bromine led them to synthesize a series of compounds $X_8Ge_{38}A_8$ (X = halogen and A = P, As, Sb), which were the first inorganic clathrates with anionic guest atoms and cationic mixed host lattices: $(X)_8\left(Ge^0_{38}A^+_8\right)$. As expected, these compounds were characterized by semiconducting properties (the gap is 0.9 eV for $I_8Ge_{38}Sb_8$).

Another example was given by Nesper et al. [93] ). As they were trying to synthesize the empty $Ge_{136}$ clathrate with iodine as transport gaseous species, they obtained a phase of composition close to $Ge_4I$, which proved to correspond to a clathrate of composition $I_8(Ge_{46-x}I_x)$ with x = 2.67 and the following ionic charge distribution: $(I^-)_8\left(Ge_{43.33}I^{3+}_{2,67}\right)$. In this compound, a same element, iodine, is present in both the guest and host lattices and at two different oxidation states. Using the same Zintl-Klemm concept, the semiconducting properties of $K_8Ge_{46}$ were explained by a deficit of 2 Ge atoms in the host lattice leading to the following charge distribution: $(K^+)_8(Ge_{44}\square_2)^{8-}$. In the $A^{II}_8B^{III}_{16}B^{IV}_{30}$ ($A^{II} \equiv$ Sr, Ba; $B^{III} \equiv$ Al, Ga; $B^{IV} \equiv$ Si, Ge, Sn) type I clathrates, the sixteen outer electrons of the eight alkaline-earth guest atoms compensate the sixteen missing electrons of the mixed host lattice $\left(B^{III}_{16}B^{IV}_{30}\right)^{16-}$ [60].

Many other examples can be given, which show the large variety of possible inorganic clathrate compounds, including light or heavier elements and leading to different physical properties.

## 1.14 Brief Introduction to the Next Chapters

In the late-eighties of the last century, the discovery of fullerene forms of carbon and high $T_C$ superconducting properties of Ba–La cuprates, followed a few years later by that of the high $T_C$ properties of $M_3C_{60}$ (M = K, Rb, Cs) fullerides, attracted the attention of a great number of scientists on the cage-like inorganic clathrates and all their interesting features [94–96]. From that time, these compounds rapidly became one of the major recent subject matter for scientists of various disciplines.

The following chapters provide a thorough and updated review on all the main topics which have been developed and still going on:

- Chapters 2, 3, 5, 9 and 10 concern the synthesis, the structure and the physical properties of numerous anionic and cationic clathrates with light or heavier elements as host lattices and various other ones as guest species. Some of them exhibit very interesting thermal properties and are promising thermoelectric materials, a point which is treated in details in Chap. 6;
- Chapters 4, 11 and 12 are dealing with three powerful characterization methods. The first one is the in situ X-ray diffraction study of the influence of pressure on the structure of clathrates, using a high pressure cell and synchrotron radiation as incident beam. The second one concerns the contribution of neutron diffraction studies to the structure of clathrates and the third one is about the mechanical properties of clathrates, which can be considered as hard materials;

- Also related to high pressure is Chap. 8, which treats of the use of this efficient technique for the synthesis of new clathrates, such as $Ba_8Si_{46}$ ($T_C$ = 8 K), $Ba_{24}Si_{100}$ ($T_C$ = 1.4 K) and $BaSi_6$. Their superconducting properties are presented and interpreted;
- Chapter 7 is about the theoretical aspect of clathrates and is of great importance. The results of the calculation give very interesting information such as the stabilization energy and band structure of the empty clathrates $Si_{46}$ and $Si_{34}$ ($Si_{136}/4$), which would be almost direct wide-gap semiconductors ($\sim$1.9 eV). Furthermore, the calculations have been extended to hypothetical clathrate structures of carbon which would be relatively stable in comparison to diamond and possibly doped by metallic species, giving rise to conducting materials almost as hard as diamond. The synthesis of such materials appears as an exciting challenge for scientists of many disciplines.

## References

1. C. Cros, M. Pouchard, P. Hagenmuller, Sur deux nouvelles phases du système silicium-sodium. C. R. Acad. Sci. Paris **260**, 4764–4767 (1965)
2. C. Cros, Le système silicium-sodium. Thèse de Troisième Cycle, Université de Bordeaux (1965)
3. J.S. Kasper, P. Hagenmuller, M. Pouchard, C. Cros, Clathrate structures of silicon. Science **150**, 1713–1714 (1965)
4. C. Cros, Sur quelques nouveaux siliciures et germaniures alcalins à structure clathrate: étude cristallographique et physique. Thèse Université de Bordeaux (1970)
5. C. Cros, M. Pouchard, Sur les phases de type clathrate du silicium et des éléments apparentés (C, Si, Ge, Sn): une approche historique. C. R. Chimie **12**, 1014–1056 (2009)
6. E. Hohmann, Silicide und germanide der alkalimetalle. Z. Anorg. Chem. **257**, 113–126 (1948)
7. R. Schäfer, W. Klemm, Das verhalten der alkalimetalle zu halbmetallen. IX- Weitere beiträge zur kenntnis der silicide und germanide der alkalimetalle. Z. Anorg. Allg. Chem. **312**, 214–224 (1961)
8. H.M. Powell, The structure of molecular compounds. Part IV. Clathrate compounds. J. Chem. Soc. **16**, 61–73 (1948)
9. H. Davy, On a combination of oxymuriatic gas and oxygene gas. Phil. Trans. R. Soc. London **101**, 155–162 (1811)
10. M. Faraday, On fluid chlorine. Phil. Trans. R. Soc. London **113**, 160–165 (1823)
11. M. Berthelot, Recherches sur les relations qui existent entre l'oxyde de carbone et l'acide formique. Ann. Chem. Phys. **3**(46), 477–491 (1856)
12. H. Le Chatelier, Sur un énoncé général des lois des équilibres chimiques. C. R. Acad. Sci. Paris **99**(19), 786–789 (1884)
13. R. De Forcrand, Recherches sur les hydrates sulphydriques. Ann. Chim. Phys. **5**(38), 5–67 (1883)
14. R. De Forcrand, Sur la composition des hydrates de gaz. C. R. Acad. Sci. Paris **135**, 959–961 (1902)
15. P. Villard, Etude expérimentale des hydrates de gaz. Ann. Chim. Phys. **11**(7), 289–394 (1897)

16. P. Villard, Sur quelques nouveaux hydrates de gaz. C. R. Acad. Sci. Paris **106**, 1602–1603 (1888)
17. P. Villard, Sur quelques nouveaux hydrates de gaz. C. R. Acad. Sci. Paris **111**, 302–305 (1890)
18. P. Villard, Sur l'hydrate carbonique et la composition des hydrates de gaz. C. R. Acad. Sci. Paris **119**, 368–371 (1894)
19. W. Schroeder, Die Geschichte der gashydrate. Sammlung Chem. Chem-Tech. Vorträge **29**, 1–98 (1927)
20. M. Von Stackelberg, Feste gashydrate. Naturwiss **36**, 327–333 (1949)
21. W.F. Claussen, Suggested structures of water in inert gas hydrates. J. Chem. Phys. **19**, 259–260 (1951)
22. M. Von Stackelberg, H.R. Müller, On the structure of gas hydrates. J. Chem. Phys. **19**, 1319–1320 (1951)
23. W.F. Claussen, A second water structure for gas hydrates. J. Chem. Phys. **19**, 1425–1426 (1951)
24. H.R. Müller, M. von Stackelberg, Zur struktur der gashydrate. Naturwiss **1**, 20–21 (1951)
25. L. Pauling, R.E. Marsch, The structure of chlorine hydrate. Proc. Nat. Acad. Sci. US. **38**, 112–118 (1952)
26. G.A. Jeffrey, in *Hydrate Inclusion Compounds*. ed. by J.L. Atwood, J.E. Davies, D.D. McNicol. Inclusion Compounds I, Chapter 5 (Academic Press Inc., London, 1984), pp. 135–190
27. J.A. Ripmeester, J.S. Tse, C.I. Ratcliffe, B.M. Powell, A new clathrate hydrate structure. Nature **325**, 135–136 (1987)
28. S.B. Roy, K.E. Sim, A.D. Caplin, The insulating-to-metal transition in Si-Na clathrate compounds: a search for superconductivity. Phil. Mag. B **65**, 1445–1450 (1992)
29. E. Reny, P. Gravereau, C. Cros, M. Pouchard, Structural characterizations of the $Na_xSi_{136}$ and $Na_8Si_{46}$ silicon clathrates using the Rietweld method. J. Mater. Chem. **8**, 2839–2844 (1998)
30. G.K. Ramachandran, J. Dong, J. Dieffenbacher et al., Synthesis and X-ray characterization of silicon clathrates. Solid State Chem. **145**, 716–730 (1999)
31. C. Cros, M. Pouchard, P. Hagenmuller, J.S. Kasper, Sur deux composés du potassium isotypes de l'hydrate de krypton. Bull. Soc. Chim. Fr. **7**, 2737–2742 (1968)
32. J. Gallmeier, H. Schäfer, A. Weiss, K8Si46, ein silicid mit käfigstruktur. Z. Naturforsch. **22b**, 1080 (1967)
33. J. Gallmeier, H. Schäfer, A. Weiss, Eine käfigstruktur als gemeinsames bauprinzip der verbindungen K8E46 (E = Si, Ge, Sn.). Z. Naturforsch. **24b**, 665–667 (1969)
34. H.G. Von Schnering, Zintl-phasen: prinzipien von struktur und bindung. Nova Acta Leopold **59**, 165–182 (1985)
35. J.V. Zaikina, K.A. Kovnir, F. Haarmann et al., First silicon-based cationic clathrate III with exceptional high thermal stability. Chem. Eur. J. **14**, 5414–5422 (2008)
36. J.-T. Zhao, J.D. Corbett, Zintl-phase in alkali-metal-Tin systems: $K_8Sn_{25}$ with condensed pentagonal dodecahedra of tin: two $A_8Sn_{44}$ phases with a defect clathrate structure. Inorg. Chem. **33**, 5721–5726 (1994)
37. E.D. Sloan Jr, Fundamental principles and applications of natural gas hydrates. Nature **426**, 353–359 (2003)
38. C. Cros, M. Pouchard, P. Hagenmuller, Sur une nouvelle famille de clathrates minéraux isotypes des hydrates de gaz et de liquides. Interprétation des résultats obtenus. J. Solid State Chem. **2**, 570–581 (1970)
39. W.M. Haynes, *Handbook of Chemistry and Physics*, 91th edn. (CRC Press, Boca Raton New York London, 2008–2011), Chap. 9, pp. 49–50
40. G.D. Holder, D.J. Manganiello, Hydrate dissociation pressure minima in multicomponent systems. Chem. Eng. Sc. **37**(1), 9–16 (1982)
41. D.W. Davidson, Y.P. Handa, C.I. Ratcliffe, J.S. Tse, B.M. Powell, The ability of small molecules to form clathrate hydrates of structure II. Nature **311**, 142–143 (1984)

42. W.L. Mao, H.K. Mao, V.V. Goncharov et al., Hydrogen clusters in clathrate hydrate. Science **297**, 2247–2249 (2002)
43. C. Cros, M. Pouchard, P. Hagenmuller, Sur deux nouvelles structures du silicium et du germanium de type clathrate. Bull. Soc. Chim. Fr. **2**, 379–386 (1971)
44. D. Weaire, R. Phelan, A counter-example to Kelvin's conjecture on minimal surface. Phil. Mag. Lett. **69**, 107–110 (1994)
45. R. Gabrielli, A.J. Meagher, D. Weaire, K.A. Brakke, S. Hutzler, An experimental realization of the Weaire-Phelan structure in monodisperse liquid foam. Phil. Mag. Lett. **92**(1), 1–6 (2012)
46. F. Liebau, H. Gies, R.P. Gunawardane, B. Marker, Classification of tectosilicates and systematic nomenclature of clathrate-type tectosilicates: a proposal. Zeolites **6**, 373–377 (1986)
47. B. Kamb, A clathrate crystalline form of silica. Science **148**, 232–234 (1964)
48. H. Gies, Studies on clathrasils III. Crystal structure of melanophlogite, a natural clathrate compound of silica. Z. Kristallogr. **164**(3–4), 247–257 (1983)
49. H. Gies, F. Liebau, Melanophlogite: composition, thermal behavior and structural refinements. Acta Cryst. **A37**, C187–C188 (1981)
50. Ch. Baerlocher, L.B. McCusker, D.H. Olson, *Atlas of Zeolite Framework Types*, 6th revised edn. (Elsevier, Amsterdam, 2007)
51. H. Gies, Studies on clathrasils VI. Crystal structure of dodecasil 3C, another synthetic clathrate of silica. Z. Kristallogr. **167**(1–2), 73–82 (1984)
52. H. Gerke, H. Gies, Studies on clathrasils IV Crystal structure of dodecasil 1H, a synthetic clathrate compound of silica. Z. Kristallogr. **166**(1–2), 11–22 (1984)
53. F.C. Frank, J.S. Kasper, Complex alloy structures regarded as sphere packings. I. Definitions and basic principles. Acta Cryst. **11**, 184–190 (1958)
54. F.C. Frank, J.S. Kasper, Complex alloy structures regarded as sphere packings II. Analysis and classification of representative structures. Acta. Cryst. **12**, 483–499 (1959)
55. M. Dutour-Sikiric, O. Delgado-Fredrichs, M. Deza, Space fullerenes: a computer search for new Frank-Kasper structures. Acta Cryst. **A66**, 1–14 (2010)
56. K. O'Keeffe, G.B. Adams, O.F. Sankey, Duals of Frank-Kasper structures as C, Si and Ge clathrates: energetics and structure. Phil. Mag. Lett. **78**(1), 21–28 (1998)
57. P. Rogl, in *Formation and Crystal Chemistry of Clathrates*, ed. by Rove D.M. Handbook of Thermoelectrics (CRC Press, Boca Raton, London, New York, 2008), Chap. 32, pp. 1–24
58. K.W. Allen, G.A. Jeffrey, On the structure of bromine hydrate. J. Chem. Phys. **38**, 2304–2306 (1963)
59. S. Bobev, S.C. Sevov, Clathrate III of group 14 exists after al. J. Am. Chem. Soc. **123**, 3389–3390 (2001)
60. B. Eisenmann, H. Schäfer, R. Zagler, Die verbindungen $A^{II}_8B^{III}_{16}B^{IV}_{30}$($A^{II} \equiv$ Sr, Ba; $B^{III} \equiv$ Al, Ga; $B^{IV} \equiv$ Si, Ge, Sn) und ihre käfigstrukturen. J. Less-Common Met. **118**, 43–55 (1986)
61. S. Paschen, W. Carrillo-Cabrera, A. Bentien et al., Structural, transport, magnetic and thermal properties of $Eu_8Ga_{16}Ge_{30}$. Phys. Rev. B **63**, 214404(1)–214404(11) (2001)
62. R. Kröner, R. Nesper, H.G. von Schnering, $Ba_6In_4Sn_{21}$- ein neuer clathrat-typ. Z. Kristallogr. **182**, 164–165 (1988)
63. H. Fukuoka, K. Ueno, S. Yamanaka, High-pressure and structure of a new silicon clathrate $Ba_{24}Si_{100}$. J. Organomet. Chem. **611**, 543–546 (2000)
64. H.W. Kroto, The stability of the fullerenes $C_n$, with n = 24, 28, 32, 36, 50, 60, 70. Nature **329**, 529–531 (1987)
65. H. Prinzbach, A. Weller, P. Landenberger et al., Gas-phase production and photoelectron spectroscopy of the smallest fullerene, $C_{20}$. Nature **405**, 60–63 (2000)
66. R.C. Haddon, Chemistry of the fullerenes: the manifestation of strain in a class of continuous aromatic molecules. Science **261**, 1545–1550 (1993)

67. R.C. Haddon, The fullerenes: powerful carbon-based electron acceptors. Phil. Trans. Roy. Soc. London **A343**, 53–61 (1993)
68. E. Kaxiras, Effect of surface reconstruction on stability and reactivity of Si clusters. Phys. Rev. Lett. **64**, 551–554 (1990)
69. J. Zhao, J. Wang, J. Jellinek et al., Stuffed fullerene structures for medium-sized silicon clusters. Eur. Phys. J. D **34**, 35–37 (2005)
70. B.X. Li, P.L. Cao, D.I. Que et al., Distorted icosahedral cage structure of $Si_{60}$ cluster. Phys. Rev. B **61**(3), 1685–1687 (2000)
71. R. Nesper, K. Vogel, P.E. Blochl, Hypothetical carbon modification derived from zeolites frameworks. Angew. Chem. Int. Edit. **32**(5), 701–703 (1993)
72. G.B. Adams, M. O'Keeffe, A.A. Demkov et al., Wide-band-gap Si in open fourfold-coordinated clathrate structures. Phys. Rev. B **49**(12), 8048–8053 (1994)
73. S. Saito, A. Oshiyama, Electronic structure of $Si_{46}$ and $Na_2Ba_6Si_{46}$. Phys. Rev. B **51**(4), 2628–2631 (1995)
74. C. Cros, J.C. Bénéjat, Préparation et propriétés d'un clathrate à très large domaine d'existence: le siliciure de sodium $Na_xSi_{136}$. Bull. Soc. Chim. Fr. **1972**, 1739–1743 (1972)
75. N.F. Mott, Properties of compounds of type $Na_xSi_{46}$ and $Na_xSi_{136}$. J. Solid State Chem. **6**, 348–351 (1973)
76. A.A. Demkov, O.F. Sankey, K.E. Schmidt et al., Theoretical investigation of alkali-metal doping in Si clathrates. Phys. Rev. B **50**, 17001–17008 (1994)
77. V.I. Smelyansky, J.S. Tse, The electronic structure of metallo-silicon clathrates $Na_xSi_{136}$ (x = 0, 4, 8, 16, 24). Chem. Phys. Lett. **264**, 459–465 (1997)
78. F.P. Bundy, J.S. Kasper, Electrical behavior of sodium-silicon clathrates at very high pressures. High Temp. High Press. **2**, 429–436 (1970)
79. J. Gryko, P.F. McMillan, O.F. Sankey, MNR studies of Na atoms in silicon clathrate compounds. Phys. Rev. B **54**(5), 3037–3039 (1996)
80. F. Shimizu, Y. Maniwa, K. Kume, NMR studies in the superconducting silicon clathrate compound $Na_xBa_ySi_{46}$. Phys. Rev. B **54**(18), 13242–13246 (1996)
81. E. Reny, M. Ménétrier, C. Cros, M. Pouchard, J. Sénégas, A $^{23}$Na NMR study of $Na_xSi_{136}$ and $Na_8Si_{46}$ silicon clathrates. C. R. Acad. Sci. Paris **1**(2c), 126–136 (1998)
82. J. Gryko, P.F. McMillan, R.F. Marzke et al., Temperature-dependent $^{23}$Na Knight shifts and sharply peaked structure in the electronic densities of states of Na-Si clathrates. Phys. Rev. B **57**(7), 4172–4179 (1998)
83. G.K. Ramachandran, P.F. McMillan, J. Diefenbacher et al., $^{29}$Si NMR study on the stoichiometry of the silicon clathrate $Na_8Si_{46}$. Phys. Rev. B **60**(17), 12294–12298 (1999)
84. G.K. Ramachandran, J. Dong, O.F. Sankey et al., 23Na and 29Si NMR shifts in the silicon clathrate $Na_{16}Cs_8Si_{136}$. Phys. Rev. B **63**, 033102(1)–033102(4) (2000)
85. S. Latturner, B.B. Iversen, J. Sepa et al., NMR Knight shifts and the electronic properties of $Rb_8Na_{16}Si_{136}$ clathrate. Phys. Rev. B **63**, 125403(1)–125403(6) (2001)
86. J. He, D.D. Klug, K. Uehara et al., NMR and X-ray spectroscopy of sodium-silicon clathrates. J. Phys. Chem. B **105**, 3475–3485 (2001)
87. M. Pouchard, C. Cros, P. Hagenmuller et al., A brief overview on low sodium content silicides: are they mainly clathrates, fullerenes, intercalation compounds or Zintl phases? Solid State Sci. **4**, 723–729 (2002)
88. H. Yahiro, K. Yamaji, M. Shiotani et al., An ESR study on the thermal electron excitation of sodium atom incorporated in a silicon clathrate compound. Chem. Phys. Lett. **246**, 167–170 (1995)
89. A. Ammar, C. Cros, M. Pouchard et al., On the clathrate form of elemental silicon, $Si_{136}$: preparation and characterization of $Na_xSi_{136}$ ($x \to 0$). Solid State Sci. **6**, 393–400 (2004)
90. H.G. Von Schnering, H. Menke, $Ge_{38}P_8I_8$ and $Ge_{38}As_8I_8$, a new class of compounds with clathrate structure. Angew. Chem. Int. Edit. **11**, 43–44 (1972)
91. H.G. Von Schnering, H. Menke, Die partielle substitution von Ge durch GaAs und GaSb in der käfigverbindungen $Ge_{38}As_8I_8$ und $Ge_{38}Sb_8I_8$. Z. Anorg. Allg. Chem. **424**, 108–114 (1976)

92. H. Menke, H.G. von Schnering, Die Käfigverbindungen $Ge_{38}A_8X_8$ mit A = P, As, Sb und X = Cl, Br. I. Z. Anorg. Allg. Chem. **395**, 223–238 (1973)
93. R. Nesper, J. Curda, H.G. von Schnering, $Ge_{4.06}I$, an unexpected germanium subiodide- a tetragermanoiodonium (III) iodide with clathrate structure $[Ge_{46-x}I_x]I_8$, x = 8/3. Angew. Chem. Int. Edit. **25**, 350–352 (1986)
94. H. Kroto, J.R. Heath, S.C. O'Brien et al., $C_{60}$: Buckminsterfullerene. Nature **318**, 162–163 (1985)
95. J.G. Bernorz, K.A. Müller, Possible high $T_C$ superconductivity in Ba-La-Cu-O system. Z. Phys. B **64**(2), 189–193 (1986)
96. A.Y. Ganin, Y. Takabayashi, P. Jeglic et al., Polymorphism control of superconductivity and magnetism in $Cs_3C_{60}$ near the Mott transition. Nature **466**, 221–225 (2010)

# Chapter 2
# Solid State Chemistry of Clathrate Phases: Crystal Structure, Chemical Bonding and Preparation Routes

**Michael Baitinger, Bodo Böhme, Alim Ormeci and Yuri Grin**

**Abstract** Clathrates represent a family of inorganic materials called cage compounds. The key feature of their crystal structures is a three-dimensional (host) framework bearing large cavities (cages) with 20–28 vertices. These polyhedral cages bear—as a rule—guest species. Depending on the formal charge of the framework, clathrates are grouped in anionic, cationic and neutral. While the bonding in the framework is of (polar) covalent nature, the guest-host interaction can be ionic, covalent or even van-der Waals, depending on the chemical composition of the clathrates. The chemical composition and structural features of the cationic clathrates can be described by the enhanced Zintl concept, whereas the composition of the anionic clathrates deviates often from the Zintl counts, indicating additional atomic interactions in comparison with the ionic-covalent Zintl model. These interactions can be visualized and studied by applying modern quantum chemical approaches such as electron localizability.

## 2.1 Clathrates as Cage Materials: From Minerals to Intermetallic Compounds

Clathrates belong to a large group of inorganic materials called cage compounds. The crystal structures of these substances are built of three-dimensional host frameworks with embedded guest species within large cavities (cages) of the framework. Representatives of this group are found among different families of inorganic materials—from silicate minerals to intermetallic compounds. In particular, minerals of the zeolite family are known for their ability to bear other

M. Baitinger · B. Böhme · A. Ormeci · Y. Grin (✉)
Max-Planck-Institut für Chemische Physik fester Stoffe,
Nöthnitzer Str. 40, 01187 Dresden, Germany
e-mail: Juri.Grin@cpfs.mpg.de

G. S. Nolas (ed.), *The Physics and Chemistry of Inorganic Clathrates*,
Springer Series in Materials Science 199, DOI: 10.1007/978-94-017-9127-4_2,

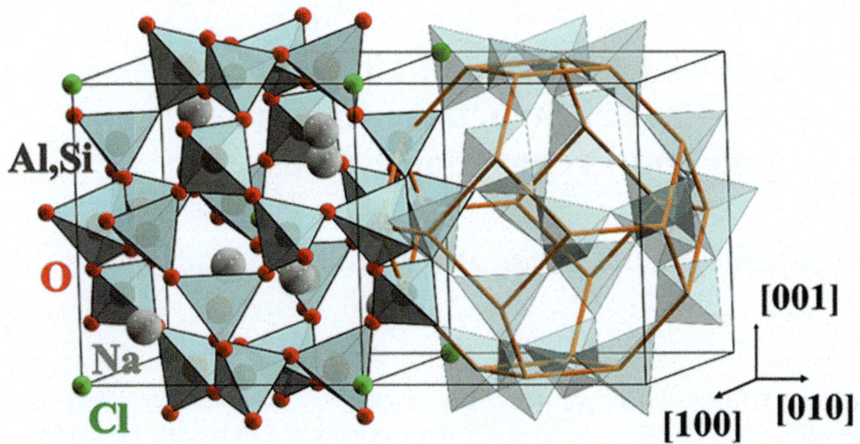


**Fig. 2.1** Crystal structure of the mineral sodalite $Na_8[Al_6Si_6O_{24}]Cl_2$ (*left*) with a structural pattern of the clathrate-VII type visualized by the *brown* graph (*right*)

substances, e.g. smaller molecules such as $CH_4$ or $CO_2$, in the cavities of their crystal structures [1]. As an example, the crystal structure of the mineral sodalite $Na_4[Al_3Si_3O_{12}]Cl$ [2, 3] constitutes one of the prototype atomic arrangements for cage compounds, the so-called clathrate-VII pattern [4, 5]. Its framework is formed by $[SiO_4]$ and/or $[AlO_4]$ tetrahedral units sharing oxygen vertices of neighboring tetrahedrons (Fig. 2.1). The formal charge of these units can be calculated from the structural formulas as $[SiO_{4/2}]^0$ and $[AlO_{4/2}]^{1-}$, where the subscript denominator denotes the number of tetrahedrons between which the given atom is shared. Because of the difference in electronegativity between silicon or aluminum on the one hand and oxygen on the other hand, the Si–O and Al–O bonds are covalent polar. The resulting three-dimensional framework carries negative charge. It comprises tetrakaidecahedral $[4^66^8]$ cavities which are filled by $[ClNa_4]^{3+}$ units, so that charge compensation is achieved according to $(Na^{1+})_8[(AlO_{4/2})^{1-}]_6[(SiO_{4/2})^0]_6(Cl^{1-})_2 = Na_8[Al_6Si_6O_{24}]Cl_2$.

A similar spatial organization is observed in the clathrate hydrates, the oldest known group of clathrate compounds. The formation of such substances was already reported by Davy [6] and Faraday [7]. The crystal structures were solved more than a 100 years later [8–13]. The prototype clathrate hydrates have been named by roman numbers or letters. In the clathrate-I crystal structure with ideal composition $X_8(H_2O)_{46}$ ($X$ stands for a small molecule), the framework is built of water molecules connected via hydrogen bonds (Fig. 2.2, left). The covalent- and hydrogen-bonded O–H–O groups are arranged in such a way that pentagondodecahedral and tetrakaidecahedral cages are formed. The whole network can be described as being formed by neutral $[OH_{4/2}]$ groups. It is therefore uncharged and the cavities can bear neutral molecules of appropriate size according to the balance $(X^0)_8[(OH_{4/2})^0]_{46} = X_8(H_2O)_{46}$. Hence this clathrate family is referred to as

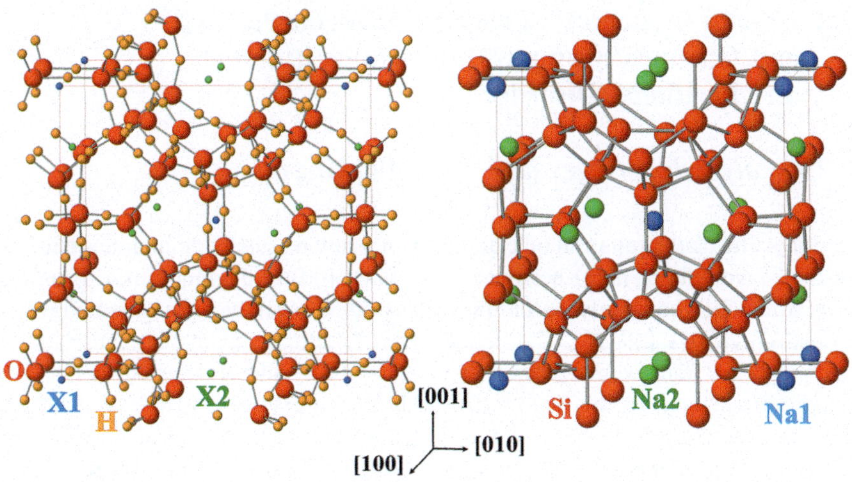


**Fig. 2.2** Relationship between the clathrate-I-type crystal structures of the clathrate hydrates ($X_8(H_2O)_{46}$, *left*) and the intermetallic clathrates ($Na_8Si_{46}$, *right*)

neutral from a chemical point of view. The framework of the mineral melanophlogite $X'_6X''_2(SiO_2)_{46}$ can be derived by substitution of $[OH_{4/2}]$ in the clathrate-I-type structure by $[SiO_{4/2}]$ units [14]. Melanophlogite is a silica modification belonging to the clathrasil family [15, 16]. In clathrasils, the framework is stabilized by covalent Si–O bonds. Unlike the aforementioned sodalite containing $[AlO_{4/2}]^-$ units, the framework in clathrasils carries no charge. As is the case for clathrate hydrates, the cavities therefore can host neutral molecules.

Formally, removing the hydrogen atoms from the clathrate hydrate structure and forming direct bonds between the tetrahedron centers yields the structural motif of intermetallic clathrates. This way (Fig. 2.2), the crystal structure of $X_8(H_2O)_{46}$ may be transformed into that of $Na_8Si_{46}$ [17]. In this case, the framework carries negative charge so that this type of clathrate is referred to as anionic. Consequently, the guest atoms are positively charged. From the crystallographic point of view the same type of crystal structure describes the so-called inverse clathrates, e.g. $Ge_{38}P_8I_8$ [18], with inverse polarity of guest and host substructure, i.e. guest iodine as anion. Despite the formal crystallographic difference in the number of atoms in the unit cell and their chemical functionality, all four crystal structures are considered to belong to the clathrate-I type. The notation originates from work on hydrate clathrates and has been adopted for the other clathrate families [4, 5].

This chapter deals with the chemical composition, crystal structures and their crystallographic features, chemical bonding and preparation routes of intermetallic clathrates.

## 2.2 Crystal Structures: From Polyhedron Packing and Chemical Composition to Substitutional and Positional Disorder

### 2.2.1 Polyhedral Cages and Their Packings

Intermetallic clathrates attract the attention of many research communities because of their favorable physical behaviors, e.g. as promising thermoelectric materials. The structural characteristics of this family have recently been summarized in extended reviews [19–23]. One of the key structural features of intermetallic clathrates are frameworks formed predominantly by four-connected atoms. The densest framework of this kind is the well-known crystal structure of diamond. This is in particular because of the staggered conformation of all neighboring atoms. In clathrate structures the neighboring framework atoms in the ecliptic conformation are also present. This local arrangement leads to a less dense framework with differently sized cavities (polyhedral cages) therein. The cavities found in intermetallic clathrates have most commonly the shape of 20-atom dodecahedron, but also the less-observed larger polyhedrons like 24-atom tetrakaidecahedron, 26-atom pentakaidecahedron and 28-atom hexakaidecahedron (Fig. 2.3). In the crystal structures of clathrates I, II, III, IV and V, only these four basic polyhedral cages occur; the space of the structures thus formed is filled completely by arranging the polyhedrons appropriately.

The host framework may be modified either by changing the ratio of staggered and ecliptic host atoms (at least by occurring of the framework atoms in an intermediate conformation) or by introducing three-bonded atoms. In both cases, the consequence is the formation of additional cavities of different size. In this way, in the crystal structure of the clathrate-VIII type, (20 + 3) polyhedrons (derived from the dodecahedron) are condensed so that additional small eight-vertices empty cavities appear [24]. The crystal structure of $Ba_6In_4Ge_{21}$ (clathrate *cP*124 or type IX) is characterized by a chiral framework of condensed dodecahedrons. The channels within this framework constitute the so-called *Y* graph. The structure represents an intermediate between clathrates and zeolites [25]. The space within this framework is filled by larger irregular 20-atom polyhedrons, filled e.g. by barium atoms in $Ba_6Ge_{25}$ [26–28]. In addition, smaller 11-vertices cavities formed mainly by three-bonded germanium atoms appear which are partially filled by potassium atoms in $K_{6+x}Sn_{25}$ [29].

The crystal structure of recently discovered $BaGe_5$ (clathrate *oP*60) [30] is formed by Ba-centered $Ge_{20}$ dodecahedrons arranged on an orthorhombic lattice and interconnected via three-bonded germanium species to two-dimensional layers. Between the layers, $Ge_{24}$ polyhedrons and additional smaller cavities are formed, which host the remaining Ba atoms, in analogy to $Ba_6Ge_{25}$.

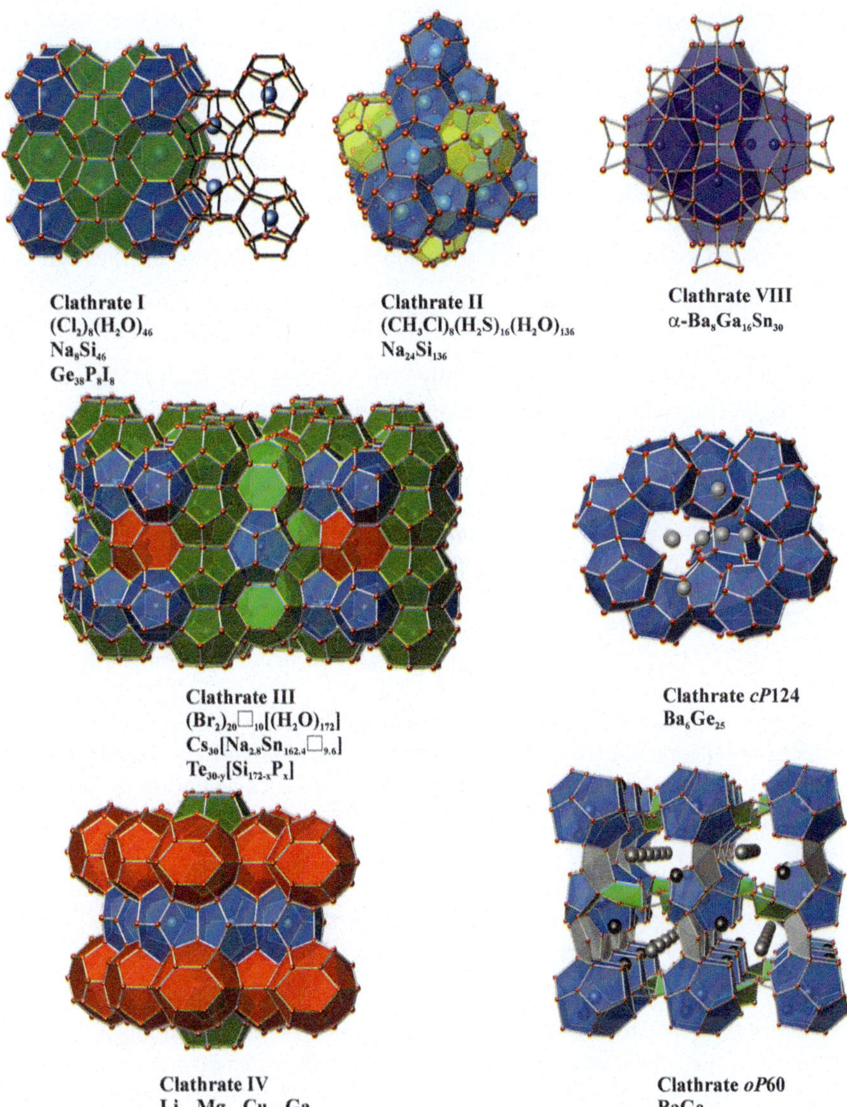


**Fig. 2.3** Crystal structures of intermetallic clathrates represented as packing of cage polyhedrons: 20-atom dodecahedron (*blue*), 24-atom tetrakaidecahedron (*green*), 26-atom pentakaidecahedron (*red*) and 28-atom hexakaidecahedron (*yellow*). The example compositions for different clathrate families (hydrate, cationic, and anionic) are given for comparison

### 2.2.2 Chemical Composition

The composition of clathrates with crystal structures built-up of space-filling polyhedrons (e.g. clathrates I, II, III and IV) can be calculated using the formalism mentioned above for the sodalite type of structure. For these clathrates, the composition results from the fact that each framework atom is shared by four polyhedral cages. For the first observed intermetallic clathrate I, $Na_8Si_{46}$ [17, 31–33] with two $NaSi_{20}$ and six $NaSi_{24}$ polyhedrons in the unit cell, the content of the unit cell is

$$2\ NaSi_{20/4} + 6\ NaSi_{24/4} = Na_8Si_{46}.$$

For the clathrate II with partially filled cavities, $Na_{24-x}Si_{136}$ [17, 31–35], the ideal composition can be obtained from the following balance:

$$8\ NaSi_{28/4} + 16\ NaSi_{20/4} = Na_{24}Si_{136}.$$

For the clathrate III $Cs_{30}(Na_{2.5}Sn_{162.6}\square_{6.9})$ [36], in fact with defects in the framework, the idealized content of the unit cell (without defects) can be calculated as

$$10\,Cs(Sn,Na)_{20/4} + 16\,Cs(Sn,Na)_{24/4} + 4\,Cs(Sn,Na)_{26/4} = Cs_{30}(Sn,Na)_{172}.$$

For the clathrates VIII, *cP*124 and *oP*60, which do not consist of space-filling centered polyhedra, the correct composition of the unit cell can be obtained by taking into account also the empty cavities and atoms contributing to more than four cages.

### 2.2.3 Crystallographic Disorder and Complexity of the Crystal Structures of Clathrates

Despite very clear structural motifs based on four-connected atoms within the frameworks, the clathrates seldom have completely ordered crystal structures. Different kinds of crystallographic disorder appear including defects in the framework and in the cages, mixed occupation by different elements, and positional disorder of the so-formed sub-sites. The resulting defect or substitution variants tend to form superstructures by ordering of defects or different atoms on the substitution positions. Most of these effects have already been studied on representatives of the clathrate-I type, thus some such examples are given below.

In the structural pattern of the clathrate-I type, the framework is formed by three positions called hereafter as position 1 (Wyckoff site 6*c*, ¼ 0 ½), 2 (Wyckoff site 16*i*, *xxx*, $x \approx 0.18$) and 3 (space group $Pm\bar{3}n$, Wyckoff site 24 *k*, 0*yz*, $y \approx 0.30$, $z \approx 0.11$). The interatomic distances are usually close to the sum of the covalent radii of the constituting atoms. The average distance between the framework atoms increases slightly with the size of the cation, e.g. *d*(Si–Si) changes from 2.37 Å for $Na_8Si_{46}$ to 2.42 Å for $Cs_{8-x}Si_{46}$ [37]. Position 2 is participating in the five-ring faces of the adjacent polyhedral cages only, i.e. it is a crossing point of six pentagonal faces. All bond angles here are close to the tetrahedral one. Position 3 is located at the crossing point of five pentagonal and one hexagonal face, having one angle within the hexagon which tends to be much larger than the tetrahedral one. Position 1 is located at the crossing point of two perpendicular six-ring faces and four five-ring faces of adjacent tetrakaidecahedrons. It has two angles tending to exceed the tetrahedral one. The deviation of the bond angles from the tetrahedral value also depends on the size ratio of host and guest atoms. With increasing guest-to-host ratio the angles approach the tetrahedral value at positions 2 and 3. For position 1 the trend is opposite (Table 2.1): for the same anion (Si) the deviation from the tetrahedral angle increases with the size of the cation.

The crystal structures try to compensate for the bonding strain, resulting at least from geometric reasons, by varying the atomic coordinates and bond lengths. Position 1 is thereby often partially occupied or occupied by substitution elements, such as transition metals, in particular at their lower concentrations. Defects and substitutional atoms (with the size different form the main host component) at the position 1 lead to a shift of the adjacent atoms at the position 3 and increase the angle deviation from the tetrahedral values which may be even less suitable for the tetrahedral environment (Table 2.1). The reduction of the total energy obtained by substitution seems to compensate the 'losses' caused by the stronger deviation from the tetrahedral environment of atoms in the framework.

In the binary type-I clathrates, the formation of vacancies at position 1 allows for the adjustment of the total electronic balance (see chapter 2.3). Crystallographically, the formation of defects at this position, e.g. in $K_8Sn_{44}\square_2$ [38] or in $Cs_8Sn_{44}\square_2$ [39] goes in parallel with an unusually large atomic displacement anisotropy of position three (Fig. 2.4, middle left). Analysis of the electronic density shows that this anisotropy increases with decreasing temperature (cf. difference electron density maps in Fig. 2.4, middle). For the case of the thermal displacement solely, the anisotropy should be also reduced. Thus two split positions were used for the description of the electron density in this region. Position 31 describes the location of the Sn3 atom when the neighboring Sn1 is present, and position 32 represents the location of the Sn3 atom next to the vacancy $\square$ at the Sn1 position (Fig. 2.4, bottom). Summing up the occupancies of the framework positions leads to the final compositions $K_8Sn_{44}\square_2$ or $Cs_8Sn_{44}\square_2$.

The formation of defects leads locally to strong changes in the electron density distribution so that four lone pairs are formed around each defect in the framework. Such regions are expected to order within the structure to achieve a more homogeneous distribution of the charge. Indeed, recent detailed studies of the crystal structure of $Ba_8Ge_{43}\square_3$ [40, 41], $Rb_8Sn_{44}\square_2$ [42] and $Cs_8Sn_{44}\square_2$ [43] reveal that the vacancies order within the hexagons along [100] and then within the tetrakaidecahedrons in a way that the distance between the nearest vacancies along the [001] direction is maximized (Fig. 2.5, middle). The symmetry remains cubic

**Table 2.1** Distances and bond angles for the framework positions in selected clathrates I

| Compound | *Distances (Å)* and Bond angles (deg) | | | Reference |
|---|---|---|---|---|
| | Position 1 | Position 2 | Position 3 | |
| $Na_8Si_{46}$ | *4 × 2.381* | *3 × 2.366* | *2 × 2.367* | [32] |
| | 4 × 108.8 | *1 × 2.292* | *1 × 2.381* | |
| | 2 × 110.9 | 3 × 108.9 | *1 × 2.396* | |
| | | 3 × 110.1 | 2 × 105.7 | |
| | | | 1 × 105.8 | |
| | | | 2 × 106.9 | |
| | | | 1 × 124.6 | |
| $Cs_{8-x}Si_{46}$ | *4 × 2.440* | *3 × 2.398* | *2 × 2.398* | [37] |
| | 4 × 107.7 | *1 × 2.321* | *1 × 2.440* | |
| | 2 × 113.1 | 3 × 109.3 | *1 × 2.498* | |
| | | 3 × 109.7 | 2 × 104.8 | |
| | | | 1 × 106.0 | |
| | | | 2 × 107.6 | |
| | | | 1 × 124.1 | |
| $Ba_{8-x}Si_{46}$ | *4 × 2.403* | *3 × 2.377* | *2 × 2.377* | [106] |
| | 4 × 108.0 | *1 × 2.329* | *1 × 2.403* | |
| | 2 × 112.4 | 3 × 109.4 | *1 × 2.509* | |
| | | 3 × 109.5 | 2 × 105.6 | |
| | | | 1 × 107.6 | |
| | | | 2 × 106.7 | |
| | | | 1 × 123.8 | |
| $Ba_8Al_{10}Si_{36}$ | *4 × 2.500* | *3 × 2.420* | *2 × 2.420* | [50] |
| | 4 × 108.5 | *1 × 2.377* | *1 × 2.440* | |
| | 2 × 111.4 | 3 × 109.2 | *1 × 2.500* | |
| | | 3 × 109.7 | 2 × 104.0 | |
| | | | 1 × 106.7 | |
| | | | 2 × 107.4 | |
| | | | 1 × 124.3 | |
| $K_8Sn_{46}$ (refinement without defects at position 1 and split at position 3) | *4 × 2.842* | *3 × 2.786* | *2 × 2.786* | [38] |
| | 4 × 109.62 | *1 × 2.810* | *1 × 2.842* | |
| | 2 × 109.18 | 3 × 110.45 | *1 × 2.717* | |
| | | 3 × 108.48 | 2 × 105.37 | |
| | | | 1 × 103.90 | |
| | | | 2 × 107.45 | |
| | | | 1 × 125.41 | |
| $K_8Sn_{44}$ (refinement with defect at position 1 and split position 3–31 and 32) | | *2 × 2.786* | *2 × 2.786* | [38] |
| | | *1 × 2.810* | *1 × 2.842* | |
| | | *1 × 3.041* | *1 × 2.717 or 1 × 3.041* | |
| | | 2 × 108.48 | | |
| | | 1 × 97.95 | 1 × 103.90 | |
| | | 1 × 110.45 | 2 × 105.37 | |
| | | 1 × 114.13 | 2 × 107.45 or 2 × 112.90 | |
| | | 1 × 116.27 | 1 × 125.41 or 1 × 115.37 | |
| | | | or | |
| | | | *2 × 3.041* | |
| | | | *1 × 3.035* | |
| | | | 1 × 92.35 | |
| | | | 2 × 93.80 | |

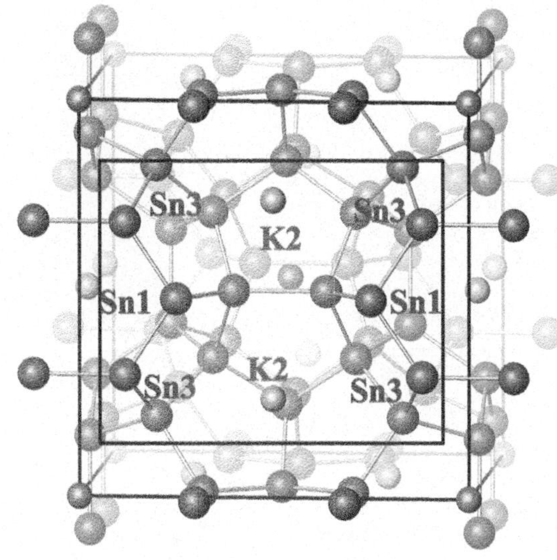


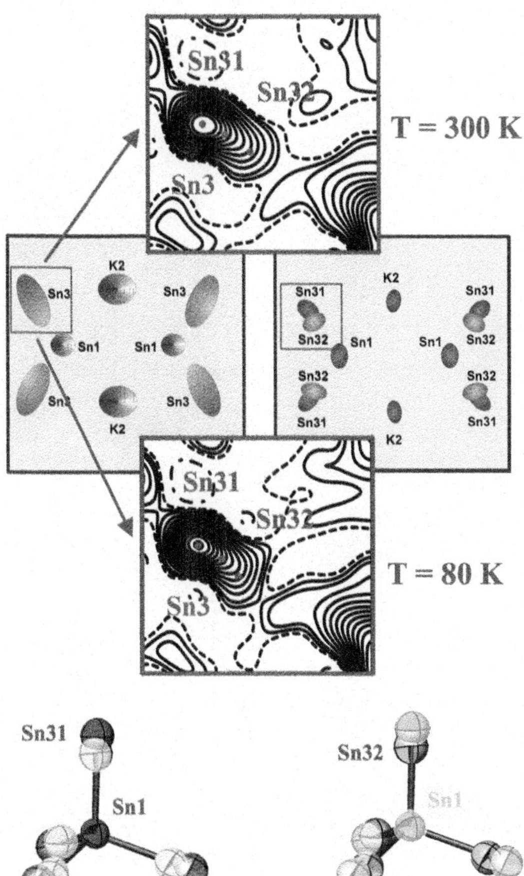


**Fig. 2.4** Defects in the crystal structure of clathrate-I $K_8Sn_{44}\square_2$: (*top*) environment of the Sn1 position in the (100) plane; (*middle*) displacement ellipsoids in the (100) plane without taking into account the splitting of the Sn3 position (*left*) and considering the splitting into Sn31 and Sn32 (*right*) together with the difference density distribution at $T = 300$ K and $T = 80$ K; (*bottom*) different local atomic arrangements without (*left*) and with (*right*) defect at the Sn1 position

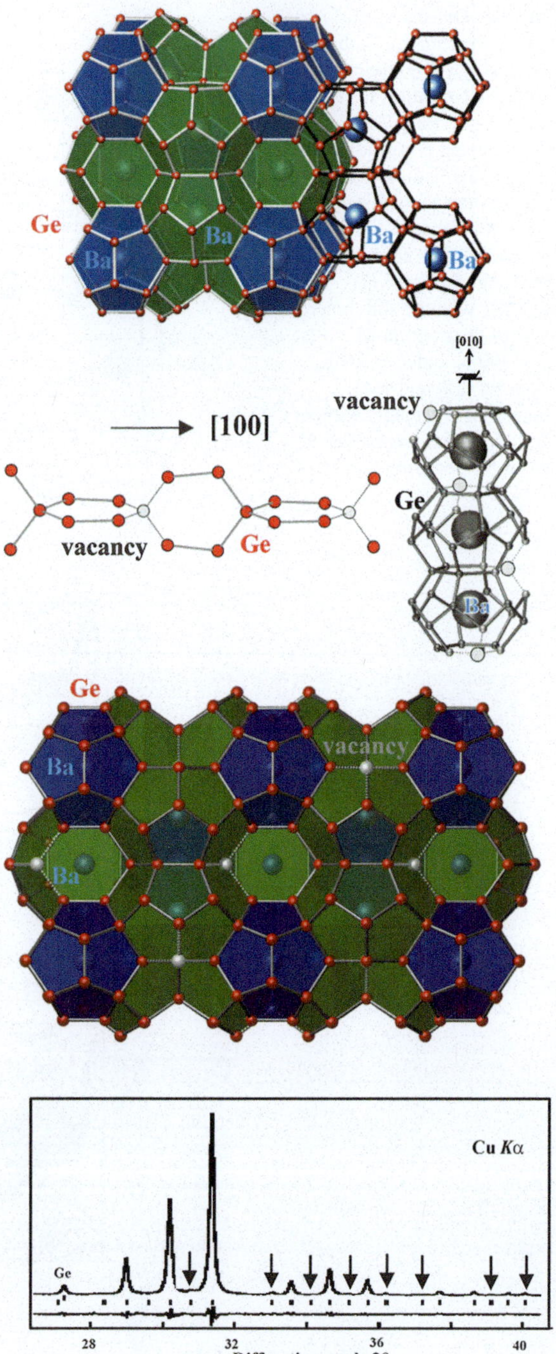


**Fig. 2.5** Ordering of defects and formation of the 2 × 2 × 2 superstructure in $Ba_8Ge_{43}\square_3$: (*top left*) packing of the polyhedral cages in the clathrate-I type of structure without defects; (*middle left*) ordering of vacancies (*grey*) within the hexagon chain along [100] direction; (*middle right*) ordering of vacancies within a column of the tetrakaidecahedrons; (*middle*) polyhedron packing in the 2 × 2 × 2 superstructure; tetrakaidecahedrons equal in the original clathrate-I structure, differ in the position of the vacancy; (*bottom left*) occurrence of the superstructure reflections in the X-ray powder diffraction pattern

(space group $Ia\bar{3}d$), the unit cell parameter doubles (cf. powder diffraction pattern in Fig. 2.5, bottom), the volume of the unit cell is eight time larger, and the resulting unit cell contains 432-$x$ atoms (e.g. for $Ba_8Ge_{43}\square_3$, $8 \times 54 - 8 \times 3 = 432 - 24$ atoms in the unit cell, Fig. 2.5, middle). For the tin clathrates, superstructure reflections appear in the diffraction patterns provided that the crystallization of the products is realized slow enough or the products were annealed for a long time at an appropriately low temperature [43]. This may explain why superstructure formation was not observed in previous investigations of $Ba_8Ge_{43}\square_3$ [44] or $Cs_8Sn_{44}\square_2$ [39] or was observed to different extent in different crystals of $Ba_8Ge_{43}\square_3$ [40].

Another way to release the bonding strain in the framework and at the same time adjust the total electronic concentration in a clathrate system is substitution of the $p$-elements in the framework by other $p$- or $s$-elements or transition metals. Studies in the system Ba–Ni–Ge show that for small amount of nickel it formally fills the vacancies of a hypothetical binary Zintl phase '$Ba_8Ge_{42}\square_4$' [45]. Only for higher Ni contents the nickel atoms start to replace germanium in the framework [46]. For the composition $Ba_8Ni_{3.5}Ge_{42.1}\square_{0.4}$ the crystal structure around the positions 1 and 3 is complex (Fig. 2.6, top). The distribution of the electron density around position three cannot be satisfactorily described applying one atom with a large anisotropic displacement. In order to describe the electron density in the vicinity of Ge3 correctly, three split positions are required. The first (31) if a vacancy is present at position one (Fig. 2.6, bottom left), the second (32) if position 1 is occupied by Ni (Fig. 2.6, bottom middle), and the third (33) for position 1 being occupied by Ge (Fig. 2.6, bottom right). The distance between the positions 1 and the split positions around 3 vary reliably from $\approx$2.2 Å (to 31) to 2.5 Å (to 33).

A similar complex disorder was observed in the study of the inverse clathrates-I compounds $Sn_{19.3}Cu_{4.7}As_{22}I_8$ [47], $Sn_{20}Zn_4P_{22-\nu}I_8$ ($\nu = 1.2$), $Sn_{17}Zn_7P_{22}I_8$, and $Sn_{17}Zn_7P_{22}Br_8$ [48]. In the latter, zinc and tin are occupying position 3, whereas phosphorus is located at positions 1 and 2. This leads to a strong crystallographic disorder around the position 3 which requires three spilt positions for correct description of the electron density in this region.

## 2.3 Chemical Bonding: From the Zintl Concept to the Electron-Localizability Approach

Considering the binary and ternary representatives of the clathrate structures in metallic systems, one immediately recognizes enormous chemical variability in the realization of the same structural motif. The framework may play the role of the anion ($Na_8Si_{46}$) or of the cation ($Ge_{38}P_8I_8$) or be neutral ($\square_{24}Ge_{136}$). In the first two cases, the cavity atoms provide charge compensation; in the third one, the cavities remain empty. Several variants for replacement of the main-group elements within the framework are found. While a substitution by another $p$ element can be expected from the chemical point of view (e.g. $Ba_8Ga_{16}Ge_{30}$ [24, 49] or $Ba_8Al_{10}Si_{36}$ [50]),

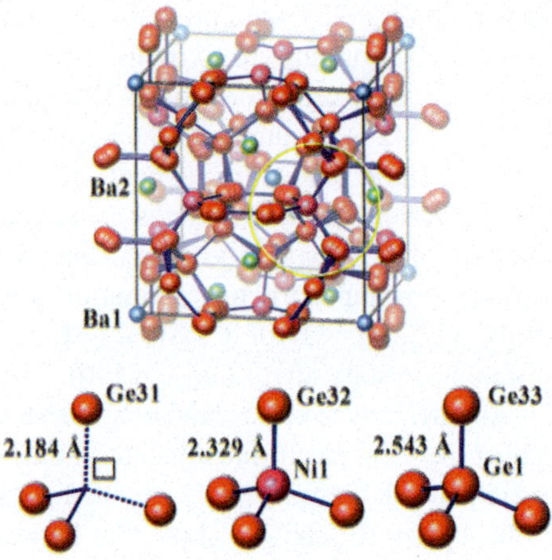


**Fig. 2.6** Crystallographic disorder in clathrate-I $Ba_8Ni_{3.5}Ge_{42.1}\square_{0.4}$: (*top*) average structure with split positions around Ge3, the *yellow circle* encloses the (Ni,Ge)1 position (*pink*) and the split positions around position three; (*bottom*) local atomic arrangements around the position (Ni,Ge)1 in case it is not occupied (*left*), occupied by Ni (*middle*), and occupied by Ge (*right*) with the resulting interatomic distances

the incorporation of transition elements [51] and—especially—alkali metals ($Cs_{30}(Na,Sn)_{172-x}$ [36], $K_8Li_xGe_{46-x}$ [52]) in the framework appears on the first glance surprising and requires additional knowledge about the atomic interactions within the framework. On the other hand, recent studies address additional questions about the role of the species filling the polyhedral cages. A typical understanding of filling atoms as charge donators for the framework assumes their ionic interaction with the framework. The example of the gold-containing clathrate $Ba_8Au_{5.3}Ge_{40.7}$ reveals that the Ba-framework relation is not limited solely to charge transfer and additional Ba–Au interactions have been found [53].

Also the non-monotonic behavior in the lattice parameter of $Na_{24-x}Si_{136}$ as a function of the sodium content cannot be explained with ionic interaction alone. One would expect a monotonic increase of the lattice parameter due to the filling of anti-bonding states of silicon. But for $x < 8$ the lattice parameter shrinks, and only for $x > 8$ does it increase with the filling [31, 34]. Furthermore, an $^7$Na ESR study for low sodium concentrations did not reveal complete charge transfer [54, 55] being in agreement with the NMR investigations [56, 57], and the semiconducting behavior is preserved for $x \leq 4$ [58]. These findings as well as information about the more or less extended homogeneity ranges for many ternary clathrates (caused by substitution in the framework and defects at the guest sites) raise the question about the general picture of atomic interactions in clathrates.

### *2.3.1 Zintl–Klemm Concept for Clathrates*

The simplest crystal structure built of four-bonded atoms is that of diamond. Each carbon atom has four shortest distances to its neighbors. Each atom has four valence electrons, which allows interpretation of all the short contacts as two center-two electron (2*c*-2*e*) bonds. The simplest heteroatomic structure built the same way is the structure of sphalerite, ZnS. Each zinc atom is surrounded by four sulfur atoms and vice versa. The average number of electrons per atom is also four, thus here all bonds can be considered as 2*c*-2*e* bonds. The diamond and the sphalerite structures can be interpreted applying the usual valence rules. The condition for the validity of this interpretation is that a sufficient number of electrons is available in the system to realize all the bonds (crystallographically the shortest distances) as 2*c*-2*e* bonds. In the unit cell of the clathrate-I motif, e.g. in $Na_8Si_{46}$, there are 92 short contacts within the framework, comparable with the sum of the covalent radii of the participating atoms ($d$(Si–Si) = 2.292 – 2.381 Å, $2r_{cov}$(Si) = 2.34 Å [59] ). Further 184 shortest contacts between the guest and the framework atoms are larger than the sum of the corresponding atomic radii ($d$(Na–Si) = 3.269 – 3.358 Å in the pentagon-dodecahedron and $d$(Na–Si) = 3.432 – 3.944 Å in the tetrakaidecahedron, $r_{at}$(Si) + $r_{at}$(Na) = 1.17 + 1.53 = 2.70 Å [59] ). The available 192 valence electrons are not sufficient to form 2*c*-2*e* bonds between Na and the surrounding Si atoms in the $Si_{20}$ and $Si_{24}$ cages. This situation is characteristic of the majority of intermetallic compounds [60]. In order to apply the Zintl–Klemm concept to clathrates only the bonds within the framework are considered as 2*c*-2*e* bonds. The average number of valence electrons per framework atom should be four, and each atom obtains a formal charge (which may be used as oxidation number) $n_{val} - n_{bonds}$, where $n_{val}$ is the number of valence electrons according to the Periodic Chart and $n_{bonds}$ is the number of bonds in which the atoms participates. The interaction between the guests and the framework is understood as charge transfer: in the anionic clathrates, the guest cations deliver their valence electrons to the framework; in case of the cationic (inverse) clathrates, the guest anions accept the excess electrons from the framework. For the cationic clathrate $Ge_{36}P_8I_8$ we therefore obtain the complete electronic balance:

$$Ge_{36}P_8I_8 = \left[(4b)Ge^0\right]_{36}\left[(4b)P^{1+}\right]_8\left[I^{1-}\right]_8 \times 0e^-,$$

where 4*b* stands for four-bonded atom.

Applying the same recipe to binary germanium and tin clathrates of alkali metals one obtains an unbalanced situation considering a complete host framework:

$$K_8Sn_{46} = \left[K^{1+}\right]_8\left[(4b)Sn^0\right]_{46} \times 8e^-.$$

To accommodate the electrons of the cation, two defects are formed per formula unit in the crystal structure (Fig. 2.4). The eight neighboring tin atoms are three-bonded, thus they obtain the oxidation number of −1, and the total balance is electron precise:

$$K_8Sn_{44}\square_2 = \left[K^{1+}\right]_8\left[(4b)Sn^0\right]_{36}\left[(3b)Sn^{1-}\right]_8 \times 0e^-.$$

Application of the Zintl–Klemm concept allows complete understanding of the composition and defects in the crystal structure of the binary germanium and tin clathrates-I of alkaline metals according to the formula $A_8E_{44}\square_2$ [61, 62]. This concept is also effective in understanding the presence of vacancies in the guest substructure of ternary alkali-metal clathrates with *p*-elements, e.g.

$$K_7Si_{39}B_7 = [K^{1+}]_{8-1}[(4b)Si^0]_{39}[(4b)B^{1-}]_7 \times 0e^-$$

[63], or for the interpretation of the ideal content of the substituting elements in many ternary clathrates with *p* elements and even transition elements as substitution components such as

$$Ba_8Ge_{30}Ga_{16} = [Ba^{2+}]_8[(4b)Ge^0]_{30}[(4b)Ga^{1-}]_{16} \times 0e^-$$

[24, 49], or

$$Ba_8Ge_{40.67}Au_{5.33} = [Ba^{2+}]_8[(4b)Ge^0]_{40.67}[(4b)Au^{3-}]_{5.33} \times 0e^-$$

[53].

The complete balance in the sense of oxidation numbers derived in such a way is not a general feature of clathrates. In particular, the defect structure of the alkaline-earth metal clathrate $Ba_8Ge_{43}\square_3$ shows a balance with excess electrons:

$$Ba_8Ge_{43}\square_3 = [Ba^{2+}]_8[(3b)Ge^{1-}]_{12}[(4b)Ge^0]_{31} \times 4e^-.$$

All binary silicon clathrates show even larger numbers of excess electrons, e.g.

$$Na_8Si_{46} = [Na^{1+}]_8[(4b)Si^0]_{46} \times 8e^-.$$

Besides the possibility to understand chemical composition and structural features, the balances above are connected to the electronic density of states, DOS (Fig. 2.7). In the case of a completely balanced situation, the Fermi level is located in a gap or pseudo gap of the calculated electronic DOS indicating a semi-conducting or bad-metal behavior in electronic transport. If excess electrons are present, the Fermi level is located in the conduction band, suggesting *n*-type metallic or metal-like electronic transport.

In contrast to the intermetallic anionic clathrates, the cationic clathrates mainly show an electron-balanced situation or very small deviations thereof. Thus the Zintl–Klemm concept provides an adequate tool to describe the general picture of atomic interactions in cationic (inverse) clathrates. In anionic clathrates deviating from the Zintl–Klemm concept mostly excess-electron situations are observed. *P*-type conduction, which can be considered as a physical 'fingerprint' for electron deficiency, is less common. Besides the known results on $Ba_8Ga_{16+x}Ge_{30-x}$ [64], such behavior was found recently for $Ba_8Au_{5.3}Ge_{40.7}$ [53] and $Ba_8Au_xSi_{46-x}$ ($x > 5.43$) [65, 66].

Assuming that the 2*c*-2*e* covalent-bond description is sufficient for the interpretation of chemical bonding within the framework of anionic clathrates, the

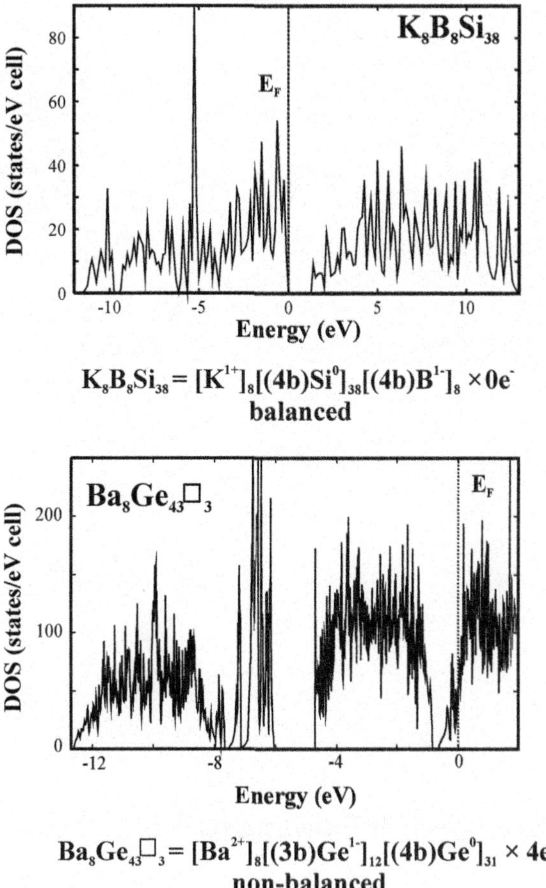


**Fig. 2.7** Electron balance and calculated electronic density of state for the clathrates-I $K_7Si_{39}B_7$ (idealized composition $K_8Si_{38}B_8$) and $Ba_8Ge_{43}\square_3$

excess electrons may be thought of as contributing to the guest-framework interactions. A clear experimental hint to the special features of these interactions was obtained for the clathrate *cP*124 $Ba_6Ge_{25}$ (Fig. 2.8) [67]. The temperature dependence of the lattice parameter exhibits two anomalies at about 180 and 230 K, caused by a structure transformation in two steps. A reconstructive nature of the structural transformation involves Ge–Ge bond breaking and barium cation displacement (Fig. 2.8, middle). Some Ge4 type atoms are so significantly displaced during cooling that Ge4–Ge6 bonds break and new three-bonded $(3b)Ge^{1-}$ species (electron acceptors) are formed. Consequently the number of charge carriers is reduced thus affecting the physical properties [68]. The driving force for this transformation is an additional Ba–Ge interaction (Fig. 2.8, bottom) which is obviously more favorable than the Ge–Ge bonding at low temperatures.

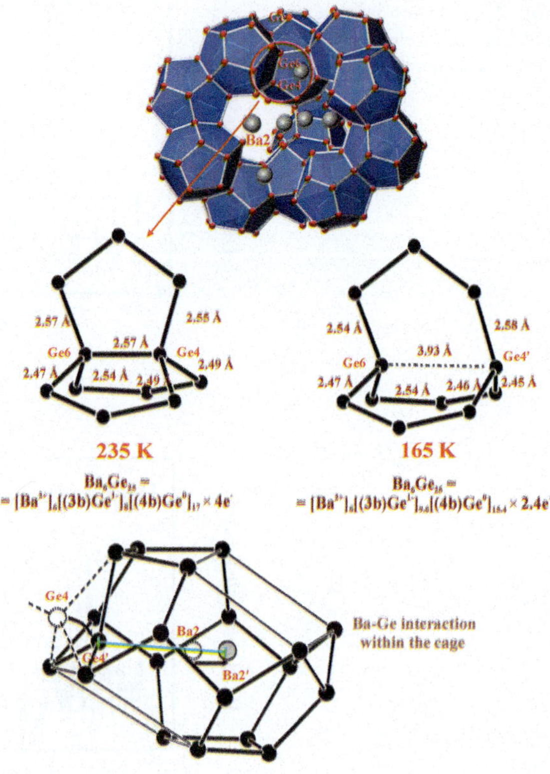


**Fig. 2.8** Structural phase transformation in clathrate *cP*124 $Ba_6Ge_{25}$: (*top*) crystal structure with the atomic arrangement relevant for the structural transformation; (*middle*) atomic arrangements and interatomic distances in the vicinity of the Ge4–Ge6 bond above (*left*) and below (*right*) the transformation temperature; (*bottom*) changes in the coordination sphere of the Ba2 atoms caused by the transformation [67]

## 2.3.2 Electron-Localizability Approach

The discussed guest-framework interactions cannot be studied within the Zintl–Klemm concept, because per definition it restricts them to the Coulomb forces caused by charge transfer. More insight may be achieved by applying quantum chemical tools for the analysis of chemical bonding in real space. The electron localizability approach based on a combined consideration of electron density and electron localizability indicator developed recently provides an efficient technique for the analysis of chemical bonding in intermetallic compounds [69]. The analysis of the topology of the electron density by means of the quantum theory of atoms in molecules (QTAIM [70] ) yields the 3D representations of atoms (atomic basins) in real space. The obtained electron density basins represent the individual atoms in the system when a nucleus is contained within the basin. The difference between the electron population of the basin surrounding the nucleus and the nuclear charge yields the charge of the atomic basin, which is often relatively close to the formal charge of the examined atomic species.

The shapes of QTAIM atoms in barium–germanium clathrates at the ideal composition $Ba_8Ge_{40}E_6$ are shown in Fig. 2.9 (top) [71]. Ba atoms show shapes close to spherical. This is in agreement with their role as cations. In such cases the

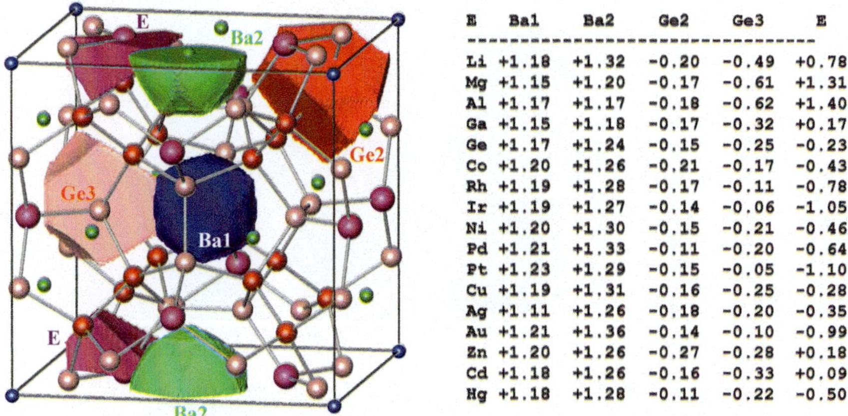


| E | Ba1 | Ba2 | Ge2 | Ge3 | E |
|---|---|---|---|---|---|
| Li | +1.18 | +1.32 | -0.20 | -0.49 | +0.78 |
| Mg | +1.15 | +1.20 | -0.17 | -0.61 | +1.31 |
| Al | +1.17 | +1.17 | -0.18 | -0.62 | +1.40 |
| Ga | +1.15 | +1.18 | -0.17 | -0.32 | +0.17 |
| Ge | +1.17 | +1.24 | -0.15 | -0.25 | -0.23 |
| Co | +1.20 | +1.26 | -0.21 | -0.17 | -0.43 |
| Rh | +1.19 | +1.28 | -0.17 | -0.11 | -0.78 |
| Ir | +1.19 | +1.27 | -0.14 | -0.06 | -1.05 |
| Ni | +1.20 | +1.30 | -0.15 | -0.21 | -0.46 |
| Pd | +1.21 | +1.33 | -0.11 | -0.20 | -0.64 |
| Pt | +1.23 | +1.29 | -0.15 | -0.05 | -1.10 |
| Cu | +1.19 | +1.31 | -0.16 | -0.25 | -0.28 |
| Ag | +1.11 | +1.26 | -0.18 | -0.20 | -0.35 |
| Au | +1.21 | +1.36 | -0.14 | -0.10 | -0.99 |
| Zn | +1.20 | +1.26 | -0.27 | -0.28 | +0.18 |
| Cd | +1.18 | +1.26 | -0.16 | -0.33 | +0.09 |
| Hg | +1.18 | +1.28 | -0.11 | -0.22 | -0.50 |

**Fig. 2.9** Atomic basins (QTAIM atoms) in the barium–germanium clathrates with the hypothetical composition $Ba_8Ge_{40}E_6$ (*left*) and the effective QTAIM charges (*right*)

QTAIM atom should envelop only the inner shells which, in turn, should show a spherical distribution of electron density. The effective charge of both Ba positions is essentially independent of the substitution elements and is much lower than the expected 2+. The shapes for the Ge2 and Ge3 positions as well as for the *E*1 position reflect their environment and are far from spherical. This implies that the interaction within the framework and between the guest and the framework is different. The effective charge of the *E*1 position is strongly dependent on the electronegativity difference between the substitution element E and germanium; the more electronegative E is the more negative charge it accommodates (Fig. 2.9, bottom). The polar character of the Ge3–*E*1 interaction is also reflected in the charge at Ge3: the more electronegative E is the more positive charge is accommodated at Ge3. The charge of Ge2 coordinated by Ge2 and Ge3 is practically independent of the kind of the substitution element *E* [71].

Further information about the chemical bonding can be obtained using the electron localizability indicator [72, 73] in its ELI-D representation as a bonding descriptor [74]. ELI-D describes the correlation of electronic motion. It is proportional to the electron population needed to form a fixed fraction of an electron pair. In analogy to QTAIM, ELI-D basins can be determined and assigned to chemically meaningful descriptors. Because ELI-D reveals regions of space that can be used as descriptors of atomic shells, bonds and lone-pairs, the examination of ELI-D may shed more light on the electronic structure of the intermetallic clathrates.

The distribution of ELI-D in the ordered superstructure $Ba_8Ge_{43}\square_3$ (Fig. 2.10, top) clearly reveals the covalent interaction between the germanium atoms within the framework [71]. Both three- and four-bonded germanium atoms form direct bonds which are visualized by ELI-D maxima (attractors) located on the shortest Ge–Ge contacts or very close to them (Fig. 2.10, middle right and bottom).

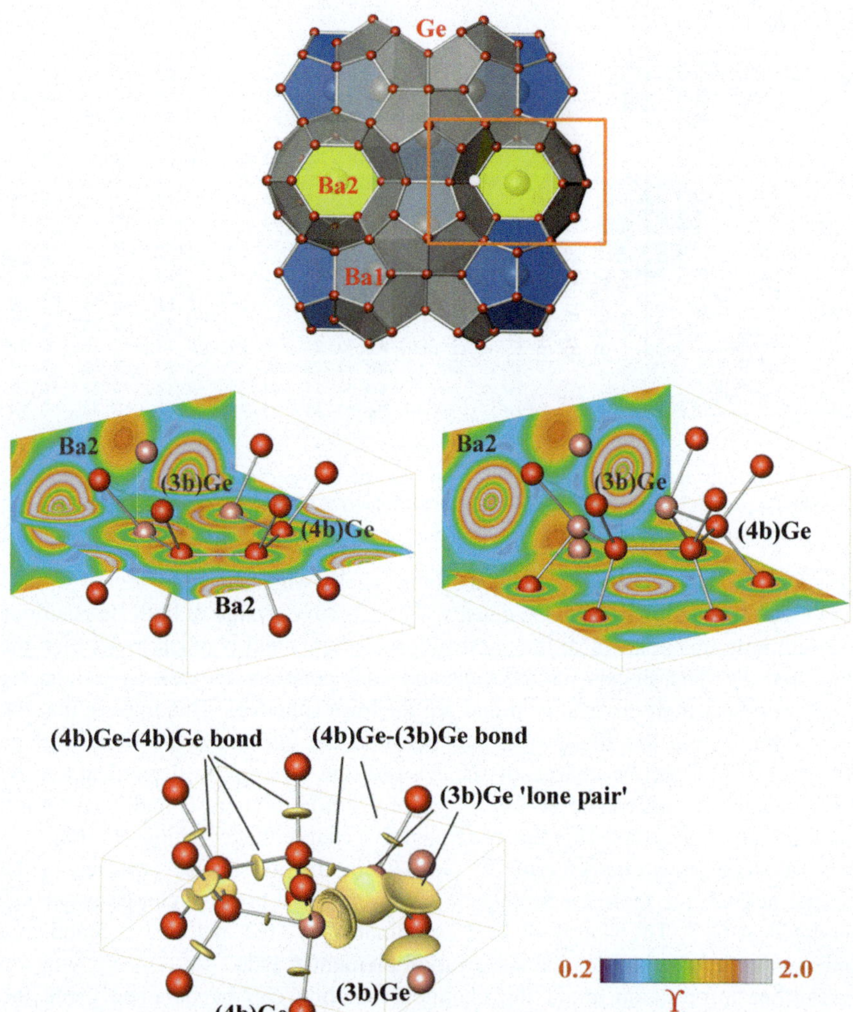


**Fig. 2.10** Electron localizability indicator in $Ba_8Ge_{43}$: (*top*) the investigated fragment of the crystal structure; (*middle*) distribution of ELI-D in the planes of Ba atoms (*left*) and in the plane of the defect hexagonal face of the tetrakaidecahedron (*right*); (*bottom*) isosurface of $Y = 1.25$ visualizing the Ge–Ge bonds and lone pairs at the three-bonded germanium atoms

Furthermore, the lone pairs at (3*b*)Ge atoms are visualized by the four ELI attractors oriented toward the vacancy position (Fig. 2.10, bottom). The distribution of ELI-D in the inner shells of the Ge atoms is spherical, indicating that, as expected, the electrons of the inner shells do not participate in valence interactions. The striking feature of the ELI-D distribution around the Ba atoms is the missing sixth shell. This suggests charge transfer to the framework and is in agreement with the idea of ionic guest-framework interaction (Fig. 2.10, middle left). The penultimate shell of Ba is slightly structured—the ELI-D distribution shows a small deviation from sphericity. This finding is a 'fingerprint' for the participation of the penultimate-shell electrons in the valence-region interactions [74, 75] revealing a first trend to some additional interactions between Ba and framework aside the ionic one.

Substitution of elements *E* with different electronegativities at the position Ge1, keeping in mind the traditional picture of chemical bonding for $Ba_8Ge_{40}E_6$, causes several changes in the details of the atomic interactions which can be resolved by application of the electron-localizability approach (Fig. 2.11). The last sixth shell of the Ba atoms is not present in all four compounds, supporting basically ionic interaction of Ba with the framework. The Ge–Ge interaction remains covalent being independent of the kind of the element *E*. The combined analysis of the electron density and ELI-D using the intersection technique [76] shows that the Li–Ge (0.04 electrons from Li and 2.3 from Ge) and Au–Ge (1.4 electron from Au and 0.4 from Ge) interactions are polar. With Li, germanium plays therefore the role of electron acceptor, and with Au germanium donates electrons. These findings are also supported by the effective QTAIM charges (cf. above). The inner shell of Au is already structured showing regions with slightly lower values on the Ba–Au contact line, signalizing an additional interaction. The same feature is observed for the inner shell of the Ba atoms. The direct Ba–Au interaction is manifested by a separate maximum of ELI-D (Fig. 2.11, bottom right). This kind of interaction is a novel feature for intermetallic compounds and was found for the first time in form of a dedicated maximum in ELI-D distribution for the gold-containing clathrates. Current studies reveal similar features for Pt, Pd and Cd compounds [71].

The application of the electron localizability approach allows for studying details of the atomic interactions in intermetallic clathrates. This may contribute to the understanding of structural features which cannot be achieved within the Zintl–Klemm concept. Another, from chemical point important, outcome of the electron localizability approach is the electron-localizability-based oxidation number (ELIBON [77]) which is the real-space equivalent of the traditional oxidation numbers. Application of oxidation numbers on intermetallic clathrates allows for new ways of understanding of experimentally observed clathrate compositions and for novel redox routes for their preparation.

## 2.4 Preparation Routes: From Direct Reaction Between Elements to Redox Processes

Various synthesis routes to intermetallic clathrates have been successfully applied covering the whole spectrum of preparation methods in solid state chemistry. The applicability of each method is dependent on the elements forming the target material and some thermodynamic properties, such as formation energy and vapor pressure of components at the reaction temperature. Although it is straightforward for many systems to prepare a clathrate phase at one certain composition,

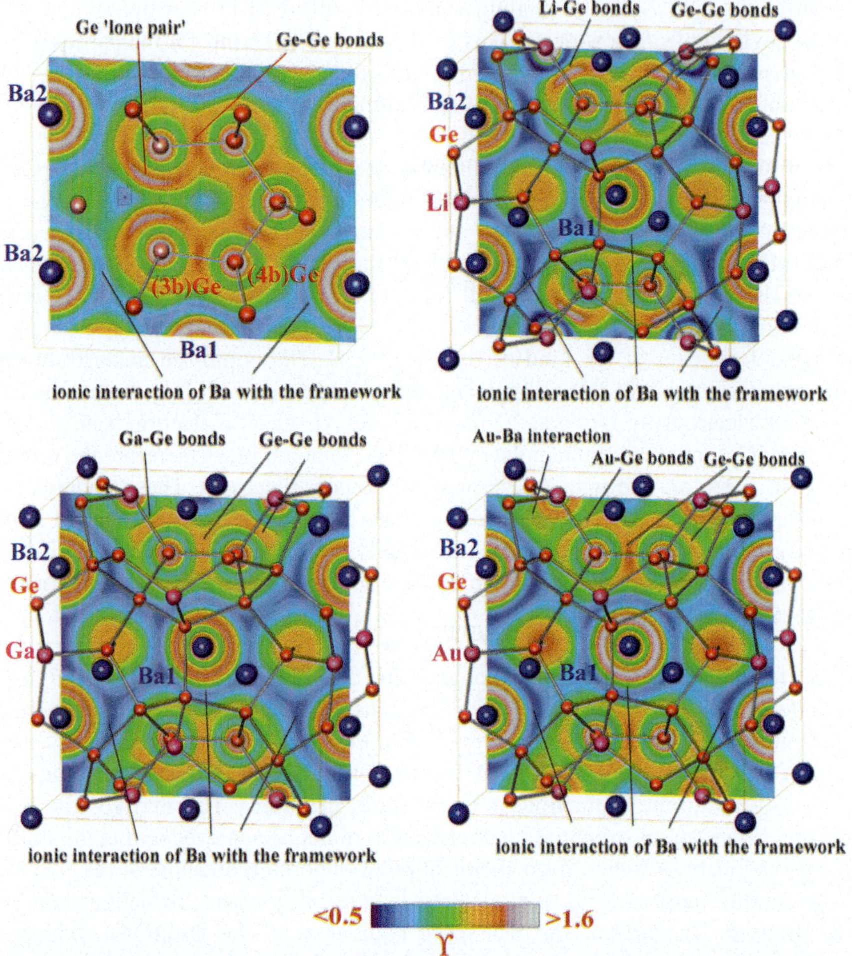


**Fig. 2.11** ELI-D distributions in the barium–germanium clathrates $Ba_8Ge_{43}\square_3$ (*top left*, $\square$ stands for vacancy), $Ba_8Li_6Ge_{40}$ (*top right*), $Ba_8Ga_6Ge_{40}$ (*bottom left*) and $Ba_8Au_6Ge_{40}$ (*bottom right*) illustrating the polar and non-polar covalent bonds in the framework, the ionic interaction of Ba with the framework and the additional covalent Ba–Au interaction

systematic investigations of electronic transport behavior require more effort. The correlation of chemical composition and structural defects with intrinsic electronic transport properties resembles the well-known challenges of semiconductor science. However, despite the fact that clathrate research in the literature has been developed into materials science and appears to be mainly focused on energy-related applications, i.e. thermoelectricity, it still holds surprising results and interesting open questions for fundamental investigations. This is also reflected by the ongoing developments of preparation methods which have resulted in new types of clathrate phases after decades of research.

### 2.4.1 Crystallization from the Melt

In favorable cases, clathrate phases show a high stability and form congruently from the melt. With respect to the clathrate-I type of crystal structure, some ternary materials are prepared with compositions tending to an electronic balance such as $Ba_8Ga_{16}Ge_{30}$ [24, 49] and $Ba_8Au_{5.33}Ge_{40.67}$ [53]. Some clathrate-II type phases with homoatomic framework, such as $Na_{16}M_8Y_{136}$ ($M$ = Rb, Cs; $Y$ = Si, Ge), may form congruently as well [78]. The respective melting behavior can be explained by considering the stability of neighboring phases. In this case mixtures, e.g. $M_{12}Si_{17}$ + Si or $M_4Ge_9$ + Ge, are less stable than, e.g., $BaSi_2$ and Si. A hindrance for crystal growth can be a large homogeneity range of clathrate phases. If the preparation starts at a composition deviating from the congruently melting one or formation reaction is peritectic, the crystals of the phase having a homogeneity range may show a marked gradient in composition along the growth direction for usual growth rates. A successful method to obtain large single crystals is the *Czochralski* technique. However, due to its technical complexity the application of this method has been restricted to systems with negligible vapor pressures such as $Ba_8Ga_{16}Ge_{30}$ [79] and $Ba_8Al_xSi_{46-x}$ [80]. For comparison, equipment and experimental procedures are much simpler for the *Bridgman* technique which allows for the use of welded crucibles. For carbon or metal crucibles with high thermal conductivity the temperature profile in the crystal growth experiment is blurred so that not all experiments are successful and several crystalline grains, instead of a single one, are typically formed. In contrast to glass ampoules, the necking of the ampoule favorable for grain selection is generally not applicable for glassy carbon and metal tubes. Instead, a simple inset with conic wholes can be placed at the bottom of the growth crucible. By using the *Bridgman* technique, mm-sized single crystals of $Ba_8Ni_{3.5}Ge_{42.1}$ [46] and even cm-sized single crystal of $Ba_8Au_{5.3}Ge_{40.7}$ [53] (Fig. 2.12) have been grown in glassy carbon crucibles.

### 2.4.2 Polycrystalline Samples by Melt Quenching

Some clathrate phases exist in the phase diagram only within a small temperature window. For example, the clathrate $Ba_8Ge_{43}\square_3$ is thermodynamically stable at ambient pressure between 770 and 810 °C [41]. By applying slow cooling rates the clathrate phase is decomposed below the stability limit, provided the temperature is still high enough to trigger a solid state reaction. Hence, rapid cooling (quenching) techniques are required for successful preparation. One possibility for quenching is the melt spinning method which is often used to influence the microstructure of clathrate phases, e.g. for $Ba_8Ga_{16}Ge_{30}$ [81]. More simple is the steel quenching technique imitating traditional forging. A portion of the melt at the clathrate composition is heated in a glassy carbon crucible, then purged on a cold steel plate and hammered by a second plate (Fig. 2.13). The cooling rate

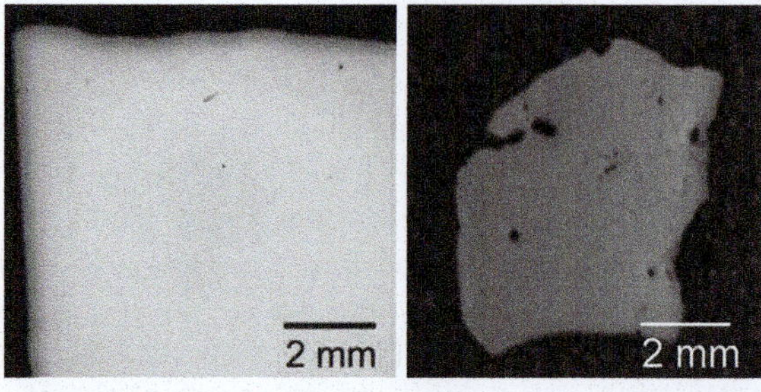


**Fig. 2.12** Microstructure of single crystals of $Ba_8Ni_{3.7}Ge_{42.1}$ (*left*) and $Ba_8Au_{5.44}Ge_{40.67}$ (*right*)

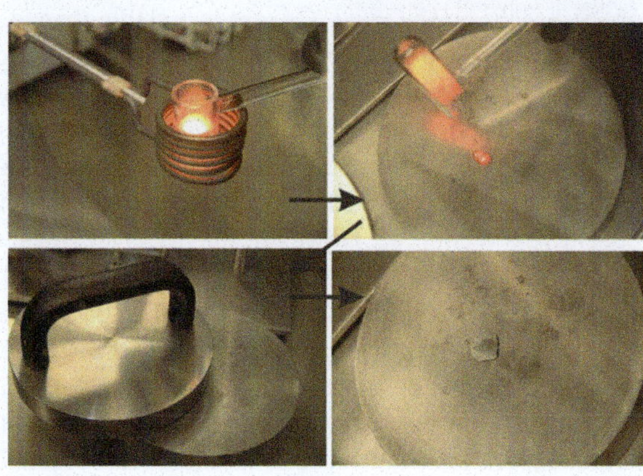

**Fig. 2.13** Steel-quenching preparation of polycrystalline clathrate phase

using this procedure is estimated to be higher than 1,000 K/s. In this way single-phase polycrystalline products have been obtained (e.g. $Ba_8Ge_{43}\square_3$ [41], $Ba_8Ir_{0.2}Ge_{43}\square_3$ [82]), and superconducting $Ba_8Ni_1Si_{45}$ has been stabilized in the metastable ambient pressure region [83].

For melts with high vapor pressure, both methods described above cannot be applied and closed ampoules must be used instead. In this case it has turned out to be useful to purge water directly on the heated metal container instead of throwing metal tubes with quartz jacket into water, as typically described in the literature. Due to the vigorous steam development of water at 1,000 °C this method requires special equipment and safety arrangements. When moderate cooling rates are sufficient, arc melting may be applied as well [84]. The method is, however, suitable only, when no marked evaporation losses appear during arc melting since these can be only roughly corrected by weighing excess amounts of the volatile components.

### 2.4.3 Solid State Reactions

To investigate homogeneity ranges and the relationship between physical properties and chemical composition, the equilibrium state for each composition has to be achieved. The ampoules or crucibles along with the starting materials are placed in vertical tube furnaces with well-controlled temperature profile (control

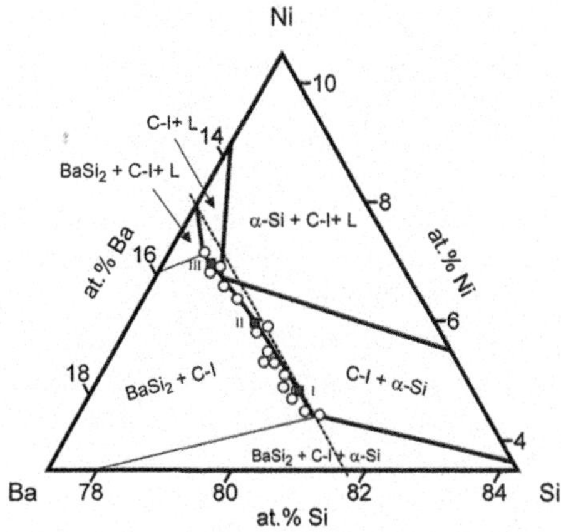


**Fig. 2.14** Phase relations for $Ba_8Ni_xSi_{46-x-y}$ at 1,000 °C. Analysis of the equilibrium phases after annealing clearly shows the deviation of the homogeneity range from the iso-concentration line describing the composition $Ba_8Ni_xSi_{46-x}$ and hence the presence of defects in the silicon framework [83]

measurements should be done before each annealing experiment). The annealing times depend strongly on the system studied. For example the clathrate $Ba_8Ni_x$-$Si_{46-x-y}$ [83] is typically obtained as a single-phase material after annealing for one week at 1,000 °C (for phase equilibria, see Fig. 2.14), and the germanium clathrate $Ba_8Ni_xGe_{46-x-y}$—after 1 week at 800 °C. The crucible material for silicon clathrates is typically corundum, for those of germanium (without alkali metals)—glassy carbon, for tin compounds—glassy carbon or metal, and for phosphorus representatives—quartz glass. An appropriate crucible material can only be identified taking into account its interaction with all components used for the synthesis. Side reactions with the container material can often be easily realized but can also lead to misconceptions. For example when a clathrate at composition $Ba_8Ni_4Ge_{42}$ is annealed in Ta ampoules, the neighboring clathrate phase $Ba_6Ge_{25}$ appears in the reaction product, which is widely inert against Ta, while $TaGe_2$ is formed at the container wall [85].

If the clathrate develops a significant vapor pressure on heating, the reaction crucible must be welded in a metal tube. For the outer metal tube the choice of material is less critical, and stainless steel is just as good as expensive tantalum. The annealing process can be shortened by careful homogenization of the starting mixture, but there are systems like clathrate phosphides for which even long-time annealing is not sufficient. Here a clathrate phase can be only obtained by several annealing steps with repeated milling and compacting [86]. More recently, our studies show that the spark plasma sintering (SPS) allows for distinctly shorter annealing time. However, the respective phase must be stable against electrochemical decomposition during SPS treatment.

### *2.4.4 Thermal Decomposition*

Thermal decomposition of alkali metal monosilicides $M_4Si_4$ and monogermanides $M_4Ge_4$ was historically the first preparation method for the respective clathrate phases in 1961 [87, 88] and it was used in 1995, when the superconducting phase $Na_2Ba_6Si_{46}$ was found. This was the first clathrate phase containing Ba atoms in a complete homoatomic silicon framework [89]. The decomposition experiments are typically performed in the temperature range 300–500 °C in which most crucible materials can be used. The method is best suitable for clathrate products which are not decomposing at the reaction conditions. A more serious problem, however, is the reproducibility of the experimental conditions. In the literature, dynamic vacuum conditions are generally used, but the actual pressure of the volatile particles at the sample position in the reactor is typically not well controlled. As a rough approximation, the counter pressure of the volatile component depends on the temperature of its precipitate forming at the colder part of the reactor. For a long time thermal decomposition experiments yielded only microcrystalline powders, and the investigation of the so-obtained very small single crystals by XRD was challenging [90]. Further development of the decomposition preparation route came up recently with the decomposition experiments of monosilicides in a closed system using a graphite acceptor in shifting the equilibrium [91–94]. The alkali metals are absorbed by the graphite providing a distinct counter pressure of the alkali-metal vapor. This way, mm-sized single crystals of silicon clathrates were obtained.

### 2.4.5 Chemical Transport

The historically first experiment of this kind for clathrates was an attempt to synthesize a noble gas clathrate $Xe_8Ge_{46}$ by chemical transport of Ge with $GeI_4$ under high Xe gas pressure. While the existence of the noble gas clathrates is still under discussion, the first "inverse" clathrate $I_8Ge_{43.33}I_{2.67}$ [95] was obtained in this way. The technique was also successfully applied for the preparation of phases such as $Ge_{38}As_8I_8$ [18]. Later it was developed into a fruitful synthetic tool with the possibility to obtain well-developed single crystals [19, 20].

### 2.4.6 High-Pressure Synthesis

The anticipation of clathrates as high-pressure phases was not natural due to their bulky host framework. In particular it is not expected that large filler atoms such as Ba and Cs form high-pressure phases. Therefore the first high-temperature high-pressure preparation of the binary silicon clathrate $Ba_{8-x}Si_{46}$ by reacting $BaSi_2$ and Si under high pressure conditions was a sensational report [96]. Later, $Ba_6Si_{25}$ [97], $Cs_{8-x}Si_{46}$ [37] and $Cs_{8-x}Ge_{46-y}$ [98] were prepared at extreme conditions. In fact, clathrates with oversized filler atoms are not stabilized by high pressure, but the competing phases are more destabilized. Generally, high-pressure reactions allow for the preparation of single crystals but can only be performed with small reaction volumes.

**Fig. 2.15** Spark-plasma electrochemical redox preparation of clathrate $Na_{24}Si_{136}$ (*top*) and the so-obtained single crystals (*bottom*) [35]

## 2.4.7 *Electrochemical Clathrate Preparation by the Spark-Plasma Technique*

Spark-plasma sintering has been known as an efficient compaction method. This also makes it of special importance for clathrate products. Microcrystalline powders can be transformed under inert atmosphere in special pressure tool to polycrystalline specimens of high density allowing for the determination of physical properties. However material problems must also be solved with the limitation that a part of the tool (dye and/or punches) should be manufactured from a conductive material. Recently, new synthetic features of SPS processing were revealed with the preparation of mm-sized single crystals of $Na_{24}Si_{136}$ [35] (Fig. 2.15). The synthesis was achieved by electrochemical transformation of $Na_4Si_4$ at the punches of the SPS tool. Since the first report [17], the preparation of a completely filled clathrate II $Na_{24}Si_{136}$ was difficult, and only microcrystalline powders were reported. Hence, this new method opens a route for investigation of the intrinsic physical properties of high-quality large single crystals [99, 100].

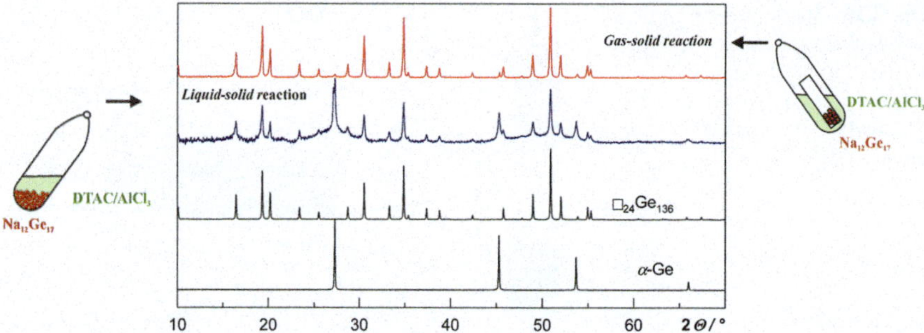


**Fig. 2.16** Schematic reaction set-ups for the oxidation of Zintl phases to clathrates (*left* and *right*) and the X-ray powder diffraction patterns of the products of the reactions (*middle*)

## *2.4.8 Clathrates by Redox Preparation*

The established preparation methods described so far need appropriate temperature conditions to achieve full conversion of the reaction educts to the clathrate phase. Alternatively, suitable precursors such as $K_4Si_4$ or $K_4Ge_4$ can be converted at relatively low reaction temperatures to clathrate phases by heterogeneous liquid–solid redox process. Due to the low reaction temperatures, metastable phases have been stabilized as well. These products are not accessible at any $(p,T)$-condition in thermal equilibrium. The method became known when the Zintl phase $Na_{12}Ge_{17}$ was reacted in the ionic liquid DTAC/$AlCl_3$ to a new allotrope of germanium with empty clathrate-II structure, Ge(*cF*136) [101]. Further research on the new products was hindered self-decomposition of the ionic liquid, leading to irreversible contamination of the products. The preparation concept was then further transformed to a gas–solid reaction [102, 103]. In a simplest procedure, suitable educts like $Na_4Si_4$ and $NH_4Cl$ are placed separately in open glass tubes. These are sealed in a bigger glass ampoule which is heated to the reaction temperature (Fig. 2.16). The reaction product consists of clathrate-I phase and NaCl which can be easily washed out [102]. Alternatively the precursors can be placed under mild conditions in a gas stream of oxidizing agent, e.g. organic chloride. This approach has provided in a short time interesting metastable intermetallic phases and promises new perspectives in the preparation not only of clathrate phases. This is testified by the preparation of elemental allotropes [101] or preparation of high-pressure phase $Ba_{8-x}Si_{46}$ at ambient pressure [104], and selective phase preparation [105].

## 2.5 Outlook: From Flamboyant Substances to Materials of Interest

Within 50 years since their discovery intermetallic clathrates have developed from flamboyant substances with esthetic crystal structures to a materials family with widely investigated properties. Despite remarkable progress in their study, the description and understanding of the crystal structures and chemical bonding in clathrates, a detailed and—at the same time general—picture of the atomic interactions is missing. For example, a prediction of the chemical composition for the homogeneity ranges is currently out of reach. While the composition and the structural features of the frameworks are usually well represented by (polar) covalent bonding between the host atoms, quantitative interpretation of the guest–host interactions remains an open and is key issue for further studies.

## References

1. S.M. Auerbach, K.A. Carrado, P.K. Dutta (eds.), *Handbook of Zeolite Science and Technology* (Marcel Dekker, New York, 2003)
2. L. Pauling, Z. Kristallogr. **74**, 213–225 (1930)
3. R. Wartchow, Z. Kristallogr. NCS **212**, 80 (1997)
4. G.A. Jeffrey, in *Inclusion Compounds*, vol. 1, ed. by J.L. Atwood, J.E.D. Davies, D.D. MacNicol (Academic Press, London, 1984)
5. P. Rogl, in *Thermoelectrics Handbook*, ed. by D.M. Rowe (CRC Taylor & Francis, Boca Raton, 2006)
6. H. Davy, Philos. Trans. R. Soc. Lond. **101**, 1–35 (1811)
7. M. Faraday, Q. J. Sci. Lit. Arts **15**, 71–74 (1823)
8. W.F. Claussen, J. Chem. Phys. **19**, 259–260, 662 (1951)
9. W.F. Claussen, J. Chem. Phys. **19**, 1425–1426 (1951)
10. M. von Stackelberg, H.R. Müller, J. Chem. Phys. **19**, 1319–1320 (1951)
11. M. von Stackelberg, H.R. Müller, Naturwiss. **38**, 456 (1951)
12. H.R. Müller, M. von Stackelberg, Naturwiss. **39**, 20–21 (1952)
13. L. Pauling, R.E. Marsh, Proc. Natl. Acad. Sci. U. S. A. **38**, 112–118 (1952)
14. B. Kamb, Science **148**, 232–234 (1965)
15. H. Gies, F. Liebau, Acta Crystallogr. **A37**, C187–C188 (1981)
16. F. Liebau, H. Gies, R.P. Gunawardane, B. Marler, Zeolites **6**, 373–377 (1986)
17. J.S. Kasper, P. Hagenmuller, M. Pouchard, C. Cros, Science **150**, 1713–1714 (1965)
18. H. Menke, H.G. von Schnering, Z. Anorg. Allg. Chem. **395**, 223–238 (1973)
19. K.A. Kovnir, A.V. Shevelkov, Russ. Chem. Rev. **73**, 923–938 (2004)
20. A.V. Shevelkov, K.A. Kovnir, Struct. Bond. **139**, 97–142 (2011)
21. M. Beekman, G.S. Nolas, J. Mater. Chem. **18**, 842–851 (2008)
22. C. Cros, M. Pouchard, Comp. Rend. Chim. **12**, 1014–1056 (2009)
23. A.J. Karttunen, T.F. Fässler, M. Linnolahti, T.A. Pakkanen, Inorg. Chem. **50**, 1733–1742 (2011)
24. B. Eisenmann, H. Schäfer, R. Zagler, J. Less-Comm. Met. **118**, 43–55 (1986)
25. H.G. von Schnering, A. Zürn, J.-H. Chang, M. Baitinger, Yu. Grin, Z. Anorg. Allg. Chem. **633**, 1147–1153 (2007)

26. W. Carrillo-Cabrera, J. Curda, H.G. von Schnering, S. Paschen, Yu. Grin, Z. Kristallogr. NCS **215**, 207–208 (2000)
27. H. Fukuoka, K. Iwai, S. Yamanaka, H. Abe, K. Yoza, L. Häming, J. Solid State Chem. **151**, 117–121 (2000)
28. S.-J. Kim, S. Hu, C. Uher, T. Hogan, B. Huang, J.D. Corbett, M.G. Kanatzidis, J. Solid State Chem. **153**, 321–329 (2000)
29. J.-T. Zhao, J.D. Corbett, Inorg. Chem. **33**, 5721–5726 (1994)
30. U. Aydemir, L. Akselrud, W. Carrillo-Cabrera, C. Candolfi, N. Oeschler, M. Baitinger, F. Steglich, Yu. Grin, J. Am. Chem. Soc. **132**, 10984–10985 (2010)
31. E. Reny, P. Gravereau, C. Cros, M. Pouchard, J. Mater. Chem. **8**, 2839–2844 (1998)
32. G.K. Ramachandran, J. Dong, J. Diefenbacher, J. Gryko, R.F. Marzke, O.F. Sankey, P.F. McMillan, J. Solid State Chem. **145**, 716–730 (1999)
33. H. Horie, T. Kikudome, K. Teramura, S. Yamanaka, J. Solid State Chem. **182**, 129–135 (2009)
34. M. Beekman, E.N. Nenghabi, K. Biswas, C.W. Myles, M. Baitinger, Yu. Grin, G.S. Nolas, Inorg. Chem. **49**, 5338–5340 (2010)
35. M. Beekman, M. Baitinger, H. Borrmann, W. Schnelle, K. Meier, G.S. Nolas, Yu. Grin, J. Am. Chem. Soc. **131**, 9642–9643 (2009)
36. S. Bobev, S.C. Sevov, J. Am. Chem. Soc. **123**, 3389–3390 (2001)
37. A. Wosylus, I. Veremchuk, W. Schnelle, M. Baitinger, U. Schwarz, Yu. Grin, Chem. Eur. J. **15**, 5901–5903 (2009)
38. M. Baitinger, Ph.D. Thesis. Technische Universität Darmstadt (2000)
39. H.G. von Schnering, R. Kröner, M. Baitinger, K. Peters, R. Nesper, Yu. Grin, Z. Kristallogr. NCS **215**, 205–206 (2000)
40. W. Carrillo-Cabrera, S. Budnyk, Yu. Prots, Yu. Grin, Z. Anorg. Allg. Chem. **630**, 2267–2276 (2004)
41. U. Aydemir, C. Candolfi, H. Borrmann, M. Baitinger, A. Ormeci, W. Carrillo-Cabrera, C. Chubilleau, B. Lenoir, A. Dauscher, N. Oeschler, F. Steglich, Yu. Grin, Dalton Trans. **39**, 1078–1088 (2010)
42. F. Dubois, T.F. Fässler, J. Am. Chem. Soc. **127**, 3264–3265 (2005)
43. A. Kaltzoglou, S.D. Hoffmann, T.F. Fässler, Eur. J. Inorg. Chem. **2007** 4162–4167 (2007)
44. W. Carrillo-Cabrera, J. Curda, K. Peters, S. Paschen, M. Baenitz, Yu. Grin, H.G. von Schnering, Z. Kristallogr. NCS **215**, 321–322 (2000)
45. U. Aydemir, M. Baitinger, Yu. Grin, Unpublished data (2013)
46. L.T.K. Nguyen, U. Aydemir, M. Baitinger, E. Bauer, H. Borrmann, U. Burkhardt, J. Custers, A. Haghighirad, R. Höfler, K.D. Luther, F. Ritter, W. Assmus, Yu. Grin, S. Paschen, Dalton Trans. **39**, 1071–1077 (2010)
47. K.A. Kovnir, A.V. Sobolev, I.A. Presniakov, O.L. Lebedev, G. van Tendeloo, W. Schnelle, Yu. Grin, A.V. Shevelkov, Inorg. Chem. **44**, 8786–8793 (2005)
48. K.A. Kovnir, M.M. Shatruk, L.N. Reshetova, I.A. Presniakov, E.V. Dikarev, M. Baitinger, F. Haarmann, W. Schnelle, M. Baenitz, Yu. Grin, A.V. Shevelkov, Solid State Sci. **7**, 957–968 (2005)
49. A. Bentien, E. Nishibori, S. Paschen, B.B. Iversen, Phys. Rev. B **71**, 144107 (2005)
50. J.H. Roudebush, N. Tsujii, A. Hurtando, H. Hope, Yu. Grin, S.M. Kauzlarich, Inorg. Chem. **51**, 4161–4169 (2012)
51. G. Cordier, P. Woll, J. Less-Common Met. **169**, 291–302 (1991)
52. Y. Liang, B. Böhme, A. Ormeci, H. Borrmann, O. Pecher, F. Haarmann, W. Schnelle, M. Baitinger, Yu. Grin, Chem. Eur. J. **18**, 9818–9822 (2012)
53. H. Zhang, H. Borrmann, N. Oeschler, C. Candolfi, W. Schnelle, M. Schmidt, U. Burkhardt, M. Baitinger, J.-T. Zhao, Yu. Grin, Inorg. Chem. **50**, 1250–1257 (2011)
54. S.B. Roy, K.E. Sim, A.D. Chaplin, Phil. Mag. **65**, 1445–1450 (1992)
55. H. Yahiro, K. Yamaji, M. Shiotani, S. Yamanaka, M. Ishikawa, Chem. Phys. Lett. **246**, 167–170 (1995)
56. J. Gryko, P.F. McMillan, O.F. Sankey, Phys. Rev. B **54**, 3037–3039 (1996)

57. E. Reny, M. Ménétier, C. Cros, M. Pouchard, J. Sénégas, Compt. Rend. Acad. Sci. Ser. **II**, 129–136 (1998)
58. V.I. Smelyansky, J. S. Tse, Chem. Phys. Lett. **264**, 459-465 (1997)
59. J. Emsley, *The Elements* (Oxford University Press Ins, New York, 1991)
60. Yu. Grin, in *Comprehensive Inorganic Chemistry II*, vol. 2, ed. by J. Reedijk, K. Poeppelmeier (Elsevier, Oxford, 2013), pp. 359–373
61. H.G. von Schnering, Nov. Acta Leopold. **264**, 165–182 (1985)
62. H.G. von Schnering, Bol. Soc. Chil. Quim. **33**, 41–57 (1988)
63. W. Jung, J. Lorincz, R. Ramlau, H. Borrmann, Yu. Prots, F. Haarmann, W. Schnelle, U. Burkhardt, M. Baitinger, Yu. Grin, Angew. Chem. Int. Ed. **46**, 6725–6728 (2007)
64. A. Bentien, M. Christensen, J.D. Bryan, A. Sanchez, S. Paschen, F. Steglich, G.D. Stucky, B.B. Iversen, Phys. Rev. B **69**, 045107 (2004)
65. N. Jaussaud, P. Gravereau, S. Pechev, B. Chevalier, M. Ménétier, P. Dordor, R. Decourt, G. Goglio, C. Cros, M. Pouchard, Compt. Rend. Chim. **8**, 39–46 (2005)
66. U. Aydemir, C. Candolfi, A. Ormeci, Y. Oztan, M. Baitinger, N. Oeschler, F. Steglich, Yu. Grin, Phys. Rev. B. **84**, 195137 (2011)
67. W. Carrillo-Cabrera, H. Borrmann, S. Paschen, M. Baenitz, F. Steglich, Yu. Grin, J. Solid State Chem. **178**, 715–728 (2005)
68. S. Paschen, V.H. Tran, M. Baenitz, W. Carillo-Cabrera, Yu. Grin, F. Steglich, Phys. Rev. B **65**, 134435 (2002)
69. F.R. Wagner, V. Bezugly, M. Kohout, Yu. Grin, Chem. Eur. J. **13**, 5724–5741 (2007)
70. R.F.W. Bader, *Atoms in Molecules* (Oxford University Press, Oxford, 1990)
71. A. Ormeci, Yu. Grin, Unpublished data (2013)
72. M. Kohout, Int. J. Quantum Chem. **97**, 651–658 (2004)
73. M. Kohout, Faraday Discuss. **135**, 43–54 (2007)
74. M. Kohout, F.R. Wagner, Yu. Grin, Int. J. Quantum Chem. **106**, 1499–1507 (2006)
75. M. Kohout, F.R. Wagner, Yu. Grin, Theor. Chem. Acc. **108**, 150–156 (2002)
76. S. Raub, G. Jansen, Theor. Chem. Acc. **106**, 223–232 (2001)
77. I. Veremchuk, T. Mori, Yu. Prots, W. Schnelle, A. Leithe-Jasper, M. Kohout, Yu. Grin, J. Solid State Chem. **181**, 1983–1991 (2008)
78. S. Bobev, S.C. Sevov, J. Solid State Chem. **153**, 92–105 (2000)
79. A. Saramat, G. Svensson, A.E.C. Palmqvist, C. Stiewe, E. Mueller, D. Platzek, S.G.K. Williams, D.M. Rowe, J.D. Bryan, G.D. Stucky, J. Appl. Phys. **99**, 023708 (2006)
80. Y. Nagatomo, N. Mugita, Y. Nakakohara, M. Saisho, M. Tajiri, R. Teranishi, S. Munetoh, J. Phys: Conf. Ser. **379**, 012008 (2012)
81. A. Prokofiev, M. Ikeda, E. Makalkina, R. Svagera, M. Waas, S. Paschen, J. Electron. Mater. **42**, 1628–1633 (2013)
82. L.T.K. Nguyen, Ph.D. Thesis, TU Wien (2010)
83. U. Aydemir, C. Candolfi, A. Ormeci, H. Borrmann, U. Burkhardt, Y. Oztan, N. Oeschler, M. Baitinger, F. Steglich, Yu. Grin, Inorg. Chem. **51**, 4730–4741 (2012)
84. W. Jung, H. Kessens, A. Ormeci, W. Schnelle, U. Burkhardt, H. Borrmann, H.D. Nguyen, M. Baitinger, Yu. Grin, Dalton Trans. **41**, 13960–13968 (2012)
85. U. Aydemir, Ph.D. Thesis, TU Dresden (2012)
86. K. Kovnir, U. Stockert, S. Budnyk, Yu. Prots, M. Baitinger, S. Paschen, A.V. Shevelkov, Yu. Grin, Inorg. Chem. **50**, 10387–10396 (2011)
87. E. Hohmann, Z. Anorg. Allg. Chem. **257**, 113–126 (1948)
88. C. Cros, M. Pouchard, P. Hagenmuller, C.R. Hebd. Seances Acad. Sci. **260**, 4764–4767 (1965)
89. S. Yamanaka, H.-O. Horie, H. Nakano, M. Ishikawa, Fullerene Sci. Technol. **3**, 21–28 (1995)
90. M. Baitinger, H.G. von Schnering, J.-H. Chang, K. Peters, Yu. Grin, Z. Kristallogr. NCS **222**, 87–88 (2007)
91. S. Stefanoski, M. Beekman, W. Wong-Ng, P. Zavalij, G.S. Nolas, Chem. Mater. **23**, 1491–1495 (2011)

92. S. Stefanoski, J. Martin, G.S. Nolas, J. Phys. Cond. Mat. **22**, 485404 (2010)
93. S. Stefanoski, G.S. Nolas, Cryst. Growth Des. **11**, 4533–4537 (2011)
94. S. Stefanoski, C.D. Malliakas, M.G. Kanatzidis, G.S. Nolas, Inorg. Chem. **51**, 8686–8689 (2012)
95. R. Nesper, J. Curda, H.G. von Schnering, Angew. Chem. Int. Ed. **25**, 350–352 (1986)
96. S. Yamanaka, E. Enishi, H. Fukuoka, M. Yasukawa, Inorg. Chem. **39**, 56–58 (2000)
97. H. Fukuoka, K. Ueno, S. Yamanaka, J. Organomet. Chem. **611**, 543-546 (2000)
98. I. Veremchuk, A. Wosylus, B. Böhme, M. Baitinger, H. Borrmann, Y. Prots, U. Burkhardt, U. Schwarz, Yu. Grin, Z. Anorg. Allg. Chem. **637**, 1281–1286 (2011)
99. M. Beekman, W. Schnelle, H. Borrmann, M. Baitinger, Yu. Grin, G.S. Nolas, Phys. Rev. Lett. **104**, 018301 (2010)
100. S. Stefanoski, M.C. Blosser, G.S. Nolas, Cryst. Growth Des. **13**, 195–197 (2013)
101. A.M. Guloy, R. Ramlau, Z. Tang, W. Schnelle, M. Baitinger, Yu. Grin, Nature **443**, 320–323 (2006)
102. B. Böhme, A. Guloy, Z. Tang, W. Schnelle, U. Burkhardt, M. Baitinger, Yu. Grin, J. Am. Chem. Soc. **129**, 5348–5349 (2007)
103. B. Böhme, S. Hoffmann, M. Baitinger, Yu. Grin. Z, Naturforsch. **66b**, 230–238 (2011)
104. Y. Liang, B. Böhme, M. Reibold, W. Schnelle, U. Schwarz, M. Baitinger, H. Lichte, Yu. Grin, Inorg. Chem. **50**, 4523–4528 (2011)
105. M.C. Blosser, G.S. Nolas, Mater. Lett. **99**, 161–163 (2013)
106. H. Fukuoka, J. Kiyoto, S.Yamanaka, Inorg. Chem. **42**, 2933–2937 (2003)

# Chapter 3
# Synthetic Approaches to Intermetallic Clathrates

**Matt Beekman and George S. Nolas**

**Abstract** Intermetallic clathrates comprise a class of unique crystalline solids displaying remarkable flexibility in chemical composition, resulting in a wide range of physical properties. Developing the scientific understanding of the unusual and potentially useful properties of these materials and their development in device applications requires a diverse synthetic toolkit. A variety of techniques have been used to prepare intermetallic clathrates, including direct solid-state reaction of the elements, flux and Czochralski crystal growth methods, and thermal decomposition of Zintl phase precursors. Recently, new approaches have been developed and applied to synthesize high quality crystals, as well as the preparation of compositions that are not easily accessible by conventional techniques. In this chapter, we provide a concise overview of both conventional and novel methods for synthesis of intermetallic clathrates, highlighting the types of compositions and characteristic products that have been prepared by each approach.

## 3.1 Introduction

Nearly 50 years have passed since the discovery that group 14 elements can adopt framework structures isotypic with the clathrate hydrates [1–3], the historical context of which is discussed in Chap. 1 of this volume. Given their significant chemical differences, it is intriguing that both water molecules and main group

M. Beekman (✉)
Department of Natural Sciences, Oregon Institute of Technology,
Klamath Falls 97601, Oregon
e-mail: matt.beekman@oit.edu

G. S. Nolas (✉)
Department of Physics, University of South Florida,
4202 E. Fowler Ave., Tampa, FL 33620, Florida, USA
e-mail: gnolas@usf.edu

G. S. Nolas (ed.), *The Physics and Chemistry of Inorganic Clathrates*,
Springer Series in Materials Science 199, DOI: 10.1007/978-94-017-9127-4_3,

elements can adopt the same crystalline arrangement, and the fundamental chemistry and physics of intermetallic clathrates have received significant attention. The technological relevance of these materials has increased in importance in recent years due to remarkable structure-property relationships that intermetallic clathrates display and the potential to use them for energy conversion, transport, and storage applications [4–11].

The ability to synthesize high quality specimens is a prerequisite for evaluating and understanding material properties as well as materials development for technological applications. Equally important, the preparation of new and difficult to synthesize compositions often requires the development of unconventional synthetic routes. The synthetic approach best suited to prepare a particular compound often depends on its composition and stability and the presence of competing phases, possibly having the same or very similar composition. This chapter aims to provide a concise overview of approaches, some conventional and some unconventional, that have been used to prepare intermetallic clathrates. Specific challenges associated with the preparation of different compositions and various advantages and disadvantages of different synthetic approaches are highlighted. Although we restrict our discussion to intermetallic clathrates, other important subsets of inorganic clathrates, e.g. "inverse" clathrates, are often also prepared by many of the techniques discussed here and are highlighted in other chapters of this volume.

## 3.2 General Considerations

Crystallographic descriptions of the known intermetallic clathrate crystal structure types have been described in several reviews [4, 12–14] as well as Chaps. 2, 5, 7, 8 of the present volume. At the time of this writing, intermetallic clathrates have been observed to crystallize in a limited number of primary structure types, although other structures can be expected and in principle new structures may be formed as intergrowths of the classic types [15]. Additional variants, e.g. superstructures of the classic types having ordered arrangements of framework vacancies, have also been reported [16, 17]. Rogl has outlined the symmetry relationships between various clathrate structure types [12]. The common structural feature of intermetallic clathrates is a covalently bonded framework (most commonly comprised of group 13, 14 and/or 15 elements, as well as some transition metals) that can encapsulate electropositive atoms (alkali, alkaline earth, and select rare earth metals) inside the various sized polyhedral cages formed by the framework (cf. Chaps. 2, 5, 7, 8).

Similar features in crystal chemistry amongst many of the known intermetallic clathrate compositions aid the prediction of new compositions. Charge balance is a key consideration in the rational design of many intermetallic clathrate compositions. A significant number of reported intermetallic clathrate-I compositions appear to adhere relatively well to two rules of thumb: (i) there is significant (often nearly complete) transfer of valence electron density from the guest to the framework (or, in the case of the so-called inverse clathrates, from framework to guest; see Chap. 5), and (ii) formally, each atom in the structure typically achieves an electron octet. These rules are not hard and fast, but for many compositions the Zintl formalism can be useful to classify and rationalize compositions [13, 18]. Simple electron counting and formal oxidation states of the guest and framework atoms have been used to guide the rational design of new clathrate compositions and qualitative prediction of their electronic properties, such that charge balanced compositions can be predicted based on these criteria [13]. While the extent to which composition deviates from this nominal charge balance is usually limited for most ternary and higher order clathrate-I systems, it can be used to optimize their thermoelectric properties by influencing the charge carrier concentration [18–21]. In spite of the above, some clathrate-I and most known clathrate-II compositions lacking framework substitution to compensate for charge transfer do not adhere to the above charge balance rules and are characterized by pronounced metallic behavior [22–24], or can be tuned from semiconductors to metals by adjusting the guest content [3, 25]. In spite of constraints imposed by crystal chemistry, Fig. 3.1 illustrates that inorganic clathrates display remarkable compositional flexibility, reflected in particular by the recent preparation of compositions that do not contain a group 14 element [26, 27]. This compositional flexibility is in part responsible for the wide range of interesting and potentially useful mechanical, electrical, magnetic, and thermal properties clathrate materials display.

Since intermetallic clathrates are crystalline guest-host systems, geometric/size relationships between guest and host are also an important consideration. The seminal work by Cros et al. [1–3] demonstrated that the relative size of the guest and cage is a simple but useful predictor for which clathrate structure type, if any, a particular composition will preferentially adopt. Since the clathrate-I and clathrate-II structures each have two different framework cavities, stable compositions can be rationalized by appropriate choice of cation for each cage, based on published ionic and covalent radii [28–31]. For example, at one extreme the relatively large Cs guest does not readily fit inside the cages of the Si clathrate-I framework, thus clathrate-II $Cs_{8-x}Si_{136}$ is preferentially formed during thermal decomposition of $Cs_4Si_4$ [2, 3].[1] At the other extreme, the Na guest appears to be too small to stabilize the Sn clathrate-I or clathrate-II structures, and to date there are no known Sn clathrate compositions containing only Na as a guest species. As pointed out by Bovev and Sevov [32], the relative size of guest and cage provides a simple explanation for the preponderance of known clathrate-I compositions versus clathrate-II compositions reported in the literature.

Many intermetallic clathrate compositions of interest are members of ternary and higher order systems. In such cases the phase equilibria can be rich and complex, while the potential for thermodynamically or kinetically stable

[1] Clathrate-I $Cs_{8-x}Si_{46}$ can be prepared by high-pressure techniques. The lattice parameter and average Si–Si bond length in this phase are the largest of any of the silicon clathrate-I framework [35].

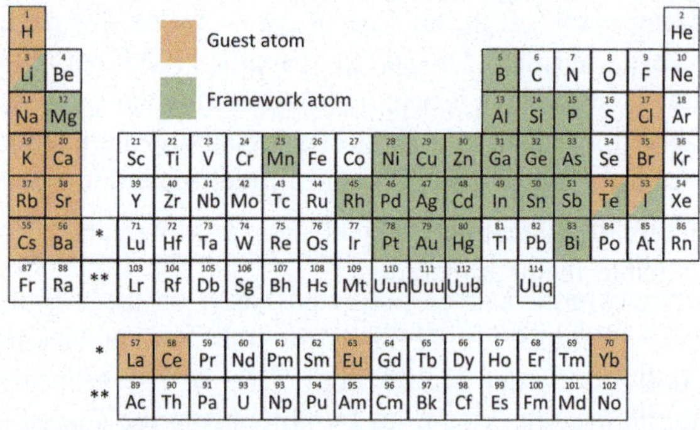


**Fig. 3.1** Known inorganic clathrate-forming elements, classified by guest atom (*orange*) and framework atom (*green*), based on the available literature

secondary (non-clathrate) phases also increases with an increase in the number of degrees of freedom. Many of the ternary and quaternary systems in which clathrates form have not been investigated in detail and published phase diagrams are not yet available, though the body of data on phase equilibria in these systems continues to grow [33, 34].

## 3.3 Conventional Preparation of Polycrystalline Materials

Arguably the most straightforward approach for the preparation of intermetallic clathrates, direct reaction of the elements and/or pre-reacted binary precursors at high temperatures via solid–solid, solid–liquid, and/or gas–solid–liquid reactions has been used to prepare a wide range of compositions. Methods include controlled heating schedules in resistive furnaces, arc melting, and radio frequency inductive melting. The products of these reactions are often polycrystalline materials, sufficient for many structural and physical properties investigations. The electrical and thermal transport properties, for example, do not depend on crystallographic direction for cubic clathrates, and suitable polycrystalline specimens of densities greater than 95 % of the theoretical (X-ray diffraction) density can be directly cut from a boule or obtained by consolidation methods such as hot-pressing or spark plasma sintering. In many cases the products contain small single crystals of sufficient size for X-ray crystallographic study. Relatively large single crystals ($\sim$mm) can also be obtained, in particular (but not exclusively) when direct access to the melt exists, e.g. for congruently melting compounds. In most cases high temperatures ($T > 900$ K) are required, thus these reactions typically only provide access to the most thermodynamically stable products.

The majority of clathrate forming elements (cf. Figure 3.1) readily form stable oxides at high temperature. To avoid oxidation, reactions are carried out under high purity inert atmospheres such as argon or nitrogen. The atmosphere and reaction vessel must be carefully chosen to avoid side reactions. Boron nitride, alumina (corundum), glassy carbon, or carbon coated fused silica crucibles are often adequate if the guest is an alkaline earth metal. For alkali metals, tantalum, niobium, and/or tungsten are better choices, although reaction of these containers with silicon can occur at high temperatures. In the case of alkali metal containing clathrates, the reaction charge is often loaded into a welded tantalum or niobium tube or sealed in a stainless steel canister to avoid the aggressive attack of the silica ampoule by the alkali metal vapor. A fused silica jacket, either evacuated or backfilled with inert gas, can then be used for secondary containment of the metal ampoule if needed.

In some cases the synthesis can be facilitated by utilizing pre-reacted binary precursors, especially in cases where diffusion and reaction kinetics remain slow, e.g. peritectic reactions, due to limitations on reaction temperature when using fused silica ampoules ($T < 1{,}100$ °C). Although the volatility of one or more of the elements and/or significant differences in melting point or vapor pressures may preclude direct formation of the clathrate composition by arc or induction melting, pre-clathrate synthesis of binary precursors of the appropriate constituents can homogenize the elements and, in some cases, create compounds or alloys that have a lower melting point than the constituent elements. For combinations of elements that are insoluble at room temperature but form a homogenous melt at elevated temperature (In-Ge, for example), pre-melting of the framework elements can still produce a precursor with finely divided mixture that can facilitate clathrate synthesis upon peritectic reaction with the guest element.

In many cases, direct reaction of the elements is not always feasible, especially when the physical characteristics for the constituents (melting point, vapor pressure, diffusion coefficients, etc.) or kinetics of the reaction are unfavorable, or if the target product is thermodynamically metastable[2] (in such cases, other synthetic routes discussed elsewhere in this chapter may be more effective). Nevertheless, direct reaction of the elements or preformed precursors has to date been the most commonly used approach to prepare intermetallic clathrates, producing the largest variety of compositions synthesized to date (cf. Table 3.1 for representative reactions).

## 3.4 Thermal Decomposition of Metal-Rich Precursors

Intermetallic clathrates were first discovered in the course of investigations on the thermal decomposition of the binary compounds $M_4Tt_4$ [48, 49], where M = alkali metal and Tt = group 14 element. Systematic investigations carried out in the seminal work by Cros et al. [1–3, 50–52] demonstrated that a variety of alkali metal silicon and germanium-based compositions are prepared via this route, which continues to be employed for preparation of clathrate materials in contemporary investigations [53–64].

---

[2] Melt-spinning has been used in some cases to produce metastable compositions via rapid quenching from the melt [46, 47]. In most cases these compositions are slight deviations from stable compositions, likely due to the broadened phase width for the compound at higher temperature.

**Table 3.1** Selected examples of intermetallic clathrate compositions prepared by direct reaction of the elements to obtain polycrystalline specimens (representative reactions)

| Composition | Structure type | Reaction vessel and atmosphere | Notes | Ref. |
|---|---|---|---|---|
| $A_8Ga_{16}Ge_{30}$ (A = Sr or Eu) | I | Pyrolytic BN crucibles, sealed in fused silica ampoules under Ar | Resistive furnace; 3 days at 1,223 K followed by 4 days at 973 K; melts congruently at 1,047 K | [36] |
| $Ba_8Ga_{16}Ge_{30}$ | I | Pryolytic BN crucibles, sealed in fused silica ampoules under $N_2$ | Induction melt; 10 min at 1,373 K followed by water quench; congruently melting compound | [20] |
| $Ba_8Au_xSi_{46-x}$ | I | Glassy carbon crucibles, Ar atmosphere | Induction melt; quenched from the melt by contacting melt with metal surfaces | [37] |
| $Ba_8Ga_{16}Si_yGe_{46-y}$ | I | Pyrolytic BN crucibles, sealed in fused silica ampoules under $N_2$ | $Si_{1-x}Ge_x$ alloy formed by arc melting first; subsequent reaction with Ba and Ga: 1 K/min heat, 72 h at 1,173 K, 2 K/min cool down; ground, compacted and sintered two additional times; congruently melting compounds, but a gap exists between the liquidus and solidus | [38] |
| $Ba_8Ga_{16}Si_{30}$ | I | Arc melting of the elements under Ar followed by spark plasma sintering | Consolidation/solid state reaction by spark plasma sintering at 1,113–1,273 K for 1–2 h | [39] |
| $A_8Hg_{3+x}Ge_{43-x}$ (A = K, Rb) | I | $A_4Ge_9$ precursor reacted with Hg in welded Nb tubes | Resistive furnace; 872 K for 24 h, 673 K for 24 h, then cooled to room temperature (rate: –4 K/min) | [40] |

(continued)

**Table 3.1** (continued)

| Composition | Structure type | Reaction vessel and atmosphere | Notes | Ref. |
|---|---|---|---|---|
| $A_8Na_{16}X_{136}$ (A = Rb or Cs; X = Si and Ge) | II | Welded Nb tubes, sealed in fused silica ampoules under Ar | Resistive furnace; 3 weeks at 923 K | [41] |
| $Cs_8Na_{16}TM_yGe_{136-y}$ (TM = Ag, Cu) | II | W crucibles in steel canisters sealed by graphite gaskets, sealed in fused silica ampoules under $N_2$ | Resistive furnace; 2 days at 1073 K followed by 6 days at 923 K | [42, 43] |
| $A_8Ba_{16-x}Ga_ySn_{136-y}$ | II | Welded Nb tubes (Ar atmosphere), sealed in fused silica ampoules under vacuum | Resistive furnace; 1,273 K (rate: 300 K/h), held for 1 h, cooled to 973 °C (rate: −150 K/h), held for 120 h, and cooled to room temperature (rate: −6 K/h) | [44] |
| $Cs_8Na_{16}In_8Ge_{128}$ | II | W crucibles in steel canisters sealed by graphite gaskets, sealed in fused silica ampoules under $N_2$ | In and Ge pre-melted and ground to form a finely divided mixture, then combined with alkali metal; resistive furnace; 2 days at 1073 K followed by 6 days at 923 K | [45] |

These compositions are thermodynamically stable (at the temperature of reaction) and various other successful reaction conditions for a given composition have often been reported in other articles in the literature

Most $M_4Tt_4$ compounds are stable to several hundred degrees Celsius or higher if a satisfactory partial pressure of alkali metal vapor is maintained over the compound.[3] However, due to the relatively high vapor pressure of the alkali metal, the $M_4Tt_4$ compounds are often unstable when heated under conditions where the partial pressure of the alkali metal is not maintained at the equilibrium vapor pressure, e.g. vacuum or controlled inert atmosphere (open container). Since the $M_4Tt_4$ precursors can explosively react with air and moisture, they are handled under inert atmosphere such as high purity $N_2$ or Ar. The apparatus used at the University of South Florida for synthesis of clathrates by thermal decomposition under vacuum is illustrated schematically in Fig. 3.2 [66, 67]. Finely ground $M_4Tt_4$ precursor is placed in an appropriate boat or crucible (tantalum or fused silica, for example), and loaded into a fused silica tube. A vacuum valve and coupling allows transfer into and out of a glove box for pre- and post-decomposition handling without exposure to air. The decomposition products are typically fine-grain clathrate powders, and often contain varying amounts of free alkali metal (at the surface of the particles), unreacted $M_4Si_4$, and various nanophase and amorphous material, all of which can be pyrophoric and can cause the entire post-decomposition product (including the clathrate) to ignite and combust uncontrollably in air. Carefully rinsing the powder with ethanol and then $H_2O$ under flowing $N_2$ or Ar can remove these impurities, allowing the stable clathrate product to be isolated [67]. Brief sonication of the suspension of powder in the ethanol-water solution aids this process. Sonication can also aid in the physical separation of clathrate from amorphous and/or nanocrystalline material, which upon completion often settles at a significantly slower rate than the clathrate and can be decanted with the ethanol-water solution. The microcrystalline clathrate powder (Fig. 3.2) can then be recovered and dried in air or vacuum.

The thermal decomposition kinetics and reaction products are highly sensitive to the temperature and atmosphere characteristics. One of the challenges (or opportunities, depending on perspective) accompanying the thermal decomposition approach is the possibility of producing various products and phases from the same precursor. The original work by Cros et al. demonstrated that clathrate-II $Na_xSi_{136}$ ($0 < x < 24$) is preferentially obtained from thermal decomposition of $Na_4Si_4$ under vacuum, whereas clathrate-I $Na_8Si_{46}$ can be preferentially obtained from thermal decomposition under 1 bar Ar atmosphere, and a correlation between the partial pressure of Na and the phase obtained was identified [3]. More recently, Horie et al. presented further evidence that the partial pressure of Na is an important factor in determining the products produced by thermal decomposition [68], in agreement with the original observations of Cros [3]. The weight ratio clathrate-I:clathrate-II ($Na_8Si_{46}$:$Na_xSi_{136}$) in specimens prepared by thermal decomposition of $Na_4Si_4$ can range from nearly 0 to nearly 100 % [53] and can be difficult to control. Ramachandran et al. [53] utilized the difference in mass densities of

[3] As an example, $Na_4Si_4$ melts congruently at 798 °C according to the reported Na–Si phase diagram [65].

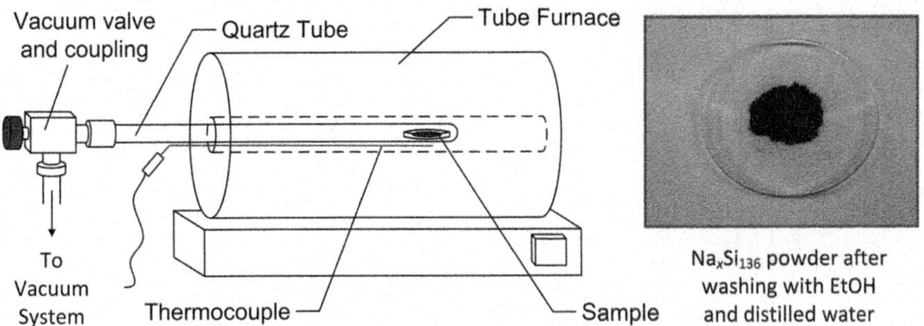


**Fig. 3.2** Schematic of the apparatus used to prepare intermetallic clathrates via thermal decomposition of $M_4Tt_4$ precursors at the University of South Florida (*left*), and an example of the resulting clathrate products formed by this approach (*right*)

$Na_8Si_{46}$ and $Na_xSi_{136}$ and appropriate solutions of dibromomethane and tetrachloroethelyne to density separate the two phases. Density separation has limitations, since the crystallites and particles are often intergrown and recovery of a completely phase pure specimen is often not possible. Gryko [69] found that very rapid heating of the $Na_4Si_4$ precursor under vacuum effectively reduces the $Na_8Si_{46}$ content when $Na_xSi_{136}$ is the target product. A modified version of this technique was used to prepare $Na_xSi_{136}$ specimens with $Na_8Si_{46}$ wt% $< 3$ % without further separation [62]. In situ studies of the thermal decomposition process are rare [70], and the detailed reaction pathway from precursor to clathrate during thermal decomposition, while clearly dependent upon the synthetic conditions, still remains not well understood. The diffusion pathways and mechanisms through which guest atoms such as Na are removed from the clathrate structure (to form partially filled $Na_xSi_{136}$ clathrates) are also still not well understood, though they may be influenced by defects in the clathrate as it nucleates and grows from the precursor.

Investigations utilizing clathrate specimens prepared by conventional thermal decomposition continue to yield important insight into the structural and physical properties of intermetallic clathrates, and a variety of compositions have been prepared by thermal decomposition (cf. Table 3.2). Nevertheless, this synthetic approach does have several limitations. The products are typically fine microcrystalline powders, and in the case of non-stoichiometric materials such as $Na_xSi_{136}$ the composition ($x$) and phase purity can be difficult to control [3, 53, 62, 68]. Comprehensive analytical studies of the microscopic homogeneity in composition (both within single grains/crystallites and between separate grains) of specimens produced by thermal decomposition have yet to be reported. The crystallite size obtained is often in the micron to submicron range with the largest crystallites typically no larger than tens of microns [60, 71]. Substantial crystal defects, including stacking faults and twin interfaces, have been reported for $Na_xSi_{136}$ specimens prepared by thermal decomposition [72]. Significant nanoparticulate material can be present in the specimens depending on preparation conditions [71], though the production of fine grain materials, possibly with substantial defects,

**Table 3.2** Selected precursor compositions, preparation conditions, and resulting products formed by thermal decomposition of metal-rich precursors

| Composition | Structure type | Precursor | High vacuum orInert atmosphere | References |
|---|---|---|---|---|
| $Na_8Si_{46}$ | I, II | $Na_4Si_4$ | 350–450 °C, high vacuum, 10 min–48 h | [1–3, 50–53, 58, 62, 66–70, 74] |
| $Na_xSi_{136}$ $(0 < x < 24)$ | | | | |
| $K_8Si_{46}$ | I | $K_4Si_4$ | 425–475 °C, dynamic vacuum | [2, 56] |
| $Rb_{8-x}Si_{46}$ | II | $Rb_4Si_4$ | 425–475 °C, dynamic vacuum | [2, 56] |
| $Cs_{8-x}Si_{136}$ | II | $Cs_4Si_4$ | 500 °C, dynamic vacuum | [2] |
| $Na_8Rb_{16}Si_{136}$ | II | $Na_2Rb_2Si_4$ | 420 °C, dynamic vacuum, 10 days | [57] |
| $Na_8Cs_{16}Si_{136}$ | II | $Na_2Cs_2Si_4$ | 425 °C, dynamic vacuum | [56] |
| $Na_2Ba_6Si_{46}$ | I | $BaNa_2Si_4$ | 427–500 °C, several seconds to 10 h | [60, 77, 78] |
| $Ba_8Na_{16}Si_{136}$ | II | | | |
| $Na_xGe_{136}$ | I | $Na_4Ge_4$ | 300–370 °C, 2 days, dynamic vacuum | [2, 79–81] |
| $Na_{1-x}Ge_{3+z}$[a] | II | | | |
| $K_8Ge_{44}$ | I | $K_4Ge_4$ | 350–380 °C, dynamic vacuum | [2, 54, 59] |
| $Rb_8Ge_{44}$ | I | $Rb_4Ge_4$ | 350–380 °C, dynamic vacuum | [66] |
| $Na_xK_yGe_{44}$ | I | $Na_2K_2Ge_4$ | 350–380 °C, dynamic vacuum | [66] |
| $Na_xK_yGe_{136}$ | II | | | |
| $Na_xRb_yGe_{44}$ | I | $Na_2Rb_2Ge_4$ | 350–380 °C, dynamic vacuum | [66] |
| $Na_xSi_{136-y}Ge_y$ | II | $Na_4Si_{4-x}Ge_x$ | 350–380 °C, dynamic vacuum | [81] |
| $Ba_{24}Ge_{100}$ | III | $BaNa_2Ge_4$ | Elevated temperature, dynamic vacuum | [82] |

[a] This phase has a different, zeolite-like structure as opposed to clathrate structure, and is a competing phase

could be beneficial for applications involving the reversible incorporation of mobile guest species [9]. In the case of silicon clathrates the small grains are typically coated by an oxide layer formed during synthesis and exposure to air [73], which can in some cases mask the intrinsic properties of the material. Consolidation of the powders is often highly challenging, and the porous nature of consolidated specimens can have a dominating effect on the measured electrical and thermal transport properties [2, 23, 25, 58, 61].

The mobile nature of Na in the clathrate-I and clathrate-II lattice allows this guest to be removed from the structure by prolonged heating under vacuum [2, 3, 53, 62, 74] or by chemical methods [75]. In this way, the Na content can be varied from $x < 1$ to close to 24 in $Na_xSi_{136}$ prepared by thermal decomposition or other techniques.[4] Further reduction of Na content was achieved by repeated density separation/centrifugation [55] and reaction with iodine [75] to obtain the allotrope $Si_{136}$ (Na content less than 100 ppm). Clathrate-II compositions such as $Cs_8Ge_{136}$ [76] and $Rb_8Ge_{136}$ [66] can be obtained by removal of Na from $Na_{16}Cs_8Ge_{136}$ and $Na_{16}Rb_8Ge_{136}$, respectively, under high vacuum at temperatures between 573 and 673 K. Removal of Na from $Na_{24}Si_{136}$ crystals allowed high quality $Na_xSi_{136}$ ($0 < x < 24$) specimens to be obtained, and as a result the intrinsic electrical and thermal transport properties could be characterized [25]. Thermal decomposition precursors and clathrate products are summarized in Table 3.2.

## 3.5 Soft Chemical Routes

Soft chemical methods for the preparation of solid-state materials possess many advantages over conventional synthesis approaches. The introduction of additional synthetic variables allows for additional ways to influence the resulting products that form, and the comparatively low temperatures used in these reactions allow access to kinetically stable (thermodynamically metastable) compounds that cannot be prepared by conventional methods. The use of low temperature chemical oxidation of precursors in heterogenous reactions was recently introduced as an effective method for synthesis of intermetallic clathrates (see also Chap. 2 of the present volume) [83, 84]. We summarize here the compositions obtained by this approach.

Guest free $Ge_{136}$ was the first reported inorganic clathrate to be prepared by chemical oxidation of reactive precursors [83]. By heating a $Na_xGe$ precursor ($x < 1$) at 300 °C under argon in a sealed reaction vessel containing a eutectic mixture of dodecyltrimethylammonium chloride (DTAC) and aluminum chloride ($AlCl_3$), the new elemental modification $Ge_{136}$ was prepared [83]. Later

[4] Fully filled (stoichiometric) $Na_{24}Si_{136}$ is not readily prepared by thermal decomposition, but can be prepared using techniques discussed elsewhere in this chapter.

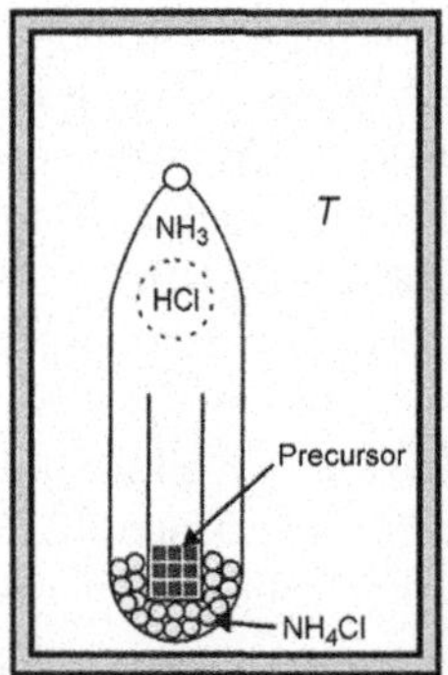

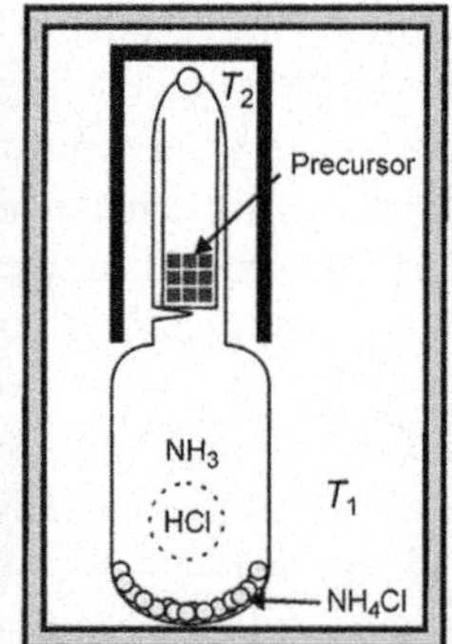

**Fig. 3.3** Schematic illustration of two methods for oxidation of $M_4Tt_4$ precursors using gaseous HCl evolved from decomposition of $NH_4Cl$: Reaction in which the precursor and HCl source are at the same temperature (*far left*), and which the precursor and HCl source are held at different temperatures (*middle pane*). The latter allows control of the reaction temperature and partial pressure of the oxidation agent (HCl) quasi-independently. *Far Right* Secondary electron SEM image of an agglomerate of $Ba_xSi_{46}$ particles, prepared by chemical oxidation of the $Ba_4Li_2Si_6$ precursor. Reprinted with permission from Ref. [86]. Copyright 2011 American Chemical Society

investigations reported by Böhme et al. [85] provided further insight into the reaction mechanism and allowed for optimization of the synthesis of $Ge_{136}$. Gaseous oxidation agents produced by the decomposition products of DTAC, DTAC/$AlCl_3$, or DTAC/$MgCl_2$, initiate heterogeneous oxidation of alkali metal tetrelide precursors resulting in clathrate products. Gaseous HCl also serves as a highly effective oxidizing agent in this approach [84], and both pure HCl and $NH_4Cl$ (HCl source upon thermal decomposition) have been used. Use of two-zone furnaces (Fig. 3.3), in which the $NH_4Cl$ is held at a lower temperature than the tetrelide precursor, allows limited control of the partial pressure of the oxidizing agent in the ampoule, independent of the reaction temperature [86].

A number of clathrate compositions have been prepared for the first time by low temperature chemical oxidation of reactive precursors, including $Ge_{136}$ [83], $K_xGe_{136}$ [87], $Na_{8-x}Si_{46}$ ($0 < x < 7$) [88], and $K_{8-x}Ge_{44+y}$ [88]. Compositions previously obtained by other methods have also been prepared by chemical oxidation, including $Na_8Si_{46}$ and $K_{8-x}Si_{46}$ [84], $Na_{24-x}Si_{136}$ [89], $Na_2Ba_6Si_{46}$ [90], and $Ba_{8-x}Si_{46}$ [86]. The composition $Ba_{8-x}Si_{46}$ is particularly significant, since it otherwise has only been obtained by high-pressure techniques [91]. Table 3.3 summarizes compositions that have been prepared by chemical oxidation methods. While the products from this synthetic approach have so far been limited to polycrystalline powders, it holds significant promise to further expand the known clathrate composition space and is expected to provide access to new compositions that may yield unique insights into structure and bonding in these materials.

**Table 3.3** Clathrate compositions that have been prepared by chemical oxidation of reactive precursors, along with selected details of the preparation

| Composition | Structure type | Precursor | Source of oxidizing agent(s) | Reaction conditions; notes | References |
|---|---|---|---|---|---|
| $Na_8Si_{46}$ | I | $Na_4Si_4$ | Gaseous HCl (produced from reaction of $NaHSO_4$ and NaCl or decomposition of $NH_4Cl$) | Argon atmosphere, 20 h at 400 °C | [84] |
| $Na_8Si_{46}$ | I | $Na_4Si_4$ | Products of decomposition of DTAC/$AlCl_3$ ionic liquid | Nitrogen atmosphere, 8 h at 210–240 °C | [89] |
| $Na_{24-x}Si_{136}$ | II | $Na_4Si_4$ | Products of decomposition of DTAC/$AlCl_3$ ionic liquid | Nitrogen atmosphere, 24 h at 210–240 °C; crystalline silicon impurity phase | [89] |
| $K_{8-x}Si_{46}$ | I | $K_4Si_4$ | Gaseous HCl (produced from decomposition of $NH_4Cl$) | Argon atmosphere, 1 h at 450 °C; amorphous impurity phase | [84] |
| $Na_2Ba_6Si_{46}$ | I | $Na_2BaSi_2$ | Gaseous HCl (produced from decomposition of $NH_4Cl$) | Argon atmosphere, 1 h at 400 °C $(Na,Ba)_xSi_{136}$ impurity phase | [90] |
| $Ba_{8-x}Si_{46}$ | I | $Li_2BaSi_2$ | Gaseous HCl (produced from decomposition of $NH_4Cl$) | Single or two zone furnace | [86] |
| $Ge_{136}$ | II | $Na_{12}Ge_{17}$ | Products of decomposition of DTAC/$AlCl_3$ ionic liquid | Best results achieved when precursor and DTAC are spatially separated as opposed to mixed. | [83, 85] |
| $K_{8-x}Ge_{44-y}$ | I | $K_4Ge_9$ | Products of decomposition of DTAC/$AlCl_3$ ionic liquid | Reaction for 24 h at 280 °C | [88] |
| $K_{24-x}Ge_{136}$ | II | $K_4Ge_9$ | Products of decomposition of DTAC/$AlCl_3$ ionic liquid | Reaction for 24 h at 300 °C. Resulting product was poorly crystallized, subsequently annealing for 48 h at 370 °C transforms product into polycrystalline $K_{24-x}Ge_{136}$ | [87] |

## 3.6 Growth of Macroscopic Single Crystals

The preparation and utility of powder and polycrystalline specimens should not be undervalued (indeed, they are the form required for numerous applications, including thermoelectric energy conversion). Nevertheless, macroscopic single crystal specimens are typically the most desirable from the perspective of understanding the intrinsic structural and physical properties of a crystalline compound. However, they can be significantly more difficult to obtain. A variety of techniques have been used to grow macroscopic single crystals of intermetallic clathrates, and high quality single crystals have been produced for a variety of compositions. For congruently melting compounds, high-quality clathrate single crystals can usually be grown using conventional techniques once the essential thermophysical characteristics of the system are understood. By slow cooling of a stoichiometric melt, relatively large single crystals ($\sim 1\ cm^3$) of clathrate-I compositions such as $Sr_8Ga_{16}Ge_{30}$, $Ba_8Ga_{16}Ge_{30}$, and $Eu_8Ga_{16}Ge_{30}$ can readily be prepared for transport properties measurements as well as neutron scattering experiments [92]. In other cases, more sophisticated methods are required. Table 3.4 provides examples of clathrate compositions and corresponding methods for which growth of macroscopic[5] single crystals has been accomplished.

The metal flux is an attractive synthetic tool [93, 94] for intermetallic phases for several reasons. In general, the equipment required for flux synthesis is relatively standard and specialized apparatus is not required. Provided the flux has a sufficiently low melting point, it can facilitate the otherwise slow diffusion of reactants and thus reactions can often be carried out at lower temperatures, and can be particularly useful for the growth of single crystals. In some cases, especially those involving peritectic reactions, it can facilitate the preparation of large single crystals when there is no direct access to the melt [93, 94]. It is rather fortuitous that many intermetallic clathrate compositions contain one or more elements that can be readily used as a molten solvent for crystal growth in a reactive or "self-flux" approach, and this fact has been utilized for successful flux crystal growth of a number of compositions [101–106]. Since the flux is comprised of one or more of the constituent elements in the compound, incorporation of foreign species from the flux into the crystal is not a concern. "Non-reactive" fluxes, i.e. fluxes that do not intentionally incorporate into the final product, have yet to be explored for crystal growth of intermetallic clathrates.

In a typical flux synthesis, a significant excess of one or more constituents is added to the reaction charge, which is loaded in an appropriate (inert) crucible and sealed under inert atmosphere either in a fused silica ampoule or welded container, depending on the composition. A plug of quartz wool is often placed above the reaction charge, either inside the crucible or directly above it, to act as a filter to separate the flux, post-crystal growth [113]. After an adequate soak at elevated

[5] Here we are concerned with single crystals large enough for physical and transport properties characterization or neutron scattering experiments.

**Table 3.4** Selected compositions and methods for clathrate single crystal growth

| Composition | Structure type | Growth Method | References |
|---|---|---|---|
| $A_8Ga_{16}Ge_{30}$ (A = Sr, Ba, and Eu) | I | Slow cooling of stoichiometric melt | [92] |
| $Ba_8Ga_{16}Ge_{30}$ | I | Czochralski method | [95, 96] |
| $Ba_8Al_{16}Ge_{30}$ | I | Czochralski method | [97] |
| $Ba_8Ga_{16}Si_{30-x}Ge_x$ | I | Czochralski method | [98] |
| $Ba_8Ni_{3.5}Ge_{42.1}\square_{0.4}$ | I | Bridgeman method | [99, 100] |
| $A_8Ga_{16}Ge_{30}$ (A = Sr, Ba, and Eu) | I | Flux synthesis | [101] |
| $Ba_8Al_xSi_{36-x}$ | I | Flux synthesis | [102, 103] |
| $Ba_8Ga_{16}Sn_{30}$ | I, VIII | Flux synthesis | [104] |
| $Sr_8Ga_{16}Si_yGe_{30-y}$ | I | Flux synthesis | [105] |
| $K_8Zn_4Sn_{42}$ | I | Flux synthesis | [106] |
| $K_8Ga_8Sn_{38}$ | I | Flux synthesis | [107, 108] |
| $Na_{24}Si_{136}$ | II | Spark plasma treatment of $Na_4Si_4$ | [109, 110] |
| $Na_{24}Si_{136}$ | II | Solid state crystal growth from $Na_4Si_4$ (KCTD) | [111] |
| $Na_8Si_{46}$ | I | Solid state crystal growth from $Na_4Si_4$ (KCTD) | [111] |
| $K_{7.5}Si_{46}$ | I | Solid state crystal growth from $K_4Si_4$ (KCTD) | [112] |
| $K_{18}Si_{136}$ | II | Solid state crystal growth from $K_4Si_4$ (KCTD) | [112] |

temperature, the temperature is slowly lowered until the crystal growth is essentially halted with the flux still molten. The reaction vessel is then typically inverted and the molten flux removed by centrifugation or spinning, with the crystals retained above the wool plug. If the resulting crystals remain coated by a thin layer of the flux, it can usually be removed by washing with an appropriate acid or base, presuming the flux is dissolved at a higher rate than the clathrate crystals [101–106].

For compositions containing more than one element that can function as a flux, the ratio of the flux constituents can be used to influence the resulting composition of the product. For example, by adjusting the Ga:Sn ratio in the reaction mixture, Suekuni et al. were able to control the carrier type in $\alpha$- and $\beta$-$Ba_8Ga_{16}Sn_{30}$ [104]. Kazlaurich et al. [102, 103] have grown single crystals of several ternary clathrate compositions, some of which are discussed in Chap. 8 of the present volume.

The Czochralski and Bridgeman methods [114] are integral crystal growth techniques widely used for growing large single crystals of a broad range of technologically important materials, in particular elemental and compound semiconductors. While the largest clathrate single crystal specimens have been grown by these two methods, they have so far only been applied to a relatively small number of compositions. The first clathrate composition to be prepared by the Czochralski technique was $Ba_8Ga_{16}Ge_{30}$ [95]. Impressive thermoelectric performance later reported for Czochralski-grown $Ba_8Ga_{16}Ge_{30}$ [96] single crystals, though the results obtained from Czochralski-grown specimens in different studies appear to vary significantly [95, 96, 115], potentially due to variation in stoichiometry (and therefore carrier concentration) of the measured sections of the crystals. The Czochralski method has recently been extended to $Ba_8Al_xSi_{46-x}$, [97] and $Ba_8Ga_{16}Si_xGe_{46-x}$ [101]. In the Si–Ge alloy clathrate, the resulting single crystals were shown to be inhomogeneous in composition along the pulling direction, evidence that a significant gap exists between the liquidus and solidus in the $Ba_8Ga_{16}Ge_{30}$–$Ba_8Ga_{16}Si_{30}$ pseudo-binary system, analogous to binary $Si_{1-x}Ge_x$ solid solutions. The Bridgeman method has been applied to grow large single crystals of $Ba_8Ni_{3.5}Ge_{42.1}\square_{0.4}$ [99].

Although single crystals of many intermetallic clathrate compositions can be prepared by standard techniques, other compositions do not lend themselves to conventional growth from the melt or using a flux. This is particularly true for the alkali metals, which have very high vapor pressures (orders of magnitude higher than Si, Ge, or Sn) and often significantly different melting points when compared to the other constituents in the reaction. Silicon, for example, melts at 1,687 K, which is well above the boiling points of the alkali metals Na, K, Rb, and Cs. Compositions that form and melt peritectically also present specific challenges. Nevertheless, such compositions are important from the perspective of understanding the fundamental chemistry and physics of intermetallic clathrates, motivating new crystal growth approaches to prepare them (Fig. 3.4).

As a result of the significant challenges alkali metal-based compositions present for conventional growth of high quality single crystal specimens, $M_4Tt_4$ precursors have played an important role in the preparation of these clathrates in single crystal form. The $Na_xSi_{136}$ clathrates are representative: although these compositions were the first intermetallic clathrates to be discovered, they eluded single crystal growth

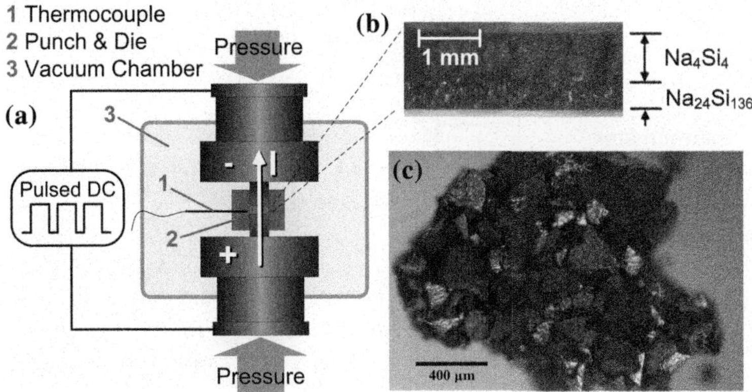


**Fig. 3.4** **a** Schematic of the apparatus used for single crystal growth of intermetallic clathrates by spark plasma processing of $M_4Tt_4$ precursors. A pulsed DC voltage is applied to the anode and cathode, resulting in **b** directional formation of the clathrate, shown here for $Na_{24}Si_{136}$ grown from $Na_4Si_4$ precursor. **c** High quality $Na_{24}Si_{136}$ single crystals are obtained by dissolution of the remaining $Na_4Si_4$ precursor in ethanol and distilled water under argon [109]. Adapted with permission from Ref. [109]. Copyright 2009 American Chemical Society

for more than four decades. The first high quality single crystals of $Na_{24}Si_{136}$ were recently prepared using pulsed electric current sintering techniques commonly referred to as field assisted sintering or spark plasma sintering (SPS) [109]. SPS is an effective and rapid sintering and consolidation technique for powder processing of a wide range of technologically important materials, in particular ceramics and nanostructured materials, but has only recently begun to be explored as a tool for materials synthesis [116–118]. During SPS crystal growth of $Na_{24}Si_{136}$, a pulsed DC current is passed through powdered $Na_4Si_4$ held under uniaxial pressure. $Na_{24}Si_{136}$ is preferentially formed at the anode of the system as shown in Fig. 3.5. The formation of the clathrate can be understood in terms of an electrochemical oxidation-reduction reaction driven by the applied potential difference and resulting electrical current that is passed through the specimen. $Na_{24}Si_{136}$ is formed at the anode, while elemental Na is formed by reduction of $Na^+$ at the cathode (Fig. 3.5). The following half reactions were proposed to explain these processes:

$$34\ Si_4^{4-} - 112\ e^- \rightarrow Si_{136}^{24-} \qquad \text{(anode)}$$

$$112\ Na^+ + 112\ e^- \rightarrow 112\ Na \qquad \text{(cathode)}$$

This requires the precursor (in this case, $Na_4Si_4$) to function as a solid electrolyte that enables $Na^+$ ion transport toward the cathode.

Temperature, pressure, and time all have significant effects on the products resulting from the reaction of $M_4Tt_4$ precursors by SPS [109, 110]. At temperatures below 550 °C, single phase clathrate-I $Na_8Si_{46}$ is obtained, whereas at 600 °C clathrate-II $Na_{24}Si_{136}$ is favored. For temperatures at or above 650 °C,

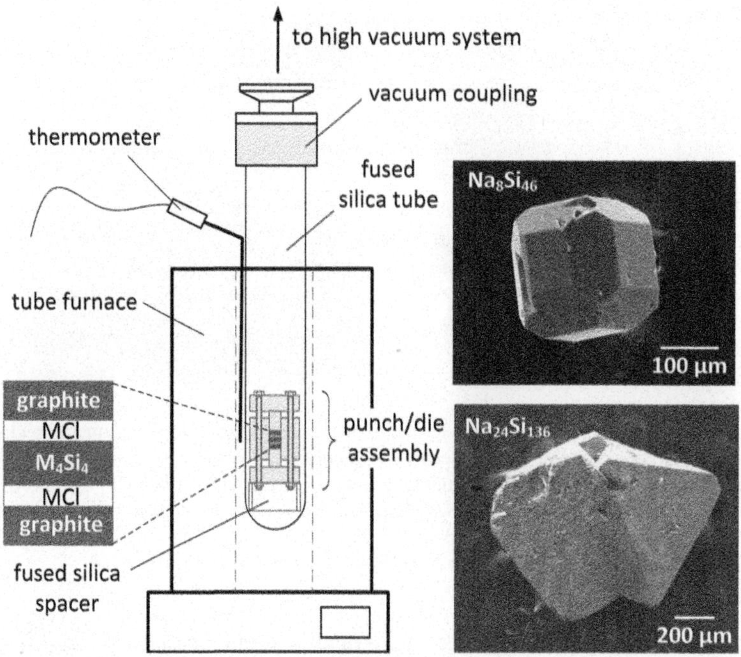


**Fig. 3.5** Apparatus (*left*) used for kinetically controlled thermal decomposition of $M_4Tt_4$ precursors [121]. The multilayer stack (sandwich) of graphite-chloride-precursor-chloride-graphite is preloaded under inert atmosphere and bolted in a steel punch and die assembly under uniaxial pressure. The assembly is heated under dynamic vacuum to initiate the reaction. As an example, high quality $Na_8Si_{46}$ and $Na_{24}Si_{136}$ clathrate single crystals (*right*) are selectively grown by adjusting the reaction temperature [111]. SEM images reprinted with permission from Ref. [111]. Copyright 2011 American Chemical Society

nanocrystalline silicon is obtained. For uniaxial pressures below 80 MPa, the products are microcrystalline, whereas above 80 MPa macroscopic single crystals are obtained [110]. By optimization of these parameters, SPS provides the ability to selectively prepare high quality, single-phase intermetallic clathrates such as $Na_8Si_{46}$ and $Na_{24}Si_{136}$ [110], which have very similar compositions and are difficult to prepare in single-phase form by other approaches (cf. Sect. 3.4). The results demonstrate the potential of this method for preparation of high quality specimens of new and synthetically challenging clathrate compounds. By appropriate reaction of $Cs_4Si_4$, $Rb_4Si_4$, or $K_4Si_4$, the respective clathrates $Cs_8Si_{136}$, $Rb_8Si_{46}$, $Rb_{24-x}Si_{136}$, and $K_{8-x}Si_{46}$ can also be synthesized [119]. The approach can likely be extended to other compositions, presuming appropriate precursors can be identified that are electrically conducting such that the electrical current passes primarily through the precursor as opposed to the surrounding die.

As discussed in Sect. 3.4, $M_4Tt_4$ precursors such as $Na_4Si_4$ are stable presuming the partial pressure of the alkali metal over the compound is maintained at the equilibrium vapor pressure at the temperature of interest. If this condition is not met, thermal decomposition ensues concurrent with evaporation of the alkali metal. For example, depending on the reaction conditions (temperature, partial pressures of Na and other gases such as $O_2$ that are present, etc.), the products of thermal decomposition of $Na_4Si_4$ can include $Na_xSi_{136}$, $Na_8Si_{46}$, elemental silicon, oxides or hydroxides, and various amorphous or nanocrystalline material [2, 3, 53, 55, 62, 66, 67, 71–75].[6] Although the thermal decomposition of $M_4Tt_4$ in open vessels can be difficult to control and in general produces microcrystalline products that can possess a high concentration of lattice defects, high quality crystal growth can be achieved by removing the alkali metal in a controlled manner. As an example, clathrate-II $M_{24-x}Si_{136}$ and clathrate-I $M_{8-x}Si_{46}$ (M = Na, K) can be selectively prepared by "kinetically controlled thermal decomposition" using the simple apparatus shown in Fig. 3.5 [121]. In this approach, finely ground $M_4Si_4$ precursor is sandwiched between graphite within a bolted punch-die-platen assembly that applies uniaxial pressure to the precursor during the reaction (NaCl acts as a barrier to spatially separate the precursor and graphite in order to prevent direct reaction and adhesion, but allows Na to be transported diffusively in the vapor phase). The system is then loaded into a fused silica tube equipped with a vacuum coupling and the reaction carried out under dynamic vacuum ($P < 10^{-5}$ torr) in a resistance furnace. A tight seal between the punches and die is important to ensure the alkali metal is slowly removed from the precursor primarily by vapor phase reaction of the alkali metal with graphite, forming alkali–carbon intercalation compounds. As the precursor composition is slowly driven Si rich, the respective clathrate nucleates and crystal growth ensues.

As with SPS crystal growth, an important attribute of kinetically controlled thermal decomposition is the selectivity it provides for crystal growth of different clathrate structure types that have very similar compositions. By merely adjusting the temperature, single phase $M_8Si_{46}$ and $M_{24}Si_{136}$ are readily prepared, in spite of the absence of these compounds from the reported equilibrium phase diagrams [65]. High quality $Na_xSi_{136}$ ($0 < x < 24$) clathrate crystals can subsequently be prepared by removal of Na from $Na_{24}Si_{136}$ by prolonged heating under high vacuum [25]. The simplicity of the method, which uses relatively standard equipment, makes it particularly attractive for the preparation of otherwise challenging compositions as well as exploratory synthesis.

## 3.7 High-Pressure Synthesis

High-pressure investigations have yielded important insights into the structural, chemical, and physical properties of clathrate materials, as well as their stability [122–124]. From a synthetic standpoint, high-pressure techniques[7] often allow

[6] Intriguing double-helical microtubules were obtained by evaporation of Na from molten $Na_4Si_4$ at 800 °C [120].

[7] We consider pressures in excess of 1 GPa to be "high".

access to solid-state phases that cannot be prepared by other techniques, and these phases often have interesting and unusual properties that allow unique insights into structure-property relationships in these materials.

The first intermetallic clathrate to be prepared by high-pressure synthesis was $Ba_8Si_{46}$, obtained by reacting a stoichiometric mixture of $BaSi_2$ and Si powders at pressures between 1 and 5 GPa and T = 800 °C [125]. Remarkably, this composition has a superconducting transition at 8.0 K, which at the time of the discovery was rather unusual given the 3-dimensional covalent silicon framework [125, 126]. New compositions, such as $NaSi_6$, have also recently been reported by high pressure synthesis [127]. Further discussion of the preparation and properties of high-pressure phases, as well as a detailed description of high-pressure synthesis and apparatus, can be found in Chap. 7.

As shown in Table 3.5, a variety of other clathrate compositions have been prepared under high-pressure conditions in the past 10 years, several of which also display superconducting transitions at low temperature. All of these clathrates are binary or pseudo-binary compounds, i.e. the framework is composed entirely of a single element. Most of the high-pressure clathrate phases reported to date are silicon-based, with no substitution of the framework atoms. Particularly noteworthy is $Cs_{8-x}Si_{46}$, the existence of which was unexpected given the large relative size of the Cs guest compared to the typical free radius of the silicon cage in $Si_{46}$ framework for the other known compounds [35].

## 3.8 Preparation of Thin Films and Nanostructures

The majority of synthetic efforts have overwhelmingly focused on the preparation of intermetallic clathrates in bulk form by the methods described above. To date, comparatively fewer efforts to prepare thin film and nanostructured materials have been reported.

$Ba_8Ga_{16}Ge_{30}$ was the first ternary intermetallic clathrate to be prepared as thin films [131, 132]. Work at the University of South Florida demonstrated that crystalline $Ba_8Ga_{16}Ge_{30}$ films can be grown on various substrates by pulsed laser ablation using a crystalline $Ba_8Ga_{16}Ge_{30}$ clathrate target [131, 132]. In these studies, the effects of incident laser wavelength and fluence on film formation and morphology were reported. Shortly thereafter, $Ba_8Ga_{16}Ge_{30}$ was prepared by radio frequency magnetron sputtering, also using a crystalline $Ba_8Ga_{16}Ge_{30}$ target [133]. The films in both cases were found to be polycrystalline, while selected area electron diffraction provided evidence for local epitaxial growth on (200) silicon substrate in the sputtered films.

Na–Si clathrates were first prepared as thin films on silicon substrates by reaction of Na with the substrate to first form a $Na_4Si_4$ layer, with subsequent decomposition to the clathrate [134]. This method has recently been revisited [135]. Sol-gel techniques for film deposition have also recently begun to be explored [136]. To date, all of these techniques require silicon substrates (as the source of silicon).

**Table 3.5** Silicon-based intermetallic clathrate compositions prepared by high-pressure reaction

| Composition | Structure type | Reaction conditions | Notes | References |
|---|---|---|---|---|
| $Ba_8Si_{46}$ | I | Reaction of stoichiometric mixture of powdered $BaSi_2$ and Si at 3 GPa, 800 °C for 1 h, quench to room temperature. | Multi-anvil press, BN cell | [125] |
| $Cs_{8-x}Si_{46}$ | I | Reaction of powdered $Cs_4Si_4$ and Si, with overall Cs:Si ratio of 1:4, at 2–10 GPa, 927 °C for 1 h | $x$ depends on pressure | [35] |
| $Ba_{24}Si_{100}$ | IX | Reaction of stoichiometric mixture of powdered $BaSi_2$ and Si at 1.5 GPa, 800 °C | Multi-anvil press, BN cell | [128] |
| $Ca_xBa_{1-x}Si_{46}$ | I | Reaction of stoichiometric mixture of powdered Si, $BaSi_2$ and $CaSi_2$ at 3 GPa, 800 °C for 1 h | Belt-type press | [129] |
| $Sr_xBa_{1-x}Si_{46}$ | I | Reaction of stoichiometric mixture of powdered Si, $BaSi_2$ and $SrSi_2$ at 3 GPa, 800 °C for 1 h | Belt-type press | [129] |
| $Na_8Si_{46}$[a] | I | Reaction of the elements at 3–6 GPa, 700–850 °C for 5.5 h | Multi-anvil press | [127] |
| $Te_{16}Si_{38}$ ($Te_8$@($Si_{38}Te_8$)) | I | Reaction of the powdered elements at 5 GPa, 825–1200 °C for 1 h | Belt-type press | [130] |

[a] $Na_xSi_{136}$ was also obtained in these experiments as a minor phase. The authors conclude from their work that the Na–Si clathrates are thermodynamically stable at high pressure

Only a limited number of reports on the preparation of nanocrystalline intermetallic clathrates are found in the literature. High energy ball milling has been used as a "top down" approach to prepare a variety of materials with nanoscale grains, and was recently used to prepare $Ba_8Ga_{16}Si_{30}$ with particle sizes as small as a few tens of nanometers [137]. Rapid quenching from the melt (melt spinning) has been explored, however for simple ternary compositions it appears difficult to obtain submicron grain sizes due to the rapid growth rate of the clathrate [138]. Recently, $K_xSi_{136}$ and $Na_xSi_{136}$ rod-shaped nanostructures were obtained via chemical oxidation of reactive precursors using methods similar to those discussed in Sect. 3.5 [139].

## 3.9 Concluding Remarks

In spite of constraints imposed by charge balance in many cases, the crystal chemistry in intermetallic clathrates affords remarkable flexibility in composition. While the overwhelming majority of known compositions have been prepared by conventional solid state and/or melt-based approaches, newly developed synthetic routes allow access to high-quality specimens and new compositions that are significant from the viewpoint of understanding the unusual and potentially useful properties of these materials. As the available synthetic toolkit for preparation of intermetallic clathrates continues to expand, it is expected that the already broad compositional space will continue to grow. Zintl precursors containing polyanionic clusters have played an important role in many of the synthetic approaches used to prepare intermetallic clathrates. As the understanding of the reaction pathways underlying these approaches improves, it is reasonable to expect that such precursors will continue to play an important role in the synthesis of intermetallic clathrates.

**Acknowledgment** The authors gratefully acknowledge the support of the U.S. Department of Energy, Basic Energy Sciences, Division of Materials Science and Engineering, under Award No. DE-FG02-04ER46145 at the University of South Florida.

## References

1. J.S. Kasper, P. Hagenmuller, M. Pouchard, C. Cros, Science **150**, 1713 (1965)
2. C. Cros, M. Pouchard, P. Hagenmuller, J. Solid State Chem. **2**, 570 (1970)
3. C. Cros, Sur quelques nouveaux siliciures et gemaniures alcalins à structure clathrate: etude cristallochimique et physique, Doctoral Thesis, University of Bordeaux, Bordeaux, France, 1970
4. G.S. Nolas, G.A. Slack, S.B. Schujman, in *Semiconductors and Semimetals*, vol. **69**, ed. by T.M. Tritt, (Academic Press, San Diego, 2001), p. 255
5. G.S. Nolas, in *Thermoelectrics Handbook: Macro to Nano*, ed. by D.M. Rowe (CRC Press, Boca Raton, 2006), pp. 33–1

6. T. Takabatake, in *Thermoelectric Nanomaterials*, ed. by K. Koumoto, T. Mori (Springer, Berlin Heidelberg, 2013)
7. M. Beekman, G.S. Nolas, J. Mater. Chem. **18**, 842 (2008)
8. A.D. Martinez, L. Krishna, L.L. Baranowski, M.T. Lusk, E.S. Toberer, A.C. Tamboli, IEEE J. Photovoltaics **3**, 1305 (2013)
9. T. Langera, S. Dupkeb, H. Trillb, S. Passerinib, H. Eckert, R. Pöttgen, M. Winter, J. Electrochem. Soc. **159**, A1318 (2012)
10. J. Yang, J.S. Tse, J. Mater. Chem. A **1**, 7782 (2013)
11. D. Neiner, N.L. Okamoto, C.L. Condron, Q.M. Ramasse, P. Yu, N.D. Browning, S.M. Kauzlarich, J. Am. Chem. Soc. **129**, 13857 (2007)
12. P. Rogl, in *Thermoelectrics Handbook: Macro to Nano*, ed.by D.M. Rowe (CRC Press, Boca Raton, 2006), pp. 32-1
13. A.V. Shevelkov, K. Kovnir, Struct. Bond. **139**, 97 (2010)
14. M. Christensen, S. Johnsen, B.B. Iversen, Dalton Trans. **39**, 978 (2010)
15. M. O'Keefe, G.B. Adams, O.F. Sankey, Phil. Mag. Lett. **78**, 21 (1998)
16. F. Dubois, T.F. Fässler, J. Am. Chem. Soc. **127**, 3264 (2005)
17. W. Carrillo-Cabrera, S. Budnyk, Y. Prots, Y. Grin, Z. Anorg, Allg. Chem. **630**, 2267 (2004)
18. X. Shi, J. Yang, S. Bai, J. Yang, H. Wang, M. Chi, J.R. Salvador, W. Zhang, L. Chen, W. Wong-Ng, Adv. Func. Mater. **20**, 755 (2010)
19. G.S. Nolas, J.L. Cohn, G.A. Slack, S.B. Schujman, Appl. Phys. Lett. **73**, 178 (1998)
20. J. Martin, H. Wang, G.S. Nolas, Appl. Phys. Lett. **92**, 222110 (2008)
21. A.F. May, E.S. Toberer, A. Saramat, G.J. Snyder, Phys. Rev. B **80**, 125205 (2009)
22. G.S. Nolas, D.G. Vanderveer, A.P. Wilkinson, J.L. Cohn, J. Appl. Phys. **91**, 8970 (2002)
23. M. Beekman, W. Schnelle, H. Borrmann, M. Baitinger, Yu. Grin, G.S. Nolas, Phys. Rev. Lett. **104**, 018301 (2010)
24. S. Stefanoski, J. Martin, G.S. Nolas, J. Phys. Cond. Matter **22**, 485404 (2010)
25. S. Stefanoski, C.D. Malliakas, M.G. Kanatzidis, G.S. Nolas, Inorg. Chem. **51**, 8686 (2012)
26. Y. Liu, L.-M. Wu, L.-H. Li, S.-W. Du, J.D. Corbett, L. Chen, Angew. Chem. Int. Ed. **48**, 5305 (2009)
27. H. He, A. Zevalkink, Z.M. Gibbs, G.J. Snyder, S. Bobev, Chem. Mater. **24**, 3596 (2012)
28. R.D. Shannon, C.T. Prewitt, Acta Cryst. B **25**, 925 (1969)
29. R.D. Shannon, C.T. Prewitt, Acta Cryst. B **26**, 1046 (1970)
30. R.D. Shannon, Acta Cryst. A **32**, 750 (1976)
31. L. Pauling, B. Kamb, Proc. Nat. Acad. Sci. USA **83**, 3569 (1986)
32. S. Bobev, S.C. Sevov, J. Solid State Chem. **153**, 92 (2000)
33. N. Melnychenko-Koblyuk, A. Grytsiv, L. Fornasari, H. Kaldarar, H. Michor, F. Röhrbacher, M. Koza, E. Royanian, E. Bauer, P. Rogl, M. Rotter, H. Schmid, F. Marabelli, A. Devishvili, M. Doerr, G. Giester, J. Phys.: Condens. Matter **19**, 216223 (2007)
34. M. Falmbigl, F. Kneidinger, M. Chen, A. Grytsiv, H. Michor, E. Royanian, E. Bauer, H. Effenberger, R. Podloucky, P. Rogl, Inorg. Chem. **52**, 931 (2013)
35. A. Wosylus, I. Veremchuk, W. Schnelle, M. Baitinger, U. Schwarz, Y. Grin, Chem. Eur. J. **15**, 5901 (2009)
36. G.S. Nolas, T.J.R. Weakley, J.L. Cohn, R. Sharma, Phys. Rev. B **61**, 3845 (2000)
37. C. Candolfi, U. Aydemir, M. Baitinger, N. Oeschler, F. Steglich, Yu. Grin, J. Appl. Phys. **111**, 043706 (2012)
38. J. Martin, G.S. Nolas, H. Wang, J. Yang, J. Appl. Phys. **102**, 103719 (2007)
39. H. Anno, H. Yamada, T. Nakabayashi, M. Hokazono, R. Shirataki, J. Phys: Conf. Ser. **379**, 012007 (2012)
40. A. Kaltzoglou, S. Ponou, T.F. Fässler, Eur. J. Inorg. Chem. **2008**, 4507 (2008)
41. S. Bobev, S.C. Sevov, J. Solid State Chem. **153**, 92 (2000)
42. M. Beekman, W. Wong-Ng, J.A. Kaduk, A. Shapiro, G.S. Nolas, J. Solid State Chem. **180**, 1076 (2007)
43. M. Beekman, J.A. Kaduk, J. Gryko, W. Wong-Ng, A. Shapiro, G.S. Nolas, J. Alloys Comp. **470**, 365 (2009)

44. M.C. Schäfer, S. Bobev J. Am. Chem. Soc. **135**, 1696 (2013)
45. M. Beekman, G.S. Nolas, Mater. Res. Soc. Proc. **1044**, 1044-U05-05 (2007)
46. S. Laumann, M. Ikeda, H. Sassik, A. Prokofiev, S. Paschen, Z. anorg. allgem. Chem. **638**, 294 (2012)
47. S. Laumann, M. Ikeda, H. Sassik, A. Prokofiev, S. Paschen, J. Mater. Res. **26**, 1861 (2011)
48. R. Schäfer, W. Klemm, Z. Anorg, Allg. Chem. **312**, 214 (1961)
49. J. Witte, H.G. von Schnering, Z. Anorg, Allg. Chem. **327**, 260 (1964)
50. C. Cros, M. Pouchard, P. Hagenmuller, C.R. Acad, Sc. Paris **260**, 4764 (1965)
51. C. Cros, M. Pouchard, P. Hagenmuller, Bull. Soc. Chim. France, **379** (1971)
52. C. Cros, M. Pouchard, P. Hagenmuller, J.S. Kasper, Bull. Soc. Chim. France, 2737 (1968)
53. G.K. Ramachandran, J.J. Dong, J. Diefenbacher, J. Gryko, R.F. Marzke, O.F. Sankey, P.F. McMillan, J. Solid State Chem. **145**, 716 (1999)
54. G.K. Ramachandran, P.F. McMillan, J. Dong, O.F. Sankey, J. Solid State Chem. **154**, 626 (2000)
55. J. Gryko, P.F. McMillan, R.F. Marzke, G.K. Ramachandran, D. Patton, S.K. Deb, O.F. Sankey, Phys. Rev. B **62**, R7707 (2000)
56. G.K. Ramachandran, J. Dong, O.F. Sankey, P.F. McMillan, Phys. Rev. B **63**, 033102 (2000)
57. S. Latturner, B.B. Iversen, J. Sepa, V. Srdanov, G. Stucky, Phys. Rev. B **63**, 125403 (2001)
58. M. Beekman, G.S. Nolas, Phys. B **383**, 111 (2006)
59. M. Beekman, G.S. Nolas, Int. J. Appl. Ceram. Technol. **4**, 332 (2007)
60. M. Baitinger, H.G. von Schnering, J.-H. Chang, K. Peters, Yu. Grin, Z. Krist, NCS **222**, 87 (2007)
61. M. Beekman, C.P. Sebastian, Yu. Grin, G.S. Nolas, J. Electron. Mater. **38**, 1052 (2009)
62. M. Beekman, E.N. Nenghabi, K. Biswas, C.W. Myles, M. Baitinger, Yu. Grin, G.S. Nolas, Inorg. Chem. **49**, 5338 (2010)
63. A.D. Ritchie, M.A. MacDonald, P. Zhang, M.A. White, M. Beekman, J. Gryko, G.S. Nolas, Phys. Rev. B **82**, 155207 (2010)
64. A.D. Ritchie, M.B. Johnson, J.F. Niven, M. Beekman, G.S. Nolas, J. Gryko, M.A. White, J. Phys.: Condens. Matter **25**, 435401 (2013)
65. H. Morito, T. Yamada, T. Ikeda, H. Yamane, J. Alloys Comp. **480**, 723 (2009)
66. M. Beekman, MSc Thesis, University of South Florida, 2006
67. M. Beekman, PhD Dissertation, University of South Florida, 2009
68. H.-O. Horie, T. Kikudome, K. Teramura, S. Yamanaka, J. Solid State Chem. **182**, 129 (2009)
69. J. Gryko, U.S. Patent No. 6,423,286
70. P.T. Hutchins, O. Leynaud, L.A. O'Dell, M.E. Smith, P. Barnes, P.F. McMillan, Chem. Mater. **23**, 5160 (2011)
71. M. Beekman, G.S. Nolas, unpublished results
72. L. Nistor, G. van Tendeloo, S. Amelinckx, C. Cros, Phys. Status Solidi A **146**, 119 (1994)
73. J. He, D.D. Klug, K. Uehara, K.F. Preston, C.I. Ratcliffe, J.S. Tse, J. Phys. Chem. **105**, 3475 (2001)
74. E. Reny, P. Gravereau, C. Cros, M. Pouchard, J. Mater. Chem. **8**, 2839 (1998)
75. A. Ammar, C. Cros, M. Pouchard, N. Jaussaud, J.-M. Bassat, G. Villeneuve, M. Duttine, M. Ménétrier, E. Reny, Sol. State Sci. **6**, 393 (2004)
76. J. Gryko, R.F. Marzke, G.A. Lamberton Jr, T.M. Tritt, M. Beekman, G.S. Nolas, Phys. Rev. B **71**, 115208 (2005)
77. H. Kawaji, H-O. Horie, S. Yamanaka, and M. Ishikawa. Phys. Rev. Lett. **74**, 1427 (1995)
78. T. Rachi, K. Tanigaki, R. Kumashiro, J. Winter, H. Kuzmany, Chem. Phys. Lett. **409**, 48 (2005)
79. M. Beekman, J.A. Kaduk, Q. Huang, W. Wong-Ng, W. Yang, D. Wang, G.S. Nolas, Chem. Commun. **2007**, 837 (2007)
80. M. Beekman, S. Stefanoski, W. Wong-Ng, J.A. Kaduk, Q. Huang, C. Reege, C.R. Bowers, G.S. Nolas, J. Solid State Chem. **183**, 1272 (2010)

81. L.L. Baranowski, L. Krishna, A.D. Martinez, T. Raharjo, V. Stevanovic, A.C. Tamboli, E.S. Toberer, J. Mater. Chem. C **2**, 3231 (2014)
82. H. Fukuoka, K. Iwai, S. Yamanaka, H. Abe, K. Yoza, L. Häming, J. Solid State Chem. **151**, 117 (2000)
83. A.M. Guloy, R. Ramlau, Z. Tang, W. Schnelle, M. Baitinger, Y. Grin, Nature **443**, 320 (2006)
84. B. Böhme, A. Guloy, Z. Tang, W. Schnelle, U. Burkhardt, M. Baitinger, Y. Grin, J. Am. Chem. Soc. **129**, 5348 (2007)
85. B. Böhme, S. Hoffmann, M. Baitinger, Y. Grin, Z. Naturforsch. **66b,** 230 (2011)
86. Y. Liang, B. Böhme, M. Reibold, W. Schnelle, U. Schwarz, M. Baitinger, H. Lichte, Y. Grin, Inorg. Chem. **50**, 4523 (2011)
87. A.M. Guloy, Z. Tang, R. Ramlau, B. Böhme, M. Baitinger, Y. Grin, Eur. J. Inorg. Chem. **2009**, 2455 (2009)
88. B. Böhme, Dissertation (Dr. rer. nat.), Technischen Universität Dresden, 2010
89. M.C. Blosser, G.S. Nolas, Mater. Lett. **99**, 161 (2013)
90. B. Böhme, U. Aydemir, A. Ormeci, W. Schnelle, M. Baitinger, Y. Grin, Sci. Tech. Adv. Mater. **8**, 410 (2007)
91. S. Yamanaka, E. Enishi, H. Fukuoka, M. Yasukawa, Inorg. Chem. **39**, 56 (2000)
92. B.C. Sales, B.C. Chakoumakos, R. Jin, J.R. Thompson, D. Mandrus, Phys. Rev. B **63**, 245113 (2001)
93. M.G. Kanatzidis, R. Pöttgen, W. Jeitschko, Angew. Chem. Int. Ed. **44**, 6996 (2005)
94. P.C. Canfield, Z. Fisk, Phil. Mag. B **65**, 1117 (1992)
95. K. Billquist, D. Bryan, M. Christensen, C. Gatti, L. Holmgren, B.B. Iversen, E. Mueller, M. Muhammed, G. Noriega, A.E.C. Palmqvist, D. Platzek, D.M. Rowe, A. Saramat, C. Stiewe, G.D. Stucky, G. Svensson, M. Toprak, S.G.K. Williams, Y. Zhang, in *Proceedings of 22nd International Conference on Thermoelectrics*, (IEEE, New York, 2003), pp. 127–130
96. A. Saramat, G. Svensson, A.E.C. Palmqvist, C. Stiewe, E. Mueller, D. Platzek, S.G.K. Williams, D.M. Rowe, J.D. Bryan, G.D. Stucky, J. Appl. Phys. **99**, 023708 (2006)
97. N. Mugita, Y. Nakakohara, R. Teranishi, S. Munetoh, J. Mater. Res. **26**, 1857 (2011)
98. M. Christensen, S. Johnsen, M. Søndergaard, J. Overgaard, H. Birkedal, B.B. Iversen, Chem. Mater. **21**, 122 (2009)
99. L.T.K. Nguyen, U. Aydemir, M. Baitinger, E. Bauer, H. Borrmann, U. Burkhardt, J. Custers, A. Haghighirad, R. Hofler, K.D. Luther, F. Ritter, W. Assmus, Yu. Grin, S. Paschen, Dalton Trans. **39**, 1071 (2010)
100. H. Euchner, S. Pailhès, L.T.K. Nguyen, W. Assmus, F. Ritter, A. Haghighirad, Y. Grin, S. Paschen, M. de Boissieu, Phys. Rev. B **86**, 224303 (2012)
101. A. Bentien, A.E.C. Palmqvist, J.D. Bryan, S. Latturner, G.D. Stucky, L. Furenlid, B.B. Inversen, Angew. Chem. Int. Ed. **39**, 3613 (2000)
102. C.L. Condron, J. Martin, G.S. Nolas, P.M.B. Piccoli, A.J. Schultz, S.M. Kauzlarich, Inorg. Chem. **45**, 9381 (2006)
103. C.L. Condron, R. Porter, T. Guo, S.M. Kauzlarich, Inorg. Chem. **44**, 9185 (2005)
104. K. Suekuni, M.A. Avila, K. Umeo, H. Fukuoka, S. Yamanaka, T. Nakagawa, T. Takabatake, Phys. Rev. B **77**, 235119 (2008)
105. K. Suekuni, M.A. Avila, K. Umeo, T. Takabatake, Phys. Rev. B **75**, 195210 (2007)
106. V. Baran, A. Fischer, W. Scherer, T.F. Fässler, Z. Anorg, Allg. Chem. **639**, 2125 (2013)
107. T. Tanaka, T. Onimaru, K. Suekuni, S. Mano, H. Fukuoka, S. Yamanaka, T. Takabatake, Phys. Rev. B **81**, 165110 (2010)
108. S. Stefanoski, Y. Dong, G.S. Nolas, J. Solid State Chem. **204**, 166 (2013)
109. M. Beekman, M. Baitinger, H. Borrmann, W. Schnelle, K. Meier, G.S. Nolas, Yu. Grin, J. Am. Chem. Soc. **131**, 9642 (2009)
110. S. Stefanoski, M.C. Blosser, G.S. Nolas, Cryst. Growth Des. **13**, 195 (2013)
111. S. Stefanoski, M. Beekman, W. Wong-Ng, P. Zavalij, G.S. Nolas, Chem. Mater. **23**, 1491 (2011)
112. S. Stefanoski, G.S. Nolas, Cryst. Growth Des. **11**, 4533 (2011)

113. R.T. Macaluso, B.K. Greve, Dalton Trans. **41**, 14225 (2012)
114. A. Borshchevsky, in *CRC Handbook on Thermoelectrics*, ed. by D.M. Rowe (CRC Press, Boca Raton, 1995), p. 83
115. E.S. Toberer, M. Christensen, B.B. Iversen, G.J. Snyder, Phys. Rev. B **77**, 075203 (2008)
116. M.J. Tokita, Soc. Powder Tech. Jpn. **790**, 30 (1993)
117. Z.A. Munir, U. Anselmi-Tamburini, M. Ohyanagi, J. Mater. Sci. **41**, 763 (2006)
118. M. Omori, Mater. Sci. Eng., A **287**, 183 (2000)
119. I. Veremchuk, M. Beekman, I. Antonyshyn, M. Baitinger, G.S. Nolas, Y. Grin, in preparation
120. H. Morito, H. Yamane, Angew. Chem. Int. Ed. **49**, 3638 (2010)
121. M. Beekman, G.S. Nolas, U.S. Patent No. 8,414,858
122. A. San-Miguel, P. Kéghélian, X. Blase, P. Mélinon, A. Perez, J.P. Itié, A. Polian, E. Reny, C. Cros, M. Pouchard, Phys. Rev. Lett. **83**, 5290 (1999)
123. J.F. Meng, N.V. Chandra Shekar, J.V. Badding, G.S. Nolas, J. Appl. Phys. **89**, 1730 (2001)
124. U. Schwarz, A. Wosylus, B. Böhme, M. Baitinger, M. Hanfland, Yu. Grin, Angew. Chem. Int. Ed. **47**, 6790 (2008)
125. S. Yamanaka, E. Enishi, H. Fukuoka, M. Yasukawa, Inorg. Chem. **39**, 56 (2000)
126. H. Kawaji, H.-O. Horie, S. Yamanaka, M. Ishikawa, Phys. Rev. Lett. **74**, 1427 (1995)
127. O.O. Kurakevych, T.A. Strobel, D.Y. Kim, T. Muramatsu, V.V. Struzhkin, Cryst. Growth Des. **13**, 303 (2013)
128. H. Fukuoka, K. Ueno, S. Yamanaka, J. Organomet. Chem. **611**, 543 (2000)
129. P. Toulemonde, A. San Miguel, A. Merlen, R. Viennois, S. Le Floch, Ch. Adessi, X. Blase, J.L. Tholence, J. Phys. Chem. Solids **67**, 1117 (2006)
130. N. Jaussaud, P. Toulemonde, M. Pouchard, A. San Miguel, P. Gravereau, S. Pechev, G. Goglio, and C. Cros, Sol. State Sci. **6**, 401 (2004)
131. S. Witanachchi, R. Hyde, M. Beekman, D. Mukherjee, P. Mukherjee, and G.S. Nolas, in *Proceedings of 25th International Conference on Thermoelectrics*, p. 44, 2006
132. S. Witanachchi, R. Hyde, H.S. Nagaraja, M. Beekman, G.S. Nolas, P. Mukherjee, Mater. Res. Soc. Symp. Proc. **886**, 401 (2006)
133. L. Miao, S. Tanemura, T. Watanabe, M. Tanemura, S. Toh, K. Kaneko, Y. Sugahara, T. Hirayama, Appl. Surf. Sci. **254**, 167 (2007)
134. P. Ecklund, S. Fang, L. Grigorian, U.S. Patent No. US6103403 A
135. T. Kume, Y. Iwai, T. Sugiyama, F. Ohashi, T. Ban, S. Sasaki, S. Nonomura, Phys. Stat. Sol. C **10**, 1739 (2013)
136. M. Gong, X.W. Zhu, T.Z. Xu, L.Z. Li, W.Z. Jiang, X. Zhou, Adv. Mater. Res. **487**, 111 (2012)
137. R. Shirataki, M. Hokazono, T. Nakabayashi, H. Anno, IOP Conf. Ser.: Mater. Sci. Eng. **18**, 142012 (2011)
138. A. Prokofiev, M. Ikeda, E. Makalkina, R. Svagera, M. Waas, S. Paschen, J. Elecron. Mater. **42**, 1628 (2013)
139. P. Simon, Z. Tang, W. Carrillo-Cabrera, K. Chiong, B. Böhme, M. Baitinger, H. Lichte, Y. Grin, A.M. Guloy, J. Am. Chem. Soc. **133**, 7596 (2011)

# Chapter 4
# Semiconductor Clathrates: In Situ Studies of Their High Pressure, Variable Temperature and Synthesis Behavior

**D. Machon, P. F. McMillan, A. San-Miguel, P. Barnes and P. T. Hutchins**

**Abstract** In situ studies have provided valuable new information on the synthesis mechanisms, low temperature properties and high pressure behavior of semiconductor clathrates. Here we review work using synchrotron and laboratory X-ray diffraction and Raman scattering used to study mainly Si-based clathrates under a variety of conditions. During synthesis of the Type I clathrate $Na_8Si_{46}$ by metastable thermal decomposition from NaSi in vacuum, we observe an unusual quasi-epitaxial process where the clathrate structure appears to nucleate and grow directly from the Na-deficient Zintl phase surface. Low temperature X-ray studies of the guest-free Type II clathrate framework $Si_{136}$ reveal a region of negative thermal expansion behavior as predicted theoretically and analogous to that observed for diamond-structured Si. High pressure studies of $Si_{136}$ lead to metastable production of the β-Sn structured Si-II phase as well as perhaps other metastable crystalline materials. High pressure investigations of Type I clathrates

D. Machon (✉) · A. San-Miguel
Institut Lumière Matière, UMR5306 Université Lyon 1-CNRS, Université de Lyon, 69622 Villeurbanne Cedex, France
e-mail: denis.machon@univ-lyon1.fr

A. San-Miguel
e-mail: alfonso.san-miguel@univ-lyon1.fr

P. F. McMillan
Christopher Ingold Laboratories, Department of Chemistry and Materials Chemistry Centre, University College London, 20 Gordon Street, London WC1H 0AJ, UK
e-mail: p.f.mcmillan@ucl.ac.uk

P. Barnes
Dept. Crystallography, School of Biological Sciences, Birkbeck College, Malet Street, London WC1E 7HX, UK
e-mail: p.barnes@mail.cryst.bbk.ac.uk

P. T. Hutchins
Infineum Business and Technology Centre, Infineum UK Ltd., P.O. Box 1 Milton Hill, Abingdon OX13 6AE, UK
e-mail: p.t.hutchins@gmail.com

G. S. Nolas (ed.), *The Physics and Chemistry of Inorganic Clathrates*, Springer Series in Materials Science 199, DOI: 10.1007/978-94-017-9127-4_4,

show evidence for a new class of apparently isostructural densification transformations followed by amorphization in certain cases.

## 4.1 Introduction

The first syntheses of the materials that became known as "semiconductor clathrates" were reported as a result of partial thermal decomposition of Zintl phase compounds such as NaSi or KGe [1–3], or by direct reaction between the elements [4]. There has been continued interest in this unusual family of solid state materials because of their unique structures and optical, electronic and thermal properties that may have technological significance [5, 6]. They were first identified as "clathrates" because of their correspondence with the open framework structures produced by H-bonded $H_2O$ networks containing guest atoms or molecules (Xe, $CH_4$ etc.) included within the large cages. The analogous materials based on tetrahedrally bonded networks of atoms within groups 13–15 of the periodic table typically contain electropositive metal atoms or ions within 4–6 Å cage sites. Although they are termed "semiconductor clathrates" for convenience, their electrical properties range from semiconducting to metallic, depending on the framework chemistry and cage site occupancy. A large number of clathrate compounds are now known with different group 13–15 elements or transition metals constituting the framework, and various alkali or alkaline-earth atoms as well as halogens, chalcogenides, lanthanides or transition metals occupying the "guest" site positions [5–7]. They are thus a chemically very diverse group of materials giving rise to many possibilities for tuning their electronic, optical, magnetic, thermal and mechanical properties.

The $sp^3$-bonded networks of $T$ atoms give rise to an assembly of $T_{20}$, $T_{24}$, or $T_{28}$ polyhedral cages that are fully or partially occupied by the guest atoms or ions. Most known clathrates fall into one of two main structure types. The Type I structure has a framework based on 46 $T$ atoms within a primitive cubic ($Pm\bar{3}m$) cell. There are two smaller dodecahedral ([$5^{12}$]) cages and 6 adjacent larger tetrakaidecahedral [$5^{12}6^2$] cage sites that are completely or partially filled by electropositive guest atoms such as Na, Rb or Ba, although clathrates are known with heavier halogens (I) or chalcogens (Se) occupying the cages. The first examples had frameworks constituted by Si or Ge, along with analogous structures based on Sn that typically contained vacancies on the framework sites to constitute polyanionic Zintl phases [1, 4]. Additional examples with alloy compositions constituting the frameworks to give rise to semiconducting series of clathrates such as $Sr_8Ga_{16}Ge_{30}$ are also known. Many such coupled substitutions among the framework and guest sites have been described with the Type I structure [5, 6]. Additional degrees of freedom are provided by the possibility of partial cage site occupancy, as in $Rb_{6.15}Si_{46}$, or vacancies on the tetrahedral framework (e.g., $Rb_8Sn_{44}$ or $Rb_8Sn_{44}\square_2$, and $K_8Ge_{44}$). Here the vacancies are typically located at the *6c* Wyckoff positions associated with the dodecahedral ($[5^{12}]$) cages. Metallic clathrates such as $Na_8Si_{46}$ are usually reported to have all framework and cage sites completely filled, but it is also possible to partially replace the metal atoms by hydrogen during synthesis [8, 9]. Type II clathrates are less common but have been equally well studied, as they were produced and identified during the first clathrate synthesis and characterization experiments [1]. They are generally found with Si or Ge atoms forming a face-centered ($Fd\bar{3}m$) structure with 136 framework atoms per formula unit (i.e., 34 $T$ species per primitive cell) to give clathrates such as $Na_xSi_{136}$ ($x = 0 - 24$) or $Cs_8Na_{16}Si_{136}$. There are 16 smaller ($[5^{12}]$) and eight larger hexakaidecahedral $[5^{12}6^2]$ cages per 136 framework atom formula unit. The cage occupancy in the $Na_xSi_{136}$ compound can be adjusted by varying the synthesis and processing conditions along with post-synthesis treatments such as vacuum heating, exposure to Na vapor or to reagents such as iodine designed to extract the alkali metal species [1, 10, 11]. Essentially "guest-free" clathrates such as $Si_{136}$ have been produced in this way. These are wide-gap semiconductors with $E_g = 1.9$ eV, much larger than that of diamond-structured Si (1.1 eV), and they provide a new allotrope of that element with an open framework structure [10]. Guest-free clathrates such as $Ge_{136}$ have now been produced by chemical precursor syntheses in ionic liquid solution [12]. A further Type III structure with a hexagonal unit cell has now been found for a few clathrate compounds such as $Ba_{24}Si_{100}$ or $Ba_{24}Ge_{100}$ [13, 14], and additional structural motifs have been predicted to exist based on first principles calculations guided by systematic prediction of the solid state structures as duals of Frank-Kasper motifs [15]. However, none of these have been observed experimentally to date. Other more exotic silicon clathrate forms related to the Type I clathrates have been synthesized with tellurium atoms as guest atoms [16].

The clathrate materials have received a great deal of attention because of their unusual and potentially useful electronic [17], optical [10], mechanical [18], thermoelectric [19–21] and electron emitter [22] properties. Depending upon the degree of guest atom filling versus framework site occupancy, their electrical properties range from semiconducting to metallic, and clathrates in systems such as $(Na_{1-x}Ba_x)_8Si_{46}$ and $\|Ba_{1-x}Sr_x|_8Si_{46}$ are superconductors with $T_c$ between 4 and 8 K [23–25]. The relationship between the electronic and vibrational properties of the framework versus guest atom sublattices is of particular interest giving rise to unique thermoelectric behavior with low phonon thermal conductivity combined with narrow gap to semimetallic electronic conductivity, essential for design of new thermoelectric materials [26–28]. Such properties are highly dependent on the electronic and vibrational coupling between the guest and the host atom sublattices. For example, the interaction between the averaged position and "rattling" vibrations of heavy guest atoms within the semiconductor cages is crucial for determining both the superconducting and thermoelectric behavior [27, 29–32].

## 4.2 Synthesis and In Situ Studies of Clathrate Formation

Two basic strategies have been developed for the synthesis of semiconductor clathrates, mainly based on the initial work that first described these unusual phases [1, 4]. Because of their useful properties and potential compatibility with existing semiconductor technology there is great interest in understanding and controlling the formation of these phases. In the Na–Si system the Type I and Type II clathrate materials are always metastable relative to the elements or liquid alloys under conditions of atmospheric pressure and above, although they develop a range of predicted thermodynamic stability in an "expanded" (negative pressure) state [33]. It has been suggested that they could be produced by crystallization from the liquid at $P = -2$ to $-4$ GPa, a range that can be achieved at semiconductor surfaces placed under tensile strain during device formation and tuning.

In one main synthesis approach, metastable thermal decomposition reactions from appropriate Zintl phase salts such as NaSi, KGe etc. are used to remove a proportion of the volatile alkali elements during heating in an inert atmosphere or under vacuum conditions to lead to Type I and Type II compounds that are mainly based on Si or Ge¯ frameworks, and direct reactions between the elements or Zintl type compounds to achieve a wide range of Type I materials with various framework and cage ion compositions. The latter approaches are well adapted to high pressure synthesis studies and new clathrate types such as superconducting $(Na,Ba)_8Si_{46}$ materials have been identified in this way [24]. These synthesis approaches have been reviewed in detail previously [5, 34]. Single crystals of Type II Si- and Ge clathrates containing Na, Cs and Rb guest atoms have been produced by metastable solid state reactions at ambient pressure [6, 35]. New synthesis strategies are emerging that involve metastable clathrate crystallization along with formation of amorphous semiconductor materials from metathesis reactions involving Zintl phases and salts such as $NH_4Cl$, and these have given rise to new $H_2$-containing clathrates that could be useful as energy storage materials [8–10, 36, 37]. Other emerging processes involving nanoparticle synthesis from molecular and Zintl phase precursors including in ionic liquids are also leading to new possibilities for clathrate formation [12]. Clathrate materials have also been identified during experiments to produce nanoscale clusters from the vapor phase following laser excitation [38]. Related experiments have produced nanofibers of potentially clathrate-structured materials by deposition from the vapor phase [39, 40]. Here we concentrate on new progress in understanding the metastable formation of clathrates from thermal decomposition of NaSi Zintl phase under vacuum conditions using newly developed synchrotron X-ray diffraction techniques with rapid data acquisition [41].

It is generally presumed that removal of Na atoms into the vapor phase accompanied by reduction of the $Na^+$ species initially contained within the ionic precursor causes formation of new Si–Si bonds that become templated around the remaining Na atoms or ions to form the clathrate cage structures. Because of the very large structural mismatch between the starting Zintl phase and the clathrate products, it has been assumed that this process must proceed via formation of an intermediate amorphous layer expressed at the surface of the starting NaSi compound, with nucleation and growth of the clathrate phase as Na removal proceeds. We have made the first in situ diffraction study of this process taking place by exploiting a time-resolved synchrotron X-ray (powder) diffraction facility in combination with an X-ray detector and in situ sample cell system tailored to the stringent conditions of synthesis.

In situ experiments to study synthesis of Type I $Na_8Si_{46}$ clathrate from NaSi Zintl phase were developed at station 6.2 of the UK Daresbury Synchrotron Radiation Source (*SRS*, now closed [42]), that was designed around needs of materials scientists to collect refinable diffraction data in real-time experiments. The beamline design concept relied on the confluence of three technological advances: a suitable high flux ten-pole wiggler X-ray source/optics within the 5 to 18 keV energy range [43, 44]; a rapid high resolution bespoke X-ray detector; and a versatile sample environmental cell. These elements have now been partly developed at other synchrotron facilities but the overall concept awaits full transfer [45]. The second condition was realized through development of *RAPID*2, an in-house-built circular position-sensitive detector [46] based on a multi(512)-wire assembly in a gas-filled proportional chamber spanning a 2θ-angular range of 60°; combined with inter-wire interpolation and multi-electronic/parallel read out features. This delivered medium-high resolution (*ca* 0.06°) at high count-rates (up to 20 MHz) such that refinable diffraction data-sets could be obtained in successive time-steps of seconds or sub-seconds from multi-component samples reacting at elevated temperatures [47]. The third essential criterion was achieved by adaptation of a capillary-based sample environment cell [41, 47, 48] that could programme and monitor chosen gas/vacuum-mixture-time profiles at high temperature.

For the $Na_8Si_{46}$ clathrate synthesis studies described here, angle-dispersive in situ X-ray diffraction experiments were carried with an incident beam energy of 8.856 keV (1.400 Å), using angle-dispersive rather than the energy-dispersive [49] rapid mode, that was preferred because of the need to engage in detailed structural refinement throughout the synthesis pathway. In order to initiate and sustain the synthesis reaction, several main requirements had to be met. The starting NaSi phase is air- and moisture-sensitive so all samples had to be prepared and loaded under dry box conditions; then the cell had to be configured to allow dynamic vacuum ($10^{-4}$ bar) under high T ramp conditions, with the furnace programmed to produce the required temperature profile. For the experiments reported here this was set to enable 20 °C → 500 °C at +8 °C/min ramp followed by 500 °C-dwell. Diffraction data were collected every 30 s during the first few hours, then at longer intervals thereafter. Typical data sets obtained are shown in Figs. 4.1 and 4.2. Notable features include:

- a broad background signal due to the glass capillary sample holder;
- the loss of diffraction after 7½ h due to the destruction of the sample capillary through attack by Na vapor evolved from the sample at high T;

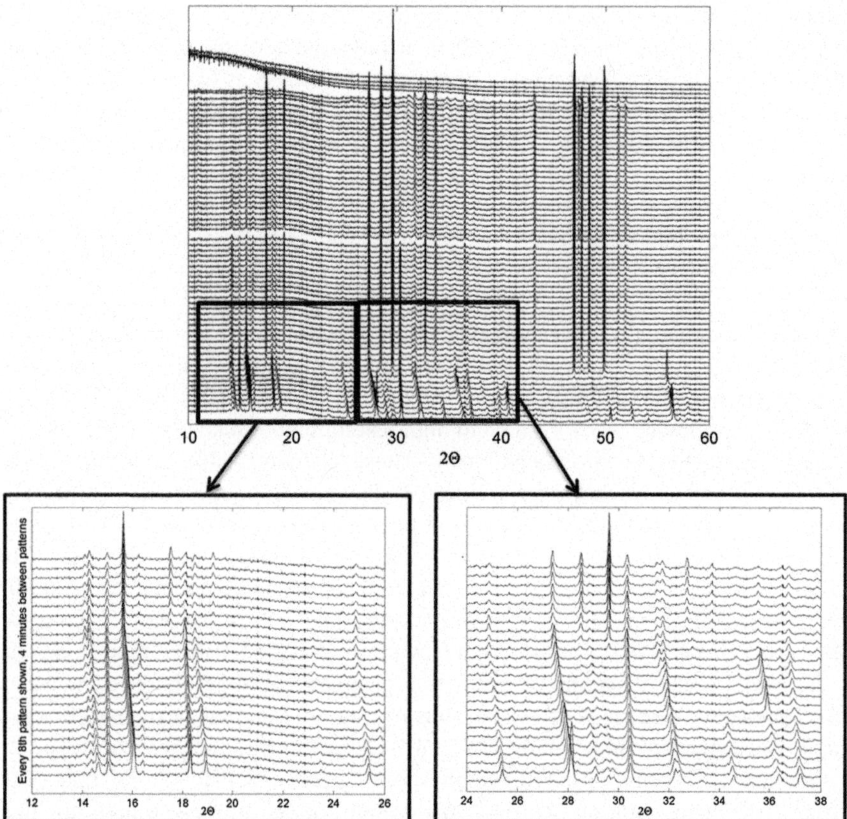


**Fig. 4.1** Time-resolved in situ powder X-ray (synchrotron) diffraction patterns obtained during the synthesis of $Na_8Si_{46}$ clathrate from NaSi in vacuum ($10^{-4}$ bar) and temperature ramped up to 500 °C then held constant (the vertical intensity scale is in arbitrary units). The main figure shows every 15th 30 s pattern such that the displayed patterns are separated by 7.5 min and cover 8 h. The two data gaps are due to, respectively, a temporary interruption of the X-ray beam at 4 h and destruction of the sample capillary by Na-vapour at 7½ h. The significant structural changes all occur during the first 10 patterns (75 min). The two boxes below show expanded regions (every 8th 30 s pattern, giving a time spacing of 4 min) to highlight the continuous peak shifting effect (from [41])

- continuous peak shifts in the low T range for the NaSi phase: this provides additional new insights on its thermal expansion behavior discussed below.

Unit cell/structural refinements were performed by successive use of Le Bail [50] and Rietveld [51] methods across the whole time/temperature-resolved diffraction data set using a multiphase refinement option assigned to the two implicated phases, precursor NaSi and clathrate-product $Na_8Si_{46}$; full details are given elsewhere [41, 48]. The $R_{Bragg}$ values were typically around 10–25 % but these could increase significantly in cases where one of the phases was not well defined.

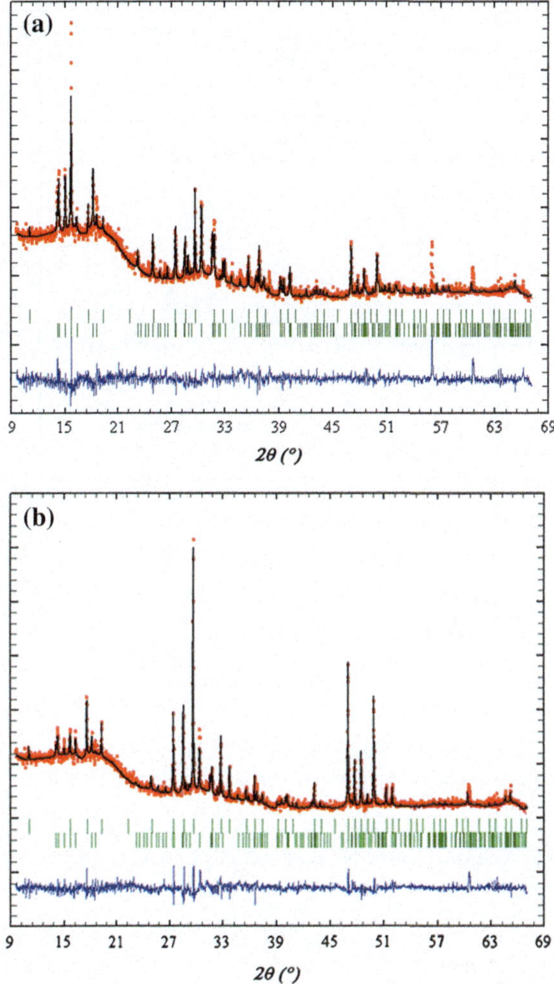


**Fig. 4.2** **a** An example of a "Rietveld" diffraction plot (intensity in arbitrary units versus 2θ (°)) for data-set No. 130, 65 min into the synthesis by which time the sample has just reached the target plateau of 500 °C and clathrate material is beginning to form [$Y_{obs}$ represents data points, the *black curve* gives the model (calculated) intensity $I_{calc}$ during refinement; and the low-lying *blue line* shows the difference between these two ($Y_{obs} - Y_{calc}$); the two rows of Bragg peak position markers relate to the NaSi (*lower*) and $Na_8Si_{46}$ (*upper*) phases]. Two unidentified contamination peaks are evident at 2θ = ~56° and ~61°, but these disappear in the further heat treatment. **b** "Rietveld" diffraction plot from the sample at 500 °C after 150 min, with this being the end point of the synthesis. The data are fitted as a two-phase model to the NaSi and Si clathrate structures. The vertical bars are the predicted positions of the Bragg diffraction peaks (as in (**a**)) showing significant peak overlap with the two structures (from [41])

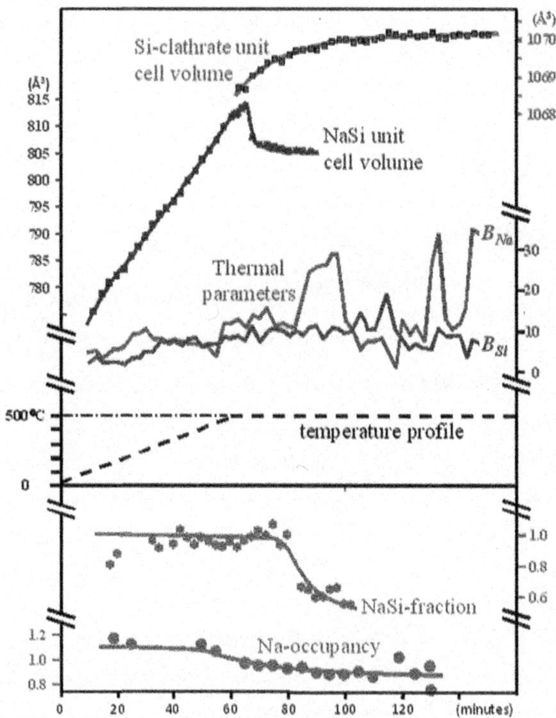


**Fig. 4.3** Behavior of NaSi and $Na_xSi_{46}$ unit cell volumes versus time/temperature, as obtained from in situ data after multi-phase refinement. The NaSi exhibits very high volume expansion up to 500 °C, noting that the $Na_xSi_{46}$-clathrate expansion is plotted on a ~7 times more sensitive scale. Also shown are the NaSi thermal-B-factors, the set temperature profile, the NaSi 'weight' fraction, the average occupancy of the Na sites in NaSi; these latter parameters are derived from standard multi-Rietveld-output (values in proportion to each other, in arbitrary units or normalized to unity). Caution must be exercised against over-interpreting such in situ data. However, the consistent trend is that, after the essential large prior NaSi expansion, other significant changes occur from around 500 °C and after (60–100 min) which indicate an increase in Na mobility leading to a discharge of structural-Na, thereby 'kick-starting' the NaSi → $Na_xSi_{46}$-clathrate decomposition reaction (from [41])

In situ experiments can never realistically match the data quality from conventional structure determination on static pure phases, and here further complications involved multi-component samples and rapidly shifting peaks. Therefore subtle structural details were not sought to be obtained but rather broader features such as thermal expansion or synthesis progression that could be justifiably extracted and interpreted from the data sets. Two typical multi-phase Rietveld refinement plots are shown in Fig. 4.2.

The output from the Le Bail/Rietveld refinements were combined to show the variation of several important structure-property indicators (unit cell, thermal, phase fraction and occupancy parameters) versus time/temperature profiles (Fig. 4.3). Given the fact that thermal *B*-factors and site occupancy parameters are highly correlated with each other during refinement procedures, the data indicate rather than confirm the existence of specific structural features and changes, but they provide essential first insights into the nature of the synthesis process as well as high-T phenomena involving the starting Zintl phase and the clathrate materials produced.

Before developing a description of the clathrate synthesis mechanism we first note the large unit cell volume expansion for the NaSi Zintl phase recorded between 60 and 500 °C (Fig. 4.3). The thermal expansion coefficient has a remarkably high value ($4.54 \times 10^{-5}$°$C^{-1}$) that is comparable with ionic conductors such as AgI. Then we note that after reaching the target temperature of 500 °C, a noticeable decrease in the NaSi fraction accompanied by growth of characteristic $Na_8Si_{46}$ clathrate peaks, with an apparent change in the Na site occupancy in the initial growth stages (Fig. 4.3). Also surprisingly after reaching 500 °C, the NaSi unit cell volume suddenly jumps back down significantly from ~815 to ~805 Å$^3$, even though the temperature remains constant at this point in the run.

The NaSi to $Na_8Si_{46}$ synthesis decomposition reaction seems to be unique amongst solid state transformations described to date, with no less than 9 reasons in support of this claim:

- Surprisingly for such dissimilar end-structures as NaSi (ionic, low symmetry) and $Na_8Si_{46}$ (covalent, high symmetry), we find no evidence anywhere during the synthesis for an intermediate amorphous or crystalline phase;
- instead, we see a continuous shifting of peaks from one structure towards the other, to an extent that certain peaks have to be labeled as having "dual ownership" (e.g. the peak moving between ~27½° and ~28° can be designated as either the $\overline{22}1$ peak of NaSi or the 222 $Na_8Si_{46}$ reflection);
- this 'dual ownership' of diffraction peaks is the direct result of the remarkably high NaSi thermal expansion that effectively bridges the structural gap between NaSi and $Na_8Si_{46}$;
- the other parameters in Fig. 4.3 ($B_{Na}$, Na-occupancy, NaSi-fraction) all indicate that the Na ions in the NaSi structure become more mobile and susceptible to loss as synthesis proceeds;
- independent high-T $^{23}$Na NMR studies [41] confirm the increased Na mobility within the structure;
- the NaSi structure displays an anomalous and significant drop in unit cell volume (~815 to ~805 Å$^3$) as the $Na_8Si_{46}$ is forming, and during a period when
- the detailed account of the experimental results and procedure [48] reports a temporary drop in the cell vacuum quality indicative of Na exiting the NaSi phase as Na vapour is formed: that result implies $Na^+$–$Na^0$ reduction has occurred in the solid, that must be accompanied by a corresponding partial oxidation of $Si^-$ ($Si_4{}^{4-}$) to $Si^0$ species along with formation of new Si-Si covalent bonds;

- the $Na_8Si_{46}$ clathrate structure (unit cell) continues to increase as synthesis continues, even though the temperature remains constant at 500 °C;
- additional studies using a format of repeated temperature-cycling experiments [41] indicate that synthesis-reversion can be achieved, appearing as a spectacular repeated pattern of partially reversible cycles, provided that a level of sublimed Na vapour is maintained inside the sample cell environment.

Our results lead us to a proposed mechanistic description of the NaSi → $Na_8Si_{46}$ clathrate synthesis process, drawing on key observations described above and in combination with essential chemical reasoning.

We start with a contextual reminder that these two phases have very dissimilar structures (i.e., isolated $Si_4^{-4}$ tetrahedra separated by $Na^+$ ions vs. open-framework Si-cages), compositions (NaSi vs. $Na_8Si_{46}$) and space groups (monoclinic *C2/c* vs. cubic $Pm\bar{3}n$). Therefore previous intuitive expectations were for some major structural re-arrangement to occur during synthesis, most likely involving an amorphous intermediate layer or bulk phase. However, we find no evidence for such amorphous formation; the in situ diffraction data point rather towards an opposite scenario, that structural coherency between the two structures becomes established by sustained high volume expansion (of NaSi) such that the diffraction peaks of NaSi shift markedly and smoothly (immediately visible in Fig. 4.1) towards their re-assignment as $Na_8Si_{46}$ peaks. As this takes place, the $Na^+$ cations exhibit (indicated by the refined diffraction/NMR data) increasing mobility and large excursions in the NaSi structure, with some distortion of the $Si_4^{-4}$ tetrahedra occurring at high temperature, eventually leaving the structure into the vacuum system. That process requires a redox reaction in which the initial $Na^+$ ions are transformed into atomic Na species, that must then be accompanied by oxidation of the $Si_4^{4-}$ (formally $Si^-$) ions of the starting Zintl phase with concomitant covalent Si-Si bond formation. Inspection of the NaSi structure shows that its *202* planes effectively constitute "Na sheets" that might aid this process via cooperative motions. Barnes et al. have observed a previous case—zeolitic Na-clinoptilolite—where Na motions display an ability to prompt a framework transformation [52]. So the structural and compositional gap between these two end members is effectively bridged by the joint action of thermal expansion and Na sublimation occurring up to 500 °C. The $Na_8Si_{46}$ clathrate phase emerges after this convergence while the remaining NaSi fraction significantly shows a sudden drop ($\sim$1 %) in unit cell volume which is associated with its loss of Na to the vacuum system.

Further to this basic description, it is then suggested that the synthesis displays characteristics of a 'quasi-epitaxial growth scenario'. First, the dual ownership of several common diffraction peaks indicates a growth process where there is structural continuity between elements of both phases at the NaSi/$Na_8Si_{46}$ interface. We noted that in the early stages of clathrate formation the unit cell volume of the clathrate is anomalously low (by $\sim$0.14 %), but with time it rises to the constant value expected for the bulk clathrate. This behavior is also consistent with a quasi-epitaxial growth scenario. New clathrate material forms at the $Na_{1-x}Si$/

$Na_8Si_{46}$ interface and so becomes structurally coherent with it and is thus constrained by the unit cell parameters of the substrate that acts as a template; by contrast, material produced later in the reaction grows out and away from the interface and therefore adopts the normal bulk clathrate unit cell value.

In conclusion we believe this is the first such exposition of the NaSi → $Na_8Si_{46}$ clathrate synthesis process, based on in situ diffraction data. The transformation is most unusual, if not unique, with structural convergence between the diverse pre-cursor and product phases attained via a very high thermal expansion of the pre-cursor throughout a crystalline continuum, followed by promotion of the structural/chemical transformation by the increasing motions of highly mobile $Na^+$-cations. Further, the unit cell—temperature profiles of these two phases are consistent with a 'quasi-epitaxial growth scenario' in which NaSi plays the role of a substrate-template at higher temperature (≥500 °C). Identifying these mechanisms should now lead to gaining more rigorous insights behind designing, developing and tuning new kinds of semiconductor clathrate materials, that can then be interfaced with existing semiconductor technology.

## 4.3 Crystallographic Studies at Low Temperature

Most work on semiconductor clathrate materials has focused on their thermal properties related to their thermoelectric behavior at temperatures within the 200–600 K regime. This is important for electrical to thermal energy conversion related to power generation or heat extraction technologies [53, 54]. In particular, the clathrates show anomalously low lattice thermal conductivity that can be traced to damping of the phonon propagation by localized "rattling" vibrations of the guest atoms in their cages, or to gaps opened up in the phonon dispersion relations because of the large framework unit cell volume compared with the parent diamond structured crystals [27, 55, 56]. The low temperature crystallographic behavior linked to the unusual anharmonic properties of the phonons in the semiconductor clathrate networks is also of great interest.

Materials exhibiting the unusual property of negative thermal expansion were first highlighted among the corner-linked tetrahedral framework structures of $SiO_2$ polymorphs, zeolites and microporous aluminophosphate networks, as well as $ZrW_2O_8$ [57, 58]. The unusual behavior was correlated with enhanced vibrational excitation of the bridging $O^{2-}$ ions that formed linkages between relatively rigid polyhedral units causing these to be drawn together as the temperature increased, thus resulting in an overall negative thermal expansion of the lattice [59, 60]. However, a completely different explanation for the classic phenomenon of negative thermal expansion observed among semiconducting materials such as diamond structured Si, Ge and Sn was based on an anharmonic analysis of the phonon behavior in these dense solid state materials. Si and Ge exhibit a well known dip in their V(T) relations with a minimum occurring near 50–100 K [61, 62]. This can be traced to the anharmonic phonon properties and phonon–phonon coupling

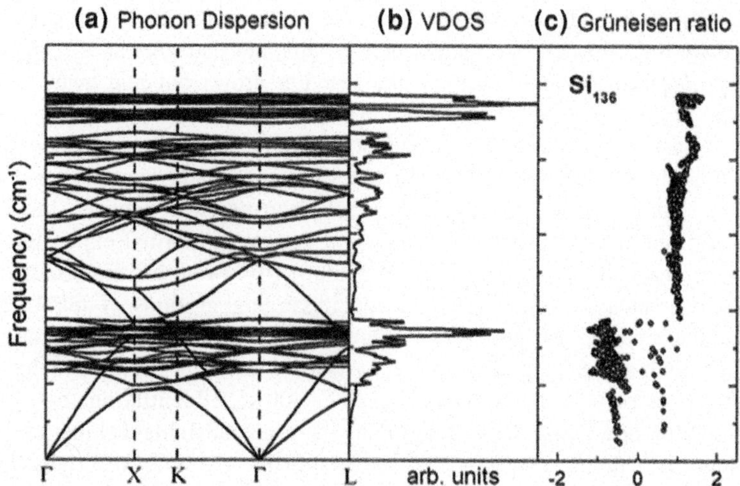


**Fig. 4.4** Phonon dispersion relations and the vibrational density of states (VDOS) calculated for guest-free clathrate $Si_{136}$ using DFT methods at T = 0 K. At right are shown the corresponding mode Gruneisen parameters calculated using a quasi-harmonic model as a function of temperature (from [66])

effects [62–65]. In the case of guest-free clathrates such as $Si_{136}$ or the hypothetical $Si_{46}$ material density-functional theoretical (DFT) calculations have revealed that these should exhibit a region of negative thermal expansion at low T due to similar phonon coupling effects. We studied this behavior that is related to the intrinsic lattice thermal conductivity for a sample of "guest-free" $Si_{136}$ Type II clathrate using a combined experimental and theoretical approach [66]. The sample was prepared by partial thermal decomposition from NaSi followed by successive vacuum extraction of Na atoms from the lattice [10]. Similar materials have been produced by iodine extraction of the alkali atoms into the vapor phase [11], or by synthesis from Zintl phase precursors in ionic liquids [12]. The lattice thermal conductivity of the $Si_{136}$ material had been predicted theoretically and studied experimentally in previous studies [67, 68]. This material provides a unique opportunity to study the intrinsic thermal and structural properties of the semiconductor clathrate lattice in the absence of guest atom rattling effects.

DFT calculations of the static lattice were performed within the local density approximation (LDA) using planewave basis sets and ultrasoft pseudopotentials, and the results were used to construct a force constant matrix within a large supercell model. The phonon spectrum was then evaluated as a function of temperature using quasi-harmonic models that allowed us to construct mode Gruneisen relationships [66] (Fig. 4.4). The results indicated a dip in the V(T) relation at ~80 K, that was slightly smaller than that observed for diamond-structured Si (Fig. 4.5).

These predictions led to an experimental study of the lattice thermal expansion at low temperature using high resolution X-ray diffraction in the laboratory. The shifts in the observed diffraction peaks are very slight throughout the temperature

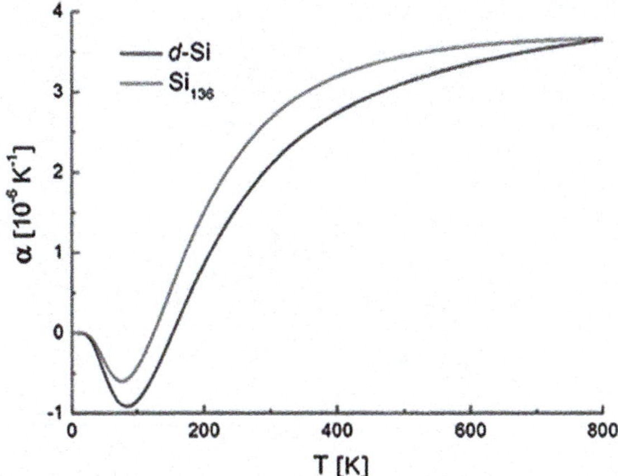


**Fig. 4.5** Lattice thermal expansion for *diamond*-structured Si and $Si_{136}$ obtained from the VDOS calculations showing the region of negative thermal expansion at low T (from [66])

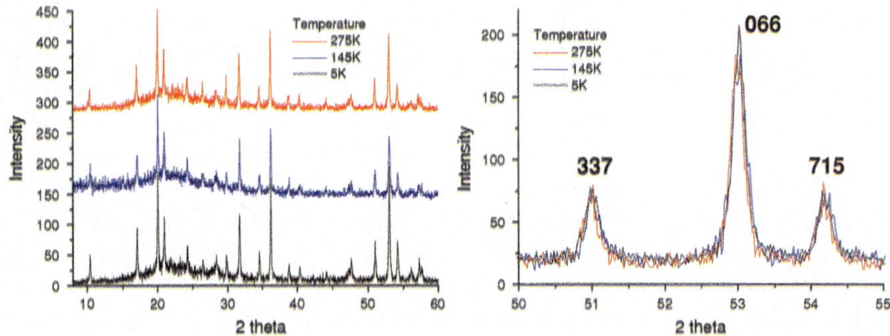


**Fig. 4.6** Selected X-ray diffraction patterns observed experimentally at low temperature for $Si_{136}$ clathrate showing the very small shifts in key reflection profiles even at high angles. The broad background near 20–25° 2Θ is due to the glass capillary used to contain the sample: this was removed from the pattern during the Rietveld fitting procedure. Careful analysis of the entire data sets provide a value for the unit cell volume at each temperature with associated errors (from [66])

range (Fig. 4.6), but by careful extraction of the volume parameter using Rietveld fitting at each temperature we could detect the presence of a dip in the V(T) relation between 80–120 K, that agreed with the DFT prediction (Fig. 4.7). We concluded that the guest-free $Si_{136}$ clathrate did in fact exhibit a range of NTE due to anharmonic phonon effects at low temperature, analogous to the behavior of the diamond-structured phase [66].

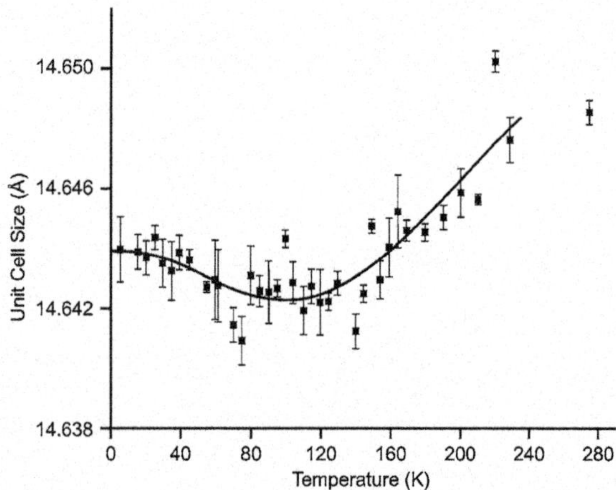


**Fig. 4.7** Unit cell volume (Å$^3$) versus temperature for the $Si_{136}$ empty clathrate structure with associated errors due to the experimental data collection and analysis. The results reveal a minimum in the V(T) relation that agrees with the DFT prediction (*solid line*) (from [66])

## 4.4 High Pressure Studies

It has been of interest to study the structural changes and metastable transformations occurring within these unique framework structures and guest-host clathrate compounds under high pressure conditions. It is important to relate the compressional behavior to the thermal properties as well as the vibrational or electronic coupling phenomena. Most studies to date have focused on Type I clathrates [5]. Upon subjecting these materials to high pressure at ambient T, a typical result is the observation of an unusual apparently isostructural volume collapse, followed by solid state pressure-induced amorphization (PIA) as the crystals become mechanically as well as thermodynamically unstable in a low T regime where nucleation and growth of stable products is kinetically hindered [5].

Clathrates containing small guest atoms ($Na_8Si_{46}$) apparently undergo pressure-induced decomposition above $P = 14$ GPa, where characteristic peaks of the metallic structures of Si-II (β-Sn) or hexagonal Si-III begin to appear in the X-ray diffraction pattern [5, 69]. Type-I clathrates with larger guest atoms (e.g., $K_8Si_{46}$ [70, 71], $Ba_8Si_{46}$ [69, 72–76], $I_8Si_{44}I_2$ [5, 77], $Rb_6Si_{46}$ [78] exhibit an unusual "volume collapse" transition in this pressure range while apparently retaining the same cubic clathrate crystal structure. This is typically followed by PIA occurring at higher pressure. The phenomena underlying the unusual structural transformation behavior and its dependence on the cage site occupancy are still not well understood [5].

The superconducting clathrate $Ba_8Si_{46}$ that is already synthesized under high-P, T conditions has been extensively studied at high pressure using Raman scattering, synchrotron X-ray scattering, electrical resistivity measurements and theoretical simulations [69, 72, 74–76]. A structural transformation occurring at 3–4 GPa was first proposed using reverse Monte Carlo (RMC) modelling based on powder diffraction data combined with resistivity measurements [75]. These conclusions

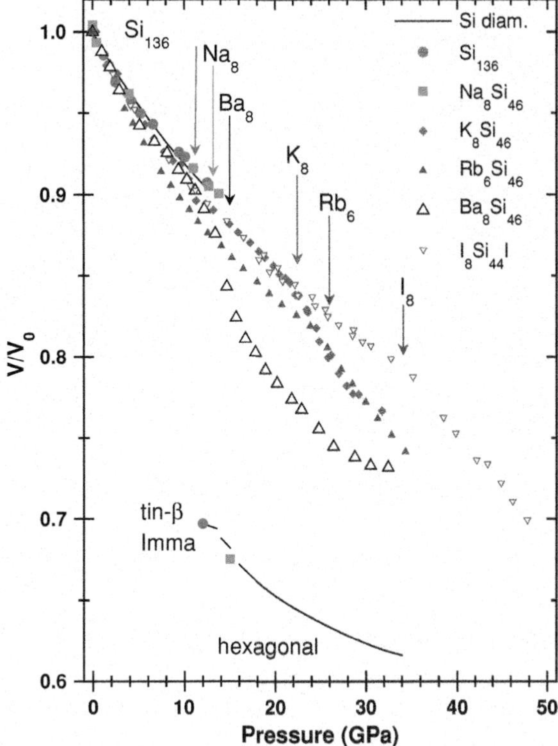


**Fig. 4.8** Pressure behavior of the reduced atomic volume for several silicon clathrates. The behavior for diamond-structured Si is also shown. *Arrows* indicate the pressure at which the isostructural phase transition occurs depending on the guest atom, except for Na for which the phase diagram of silicon in the diamond structure is followed

were then used to tentatively reinterpret changes observed in the Raman spectra observed at high pressure [72], and it was suggested that a structural transition could be occurring related to changes in the electronic state of Ba atoms in the large cage site. At around 7 GPa, Kume et al. observed a further large change in the Raman spectra [72]. The main effect was the attenuation of a low-energy phonon mode attributed to the Ba@$Si_{24}$ coupling with the host lattice. X-ray absorption spectroscopy (XAS) experiments including XANES data at the Ba L-III edge indicated a change of hybridization of the Ba 5d-electrons at P $\sim$ 5 GPa [73] and X-ray diffraction studies during near-hydrostatic compression using He as the pressure-transmitting medium revealed that certain thermal parameters changed their behavior at around 7 GPa [74]. Below this pressure, the thermal parameters for the Ba (6d), Si (24 k), and Si (6c) sites all decrease, whereas above 7 GPa an increase is noted [74]. Using Rietveld and maximum-entropy analysis techniques Tse et al. deduced the presence of pressure-induced changes in the electron density distribution [75]. These authors showed that the Ba, Si (24 k), and Si (6c) atomic positions became disordered above 9 GPa, with a further pressure-induced electronic change occurring at lower pressure ($\sim$7 GPa). Those conclusions were supported by RMC simulations and electrical resistivity measurements under pressure [76].

Analysis of in situ synchrotron diffraction and spectroscopy data first revealed the presence of the unusual isostructural phase transition occurring within semiconductor clathrate materials, that was associated with a large change in the compressibility but with no apparent change in the crystal structure (Fig. 4.8) [69, 74]. The pressure range associated with this unexpected transformation occurs between 13–16 GPa, but it also appears to depend upon the hydrostatic nature of the pressurization conditions. Various possible explanations for the transition have been advanced, including the presence of electronic [73, 79] or phonon instabilities [71], or potential vacancy formation occurring on the framework silicon sublattice [80].

XANES, X-ray diffraction and RMC studies have indicated that a modification of the hybridization state or a more disordered environment around the guest Ba atoms occurs and could lead to this "homoleptic" structural transformation, along with possible disordering in the Si framework [73, 74]. An electronic topological transition has been proposed as the driving force of this transformation [74, 76]. Tse et al. [70] have suggested an interpretation in which entire phonon branches associated with K guest atom displacements become unstable in $K_8Si_{46}$ at the transition pressure, resulting in a positionally disordered form of the clathrate. Other recent results using a combined Rietveld and maximum-entropy approach to analyze the X-ray diffraction data suggest that the volume collapse transition of $Ba_8Si_{46}$ could be associated with s-p-d rehybridization of the Si atoms at high pressure, leading to a weakening of the Si–Si bonds and a collapse of the unit-cell volume without changing the crystalline structure [75]. In this interpretation the volume collapse could be traced to an electronic topological transition involving a substantial rearrangement of the valence electron distribution accompanied by positional disordering in the Si sites [73, 75]. The Si–Si bonding electrons are removed from the bonding region leading to the weakening of the Si–Si bonds, resulting in a displacive but isostructural phase transition taking place with a large volume reduction. Iitaka also recently proposed a model in which vacancies could occur on silicon framework sites to yield a defective clathrate $Ba_8Si_{43}$, analogous to the family of isomorphous $E_8Ge_{43}$ (E: electropositive element) materials in which there are always Ge vacancies [80]. However, recent accurate X-ray measurements performed on $Ba_8Si_{46}$ in highly hydrostatic conditions did not detect the formation of any amorphous or crystalline high-pressure Si phases after the collapse [74].

It is noted that the pressure at which the volume collapse occurs is related to the valency of the cation. $K_8Si_{46}$ and $Rb_{6.15}Si_{46}$ clathrates containing alkali guest atoms exhibit the homoleptic transition at $23 \pm 3$ and $24 \pm 1$ GPa, respectively, whereas $Ba_8Si_{46}$ collapses at a significantly lower pressure $13 \pm 2$ GPa [5].

A new approach based on the phenomenological Landau theory of phase transitions has now been developed to identify physical quantities associated with order parameters ($\eta$) that might drive the isostructural transformation according to the free energy expansion ($F(\eta)$) [81]:

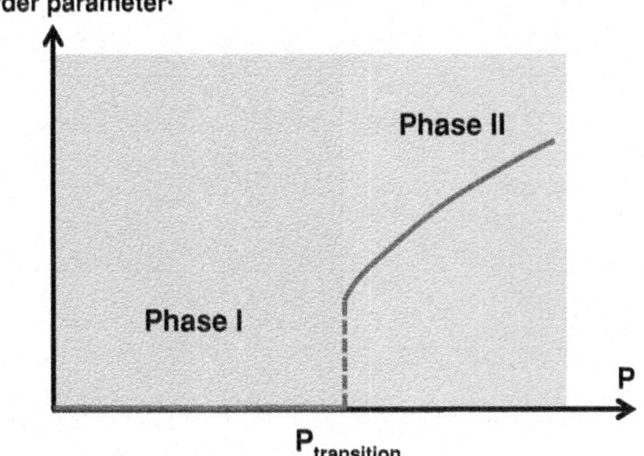


**Fig. 4.9** Evolution of the order parameter predicted for an isostructural phase transition based on the Landau theory of transitions through expansion of the free energy in terms of an order parameter that is consistent with the symmetry constraints of the phase transformation

$$F = F_0 + \frac{\alpha}{2} \cdot \eta^2 + \frac{B}{3} \cdot \eta^3 + \frac{\beta}{4} \cdot \eta^4$$

As no symmetry breaking occurs during the process $\eta$ should transform as the totally symmetric irreducible representation of the group. The physical quantity $\Delta V/V_0$ is a good candidate for use as a macroscopic order parameter compatible with the symmetry requirement. Minimization of F leads to two solutions (Fig. 4.9):

(1) $\eta = 0$ that corresponds to the low-pressure phase (phase I)
(2) $\eta = \frac{-B+(B^2-4\alpha\beta)^{1/2}}{2\beta}$ that corresponds to the high-pressure phase (phase II).

The stability conditions for the crystalline state require that $B < 0$ and $\beta > 0$, and the usual requirement on $\alpha$ is that it changes sign at the transition according to $\alpha = a(P - P_c)$, where $P_C$ is the pressure where phase I becomes unstable and $a$ is a constant to be determined by experiment. By definition $\eta = 0$ in phase I. The evolution of $V/V_0$ is obtained by subtraction from the compression data to yield the order parameter, that we term $\Delta_{Spontaneous}$ associated with the volume change (Fig. 4.10). The evolution of this quantity under pressure fits well with the form of the expression predicted by the theory. The jump in $\Delta_{Spontaneous}$ is small and this observation has led to the suggestion that the isostructural phase transition could be second-order in nature [5]. However that proposal is incompatible with the symmetry arguments used to derive the Landau potential expansion [81]. The normalized atomic displacement parameters ($U^{iso}$) of the Si framework atoms also provide a microscopic physical quantity that can be correlated with the pressure variation of $\Delta_{Spontaneous}$. It appears that this phenomenological analysis can support the idea that the isostructural phase transition is related to a disordering phenomenon within the Si clathrate sublattice.

PIA typically occurs within Type I clathrates pressurized to 30–40 GPa at room temperature, depending on the guest atoms [5]. The nature of the amorphous

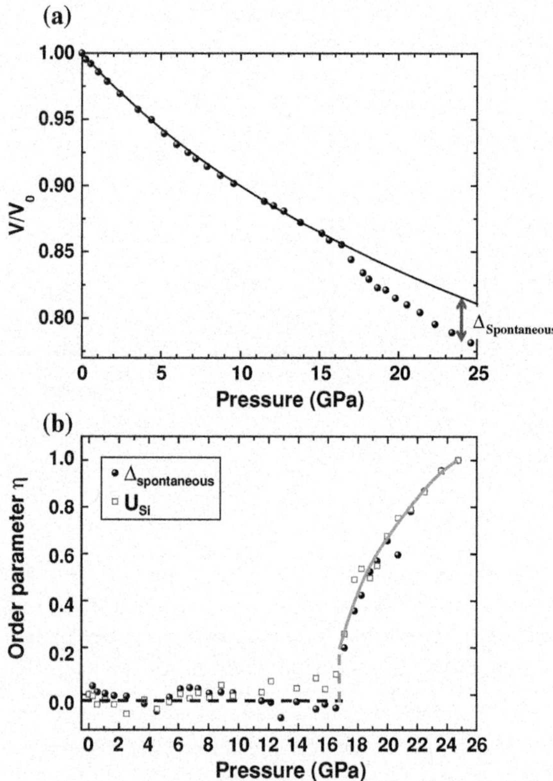


**Fig. 4.10** **a** $V/V_0$ plot for $Ba_8Si_{46}$ as a function of pressure (*points*) along with a third-order Birch-Murnaghan equation of state (*line*). The macroscopic order parameter corresponds to $\Delta_{spontaneous}$ i.e., the variation of the volume corrected from the compressibility. **b** $\Delta_{spontaneous}$ as a function of pressure fits with the theoretical analysis based on the Landau theory of phase transition: null before the transition pressure, a jump at the transition and a square-root evolution after the transition. Such a behavior is also observed for Si atomic displacement parameters that can be used as microscopic order parameter [82]. The correlation between these physical quantities underlines the relationships between the isostructural transition and disordering of the Si sub-lattice

solids, along with their relationship to thermal glasses or other amorphous materials formed by other means, is not yet well understood [83]. Evidence for a possible polyamorphic transition in the amorphous solid produced by PIA in $Rb_{6.15}Si_{46}$ clathrate was first obtained by Raman spectroscopy [78].

The expected Raman-active modes for a Type I clathrate with fully filled guest atom sites are:

$$\Gamma_{raman} = 3A_{1g} + 8E_g + 9\,T_{2g}$$

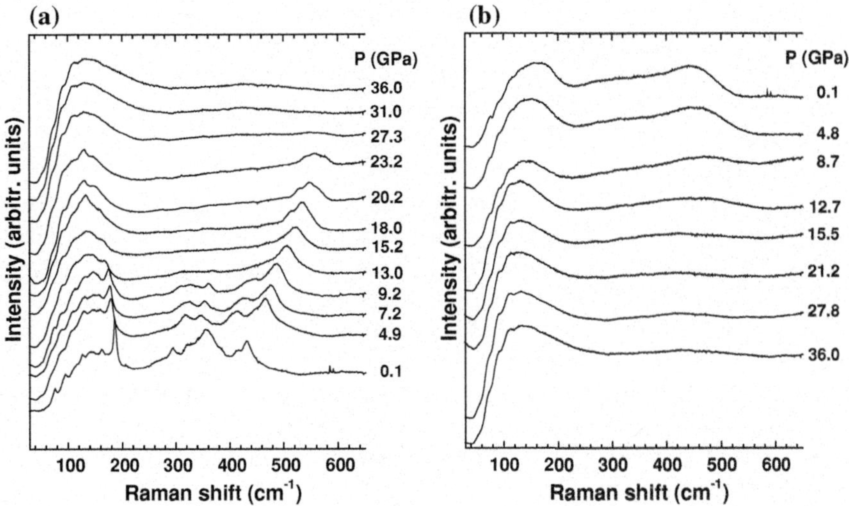


**Fig. 4.11** **a** Raman spectra of $Rb_{6.15}Si_{46}$ during quasi hydrostatic compression (using argon as pressure-transmitting medium). **b** Raman spectra of $Rb_{6.15}Si_{46}$ during decompression after quasi hydrostatic compression to 36 GPa (from [78])

Two modes of symmetry $E_g$ and $T_{2g}$ occur at low frequency (<200 cm$^{-1}$), mainly involving the encapsulated atoms in the large (6d) guest sites [84–87]. These play an important role in determining the anomalously low lattice thermal conductivity of the clathrate structures [56, 88]. The remaining modes are associated with Si–Si stretching and bending vibrations of the clathrate framework. Many of the Raman bands in reported spectra are often highly broadened for reasons that are not yet clear, but this observation could imply disordering, vacancies on the framework and guest atom sites, or electron–phonon coupling processes occurring (Fig. 4.11) [85].

During compression of $Rb_{6.15}Si_{46}$ up to 9.2 GPa, all of the high-frequency peaks (>300 cm$^{-1}$) shifted to higher wave number with increasing pressure as expected. However, the most intense Raman mode initially occurring at 186 cm$^{-1}$ shifts to lower frequency, as does the shoulder at 164 cm$^{-1}$, indicating softening of these vibrational modes as predicted theoretically [78]. Other low-frequency modes instead stiffen slightly with pressure, but less than the high-frequency vibrations. Similar observations were made for $Ba_8Si_{46}$ and $K_8Si_{46}$ Type I clathrates [71, 72]. Kume et al. noted an analogy with softening in low-lying TA modes at the Brillouin zone boundary in diamond-structured Si that could be related to the clathrate behavior, and that could thus be intrinsic to such sp$^3$-bonded tetrahedral lattices.

The appearance of the Raman spectrum is affected significantly by increasing pressure. Important changes in the entire spectrum are observed starting from 9.2 GPa but some of the initial peaks can be followed to 13.0–15.2 GPa, although with a strong attenuation beyond 9.2 GPa. By 18 GPa an entirely new Raman spectrum was recorded for the $Rb_{6.15}Si_{46}$ phase, indicating that a major change in the clathrate structural parameters affecting the Raman selection rules had occurred. These changes could be associated with the homoleptic "volume collapse transition". Above 31 GPa, the Raman spectrum contained only two very broad and weak bands at ~450 and ~160 $cm^{-1}$ indicating the presence of an amorphous structure. This is correlated with the PIA event recorded by X-ray diffraction [78]. The data indicate that the amorphous material is in a high-density form, similar to the HDA metallic polyamorph of a-Si produced via PIA from nanocrystalline porous-Si, or by compressing the low pressure (LDA) amorphous semiconductor [89, 90]. During decompression, the Raman spectra show changes analogous to those observed during the HDA–LDA transformation in amorphous Si, indicating that the polyamorphic transition extends into the Si–Rb system, and that the behavior may be general among amorphous semiconductor-metal alloys that do not undergo compositional phase separation. The predominance of pentagonal rings in the clathrate structure has led some authors to use clathrates as crystalline prototypes of amorphous Si or Ge [91].

Interesting results from clathrate compression studies is the observation of unusual mechanical properties that can be compared with those of the dense (diamond structured) $sp^3$-bonded phases. For instance, Benedek et al. [92] suggested that $Na_2Ba_6Si_{46}$, first synthesized by Yamanaka et al. [24], could exhibit a bulk modulus about 15 % higher than that of the Si-I diamond structured material. That interesting result was supported by simulation studies [5]. However, the determined bulk modulus ($K_O$) values depend on the equation of state formalism used as well as the fixing of parameters such as the zero pressure volume ($V_o$) and the pressure derivative of the bulk modulus $K_O$. Improvements in high-pressure techniques leading to a better control of hydrostatic versus non-hydrostatic compression conditions along with synchrotron X-ray diffraction and advanced data analysis methods are leading to a re-evaluation of earlier results. In particular we note the advantage of refining the $K_O$, $K_O'$ and $V_O$ parameters using a linearized stress-strain relation applied to the observed V(P) data (Fig. 4.12) [93]. The Eulerian strain parameter (f) and the normalized pressure F are defined by:

$$f = \frac{1}{2}\left[\left(\frac{V}{V_0}\right)^{-2/3} - 1\right]$$

$$F = P[3f(1+2f)^{2.5}]^{-1}$$

This formalism yields the second-order finite strain equation

$$F = K_0[1 - 1.5(4 - K_0')f]$$

Using this relation we can conduct a linear fit to the resulting F-f plot and apply well accepted goodness of fit criteria, and likely versus unlikely ranges of values for $K_0'$ can be readily tested against the data within error limits. $K_O$ and $K_O'$ values

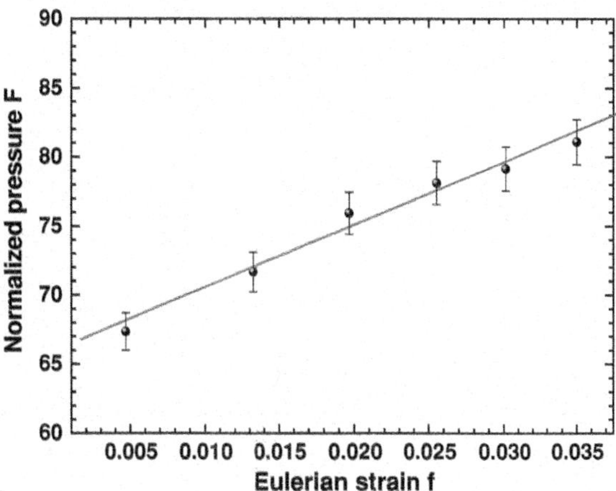


**Fig. 4.12** Plot of the normalized pressure (F) as a function of the Eulerian strain variable (f) for $Ba_8Si_{46}$ using neutron diffraction data [94]

**Table 4.1** Systematic changes in cell parameter $a_0$, the bulk modulus ($K_0$) and its pressure derivative ($K_0'$) with the progressive filling of the nanocages (x = 0, 2, 6 and 8) in the Type I clathrate structure obtained from ab initio calculations (from [78])

| $a_0$ | | $K_0$ | $K_0'$ |
|---|---|---|---|
| $Si_{46}$ | 10.2296 | 76.055 | 4.0886 |
| $Ba_2Si_{46}$ | 10.2781 | 77.59 | 4.0516 |
| $Ba_6Si_{46}$ | 10.3092 | 77.59 | 4.4531 |
| $Ba_8Si_{46}$ | 10.395 | 76.813 | 4.2288 |
| $Rb_2Si_{46}$ | 10.28 | 76.185 | 3.9799 |
| $Rb_6Si_{46}$ | 10.3335 | 74.554 | 4.1928 |
| $Rb_8Si_{46}$ | 10.4148 | 69.341 | 4.6828 |

are obtained as the intercept as f → 0 and the slope of the F-f-plot, respectively. As an example of the importance of this procedure, the first experimental determination of the $Ba_8Si_{46}$ bulk modulus using a Murnaghan equation of state gave $K_0 = 93 \pm 5$ GPa, using a fixed value of $K_0' = 3.6$. Instead, recent neutron diffraction experiments on $Ba_8Si_{46}$ compressed with an ethanol-methanol mixture and analyzed using the F-f plot technique gave $K_0 = 67(2)$ GPa and $K_0' = 8(1)$ (Fig. 4.12) [94]. In determination and comparisons of the mechanical response to compressive stress it is obviously important to use a consistent set of analysis techniques, as well as considering the hydrostatic versus non-hydrostatic nature of the pressurization environment.

Ab initio methods have also been used to calculate the cell parameter, the bulk modulus $K_0$, and its derivative $K_0'$ for $Ba_xSi_{46}$ and $Rb_xSi_{46}$ with varying metal cage site occupancies of x = 0, 2, 6, and 8 [78]. The results are summarized in Table 4.1. Filling the cages with guest atoms increases the cell parameter by up to 1.6 and 1.8 % for $Ba_8Si_{46}$ and $Rb_8Si_{46}$, respectively. The mechanical properties are not strongly affected (typically less than 2–3 %) by cage occupancy, except when

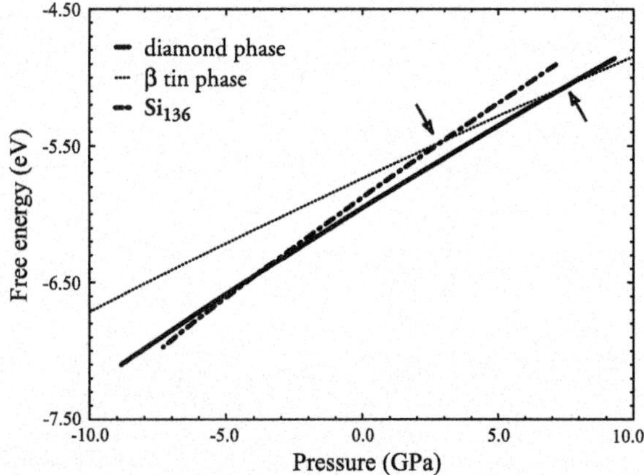


**Fig. 4.13** Gibbs free energy versus pressure plot showing the stable and metastable relationships between diamond structured Si–I, β-Sn structured Si-II and Type I empty cage clathrate structure $Si_{136}$. Parameters used to construct the plot are derived from DFT calculations within the LDA. The slope of the plot at any pressure for a given phase is its atomic volume (per atom), and the curvature is proportional to its compressibility. Theory predicts a transition pressure of 8 GPa for the Si-I to Si-II transition (*arrow pointing up*) in excellent agreement with experimental determinations. Theory also predicts a much lower (3–4 GPa, *arrow pointing down*) pressure for the $Si_{136}$ to Si-II metastable transition, but this is observed experimentally between 8.0 and 10.3 GPa. This occurs due to the absence of a convenient low-energy pathway for the $Si_{136}$ to β-tin transition (from [96])

the small $Si_{20}$ cages start to be filled with Rb atoms. The filling of all the $Si_{20}$ cages leads to a decrease of $K_O$ by 10 % and an increase of $K_O'$ by 15 %.

There have been fewer studies of the high pressure properties of Type II clathrates, and these have mostly concerned the metastable phase transformation properties of the guest-free $Si_{136}$ material compressed at ambient T conditions. Based on the observation that this initially metastable elemental polymorph had a larger molar volume than the diamond-structured Si-I phase at ambient P, T conditions, the thermodynamic transformation into the metallic β-Sn structured Si-II polymorph could be expected to occur at lower pressure during ambient T compression (Fig. 4.13).

However, that predicted result was not obtained during energy-dispersive X-ray diffraction experiments on polycrystalline samples conducted in a diamond anvil cell [95, 96] (Fig. 4.14). Instead a rapid transformation was observed to occur between the metastably compressed $Si_{136}$ phase and the β-Sn structured Si-II metallic phase at approximately 10 GPa, i.e. in a similar pressure range to that observed for the diamond-structured Si-I to Si-II transition that occurs at ambient T [96]. The variation of the cell volume with pressure using F-f analysis yields an estimated bulk modulus $K_O = 90(3)$ GPa with $K_O' = 5.2(8)$ for $Si_{136}$ clathrate, in good agreement with LDA results [67] that predict $K_O = 81.2$ GPa and $K_O' = 4.45$ (Figs. 4.15 and 4.16). These results are also in agreement with a second study

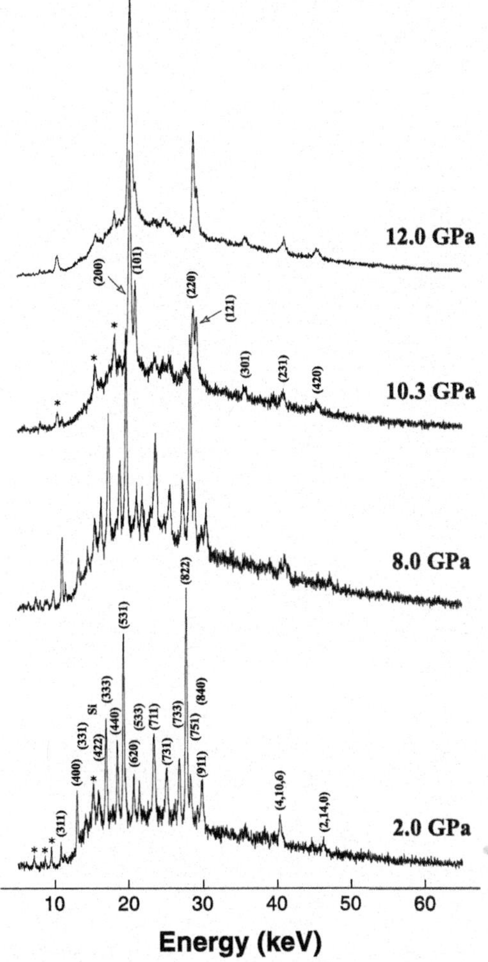


**Fig. 4.14** Selected X-ray diffraction patterns showing the structural evolution of $Si_{136}$ Type I clathrate as a function of pressure. The sample was loaded with 4:1 methanol-ethanol mixture as pressure-transmitting medium in a Mao-Bell piston cylinder type DAC with 350 μm diamond culets. The sample chamber was provided by a 120 μm hole drilled in a T301 stainless steel gasket. Pressure was measured using the ruby fluorescence scale. Energy dispersive X-ray diffraction patterns were collected at beamline X17C of the National Synchrotron Light Source at Brookhaven National Laboratory (USA). The diffracting angle was $2\Theta = 15.00(5)°$. Transformation into the dense metallic β-tin structure of silicon (Si-II) occurs between 8.0 and 10.3 GPa. The strongest lines at 10.3 GPa can be attributed to the (200), (101), (220) and (211) reflections of the tetragonal structure with a = 4.6813(1) Å and c = 2.5884(1) Å. Similar data collected at 12.0 GPa (*top most curve*) could also be indexed to the β-tin structure. Small peaks identified by stars at 2 GPa correspond to the (111) and (110) reflections of diamond structured Si–I. Peaks below 10 keV correspond to Ge fluorescence lines and "escape" peaks that are observed in the pattern at 10.3 GPa (from [96])

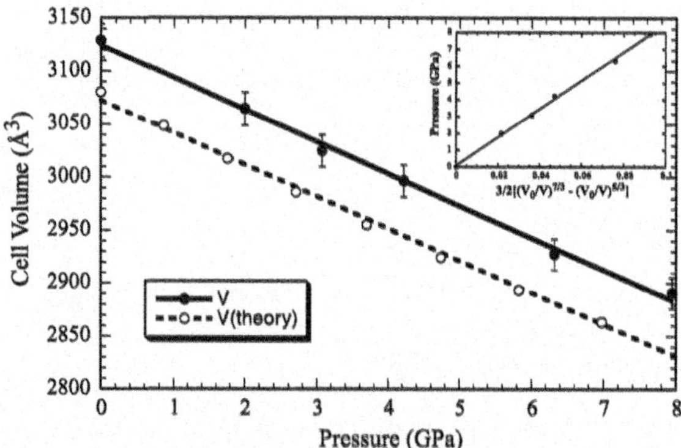


**Fig. 4.15** Pressure dependence of the unit cell volume in $Si_{136}$ ($Fd\bar{3}m$ structure). The experimental data are compared with results of density functional calculations carried out within the local density approximation. Over the pressure range shown, $Si_{136}$ undergoes a ~7.5 % volume change. In the inset, pressure is plotted against a reduced volume scale (from [96])

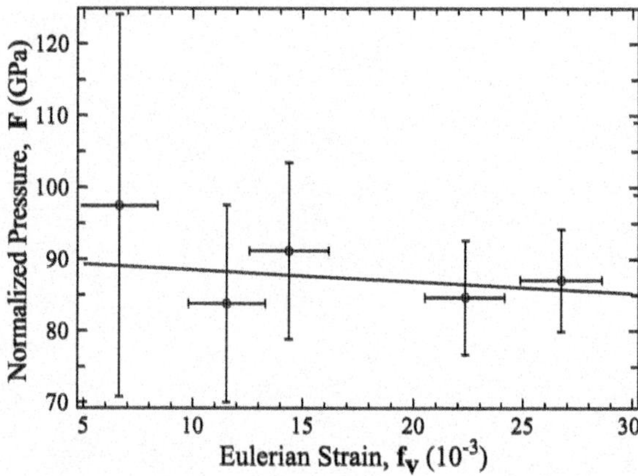


**Fig. 4.16** More detailed examination of the V(P) relations for $Si_{136}$ clathrate using a third-order Birch-Murnaghan equation of state and linearized Eulerian strain variables (F-f plot) gave a bulk modulus $K_0 = 90(3)$ GPa and its pressure derivative $K_0' = 5.2(8)$ (from [96])

combining X-ray diffraction at high-pressure with ab initio calculations that found $K_0 = 90(5)$ GPa with $K_0' = 3.6$ [95].

In a further study of the compression behavior of $Si_{136}$ sample we used Raman spectroscopy in the DAC [66]. The guest-free elemental polymorph with space group $Fd\bar{3}m$ and 34 atoms in the corresponding primitive unit cell gives rise to zone center vibrational modes:

$$\Gamma = 3\,A_{1g} + A_{1u} + A_{2g} + 3\,A_{2u} + 4\,E_g + 4\,E_u + 5\,T_{1g} + 7\,T_{1u} + 8\,T_{2g} + 5T_{2u}$$

of which $\Gamma_R = 3A_{1g} + 4E_g + 8T_{2g}$ are Raman active. At ambient conditions, we could assign all of the expected one-phonon Raman peaks by comparing with DFT results (Fig. 4.17 and Table 4.2). Second order features (two-phonon peaks) occur

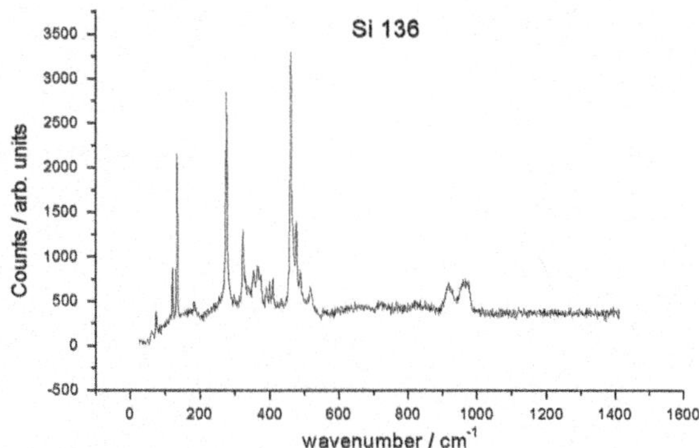


**Fig. 4.17** Raman spectrum obtained for $Si_{136}$ at ambient P and T. (from [66])

**Table 4.2** Calculated and experimentally measured Raman frequencies and mode Grüneisen parameters ($\gamma_i$) for clathrate-structured $Si_{136}$ (from [66])

| Mode | Frequency (cm$^{-1}$) (theory) | Frequency (cm$^{-1}$) (experiment) | Grüneisen parameters ($\gamma_i$) (theory) | Grüneisen parameters ($\gamma_i$) (experiment) |
|---|---|---|---|---|
| $T_{2g}$ | 121 | 117 | −1.17 | −1.32 ± 0.04 |
| $E_g$ | 130 | 130 | −0.71 | −0.86 ± 0.04 |
| $T_{2g}$ | 176 | 184 | −1.22 | – |
| $T_{2g}$ | 267 | 271 | 0.94 | 1.01 ± 0.04 |
| $A_{1g}$ | 316 | – | 1.04 | – |
| $T_{2g}$ | 325 | 324 | 0.93 | −1.17 ± 0.04 |
| $E_g$ | 360 | 360 | 1.18 | −0.66 ± 0.04 |
| $A_{1g}$ | 397 | 387 | 1.08 | −0.89 ± 0.05 |
| $T_{2g}$ | 406 | 401 | 1.49 | 0.97 ± 0.04 |
| $A_{1g}$ | 458 | 454 | 1.22 | 1.33 ± 0.04 |
| $E_g$ | 463 | 466 | 1.26 | – |
| $T_{2g}$ | 466 | | 1.09 | |
| $T_{2g}$ | 473 | 472 | 1.42 | – |
| $E_g$ | 483 | 480 | 1.33 | 1.21 ± 0.07 |
| $T_{2g}$ | 487 | 488 | 1.04 | 1.39 ± 0.07 |
| (Si) | 516 | 520 | 0.94 | 0.95 ± 0.005 |
| 2- | | phonon | | |
| 920 | | | | |
| 2- | phonon | 965 | | |

at 919 and 965 cm$^{-1}$. A very weak feature at 518 cm$^{-1}$ could be due to a trace of nanocrystalline diamond-structured Si-I within the sample, but a peak for $Si_{136}$ clathrate is also expected near this position (Table 4.2).

During compression, both positive and negative frequency shifts were observed to occur (Fig. 4.18). In particular, the two lowest frequency modes exhibited a frequency decrease with increasing pressure, as predicted theoretically by Dong

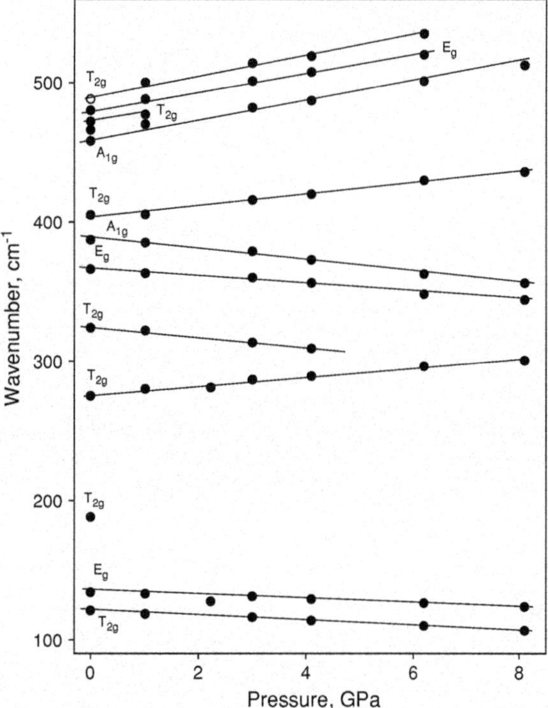


**Fig. 4.18** Pressure dependence of Raman shifts of $Si_{136}$ (from [66])

et al. [67]. Continued mode softening of this type would result in a second order phase transition above P ~ 25–30 GPa; however, structural collapse and the onset of metastable first-order phase transitions occur well before this potential event.

Between 8.1 and 9.7 GPa, the spectrum changes abruptly (Fig. 4.19). This is the pressure at which the onset of transformation into the β-Sn structured metallic phase was observed in synchrotron X-ray diffraction studies [95, 96]. At 9.2 GPa, only very weak features due to remnant traces of the $Si_{136}$ polymorph are observed on top of a broad weak background, that also contains the two Raman bands at 120 and 386 $cm^{-1}$ that are characteristic of the β-Sn structured Si-II phase [97]. Those peaks are also present in the spectrum obtained at 8.1 GPa, indicating the onset of the metastable $Si_{136}$-SiII phase transition, in agreement with the X-ray diffraction results [96]. However, continued compression to 9.7 GPa yields a completely new spectrum. This as-yet unidentified polymorph only exists during compression between 9.2–10.0 GPa. An additional weak band at about 570 $cm^{-1}$ is detected in spectra starting from 10 GPa that might be assigned to a Si-IV phase, known to be a metastable form of silicon possessing the hexagonal diamond (lonsdaleite type) structure and that is normally synthesized at high pressure (above 18 GPa) from the cubic diamond-structured Si-I. The Si-IV phase has also been obtained metastably upon heating Si-III (BC8 structure) above 200 C at ambient P [98–101]. Raman spectra obtained above 13 GPa show only signatures of metallic β-Sn

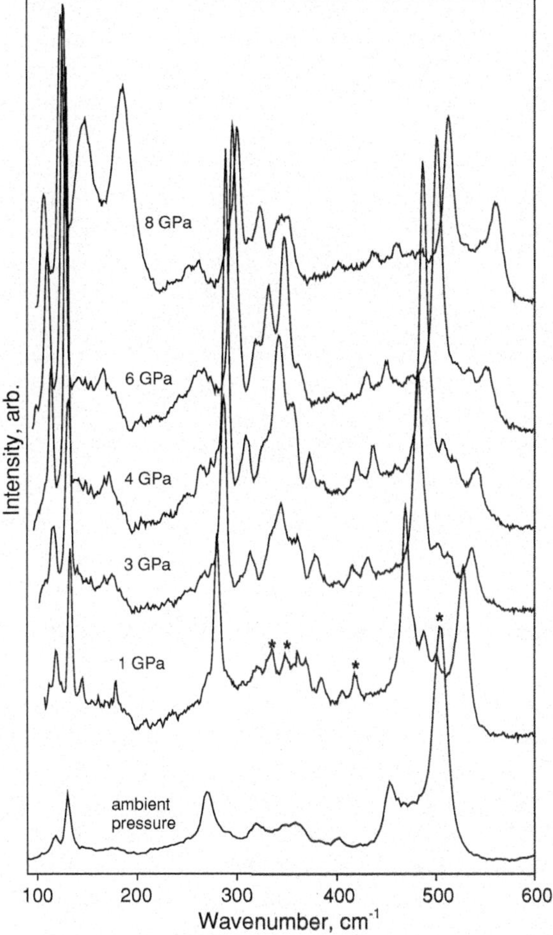


**Fig. 4.19** Raman spectra of $Si_{136}$ collected during compression up to 8 GPa. *Asterisks* mark peaks that are not identified as part of the $Si_{136}$ Raman spectrum and may be due to impurities within the sample compressed within the diamond anvil cell (from [66])

(Si-II) and hexagonal Si-III structures. These results indicate that low T compression of semiconductor clathrate materials can lead to identification of new crystalline polymorphs and elemental structure types.

Only very few studies on Type III clathrates under high pressure have been conducted to date. Grosche et al. studied the superconducting behavior of Type III $Ba_{24}Ge_{100}$ and $(Ba_4Na_2)Ge_{100}$ germanium-based clathrates up to 3.4 GPa and at low temperature [102–104] while Shimizu et al. have reported a Raman spectroscopy study of $Ba_{24}Ge_{100}$ up to 26 GPa at room temperature [105]. Three studies have recorded the high pressure behavior of $Ba_{24}Si_{100}$: a low temperature resistivity study up to 1.15 GPa [106] in its superconducting regime, a Raman spectroscopy study by Shimizu et al. at room T up to 27 GPa [107], and a synchrotron X-ray diffraction study up to ~37 GPa [108].

The high pressure Raman spectroscopy study of $Ba_{24}Si_{100}$ by Shimizu et al. [107] showed evidence for two structural transitions. The first occurs between 3.9 and 6.5 GPa and is characterized by the splitting of the 126 $cm^{-1}$ Raman mode along with a general decrease of the intensity of the modes associated with vibrations of Ba atoms. The second transformation at 20.7–23.2 GPa is associated with the disappearance of all the observable Raman features, and this was initially attributed to a pressure-induced amorphization. However, subsequent X-ray diffraction studies showed that the transition at ~23 GPa in fact corresponds a volume collapse similar to Type I silicon clathrates $M_8Si_{46}$ [108].

In the lower pressure phase transition, the Si network is not affected, but the Ba atoms show a pronounced discontinuous positive variation of their isotropic thermal factors, which was interpreted as a sign of random disordering of the encaged atoms with an associated change of their hybridization with the Si network. During the second transition at 23 GPa, the host network and guest atoms are more strongly affected, probably with the creation of a very distorted and disordered structure. This disordered but nanocage-based structure could explain the disappearance of the Raman peaks for the $Ba_{24}Si_{100}$ phase.

## 4.5 Properties and Potential Existence of C-Clathrates

So far, it has not been found possible to prepare or definitively identify carbon clathrates analogous to the Si, Ge and other group 13–15 framework members. However extrapolations of the properties of the other clathrate compounds especially in the area of mechanical properties and superconductivity, make the formation of such hypothetical carbon clathrate phases very attractive for potential synthesis studies [18]. Already in 1995, Saito and Oshiyama [109] had noted $C_{46}$ or its metal-doped forms as good candidates for high-$T_c$ superconductors due to the high phonon frequency and the strong electron–phonon interaction generally expected for carbon-based materials. The ideal tensile strength of the hypothetical $C_{46}$ phase is predicted to be larger by at least 25 % than that of diamond in its <111> direction [18]. Carbon clathrates are also predicted to be superconductors with high $T_c$ (>100 K) values [110], as well as potentially leading to new wide band gap semiconducting materials [10]. Contrary to diamond, n-doping in C clathrates could be achieved by incorporating donor atoms inside the cages as well as by substitutional or interstitial mechanisms within the dense $sp^3$-bonded networks [111]. As observed for other group 14 elements, carbon clathrates are expected to be metastable with respect to graphite and dense $sp^3$-bonded phases, but they are predicted to lie quite close in energy to graphitic materials and in particular, they might be more thermodynamically stable than solid-state fullerites [112]. Potential high-P, T synthesis pathways have been suggested but none of the methods suggested are currently viable possible because of the lack of suitable precursors and availability of the necessary experimental conditions for synthesis and recovery [113]. Also, the potential C-clathrates occur in competition with other complex low-density carbon forms that are known to form including onions, nanotubes, foams and schwarzites [114, 115]. One interesting possible route to formation of carbon clathrates is the 3D-polymerization of fullerene samples, and one proposed 3D-form of a clathrate material (but not Type I or Type II materials described to date) has been observed experimentally [116].

## 4.6 Conclusions

Semiconductor clathrates form an unusual class of open framework compounds with structural analogies to H-bonded $H_2O$ ice clathrates as well as zeolitic materials but lacking inter tetrahedral oxygen linkages. The result is a fascinating class of new materials with interesting and potentially useful electronic, optical, thermal and magnetic properties. The frameworks based on elemental Si and Ge are typically thought be metastable relative to the diamond structured polymorphs at ambient pressure but they could potentially crystallize from the molten state under negative pressure conditions, and they are observed to grow from the vapor phase or during metastable synthesis from precursors. The growth processes are not well understood and here we have described pioneering work using synchrotron X-ray diffraction to elucidate the early stages of growth of a Type I $Na_8Si_{46}$ clathrate from the Zintl phase NaSi under dynamic vacuum conditions at high temperature. Unexpectedly this process appears to involve a quasi-epitaxial mechanism in which the clathrate phase nucleates and grows in structural registry with the underlying $Na_{1-x}Si$ Zintl compound that has undergone anomalously large thermal expansion and then loses part of its Na component to the vapor phase, accompanied by $Na^+$–$Na^0$ reduction that implies Si–Si bond formation in the underlying solid surface region. Semiconductor clathrates are interesting materials for their thermoelectric properties that are associated with an unusually low lattice thermal conductivity. These behaviors are typically associated with localized "rattling" vibrations of the cage site atoms as well as gaps developed in the phonon dispersion relations of the framework structure. Additional low temperature structural properties include negative thermal expansion that has been predicted theoretically and demonstrated experimentally. High pressure studies have revealed the existence of an unusual new type of "homoleptic" phase transition involving an abrupt change in unit cell volume while maintaining the cubic crystalline symmetry, that can be related to electronic structure changes around the guest atoms and perhaps associated with defects and ordering in the clathrate framework structure. Further pressurization results in metastable phase transitions or pressure-induced amorphization phenomena, into a high-density amorphous material that may undergo a polyamorphic transformation into a low-density form following decompression.

## References


1. C. Cros, M. Pouchard, P. Hagenmuller, C.R. Hebd, Seances Acad. Sci. **260**, 4764 (1965)
2. J.S. Kasper, P. Hagenmuller, M. Pouchard, C. Cros, Science **150**, 1713 (1965)
3. C. Cros, M. Pouchard, P. Hagenmuller, J.S. Kasper, Bull. Soc. Chim. Fr. **7**, 2737 (1965)
4. J. Gallmeier, H. Schäfer, A. Weiss, Z. Naturforsch. B **24b**, 665 (1969)
5. A. San Miguel, P. Toulemonde, High Press. Res. **25**, 159 (2005)
6. S. Bobev, S.C. Sevov, J. Solid State Chem. **92**, 153 (2000)
7. N. Jaussaud, P. Toulemonde, M. Pouchard, A. San Miguel, P. Gravereau, S. Pechev, G. Goglio, C. Cros, Solid State Sci. **6**, 401 (2004)
8. D. Neiner, N.L. Okamoto, P. Yu, S. Leonard, C.L. Condron, M.F. Toney, Q.M. Ramasse, N.D. Browning, S.M. Kauzlarich, Inorg. Chem. **49**, 815 (2010)
9. D. Neiner, N.L. Okamoto, C.L. Condron, Q.M. Ramasse, P. Yu, N.D. Browning, S.M. Kauzlarich, J. Am. Chem. Soc. **129**, 13857 (2007)
10. J. Gryko, P.F. McMillan, R.F. Marzke, G.K. Ramachandran, D. Patton, S.K. Deb, O.F. Sankey, Phys. Rev. B **62**, R7707 (2000)
11. A. Ammar, C. Cros, M. Pouchard, N. Jaussaud, J.M. Bassat, G. Villeneuve, M. Duttine, M. Ménétrier, E. Reny, Solid State Sci. **6**, 393 (2004)
12. A.M. Guloy, R. Ramlau, Z. Tang, W. Schnelle, M. Baitinger, Y. Grin, Nature **443**, 320 (2006)
13. H. Fukuoka, K. Ueno, S. Yamanaka, J. Organomet. Chem. **611**, 543 (2000)
14. H. Fukuoka, K. Iwai, S. Yamanaka, H. Abe, K. Yoza, L. HaKming, J. Solid State Chem. **151**, 117 (2000)
15. M. O'Keeffe, G.B. Adams, O.F. Sankey, Philosoph. Mag. Lett. **78**, 21 (1998)
16. N. Jaussaud, P. Toulemonde, M. Pouchard, A. San Miguel, P. Gravereau, S. Pechev, G. Goglio, C. Cros, Solid State Sci. **6**, 4001 (2004)
17. G.B. Adams, M. O'Keeffe, A.A. Demkov, O.F. Sankey, Y.-M. Huang, Phys. Rev. B **49**, 8048 (1994)
18. X. Blase, P. Gillet, A. San Miguel, P. Mélinon, Phys. Rev. Lett. **92**, 215505 (2004)
19. D. Kahn, J.P. Lu, Phys. Rev. B **56**, 13898 (1997)
20. J.L. Cohn, G.S. Nolas, V. Fessatidis, T.H. Metcalf, G.A. Slack, Phys. Rev. Lett. **82**, 779 (1999)
21. J.S. Tse, K. Uehara, R. Rousseau, A. Ker, C.I. Ratcliffe, M.A. White, G. Mackay, Phys. Rev. Lett. **85**, 114 (2000)
22. V. Timoshevskii, D. Connétable, X. Blase, Appl. Phys. Lett. **80**, 1385 (2002)
23. H. Kawaji, H.O. Horie, S. Yamanaka, M. Ishikawa, Phys. Rev. Lett. **74**, 1427 (1995)
24. S. Yamanaka, E. Enishi, H. Fukuoka, M. Yasukawa, Inorg. Chem. **39**, 56 (2000)
25. P. Toulemonde, Ch. Adessi, X. Blase, A. San Miguel, J.L. Tholence, Phys. Rev. B **71**, 094504 (2005)
26. G.S. Nolas, G.A. Slack, Am. Sci. **89**, 136 (2001)
27. G.S. Nolas, J.L. Cohn, G.A. Slack, S.B. Schujman, Appl. Phys. Lett. **73**, 178 (1998)
28. V.L. Kuznetsov, L.A. Kuznetsova, A.E. Kaliazin, D.M. Rowe, J. Appl. Phys. **87**, 7871 (2000)
29. T. Kume, H. Fukuoka, T. Koda, S. Sasaki, H. Shimizu, S. Yamanaka, Phys. Rev. Lett. **90**, 155503 (2003)
30. B.B. Iversen, A.E.C. Palmqvist, D.E. Cox, G.S. Nolas, G.D. Stucky, N.P. Blake, H. Metiu, J. Solid State Chem. **149**, 455 (2000)
31. G.S. Nolas, C.A. Kendziora, Phys. Rev. B **62**, 7157 (2000)
32. J. Dong, O.F. Sankey, G.K. Ramachandran, P.F. McMillan, J. App. Phys. **87**, 7726 (2000)
33. M. Wilson, P.F. McMillan, Phys. Rev. Lett. **90**, 135703 (2003)
34. S. Yamanaka, Dalton Trans. **39**, 1901 (2010)
35. S. Bobev, S.C. Sevov, J. Am. Chem. Soc. **121**, 3795 (1999)

36. P.F. McMillan, J. Gryko, C. Bull, R. Arledge, A.J. Kenyon, B.A. Cressey, J. Solid Stat. Chem. **178**, 937 (2005)
37. S. Kauzlarich, Light element group 13–14 clathrate phases, in *The Physics and Chemistry of Inorganic Clathrates*, ed. by G. Nolas (Springer, Heidelberg) This volume
38. P. Mélinon, P. Kéghélian, B. Prével, A. Perez, G. Guiraud, J. LeBrusq, J. Lermé, M. Pellarin, M. Broyer, J. Chem. Phys. **108**, 4607 (1998)
39. R. Kamalakaran, A.K. Singh, O.N. Srivastava, J. Phys. Condens. Matter **7**, L529 (1995)
40. R. Kamalakaran, A.K. Singh, O.N. Srivastava, J. Phys. Condens. Matter **12**, 2681 (2000)
41. P.T. Hutchins, O. Leynaud, L.A. O'Dell, M.E. Smith, P. Barnes, P.F. McMillan, Chem. Mater. **23**, 5160 (2011)
42. G.N. Greaves, C.R.A. Catlow, G.E. Derbyshire, M.I. McMahon, R.J. Nelmes, G. Van Der Laan, Nat. Mater. **7**, 827 (2008)
43. N.J. Terrill, G. Bushnell-Wye, R.J. Cernik, A.J. Dent, G.P. Diakun, C.J.V. Flaherty, J. Kay, C. Tang, P. Barnes, G.N. Greaves, T. Rayment, A. Ryan, Fibre Diffr. Rev. **11**, 20 (2003)
44. R.J. Cernik, P. Barnes, G. Bushnell-Wye, A.J. Dent, G.P. Diakun, J.V. Flaherty, G.N. Greaves, E.L. Heeley, W. Helsby, S.D.M. Jacques, J. Kay, T. Rayment, A. Ryan, C.C. Tang, N.J. Terrill, J. Synchr. Rad. **11**, 163 (2004)
45. P. Abdala, O. Safonova, G. Wiker, W. Van Beek, H. Emerich, J. Van Bokhoven et al., Chimia Int. J. Chem. **66**, 699 (2012)
46. A. Berry, W.I. Helsby, B.T. Parker, C.J. Hall, P.A. Buksh, A. Hill, N. Clague, M. Hillon, G. Corbett, P. Clifford, A. Tidbury, R.A. Lewis, R.J. Cernik, P. Barnes, G.E. Derbyshire, Nucl. Instrum. Methods Phys. Res. A **513**, 260 (2003)
47. S.D.M. Jacques, O. Leynaud, D. Strusevich, A.M. Beale, G. Sankar, C.M. Martin, P. Barnes, Angew. Chem. Int. Ed. **45**, 445 (2006)
48. P.T. Hutchins, In Situ Synthesis Studies of Silicon Clathrates, Ph.D. thesis, University of London, London, 2007
49. A.C. Jupe, X. Turrillas, P. Barnes, S.L. Colston, C. Hall, D. Häusermann, M. Hanfland, Phys. Rev. B **53**, R14697 (1996)
50. A. Le Bail, Powder Diffr. **20**, 316 (2005)
51. H.M.J. Rietveld, Appl. Crystallogr. **2**, 65 (1969)
52. M. Johnson, D. O'Connor, P. Barnes, C.R.A. Catlow, S.L. Owens, G. Sankar, R. Bell, S.J. Teat, R. Stephenson, J. Phys. Chem. B **107**, 942 (2003)
53. G.S. Nolas, J.W. Sharp, H.J. Goldsmid, *Thermoelectrics: Basics Principles and New Materials Developments* (Springer, Berlin, 2001)
54. D.M. Rowe (ed.), *Thermoelectrics Handbook: Macro to Nano-Structured Materials* (CRC Press, Boca Raton, FL, 2005)
55. G.A. Slack, in *CRC Handbook of Thermoelectrics*, ed. by D.M. Rowe (CRC Press, Boca Raton, 1995), p. 407
56. J. Dong, O.F. Sankey, C.W. Myles, Phys. Rev. Lett. **86**, 2361 (2001)
57. T.A. Mary, J.S.O. Evans, T. Vogt, A.W. Sleight, Science **272**, 90 (1996)
58. D.A. Woodcock, P. Lightfoot, L.A. Villaescusa, M. Diaz-Cabanas, M.A. Camblorb, D. Engberg, Chem. Mater. **11**, 2508 (1999)
59. M.T. Dove, K.O. Trachenko, M.G. Tucker et al., Rev. Mineral. Geochem. **39**, 1 (2000)
60. V. Heine, P.R.L. Welche, M.T. Dove, J. Am. Ceram. Soc. **82**, 1793 (1999)
61. C.A. Swenson, J. Phys. Chem. Ref. Data **12**, 179 (1983)
62. T.H.K. Barron, J.G. Collins, G.K. White, Adv. Phys. **9**, 609 (1980)
63. G. Dolling, R.A. Cowley, Proc. Phys. Soc. **88**, 463 (1966)
64. J.A. Reissland, *The Physics of Phonons* (Wiley, Toronto, 1972)
65. S. Biernacki, M. SchefBer, Phys. Rev. Lett. **63**, 290 (1989)
66. X. Tang, J. Dong, P. Hutchins, O. Shebanova, J. Gryko, P. Barnes, J.K. Cockcroft, M. Vickers, P.F. McMillan, Phys. Rev. B **74**, 014109 (2006)
67. J. Dong, O.F. Sankey, G. Kern, Phys. Rev. B **60**, 950 (1999)
68. G.S. Nolas, M. Beekman, J. Gryko, G.A. Lamberton, T.M. Tritt, P.F. McMillan, Appl. Phys. Lett. **82**, 910 (2003)

69. A. San Miguel, P. Mélinon, D. Connétable, X. Blase, F. Tournus, E. Reny, S. Yamanaka, J.P. Itié., Phys. Rev. B **65**, 054109 (2002)
70. J.S. Tse, S. Desgreniers, Z.Q. Li, M.R. Ferguson, Y. Kawazoe, Phys. Rev. Lett. **89**, 195507 (2002)
71. T. Kume, T. Koda, S. Sasaki, H. Shimizu, J.S. Tse, Phys. Rev. B **70**, 052101 (2004)
72. T. Kume, H. Fukoaka, T. Koda, S. Sasaki, H. Shimizu, S. Yamanaka, Phys. Rev. Lett. **90**, 155503 (2003)
73. A. San Miguel, A. Merlen, P. Toulemonde, T. Kume, S. LeFloch, A. Aouizerat, S. Pascarelli, G. Aquilanti, O. Mathon, T. LeBihan, J.P. Itié, S. Yamanaka, Europhys. Lett. **69**, 556 (2005)
74. L. Yang, Y.M. Ma, T. Iitaka, J.S. Tse, K. Stahl, Y. Ohishi, Y. Wang, R.W. Zhang, J.F. Liu, H.-K. Mao, J.Z. Jiang, Phys. Rev. B **74**, 245209 (2006)
75. J.S. Tse, R. Flacau, S. Desgreniers, T. Iitaka, J.Z. Jiang, Phys. Rev. B **76**, 174109 (2007)
76. J.S. Tse, L. Yang, S.J. Zhang, C.Q. Jin, ChJ Sahle, C. Sternemann, A. Nyrow, V. Giordano, J.Z. Jiang, S. Yamanaka, S. Desgreniers, C.A. Tulk, Phys. Rev B **84**, 184105 (2011)
77. H. Shimizu, T. Kume, T. Kuroda, S. Sasaki, H. Fukuoka, S. Yamanaka, Phys. Rev. B **68**, 212102 (2003)
78. D. Machon, P. Toulemonde, P.F. McMillan, M. Amboage, A. Muñoz, P. Rodríguez-Hernández, A. San Miguel, Phys. Rev. B **79**, 184101 (2009)
79. J.S. Tse, R. Flacau, S. Desgreniers, T. Iitaka, J.Z. Jiang, Phys. Rev. B **76**, 174109 (2007)
80. T. Iitaka, Phys. Rev. B **75**, 012106 (2007)
81. P. Tolédano, V.P. Dmitriev, *Reconstructive Phase Transitions* (World Scientific, Singapore, 1996)
82. R. Debord, D. Machon, S. Pailhes, V. Pischedda, M. Hanfland, A. San-Miguel (unpublished)
83. D. Machon, F. Meersman, M. Wilding, M. Wilson, P.F. McMillan, Progr. Mater. Sci. **61**, 216 (2014)
84. E. Reny, A. San Miguel, Y. Guyot, B. Masenelli, P. Mélinon, L. Saviot, S. Yamanaka, B. Champagnon, C. Cros, M. Pouchard, M. Borowski, A.J. Dianoux, Phys. Rev. B **66**, 014532 (2002)
85. S.L. Fang, L. Grigorian, P.C. Eklund, G. Dresselhaus, M.S. Dresselhaus, H. Kawaji, S. Yamanaka, Phys. Rev. B **57**, 7686 (1998)
86. G.S. Nolas, C.A. Kendziora, J. Gryko, C.W. Myles, A. Poddar, O.F. Sankey, J. Appl. Phys. **92**, 7225 (2002)
87. J.S. Tse, T. Iitaka, T. Kume, H. Shimizu, K. Parlinski, H. Fukuoka, S. Yamanaka, Phys. Rev. B **72**, 155441 (2005)
88. G.S. Nolas, T.J.R. Weakley, J.L. Cohn, R. Sharma, Phys. Rev. B **61**, 3845 (2000)
89. S.K. Deb, M.C. Wilding, M. Somayazulu, P.F. McMillan, Nature **414**, 528 (2001)
90. P.F. McMillan, M. Wilson, D. Daisenberger, D. Machon, Nat. Mater. **4**, 680 (2005)
91. F.C. Weinstein, E.A. Davis, J. Non-Cryst. Solids **13**, 153 (1973)
92. G. Benedek, E. Galvani, S. Sanguinetti, Nuovo Cimento D **17**, 97 (1995)
93. R.J. Angel, Equations of State, in *High-Pressure, High-temperature Crystal Chemistry* ed. by R.M. Hazen, R.T. Downs. Reviews in Mineralogy and Geochemistry, 41 (Mineralogical Society of America and the Geochemical Society, Washington, D.C., 2000), pp. 35–60
94. P. Toulemonde, D. Machon, M. Tucker, A. San Miguel (unpublished)
95. A. San-Miguel, P. Kéghélian, X. Blase, P. Mélinon, A. Perez, J.P. Itié, A. Polian, E. Reny, C. Cros, M. Pouchard, Phys. Rev. Lett. **83**, 5290 (1999)
96. G.K. Ramachandran, P.F. McMillan, S.K. Deb, M. Somayazulu, J. Gryko, J.J. Dong, O.F. Sankey, J. Phys. Condens. Matter **12**, 4013 (2000)
97. H. Olijnyk, Phys. Rev. B **46**, 6589 (1992)
98. R.H. Wentorf, J.S. Kasper, Science **139**, 338 (1963)
99. J.M. Besson, E.H. Mokhtari, J. Gonzalez, G. Weill, Phys. Rev. Lett. **59**, 473 (1987)
100. R.J. Kobliskat, S.A. Sojin, Phys. Rev. B **8**, 3799 (1973)

101. G. Weill, J.L. Mansot, G. Sagon, C. Carlone, J.M. Besson, Semicond. Sci. Technol. **4**, 280 (1989)
102. F.M. Grosche, H.Q. Yuan, W. Carrillo-Cabrera, S. Paschen, C. Langhammer, F. Kromer, G. Sparn, M. Baenitz, Yu. Grin, F. Steglich, Phys. Rev. Lett. **87**, 247003 (2001)
103. H.Q. Yuan, F.M. Grosche, W. Carrillo-Cabrera, S. Paschen, G. Sparn, M. Baenitz, Yu. Grin, F. Steglich, J. Phys. Condens. Matter **14**, 11249 (2002)
104. H.Q. Yuan, F.M. Grosche, W. Carrillo-Cabrera, V. Pacheco, G. Sparn, M. Baenitz, U. Schwarz, Yu. Grin, F. Steglich, Phys. Rev. B **70**, 174512 (2004)
105. H. Shimizu, T. Fukushima, T. Kume, S. Sasaki, H. Fukuoka, S. Yamanaka, J. Appl. Phys. **101**, 113531 (2007)
106. T. Rachi et al., Phys. Rev. B **72**, 144504 (2005)
107. H. Shimizu, T. Kume, T. Kuroda, S. Sasaki, H. Fukuoka, S. Yamanaka, Phys. Rev. B **71**, 094108 (2005)
108. P. Toulemonde, D. Machon, A. San Miguel, M. Amboage, Phys. Rev. B **83**, 134110 (2011)
109. S. Saito, A. Oshiyama, Phys. Rev. B **51**, 2628 (1995)
110. D. Connétable, V. Timoshevskii, B. Masenelli, J. Beille, J. Marcus, B. Barbara, A.M. Saitta, G.-M. Rignanese, P. Mélinon, S. Yamanaka, X. Blase, Phys. Rev. Lett. **91**, 247001 (2003)
111. M. Bernasconi, S. Gaito, G. Benedek, Phys. Rev. B **61**, 12689 (2000)
112. G. Benedek, E. Galvani, S. Sanguinetti, S. Serra, Chem. Phys. Lett. **244**, 339 (1995)
113. N. Rey, A. Múñoz, P. Rodríguez-Hernández, A. San-Miguel, J. Phys. Condens. Matter **20**, 215218 (2008)
114. P. Mélinon, A. San Miguel, in *Clusters and Fullerenes—The Handbook of Nanophysics*, ed. by K. Sattler (Taylor & Francis, Boca Raton)
115. I. Spagnolatti, M. Bernasconi, G. Benedek, Eur. Phys. J. B **34**, 63 (2003)
116. S. Yamanaka, A. Kubo, K. Inumaru, K. Komaguchi, N.S. Kini, T. Inoue, T. Irifune, Phys. Rev. Lett. **96**, 076602 (2006)

# Chapter 5
# Chemistry and Physics of Inverse (Cationic) Clathrates and Tin Anionic Clathrates

**Andrei V. Shevelkov, Kirill A. Kovnir and Julia V. Zaikina**

**Abstract** Clathrates are a family of cage compounds featuring complete sequestering of guests inside polyhedral cages formed by host frameworks. Compounds of this type receive rapt attention because it is expected that they can be a base for creating thermoelectric materials for heat-to-power conversion. This chapter is primarily devoted to a group of clathrates which frameworks are built on group 14 atoms and carry a positive charge compensated by guest halide or chalcogenide anions. It summarizes peculiarities of their crystal and electronic structure and transport properties and outlines ways of optimization of their thermoelectric performance. Tin clathrates with anionic frameworks are also discussed.

**Keywords** Clathrates · Cage compounds · Crystal structure · Thermoelectric materials · Tin · Germanium · Silicon · Pnicogens

## 5.1 Introduction

As early as 1965 Kasper et al. showed that the products of the decomposition of sodium monosilicides in vacuum, $Na_8Si_{46}$ and $Na_xSi_{136}$, possess the structures already known as type-I and type-II clathrates [1, 2]. The latter were presented by hydrates of various gases and liquids and had very little in common with the newly

A. V. Shevelkov (✉)
Department of Chemistry, Lomonosov Moscow State University, Moscow, Russia
e-mail: shev@inorg.chem.msu.ru

K. A. Kovnir · J. V. Zaikina
Department of Chemistry, University of California, Davis, Davis, CA, USA
e-mail: kkovnir@ucdavis.edu

J. V. Zaikina
e-mail: yzaikina@ucdavis.edu

G. S. Nolas (ed.), *The Physics and Chemistry of Inorganic Clathrates*,
Springer Series in Materials Science 199, DOI: 10.1007/978-94-017-9127-4_5,

discovered cage-like sodium silicides save for the type of the crystal structure. This discovery was the first indication of the fact that the similarity of the crystal structures of cage-like compounds rests on host-guest complementarity and largely ignores the nature of the atoms involved. Moreover, compounds of the same structure displayed different chemical and physical properties. Clathrate-hydrates were white, transparent and hardly stable at room temperature whereas Na–Si clathrates were found to be air-stable black solids with metal-like electrical conductivity [3, 4].

In 1972, von Schnering and Menke reported first clathrates with the host-guest polarity reversed compared to sodium silicides [5]. In $Ge_{38}P_8I_8$ and $Ge_{38}As_8I_8$ the framework composed of germanium and pnicogen atoms carry a positive charge compensated by the iodide guests. 27 years after the discovery of von Schnering the first tin-based cationic clathrate was obtained [6]; since then the family of the cationic clathrates has grown rapidly and now includes over 40 representatives, the majority of which are the tin-based derivatives.

Many clathrate compounds attract attention as potential thermoelectric materials of new generation [7]. They conform to the Slack's concept of "phonon glass—electron crystal (PGEC)" [8], having a structure with a spatial separation of host and guest substructures and thus providing possibilities for almost independent tuning of charge carrier transport through the host framework and of heat transport through pseudo localized vibrations of guests. The same property concerns the cationic clathrates and tin-based anionic clathrates treated in this chapter which summarizes details of their crystal and electronic structures and transport properties and addresses the issues of their thermoelectric potential.

## 5.2 Crystal Structure of Cationic Clathrates

### 5.2.1 *Proper Type-I Structure*

Despite the variety of the clathrates crystal structures, the majority of cationic clathrates belong to the most widespread clathrate-I structure type. This structure is described in the space group $Pm\bar{3}n$ with just five atomic positions. Three of them, $24k$, $16i$, and $6c$, belong to the host framework whereas the $2a$ and $6d$ positions are occupied by the guest atoms sitting inside the polyhedral cages formed in the framework (Fig. 5.1). The guest positions are occupied by halogen or chalcogen anions, while the framework is based on the group 14 atoms (silicon, germanium, or tin) always with addition of the group 15 (phosphorus, arsenic, antimony) or group 16 or 17 (tellurium, iodine) elements. The $2a$ position centers the smaller 20-vertex pentagonal dodecahedron and the $6d$ position centers the bigger 24-vertex tetrakaidecahedron. The bigger polyhedra form a three-dimensional framework, sharing their hexagonal faces, whereas the smaller polyhedra do not touch each other but share pentagonal faces with bigger polyhedra. Such a structure can be named the proper structure of type-I clathrates. However, only a limited number of cationic

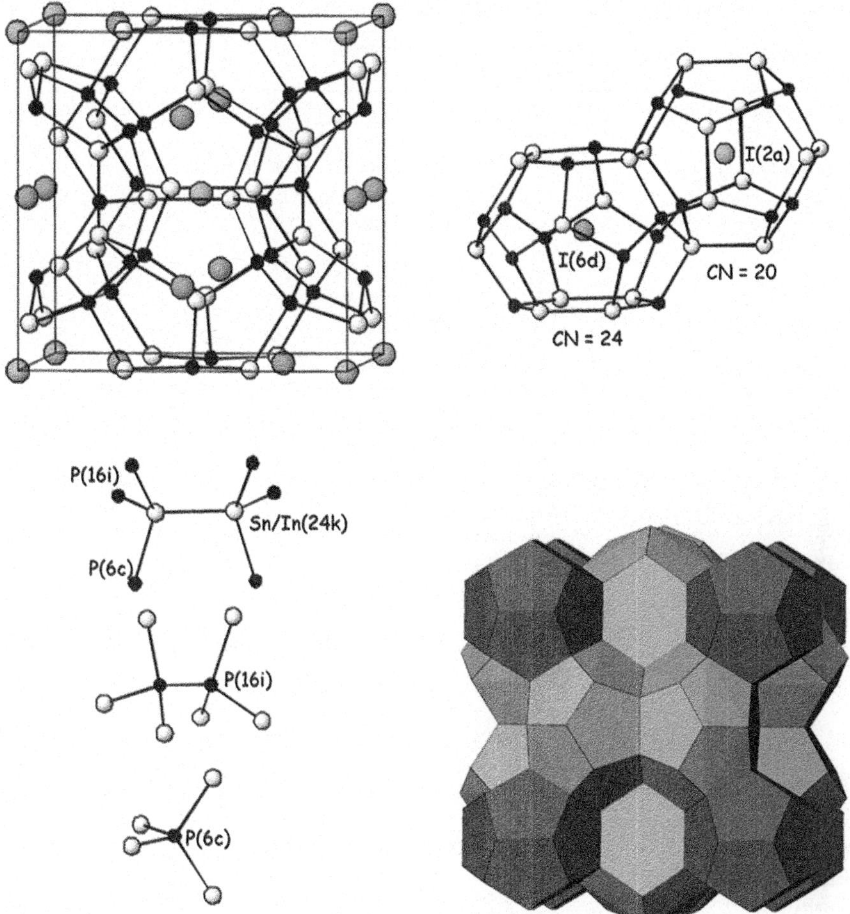


**Fig. 5.1** Crystal structure of $Sn_{10}In_{14}P_{22}I_8$: (clockwise from *top left*) a view of the unit cell; two adjacent cages with guest atoms inside; polyhedral presentation; coordination of the framework atoms

clathrates adopt it without introducing various irregularities (Table 5.1). For instance, $Sn_{10}In_{14}P_{22}I_8$ [9] displays the proper clathrate-I crystal structure, although having dissimilar distribution of atoms over the framework positions. In its crystal structure, tin and indium atoms jointly occupy the 24*k* position leaving the 16*i* and 6*c* sites for phosphorus. In $Ge_{30}P_{16}Te_8$ [10] germanium and phosphorus atoms jointly occupy all three positions within the framework with some preference of phosphorus to the 16*i* site. $Sn_{38}Sb_8I_8$ is believed to have an even distribution of tin and antimony atoms over all three framework positions [11]. However, taking into account a one electron difference between Sn and Sb, neighbors in the periodic table, the Sn/Sb distribution over the framework positions based on X-ray data looks rather ambiguous.

**Table 5.1** Structural data for the cationic clathrates

| Composition | Unit cell parameters, Å | Remarks[a] |
|---|---|---|
| *Proper type-I structures* | | |
| $Si_{44.5}I_{1.5}I_8$ [12] | $a = 10.4195$ | Si and I mix at 16*i* site |
| $Si_{30}P_{16}Te_8$ [13] | $a = 9.9646$ | Si and P mix at all framework sites |
| $Si_{30}P_{16}Se_xTe_{8-x}$ [14] | $a = 9.929$ ($x = 1.53$) $a = 9.938$ ($x = 1.26$) $a = 9.9411$ ($x = 1.06$) | Si and P mix at all framework sites, Se prefers 2*a* site |
| $Ge_{43.33}I_{2.67}I_8$ [15] | $a = 10.814$ | Ge and I mix in 6*c* |
| $Ge_{38}Sb_8I_8$ [11] | $a = 10.8892$ | Ge and Sb mix at all framework sites |
| $Ge_{30}P_{16}Te_8$ [10] | $a = 10.3376$ | Uneven distribution of Ge and Sb within the framework sites |
| $Sn_{10}In_{14}P_{22}I_8$ [9] | $a = 11.0450$ | Sn and In mix at 24*k* site |
| $Sn_{38}Sb_8I_8$ [11] | $a = 12.0447$ | Even distribution of Sn and Sb within the framework sites |
| *Irregular type-I structures, no superstructures* | | |
| $Si_{40}P_6I_{6.5}\square_{1.5}$ [16] | $a = 10.1293$ | Vacancies at 2*a* |
| $Si_{38}Te_8Te_8$ [17, 18] | $a = 10.456$ | Space group $P\bar{4}3n$, complete separation of Si and Te over framework positions |
| $Si_{31.3}P_{14.7}Te_{7.35}$ [19] | $a = 9.9702$ | Space group $Pm\bar{3}$, partial ordering of guest position vacancies; preferential occupation of one of the positions by P, other positions are mixed occupied by Si/P or Si solely |
| $Si_{32.4}P_{13.6}Te_{6.98}$ [19] | $a = 9.9794$ | Same as above |
| $Si_{33.0}P_{13.0}Te_{6.88}$ [19] | $a = 9.9808$ | Same as above |
| $Si_{31.5}P_{14.5}Te_{6.0}Br_{1.5}$ [14] | $a = 9.9863$ | Space group $Pm\bar{3}$, vacancies in smaller cages |
| $Si_{35}P_{11}Te_{2.27}Br_{5.04}$ [14] | $a = 9.9808$ | Vacancies at 2*a* |
| $Ge_{38}P_8I_8$ [5, 20] | $a = 10.5067$ | Space group $P\bar{4}3n$, segregation of P atoms in one eightfold position |
| $Ge_{38}P_8Br_8$ [20] | $a = 10.4074$ | Same as above |
| $Ge_{38}P_8I_2Br_6$ [20] | $a = 10.430$ | Same as above |
| $Ge_{38}As_8I_8$ [5, 20] | $a = 10.6158$ | Space group $P\bar{4}3n$, segregation of As atoms in one of eightfold position |
| $Ge_{32}Ga_3As_{11}I_8$ [21] | $a = 10.616$ | Space group $P\bar{4}3n$, Ge, Ga and As mix in 3 framework positions |
| $Ge_{22}Ga_8Sb_{16}I_8$ [21] | $a = 11.273$ | Space group $P\bar{4}3n$, Ge, Ga and Sb mix in 3 framework positions |
| $Ge_{30.5}Sn_{7.7}P_{7.75}I_{7.88}$ [22] | $a = 10.7210$ | Vacancies at 2*a* site, Sn mixes with Ge in 24*k* |
| $Ge_{40.0}Te_{5.3}I_8$ [23] | $a = 10.815$ | Uneven distribution of Ge and Te within the framework sites; bulk material is not uniform containing domains with segregation of Te and Ge within the fourfold positions (S.G. *P*23) |
| $Sn_{19.3}Cu_{4.7}P_{22}I_8$ [24] | $a = 10.852$ | 24*k* site splits into 2 |
| $Sn_{20}Cu_4As_{22}Br_8$ [25] | $a = 11.068$ | 24*k* site splits into 2 |
| $Sn_{24}Cu_{4.7}As_{22}I_8$ [26] | $a = 11.1736$ | 24*k* site splits into 3 |

(continued)

**Table 5.1** (continued)

| Composition | Unit cell parameters, Å | Remarks[a] |
|---|---|---|
| $Sn_{17}Zn_7P_{22}Br_8$ [27] | $a = 10.7449$ | 24*k* site splits into 3 |
| $Sn_{17}Zn_7P_{22}I_8$ [27] | $a = 10.8458$ | 24*k* site splits into 3 |
| $Sn_{20}Zn_4P_{20.8}I_8$ [27] | $a = 10.883$ | Splitting and vacancies coexist |
| $Sn_{17}Zn_7As_{22}I_8$ [28] | $a = 11.157$ | 24*k* site splits into 3 |
| $Sn_{17}Cd_7As_{22}I_8$ [28] | $a = 11.064$ | Only unit cell parameters determined |
| $Sn_{20}Ga_4P_{20}I_8$ [28] | $a = 10.949$ | Only unit cell parameters determined |
| $Sn_{10}Ga_{14}As_{22}I_8$ [28] | $a = 11.083$ | Only unit cell parameters determined |
| $Sn_{24-x}In_xAs_{19.2+x/5}I_8$ [29] | $a = 11.1680$ ($x = 4.1$), $a = 11.1819$ ($x = 4.8$), $a = 11.1970$ ($x = 5.0$), $a = 11.2590$ ($x = 6.9$), $a = 11.3480$ ($x = 9.5$) | Complex composition dependent pattern of splitting and vacancy formation |
| $Sn_{15.6(9)}In_{7.9(8)}P_{21.50(3)}Br_{8-}$ [30] | $a = 10.916$ | 24*k* site splits into 3, vacancy forms in 6*c* site |
| $Sn_{24}P_{19.3}I_8$ [6] | $a = 10.9540$ | 24*k* site splits into 2, vacancy forms in 6*c* site |
| $Sn_{24}P_{19.6}Br_8$ [31] | $a = 10.8142$ | 24*k* site splits into 2, vacancy forms in 6*c* site |
| $Sn_{24}P_{19.3}Br_xI_{8-x}$ [31] | $a = 10.9200$ ($x = 2.35$), $a = 10.9110$ ($x = 3.14$), $a = 10.8860$ ($x = 4.62$), $a = 10.8440$ ($x = 6.1$) | 24*k* site splits into 2, vacancy forms in 6c site, Br prefers 2*a* sites |
| $Sn_{24}P_{19.3}Cl_xI_{8-x}$ [32] | $a = 10.948$ ($x = 0.25$), $a = 10.9408$ ($x = 0.5$), $a = 10.9331$ ($x = 0.8$) | 24*k* site splits into two, vacancy forms in 6*c* site, Cl sits only on 2*a* sites |
| *Type-I superstructures* | | |
| $Sn_{14}In_{10}P_{21.2}I_8$ [9] | $a = 24.745$, $c = 11.067$ | Space group $P4_2/m$ Partial ordering of vacancies in P positions |
| $Sn_{20.5}As_{22}I_8$ [33] | $a = 22.1837$ | Space group *F*23 Partial ordering of vacancies in Sn positions |
| $Ge_{30-x}P_xTe_y$ [34]($x = 13.9–15.6$; $y = 5.92–7.75$) | $a = 20.544–20.698$ | Space group $Fm\bar{3}$, semiclathrate |
| $Ge_{30.5}P_{15.5}Se_8$ [35] | $a = 20.406$ | Space group $Fm\bar{3}$, semiclathrate |

(continued)

**Table 5.1** (continued)

| Composition | Unit cell parameters, Å | Remarks[a] |
|---|---|---|
| $Ge_{30.7}P_{15.3}Se_{5.38}$ [35] | $a = 20.310$ | Space group $Fm\bar{3}$, semiclathrate, vacancies in 3 out of 4 Se positions |
| *Other types* | | |
| $Si_{130}P_{42}Te_{21}$ [36] | $a = 19.2632$, $c = 10.0706$ | Space group $P4_2/mnm$ type-III |
| $Ge_{130}P_{42}Te_{21}$ [37] | $a = 19.948$, $c = 10.440$ | Space group $P4_2/mnm$ type-III |
| $Si_{17.5}Te_{9.5}$ [38] | $a = 21.136$ | Space group $Fd\bar{3}c$ Unique clathrate-like structure |
| $Ge_{79}P_{29}S_{18}Te_6$ [39] | $a = 19.2632$, $c = 10.0706$ | Space group $R\bar{3}m$, type-X |

## 5.2.2 *Irregular Type-I Structures*

All compounds of the $Ge_{38}E_8X_8$ family (E = P, As, Sb; X = Cl, Br, I) were reported [5, 20] to crystallize in the acentric space group $P\bar{4}3n$ owing to a complete segregation of the phosphorus atoms in one of two positions formed from the 16*i* position upon transformation from the space group $Pm\bar{3}n$. This finding signals that the ideal structure of a cationic clathrate can be lost in favor of fulfilling coordination demands of particular atoms composing the clathrate framework. Interestingly, variation of the synthetic conditions, e.g. different synthesis temperature and absence of the temperature gradient leads to the variation in composition and, consequently, in different unit cell parameter for clathrate phases. Vacancies in iodine positions have been reported together with mixing of Ge/Sb over all positions of the framework [22]. More detailed investigations to relate synthetic conditions, stoichiometry ranges and structure of Ge–Pn–X clathrates are necessary.

In many cases the space group $Pm\bar{3}n$ is preserved despite of a noticeable variation of the ideal type-I crystal structure. A combination of metal atoms with significantly different atomic radii within the clathrate framework causes splitting of the 24*k* position in various tin-based cationic clathrates. For instance, in the crystal structure of $Sn_{17}Zn_7P_{22}I_8$ the tin and zinc atoms should occupy the same 24-fold position of the framework to keep the proper clathrate-I crystal structure [27]. This would lead to the formation of the Zn–Sn and Sn–Sn bonds of the same interatomic distance, which is unreasonable because in intermetallic compounds these distances may differ by more than 0.2 Å. As a consequence, the 24*k* position splits into three closely lying positions, of which one tin position is responsible for forming the Sn–Sn bonds whereas other two positions are occupied by tin and zinc forming the Zn–Sn bonds. A similar type of 24*k* site splitting without alternation of the space group $Pm\bar{3}n$ is observed in some other crystals structures, for instance, $Sn_{19.3}Cu_{4.7}As_{22}I_8$ (Fig. 5.2) [26].

$Sn_{24}P_{19.3}I_8$ has vacancies within the host framework, as stressed by the formulation $Sn_{24}P_{19.3}\square_{2.7}I_8$, where □ denotes a vacancy [6]. This causes the split of the 24*k* position into two closely-lying ones to satisfy coordination demands of tin. Two new positions, Sn1 and Sn2, sit such that if the phosphorus atom is present in the 6c position it forms a single Sn1–P bond, but if there is a vacancy in the 6c position it is neighbored by the Sn2 atom. Consequently, the environment of two tin atoms is different. The Sn1 atom has three phosphorus neighbors and forms one Sn1–Sn2 bond, whereas the Sn2 atom forms two short bonds with the phosphorus atoms, one bond to Sn1 and three much longer bonds (d $\approx$ 3.3 Å) with other Sn2 atoms (the so-called 3 + 3 coordination). In this way each vacancy in the 6*c* position is surrounded by four Sn2 atoms (Fig. 5.2).

A combination of the site splitting due to dissimilar coordination of metal atoms and vacancy formation is possible and actually observed in several tin-based cationic clathrates. Probably the most interesting case is presented by the clathrates with the general formula $Sn_{24-x}In_xAs_{19.2+x/5}I_8$ [29]. With the indium concentration varying from $x = 4.1$ to $x = 9.5$ the details of the crystal structure change twice within the space group $Pm\bar{3}n$. For $x = 9.5$ the splitting of the 24*k* position into three partially filled (four-bonded Sn and In plus three-bonded Sn) ones coexists with the vacancy formation at the 6*c* site. Upon lowering the In content to $x = 5$–7 the vacancy formation is accompanying by the shift of the As2 from the ideal 6*c* to the 12*g* site with 50 % occupancy. A further decrease of the In content leads to the increase of the degree of disorder in the clathrate crystal structure: atom As2 is shifted to the 24*k* site with a concomitant displacement of the As1 atom from the ideal 16*i* position towards the 48*l* site with the partial occupancy of 33 %.

Noticeably, the tin-based cationic clathrates never possess vacancies in the guest positions, which is a frequent feature in silicon and germanium-based anionic and cationic clathrates [7]. For instance, in the crystal structure of $Si_{40}P_6I_{6.5}$ the smaller cage is only 25 % occupied by the iodine atoms, which, however, does not lead to altering the space group $Pm\bar{3}n$ [16]. On contrary, the existence of vacancies in the guest positions of the crystal structure of $Si_{46-x}P_xTe_y$ ($y \leq 8$) is associated with the uneven distribution of the phosphorus and silicon atoms within the framework, altogether leading to lowering the space group symmetry down to $Pm\bar{3}$ [40]. The resulting crystal structure features two independent tellurium positions inside the 20-vertex polyhedra. The one preferably surrounded by silicon is 100 % occupied by tellurium; the remaining position is only partially filled by tellurium and has much more phosphorus atoms around it.

Distribution of atoms of different nature over the framework positions deserves more discussion. It is noticed that the 16*i* site of the space group $Pm\bar{3}n$ is always occupied by atoms of the most electronegative elements—phosphorus or arsenic, whereas the most electropositive elements prefer the 24*k* site. The latter position is always occupied by various metal atoms, including transition metals copper and zinc and *p*-metals, such as tin and indium. Another important feature is that vacancies form at the 6*c* or 24*k* sites but never at the 16*i* sites. This pattern is remarkably different from that observed for anionic clathrates. In them, electropositive substituents, including group 13 and some transition metals, either exclusively sit on the 6*c* site or also occupy the 24*k* site but always avoid the

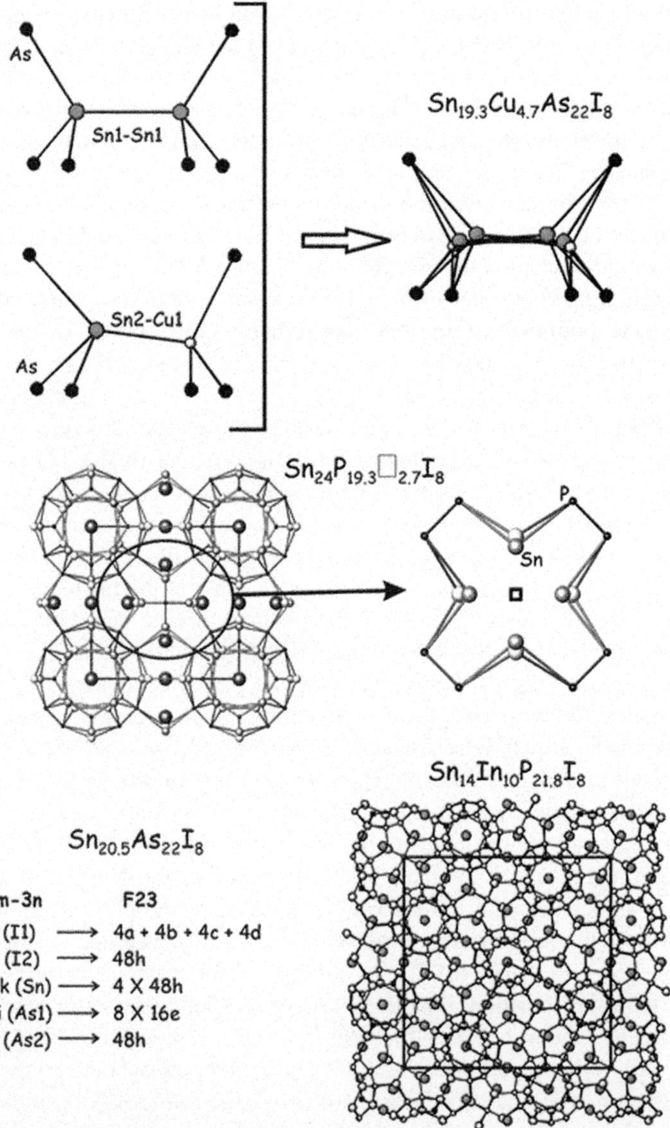


**Fig. 5.2** *Top* Splitting of the 24 k position in the crystal structure of $Sn_{19.3}Cu_{4.7}P_{22}I_8$ and coordination of the metal atoms. *middle* Environment of a vacancy in the crystal structure of $Sn_{24}P_{19.3}I_8$. *bottom left* Wickoff positions transformation upon transition from the space group $Pm\bar{3}n$ to *F*23; *bottom right* Choice of a unit cell in the structure of $Sn_{14}In_{10}P_{21.2}I_8$ in comparison with the cubic unit cell. *Dark grey*, iodine; *open circles*, framework atoms; *black*, 50 % occupied positions of the phosphorus atoms

16*i* site [41, 42]. Contrastingly, relatively electronegative antimony prefers the 24*k* site but avoids the 6c site [43]. Probably, the recently reported $Ae_8Al_{10}Si_{36}$ (Ae = Sr, Ba) clathrates provide exceptions to these rules [44].

Distribution of guest atoms of different nature is also element specific. In the tin-containing cationic clathrates only halide atoms serve as guests; partial substitution in the anionic substructure is possible. In $Sn_{24}P_{19.3}I_8$ iodine can be replaced by bromine to form the unlimited solid solution, whereas substitution of chlorine for iodine is possible only up to $Sn_{24}P_{19.3}Cl_{0.8}I_{7.2}$ [32]. Chlorine occupies only the 2*a* position but the distribution of iodine and bromine is more complex. Clearly, bromine shows some preference for the 2*a* site with the smaller effective diameter, however, even in $Sn_{24}P_{19.3}Br_{2.35}I_{5.65}$ both 2*a* and 6*d* sites exhibit joint occupation by the Br and I guest atoms. In a similar way, the crystal structure of $Si_{30+x}P_{16-x}Te_{8-x}Br_x$ shows some preference of bromine atoms to occupy the centers of the smaller cages; however, only for $x \geq 5$ tellurium is no longer present in the smaller cages [14]. Such a behavior is not typical for anionic clathrates. For example, $Eu_2Ba_6Ga_8Si_{36}$ features a complete separation of europium and barium occupying, respectively, the 2*a* and 6*d* positions aloof [45]. Probably, this reflects that the framework is much more rigid in anionic than in cationic clathrates.

### 5.2.3 Superstructures of Type-I Clathrate

Five different types of superstructure that lead to an increase in the unit cell volume have been documented for the type-I clathrates in the literature [9, 33, 46–49]. Two of these types were observed for the cationic clathrates.

The crystal structure of $Sn_{20.5}As_{22}I_8$ presents a very complicated and disordered variant of the type-I clathrate superstructure that appeared as a result of partial ordering of vacancies in positions of the tin atoms and a non-concerted shift of arsenic and tin atoms away from their proper positions [33]. The main feature of this crystal structure is a random ordering of vacancies, resulting in a doubling of the unit cell parameter of the cubic unit cell (Fig. 5.2). It was shown that only four variants of vacancy ordering would be possible for the hypothetical composition $Sn_{21}As_{22}I_8$, which is very close to the actual composition $Sn_{20.5}As_{22}I_8$. Each of four models requires that four out of eight independent positions of the tin atoms of the *F*23 space group be 50 % occupied. However, the difference between the four models is very small and there is no surprise that they coexist in the crystal structure. Moreover, the total amount of vacancies and, consequently, the amount of three-coordinated tin atoms is slightly different for the experimentally determined and valence-precise composition $Sn_{20.5}As_{22}I_8$. These "excess" vacancies in the tin atom positions may be an additional factor facilitating the intergrowth of differently ordered domains. The HRTEM study of this compound showed that though the areas of distinct ordering can be observed, their overlap is a more common picture, which leads to an averaged crystal structure determined from the X-ray single crystal experiment.

Less complicated superstructure is observed in $Sn_{14}In_{10}P_{21.2}I_8$, where a fivefold enlargement of the unit cell appears due to the transition from the space group $Pm\bar{3}n$ to $P4_2/m$: $a\sqrt{5} \times a\sqrt{5} \times a$ [9]. The origin of the superstructure formation is the partial ordering of vacancies in the positions of the phosphorus atoms (Fig. 5.2). The ordering is incomplete, and two out of 17 phosphorus atom positions are 50 % occupied. In general, this provides separation of tin atoms with two distinct coordination modes in 20 positions of the unit cell.

#### 5.2.3.1 Semiclathrates

A remarkable variation of the type-I clathrate superstructure is associated with the formation of semiclathrates in the Ge–P–Q (Q = Se, Te) systems [34, 35]. The term "semiclathrate" was first introduced for hydrates of quaternary ammonium salts, which show general structural and physical properties of normal clathrates but in addition feature a single hydrogen bond between host and guest substructures [50]. Up to now quite a number of semiclathrates-hydrates have been reported in the literature [51] compared to a very limited number of inorganic semiclathrates, which are exclusively cationic semiclathrates.

Compounds with the general formula $Ge_{46-x}P_xQ_{8-y}(13.9 \leq x \leq 15.6; y \leq 2.65;$ Q = Se, Te) crystallize in the cubic space group $Fm\bar{3}$ with the unit cell parameter varying from 20.31 to 20.70 Å depending on the nature of the chalcogen and the values of the coefficients $x$ and $y$ [34, 35]. Their crystal structures can be considered as a peculiar deviation from the crystal structure of a proper type-I clathrate formed as a result of a concerted rotation of atomic pairs (dumbbells) within a framework, which keeps the cubic symmetry intact. Such a rotation breaks some bonds within the framework and causes formation of bonds between the guest in the larger cage and a framework atom, distorting the larger cage such that it is no longer can be described by a tetrakaidecahedron. As a consequence, two features not typical for proper clathrates emerge. They are the existence of three-bonded phosphorus atoms not associated with vacancies in the framework and a single covalent bond between the guest chalcogen atom and the framework germanium atom (Fig. 5.3). This bond is quite short, d(Ge–Te) = 2.44–2.52 Å and d(Ge–Se) = 2.24–2.33 Å, and makes part of guest chalcogen atoms one-coordinate anions. The three-bonded framework atoms can be considered as forming three 2c–2e bonds and having a lone electron pair. The lone pair of atoms arrange in such a way that the doubling of the unit cell is necessary to describe the symmetry of the crystal structure. Consequently, the unit cell of the semiclathrate can be seen as corresponding to a 2 × 2 × 2 superstructure of the clathrate-I type.

The crystal structure of the semiclathrate shows definite changes upon varying the chalcogen content in $Ge_{46-x}P_xQ_{8-y}$ from $y = 2.65$ to $y = 0$. For $y = 0$ all positions of the chalcogen atom that forms a single covalent bond with a framework atom are occupied. Upon decreasing the chalcogen content this position is becoming less populated (Fig. 5.4). As a result, the neighboring germanium atom becomes

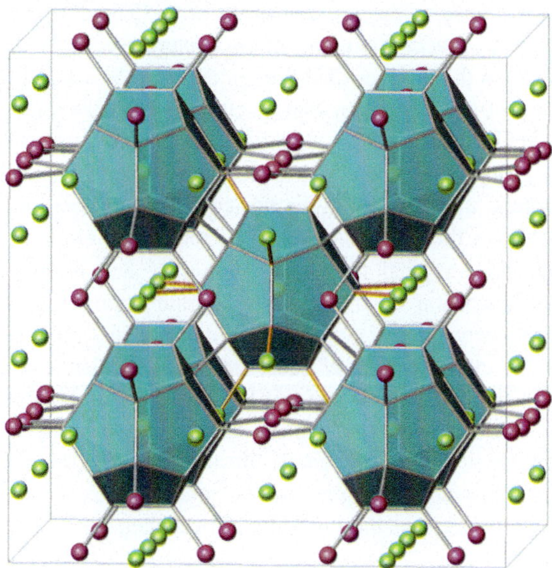

**Fig. 5.3** (color online) Crystal structure of semiclathrate-I. A unit cell content is shown. Te-filled dodecahedra, *cyan*; tellurium, *green*; germanium/phosphorus, *magenta*; bonds within the framework, *gray*; host-guest bonds, *red*

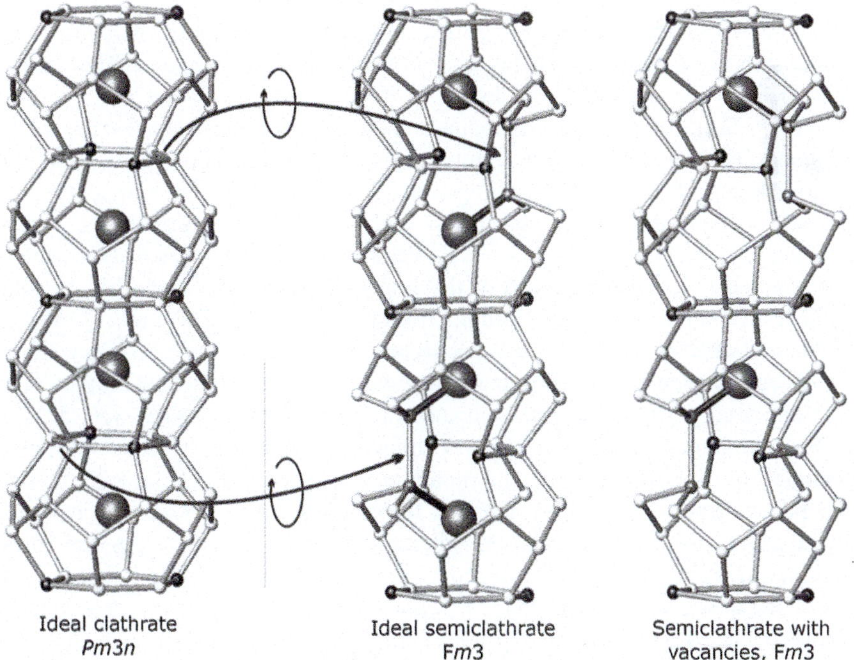


**Fig. 5.4** Transformation of the crystal structure from ideal clathrate-I to semiclathrate. Reprinted with permission from *Inorg. Chem.* 52 (2013) 577–588. Copyright 2013 American Chemical Society

three-bonded in response to the missing guest atom. Therefore, the homogeneity range for $Ge_{46-x}P_xQ_{8-y}$ is accompanied by a change in the number of vacancies in the guest positions and in the number of three-bonded framework atoms.

Two polymorphs of $Si_{38}Te_{16}$ can also be classified as semiclathrates accounting for rather short Si–Te host-guest bonds [17]. This compound crystallizes in either the cubic space group $P\bar{4}3n$ or the rhombohedral space group $R3c$ with similar unit cell parameters, $a = 10.457$ Å for the cubic polymorph and $a = 10.465$ Å and $\alpha = 89.88$ deg for the rhombohedral one. Both space groups are subgroups of the space group $Pm\bar{3}n$ typical for type-I clathrates, and both crystal structures show substantial distortion of the type-I clathrate framework, which includes formation of bonds between the guest tellurium atoms and the host silicon atoms. Interestingly, in the cubic form only guest atoms in the distorted 24-vertex cages are involved in such bonding, whereas in the rhombohedral polymorph all guest atoms form one bond of this type. Unlike the crystal structure of $Ge_{46-x}P_xQ_{8-y}$, these compounds show rather long Si–Te host-guest bonds ranging from 2.63 to 2.71 Å, which is significantly longer than a typical Si–Te covalent bond (2.5 Å), especially taking into account the coordination number 1 for tellurium.

### *5.2.4 Silicon and Germanium Clathrates with Other Structure Types*

Structure types other than type-I are exceptions for the cationic clathrates. Two compounds crystallizing with the type-III crystal structures are known. They can be described by a general formula $Tt_{172-x}P_xTe_{20+y}$ (Tt = Si or Ge; $x = 2y$; $20 \leq y \leq 22$) [36, 37]. These compounds crystallize in the tetragonal space group $P4_2/mnm$. Their crystal structure features three types of polyhedral cages composed of 20, 24, and 26 atoms within the vacancy-free framework. The respective pentagonal dodecahedra, tetrakaidecahedra and pentakaidecahedra fill the space, alternating in the 10:16:4 ratio (Fig. 5.5). Tellurium atoms are located at the centers of those cages, leaving about 4/5 of the smaller cages vacant, such that the total number of tellurium atoms per formula does not exceed 22. There are 17 crystallographic positions for the framework atoms. A combination of the X-ray single crystal and neutron powder diffraction methods enabled to prove that only three of them are jointly populated by silicon and phosphorus in $Si_{130}P_{42}Te_{21}$ [36, 52].

The compound $Ge_{79}P_{29}S_{18}Te_6$ is a unique representative of a new type of clathrates [39]. It crystallizes in the rhombohedral space group $R\bar{3}m$ with the unit cell parameters $a = 17.120(3)$ and $c = 10.608(2)$ Å. In its crystal structure, Ge and P atoms form the framework, whereas tellurium atoms occupy the guest positions inside the 24-vertex polyhedra. The polyhedra can be described as distorted tetrakaidecahedra, in which a regular hexagonal face sits opposed to a distorted one. These polyhedra, bonded through the alternation of regular and distorted hexagons, are packed into infinite columns, which, in turn, are bridged by sulfur atoms (Fig. 5.6). This is a new polyhedral arrangement, and according to the

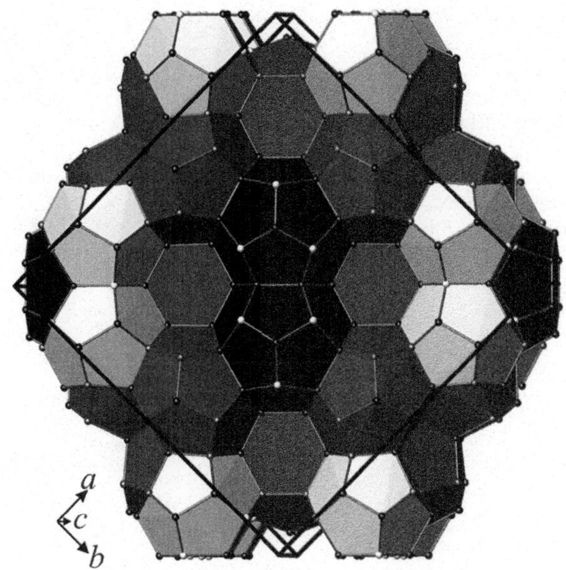


**Fig. 5.5** Polyhedral representation of the clathrate-III crystal structure. Te-filled polyhedra are drawn in *black* (26-vertex cage), *grey* (24-vertex cage) and *white* (20-vertex cage). Si and P atomic positions are denoted as *black* and *white circles*, respectively; positions with mixed Si/P occupation are drawn in *grey*

classification by Jeffrey [53] the label type-X was assigned to this clathrate crystal structure. The crystal structure of the type-X clathrate has two distinctive features. First is the alternation of three- and four-bonded Ge and P atoms within the framework. The second is the presence of sulfur atoms playing a dubious role as they can be considered as additional guest atoms with the 2 + 8 coordination or as bridges between the columns of the distorted tetrakaidecahedra. These features point at certain analogy of the type-X and type-IX clathrates.

Although no other crystal structure is classified as cationic clathrates, the compound $Si_{20-x}Te_{7+x}$ ($x \sim 2.5$) deserves mentioning as having a clathrate-like crystal structure with a positively charged framework [38]. Only one type of polyhedral cages—pentagonal dodecahedra—is present in its structure. These polyhedra are assembled into a three-dimensional framework through additional two-coordinated tellurium atoms (Fig. 5.6). However, considering these tellurium atoms as forming an alternative framework, the analogy between the crystal structures of $Si_{20-x}Te_{7+x}$ and the type-VII clathrate can be drawn [18, 38].

## 5.3 Electronic Structure and Bonding in Cationic Clathrates

### *5.3.1 Application of the Zintl Counting Scheme*

Qualitatively, the semiconducting behavior of the germanium- and tin-based cationic clathrates can be rationalized within the frames of the Zintl counting scheme, assuming the reversed host-guest polarity. In traditional Zintl phases of a general

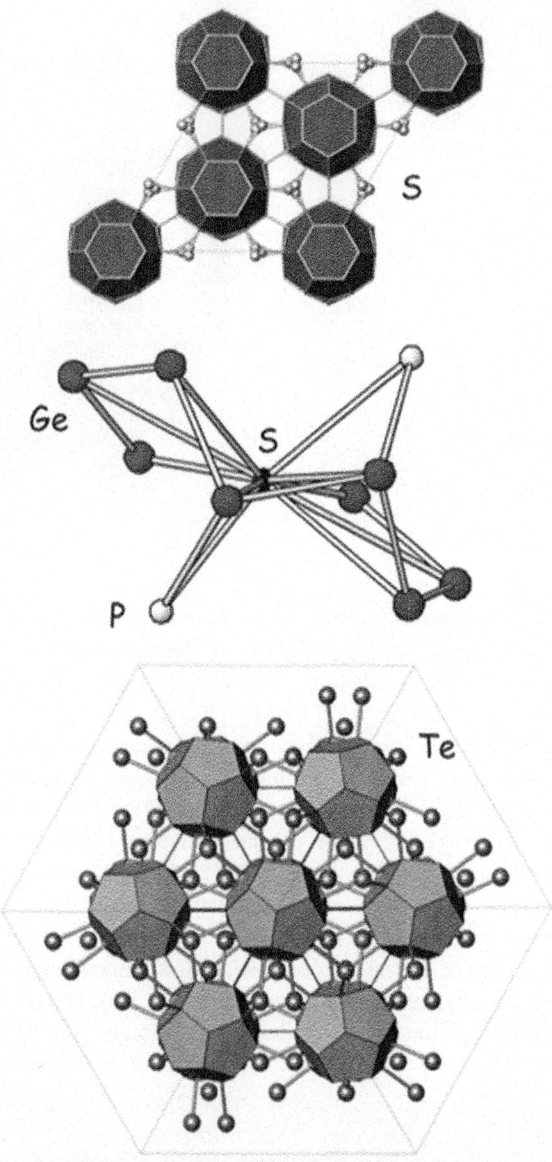


**Fig. 5.6** Crystal structure of $Ge_{78}P_{29}S_{18}Te_8$. *Top* Polyhedral presentation; shown are Te-filled tetrakaidekahedra linked by sulfur atoms. *middle* Coordination of sulfur. *bottom* Crystal structure of $Si_{20-x}Te_{7+x}$. Shown are Te@$Si_{20}$ dodecahedra linked by tellurium atoms

formula $A_nE_m$, it is assumed that the electropositive cations A donate their electrons to E atoms which achieve their 8-electron shell by forming two-center, two-electron E–E bonds and (if necessary) localizing lone electron pairs. In Zintl phases with reversed polarity of a general formula $E_mX_n$, the X anions accept electrons from the E elements rendering them forming an electron octet (Fig. 5.7).

Let us consider $Ge_{38}P_8I_8$ first. The guest iodine atoms behave as anions and should be assigned a formal charge of −1. Each atom of the framework is four-bonded, and if it has four valence electrons, as germanium, it is assumed to have a

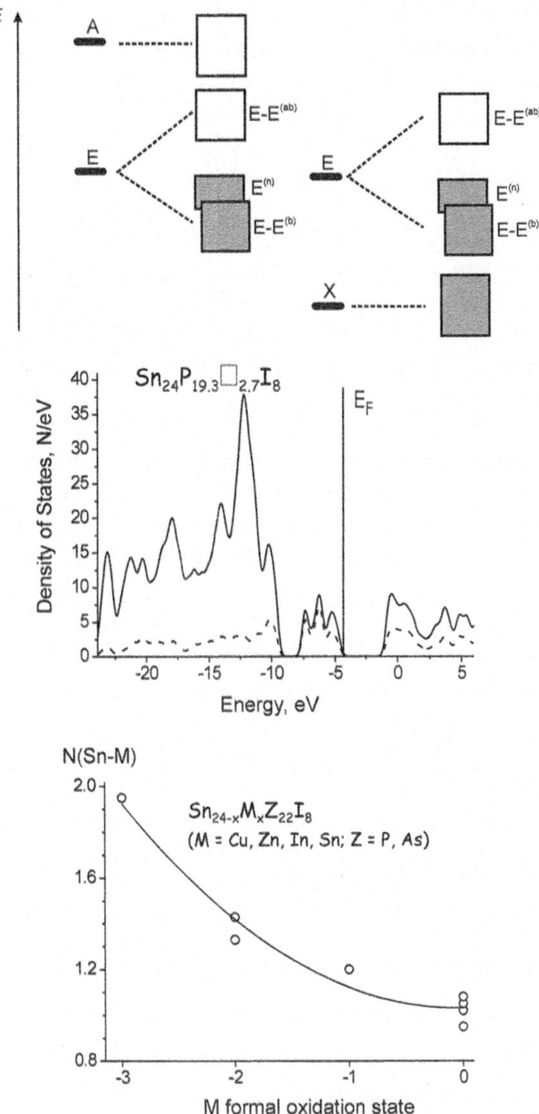


**Fig. 5.7** *Top* Comparison of band structures for traditional $A_mE_n$ and inversed $E_nX_m$ Zintl phases. *middle* DOS plot for $Sn_{24}P_{19.3}I_8$, solid line, with the contribution of the lone pairs of tin atoms, dashed line. *bottom* Population of the tin hybrid orbitals in Sn–M bonding as a function of the formal oxidation state of M

zero oxidation state since there is no need to accept or donate electrons to complete the electron octet. A phosphorus atom having five valence electrons has to surrender one electron to form four 2*c*–2*e* bonds; therefore, it is assigned an oxidation state of +1. The resulting formula $\left[Ge^{0}_{38}P^{1+}_{8}\right]I^{1-}_{8}$ shows that the clathrate is electronically balanced. In the case of $Sn_{24}P_{19.3}I_8$ the remaining question is the role of vacancies. Each vacancy is surrounded by four tin atoms, making each of them three-bonded, that is, with the oxidation state of –1. Taking into account the numbers of three-bonded tin atoms, phosphorus atoms and guest atoms, the electron balance can be achieved for the composition $Sn_{24}P_{19.2}I_8$, which coincides with the experimentally observed within the accuracy of determination (Table 5.1). There are no vacancies in the structure of $Sn_{10}In_{14}P_{22}I_8$ and all atoms are four-bonded. As the four-bonded indium atoms can be assigned an oxidation state of –1, the electrical balance is achieved as follows: eight iodine atoms and 14 indium atoms constitute a total negative charge of –22, compensated by 22 phosphorus atoms having a +1 oxidation state.

Another way of electron counting is realized for the ternary $Sn_{20.5}As_{22}I_8$ clathrate. In this case the vacancies are observed not in the pnicogen positions as in the phosphorous analogue, but rather in the tin position. Here the formation of each vacancy leads to the formation of three 3-coordinated As atoms, thus to three $As^0$, together with the formation of one 3-coordinated tin atom. Three-bonded tin atoms should be assigned the formal $Sn^-$ state, whereas four-coordinated arsenic atoms are $As^+$. The total number of vacancies is 3.5, thus 10.5 arsenic atoms are $As^0$, 11.5 arsenic atoms are $As^+$, 3.5 tin atoms are $Sn^-$, and 8 iodine atoms are $I^-$. This leads to the electron balanced composition $Sn_{20.5}As_{22}I_8$ (Table 5.1). No surprise that all tin-based clathrates are not only semiconductors but also diamagnets.

The silicon-based clathrates do not necessarily follow the Zintl rule. The type-III clathrate $Si_{130}P_{42}Te_{21}$ follows this rule and displays properties of a typical semiconductor. On contrary, the $Si_{46-x}P_xTe_y$ ($y \geq 6.6$) type-I clathrates behave as semimetals except for the composition with $x = 16$ and $y = 8$ [13, 40], for which the semiconducting properties were reported. The latter case confirms that to follow the Zintl rule, the amount of phosphorus in the framework should be twice the amount of tellurium guest atoms according to the formula $\left[Si^{0}_{30}P^{1+}_{16}\right]Te^{2-}_{8}$. However, for all compositions other than $Si_{30}P_{16}Te_8$ the relation $x < 2y$ was found. As a result these clathrates are semimetals or heavily-doped semiconductors and temperature-dependent diamagnets. A probable reason for the non-Zintl behavior of the silicon-base clathrates, both cationic and anionic, is the strength of the Si–Si bond (E = 226 kJ mol$^{-1}$), which makes filling of the conduction band preferable over breaking the homoatomic bond thus forming vacancies in the framework [7].

## 5.3.2 Band Structure

Band structure calculations were performed for various type-I cationic clathrates and semiclathrates [9, 22, 35]. Several important features were found in these calculations. All compounds were calculated to be semiconductors with a band gap of 0.01–0.3 eV depending on the phase composition and the method used for calculations. Guest atoms do not contribute to the states in vicinity of the Fermi level. Typically, they show very little dispersion, suggesting that their interaction with the framework atoms is quite weak, and carry a negative charge which, in the case of iodine, was found to be close to the formal oxidation state –1. Although in semiclathrates part of the guest atoms form a covalent bond, its contribution to the band structure is observed far below the Fermi level. The states just below the Fermi level are composed of different orbitals of the atoms of the framework, depending on the peculiarities of the crystal structure. In the case of vacancy formation, the dominating orbitals are those that can be considered as lone electron pairs on the group 14 atoms surrounding a vacancy (Fig. 5.7). If the vacancies are not present the states near the Fermi level are composed of the orbitals of the most electropositive metal atom, for instance, indium in $Sn_{10}In_{14}P_{22}I_8$. Therefore, the nature of the metal atom and the vacancy concentration play an important role in the band structure near the Fermi level and influence greatly the electronic properties of the compounds.

### *5.3.3 Bonding*

Experimentally, bonding was investigated only for the tin-based cationic clathrates owing to a facile application of Mössbauer spectroscopy with the $^{119}$Sn nucleus [6, 9, 26, 29, 33]. It is found that for the tin atoms having a tetrahedral coordination of one metal and three phosphorus atoms the isomer shift falls in the range of 1.77–1.99 mm $s^{-1}$, and if phosphorus is replaced with arsenic these values move to 2.05–2.23 mm $s^{-1}$. The application of a semi-quantitative Towns-Daily model enables analyzing the population of the corresponding tin hybrid orbitals. It was calculated that this population in the case of a Sn–Sn bond corresponds to a single covalent bond. However, in the case of the tin atom bonded to other metal atoms a more complex picture was found. Surprisingly, it was proved that the population of the tin orbitals forming a Sn–M bond does not depend on the relative electonegativity of M but rather reflects the formal charge of the M atom [26]. The extreme situation is observed for the 2*c*–2*e* Sn–Cu bond, for which the hybrid Sn population achieves 1.91, meaning that the Sn–Cu bonding electron pair occupies an orbital which is predominantly tin orbital in nature. Clearly, copper with the $4s^1$ configuration requires three electrons to form four directional bonds, which is one electron less than required by Zn or Cd and two electrons less than required by Ga or In. Accordingly, the tin hybrid population diminishes on going from the Sn–Cu to Sn–Zn(Cd) to Sn–In(Ga) and finally to Sn–Sn bond (Fig. 5.7).

Other important information that can be extracted from the $^{119}$Sn Mössbauer spectroscopy is the role of the lone electron pairs on the tin atoms having the 3 + 3 coordination. Opposing all expectations, the tin atoms having this type of environment do not show large values of the quadrupole splitting that would show great asymmetry due to localization of those lone pairs. Instead, the values of the quadrupole splitting do not exceed 0.55 mm $s^{-1}$, indicating a high degree of the lone pair suppression [6, 9, 29]. In turn, this means that these apparent lone pairs are involved in bonding and form compact regions in the density of electronic states just below the Fermi level.

Attempts to gain further information on bonding in Si- and Sn-based cationic clathrates from the $^{31}P$, $^{29}Si$ and $^{119}Sn$ NMR spectra were less informative because of the broadening of the spectral lines due to statistical disorder of atoms in the crystal structure [27, 36, 40]. Although this method is unable to probe the local environment of a target atom, it can be used for investigating the extended electronic structure of clathrates. In particular, the analysis of the chemical shifts of type-I and type-III clathrates in the Si–P–Te system confirmed that the latter clathrate is charge balanced whereas the former is electron-deficient [36, 40].

Beyond the experimental methods, bonding in semiclathrate $Ge_{32}P_{14}Se_8$ was analyzed by means of spatial distribution of Electron Localizability Indicator (ELI) [35]. It was confirmed that the cationic framework is formed by covalent Ge–Ge and polar covalent Ge–P bonds, while there are two types of Se guest anionic species playing different roles. The guest atoms found in the center of the smaller 20-vertex of the parent clathrate-I structure have Coulomb interactions with the framework when the Se-Ge/P distance exceeds of 3.3 Å. Additionally direct covalent bonds are found between the germanium framework atom and the Se guest located in the large 24-vertex cage of the parent clathrate-I structure with the Ge–Se distance of about 2.3 Å. As a result of the newly formed host-guest Ge–Se bond, breaking of one of the Ge–P bonds within the framework and formation of the lone electron pairs on the adjacent phosphorus atoms was found.

## 5.4 Notes on Synthesis and Thermal Stability of Cationic Clathrates

Tin is a low-melting metal, it liquefies already at 505 K and is quite reactive above this temperature. This opens a way for synthesizing tin-based cationic clathrates at as low temperature as 600 K in sealed silica tubes. The reactions normally require two-step annealing with an intermediate regrinding. In most cases there is no need in preparation of precursors as the elements can serve as starting materials. However, it should be noted that in most cases the reaction is facilitated by formation of volatile intermediate products serving as transport agents, which sometimes requires a proper choice of starting chemicals. For instance, for the synthesis of $Sn_{17}Zn_7P_{22}I_8$ elemental zinc, tin, and phosphorus react with tin tetraiodide, which serves both as a reagent and a chemical transport agent, making the reaction very easy. On contrary, preparation of $Sn_{17}Zn_7P_{22}Br_8$ is a more difficult process. In this case tin dibromide is used as a source of bromine. This compound is much less volatile and the target compound is frequently contaminated by a side phase, $SnZnP_2$ [27].

Paradoxically, liquid tin itself may cause difficulties if in the course of reaction it becomes covered by a solid intermediate. The situation gets worse if two liquids appear under synthetic conditions as in the case of Sn and In. The synthesis of the clathrates in the Sn–In–As–I system was always incomplete if elemental tin and indium were used as the starting materials. It was found that the reaction products always contained small portions of unreacted metals and binary impurities, including SnAs and InAs. It was assumed that these arsenides formed on the liquid metal surface precluding further reaction of metals towards target clathrates. To avoid this impediment, finely ground SnAs and InAs were chosen as the starting materials together with $SnI_4$ and elemental arsenic, which led to single-phase products after annealing for 5 days at 823 K [29].

Compared to tin, germanium and silicon have much higher melting points and are less reactive. This calls for higher reaction temperatures and sometimes for higher pressures for the preparation of single-phase cationic clathrates. For instance, $Si_{38}Te_{16}$ ($[Si_{38}Te_8]Te_8$ emphasizing the host and guest substructures) was prepared from the mixture of the elements at a temperature of 1,473 K and a pressure of 5 GPa [17]. Evidently, high-pressure conditions should be applied carefully because the crystal structure of clathrates is not dense, and excessive pressure will crush the tracery framework structure. The role of high pressure is to ensure the tight contact of reacting precursors and to prevent any volatile precursor or intermediate from leaving the reaction site as provided by the example of clathrate-I $[Si_{44.5}I_{1.5}]I_8$ [12]. Lower pressure yet high temperature (up to 1,273 K) are required when the spark plasma sintering (SPS) is used for synthesizing silicon or germanium based clathrates. In this case, pressure of about 30–50 MPa appears to be sufficient [10, 11, 13].

Other synthetic methods towards silicon and germanium based clathrates are rare. $Si_{40}P_6I_{6.5}$ was synthesized using iodine as a transport agent [16], whereas the decomposition of $GeI_4$ in an argon atmosphere led to the preparation of $[Ge_{43.33}I_{2.67}]I_8$ [15].

Tin-based cationic clathrates are stable in moist air for months. They readily oxidize upon heating in air. The destruction temperature varies depending on the chemical composition, achieving the maximum of 650 K for the clathrates of the Sn–In–As–I system [29]. Also, they decompose upon heating in vacuum between 730 and 970 K, liberating volatile tin halides.

Germanium-based clathrates show similar stability, whereas silicon-based compounds are remarkably different. It is shown that the type-I and type-III clathrates in the Si–P–Te system are stable in vacuum up to 1,300 K and in air up to 1,150 K owing to the formation of a thin oxide layer on its surface. This layer, about 3 nm thick, is composed of a phosphorus-doped silicon dioxide. It forms upon heating the compound in air and prevents further oxidation up to very high temperatures [52].

## 5.5 Thermoelectric Properties of Cationic Clathrates

Clathrates, both anionic and cationic, are treated as a base for new thermoelectric materials mainly because they display cage-like structures with guest atoms rattling inside the oversized polyhedral voids. As a rule, there are no covalent bond formed between the guest atoms and the atoms of the framework, suggesting that the concerted motion of the guest atoms can scatter heat-carrying phonons, not impeding transport of the charge carriers through the covalently-bonded framework. This property is believed to lead to very low, glass-like thermal conductivity of a narrow-gap semiconductor, which is one of prerequisites of a high thermoelectric performance of a material [54, 55].

### 5.5.1 Guest Rattling and Lattice Dynamics

The detailed analysis of the crystal structures of all cationic clathrates shows that the guest atoms exhibit greater atomic displacement parameters (ADP's) than the framework atoms. In the type-I clathrates the ADP's for the guest atoms occupying the 24-vertex cages are nearly twice larger than those for the guests inside the 20-vertex cages irrespective of the fine details of the crystal structure. The ADP values decrease linearly with the temperature for all atoms (Fig. 5.8). This provides an opportunity to estimate characteristic Debye and Einstein temperatures, $\theta_D$ and $\theta_E$, describing the collective motion of the framework and the localized motion of guest atoms, respectively. The linear fits of the $<u^2> = f(T)$ and $<u_{ii}> = f(T)$ functions, where $<u^2>$ is the mean square atomic displacement for all atoms and $<u_{ii}>$ is the directional atomic displacement parameter, were used to calculate $\theta_D$ and $\theta_E$ for various cationic clathrates. It was found that for germanium- and tin-based compounds the Debye temperature falls in the range of 185–238 K [11, 31, 56–58], which is comparable with the respective anionic clathrates [59]. Si-based cationic clathrates show much greater values, for instance $\theta_D = 491$ K was found for $Si_{30}P_{16}Te_{6.8}Se_{1.2}$ [14]. The $\theta_E$ values are more diverse; they are reported to vary from 50 to 79 K for various Ge- and Sn-based compounds [11, 31, 56, 57] and from 108 to 153 K for Si-based cationic clathrates [14].

Anisotropy of the guest motion is also reflected by the Einstein temperature. It was found that the pseudo-localized vibrations of the guest iodine atoms in $Sn_{24}P_{19.3}I_8$ can be described by three different Einstein temperatures [56]. The guests occupying the centers of the 20-vertex voids show isotropic motion characterized by $\theta_E = 76$ K. Vibrations of the guests in the 24-vertex cages are anisotropic and described by two terms, $U_{11} = U_{22}$ and $U_{33}$. The former term describes equatorial movements of the guest atom, that is, along the six-member ring of the tetrakaidekahedra, whereas the latter term describes axial vibrations. The Einstein temperature associated with $U_{33}$ was found to be 79 K but the corresponding value for the equatorial vibrations was calculated as 63 K (Fig. 5.8). These values are similar to those for the germanium anionic clathrate-I $Ba_8Ge_{43}$ and imply that the in-plane vibrational amplitude characterized by lower Einstein temperature is larger compared to the out-of-plane motions [60].

Guest rattling was also analyzed by means of the Raman spectroscopy. It was shown that in $Si_{44}I_{10}$ (= $[Si_{44}I_2]I_8$) two peaks at 75 and 101 $cm^{-1}$ can be assigned to vibrations of guest iodine anions inside the cationic framework [61].

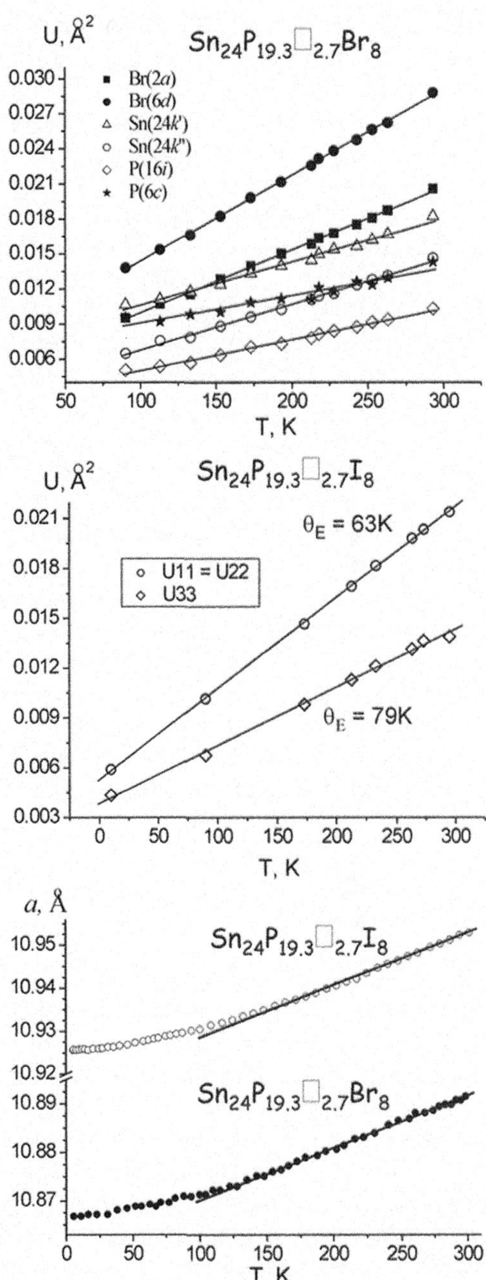


**Fig. 5.8** *Top* Atomic displacement parameters as functions of temperature for $Sn_{24}P_{19.5}Br_8$. *middle* Anisotropy of the atomic displacement parameters for the I2 atom in the crystal structure of $Sn_{24}P_{19.3}I_8$. *bottom* Thermal expansion/contraction of tin-based cationic clathrates $Sn_{24}P_{19.3}X_8$ (X = I or Br)

The crystal structures of tin-based cationic clathrates readily contract upon cooling [31, 56, 58]. The dependence is linear with temperature down to about 100 K and then shows lesser increment with further cooling (Fig. 5.8). Thermal contraction is constant in the range of 100–300 K and similar for all investigated compounds, $\alpha = 1.0–1.1 \times 10^{-5}$ $K^{-1}$. These values are typical for all clathrates with the Sn-based and Ge-based frameworks. Contrastingly, the Si-based clathrates behave differently. For instance, $Si_{30}P_{16}Te_{8-x}Se_x$ display thermal expansion constants of $6 \times 10^{-6}$ $K^{-1}$, which falls in the range of $6–8 \times 10^{-6}$ $K^{-1}$ typical for Si-based anionic clathrates [59]. Lower values observed for the latter clathrates reflect that the Si–Si bonds are stronger than the Ge–Ge and Sn–Sn ones and that the Si-based clathrates form more rigid frameworks.

Heat capacity as a function of temperature was investigated in detail for $Sn_{24}P_{19.3}I_8$ [56]. It was found that the heat capacity increases smoothly up to about 300 K and then saturates reaching the Dulong-Petit value of about 1,280 J $mol^{-1}$ $K^{-1}$. The characteristic Debye temperature of 265 K calculated from the experimental data describe the collective motion of the framework atoms. Its value is typical for most of clathrates compounds based on tin or germanium but different from typical values of the Si-based anionic clathrates [54]. The heat capacity temperature dependence was fitted using a model with two Einstein and one Debye terms, showing two modes of guest atoms rattling inside the oversized cages of the host framework (Fig. 5.9). The characteristic Einstein temperatures of 60 and 72 K reflect different energy of pseudo-localized motion of the guest atoms; however, their values confirm that the host-guest interaction in this compound is weak and that the dynamics of the guest atoms can be treated independently of the dynamics of the host framework. It should be noted that the Debye temperatures extracted from the heat capacity and structural data differ by almost 20 %. The latter data seem to be less reliable because the model used implies that the cubic unit cell is composed of atoms of a single kind, which is far from being true for the tin cationic clathrates.

### 5.5.2 Thermal Conductivity

Experimentally observed thermal conductivity $\kappa$ of the cationic clathrates is characterized by low values. Among this group of compounds, tin-base cationic clathrates display the lowest thermal conductivity ranging from 0.36 to about 2.0 W $m^{-1}$ $K^{-1}$ at room temperature [11, 32, 33, 57, 58]. The temperature dependence of thermal conductivity is typical of semiconductors (Fig. 5.9). At low temperature it increases, reaching a maximum around 30–70 K, and then decreases down to room temperature, following for many compounds the $T^{-1}$ dependence which corresponds to the predominant phonon-phonon scattering. In some cases a shallow minimum followed by a slight upturn towards room-temperature values was registered, however, this was attributed to the radiation losses inherent in measurement of the samples with low thermal conduction employing the steady-state method.

Very low thermal conductivity, $\kappa = 0.4$ W $m^{-1}$ $K^{-1}$ was found for $Sn_{20.5}As_{22}I_8$ at 210 K [33], which is among the lowest for all clathrates [7, 29] and at the minimal values for low-gap semiconductors [62, 63]. It is easy to notice that the lowest values of the thermal conductivity are peculiar to compounds with the

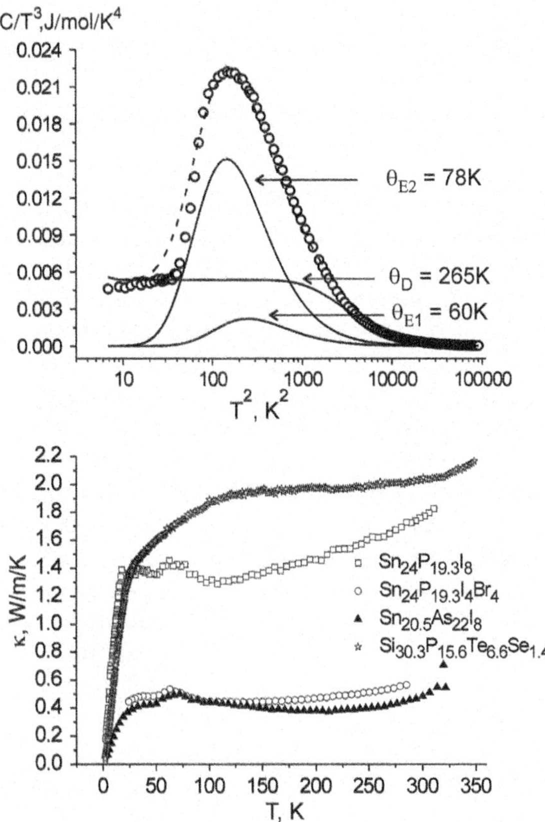


**Fig. 5.9** *Top* A $C/T^3$ versus $T^2$ plot for $Sn_{24}P_{19.3}I_8$ with the contributions of one Debye and two Einstein modes. *bottom* Thermal conductivity versus temperature for some cationic clathrates

distortions in their crystal structure. $Sn_{20.5}As_{22}I_8$ shows incomplete ordering of the vacancies accompanied with the non-concerted shift of the arsenic atoms away from their proper positions. Having the most disordered crystal structure, this compound displays almost the lowest thermal conductivity. However, substituting partially indium for tin leads to even lower thermal conductivity for $Sn_{24-x-\delta}In_xAs_{22-y}I_8$, reaching the absolute room-temperature minimum of 0.36 W $m^{-1}$ $K^{-1}$ for $x = 4.0$–$6.5$. This phase features less disorder of the crystal structure compared to the parent ternary phase; however, the effect of alloying within the framework may add to the lowering of the thermal conductivity. In $Sn_{24}P_{19.3}Br_xI_{8-x}$, the framework substructure is much less disordered. The disorder is manifested by a split of the 24k position of tin into two having different environment. Besides, the bromine and iodine guests having a different atomic weight alternate almost randomly, introducing another constituent to the total disordering. As a result, the thermal conductivity of this solid solution for $x = 4$ is 0.45 W $m^{-1}$ $K^{-1}$, which is slightly higher than for $Sn_{20.5}As_{22}I_8$. Remarkably, the solid solution $Sn_{24}P_{19.3}Br_xI_{8-x}$ with $x = 2$ exhibits three times lower thermal conductivity compared to the parent ternary phase $Sn_{24}P_{19.3}I_8$, $\kappa = 0.5$ W $m^{-1}$ $K^{-1}$ versus 1.5 W $m^{-1}$ $K^{-1}$ [32].

A comparison of the thermal conductivity observed for $Sn_{24}P_{19.3}I_8$ [32] and $Sn_{24}P_{19.3}Br_8$ [58] indicates that although the degree of disorder in isostructural compounds is apparently the same their thermal conductivity differs almost by 50 %. The fact that the thermal conductivity of $Sn_{24}P_{19.3}Br_8$ is significantly higher may be explained by two reasons. First, bromine has a smaller atomic mass than iodine, making the iodine-containing analog ~10 % heavier. The second reason is that the host-guest mismatch is lower in $Sn_{24}P_{19.3}Br_8$, leading to less pronounced vibrations of the guest bromine atoms and thus to higher values of the thermal conductivity.

Total thermal conductivity is a sum of the lattice and electronic parts, $\kappa = \kappa_L + \kappa_e$. The lattice part of the thermal conductivity describes the scattering of phonons on the vibrations of atoms, whereas the electronic part describes thermal conductivity appearing due to conduction electrons and is related to the electrical conductivity $\sigma$ of a material by the Wiedemann-Franz equation, $\kappa_e = \sigma T L_0$, where T is the absolute temperature and $L_0$ is the ideal Lorenz number, $2.45 \times 10^{-8}$ W$\Omega$K$^{-2}$ [64]. The electronic part of the thermal conductivity is typically low for low-gap semiconductors. For the tin-based cationic clathrates it was calculated to contribute less than 1% to the total thermal conductivity. The lattice part of the thermal conductivity can be estimated based on the Debye equation $\kappa_L = 1/3(C_v \lambda v_s)$, where $C$ is the volumetric heat capacity, $\lambda$ is the mean free path of phonons and $v_s$ is the velocity of sound [64]. The latter is related to the Debye characteristic temperature $\theta_D$ as $v_s = (k_B\theta_D/\hbar) \cdot [6\pi^2(N/V)]^{-1/3}$. Extracting the Debye temperature from either heat capacity or structural data and assuming that the mean free path of phonons is the average distance between the guest atoms, the lattice part of the thermal conductivity was calculated for several tin-based cationic clathrates to be in the range of 0.7–0.9 W m$^{-1}$ K$^{-1}$, which is in good agreement with the experimental data [31, 32, 56, 58].

In general, the thermal conductivity of the tin-based cationic clathrates is low, which is explained by a combination of different factors, including pseudo-localized guest vibrations, mass alternation, and framework disorder. Depending on the structure and composition of a given clathrate these factors may contribute differently.

A different picture is observed for the silicon-based cationic clathrates. The values of the total thermal conductivity are greater than those for the tin-based clathrates. Typical room-temperature values for these compounds lie between 3 and 5 W m$^{-1}$ K$^{-1}$ [13, 52]. This is explained by two facts. First, the silicon-based clathrates possess very rigid frameworks. The concerted motion of the framework atoms is not impeded by heavy distortions of the crystal structure. As a result, the phonon transport is more efficient than in the case of the Ge- and Sn-based clathrates. Second, they behave like degenerate semiconductors or bad metals rather than normal semiconductors; this increases the contribution of the electronic part to the total thermal conductivity of the Si-based cationic clathrates, which was estimated using the Wiedemann-Franz relation to constitute 15 % of the total thermal conductivity for type-I clathrates in the Si–P–Te system [52].

The thermal conductivity of $Si_{30.3}P_{15.6}Te_{6.6}Se_{1.4}$ shows additional features and deserves more discussion. It was shown [14] that the total thermal conductivity of this clathrate is much lower than for other Si-based compounds. At room temperature it reaches only 2.0 W $m^{-1}$ $K^{-1}$. Moreover, the temperature dependence of the thermal conductivity is typical for glasses. It increases sharply from 2 to 50 K, then slowly reaches 2.0 W $m^{-1}$ $K^{-1}$ at about 130 K and remains constant up to room temperature (Fig. 5.9). The glass-like behavior of this semimetallic compound is associated with the mass alteration of two types of the guest atoms, tellurium and selenium, coupled to a slight disorder in distribution of silicon and phosphorus atoms within the cationic framework.

$Si_{132}P_{40}Te_{21.5}$ is the only type-III cationic clathrate for which the thermal conductivity was measured [52]. It takes the value of 1.5 W $m^{-1}$ $K^{-1}$ at room temperature and decreases with increasing temperature until a broad maximum of 1 W $m^{-1}$ $K^{-1}$ is reached at about 600 K; it increases slightly with further increase of temperature reaching again 1.5 W $m^{-1}$ $K^{-1}$ at 1,023 K. Noticeably, the thermal conductivity of the type-III cationic clathrates is significantly lower than for the type-I counterparts (Table 5.2). Two factors can be considered. First, the crystal structure of the type-III clathrate is much more complex than that of the type-I clathrate; second, the type-III clathrate is a semiconductor, and the electronic part of the thermal conductivity is low.

Very little is known about the thermal conductivity of germanium-based cationic clathrates. $Ge_{30}P_{16}Te_8$ is reported to have a room-temperature thermal conductivity of 0.9 W $m^{-1}$ $K^{-1}$ [10]. This compound is a semiconductor; probably, there is a similarity in the thermal conductivity of the Ge- and Sn-based cationic clathrates.

### *5.5.3 Transport of Charge Carriers*

Most of silicon-based cationic clathrates demonstrate properties alien to Zintl phases. As a rule, they are semimetals or degenerate semiconductors and exhibit weakly temperature dependent diamagnetism [13, 14, 40, 52]. Their electrical conductivity slightly decreases with increasing temperature or remains constant. The magnetic susceptibility of clathrates includes two terms. One of them is a standard term, which accounts for the core diamagnetism and is calculated as a sum of the atomic diamagnetic increments. The second term accounts for the structural diamagnetism, which is most likely caused by molecular ring currents on the five and six-member rings of the framework [65]. This kind of contribution is temperature dependent, and its room temperature value for the Si-based cationic clathrates lie in the range of 5–10 $10^{-5}$ emu $mol^{-1}$ [40].

The examples of the semiconducting behavior are rare for the Si-based clathrates. Among the type-I clathrates only $Si_{30.8}P_{15.1}Te_{7.50}$ [40] and $Si_{30}P_{16}Te_8$ [13] display typical behavior of thermally activated semiconductors. Their band gap is

**Table 5.2** Physical properties of the cationic clathrates at room temperature

| Phase | $E_g$, eV | $\sigma$, S m$^{-1}$ | $S$, µV K$^{-1}$ | $\kappa$, W m$^{-1}$ K$^{-1}$ | Ref. |
|---|---|---|---|---|---|
| $Sn_{24}P_{19.3}I_8$ | 0.03; 0.04 | 1650 | +80 | 1.6 | 6, 32, 66 |
| $Sn_{24}P_{19.3}Br_8$ | 0.14 | 333, 6650 | +140 | 1.9 | 32, 66 |
| $Sn_{24}P_{19.3}Br_xI_{8-x}$ ($x = 2$–7) | 0.056–0.138 | 405–1340 | +80 ÷ +180 | 0.5–2.0 | 32, 66 |
| $Sn_{17}Zn_7P_{22}I_8$ | 0.25 | 0.4 | | | 27 |
| $Sn_{17}Zn_7P_{22}Br_8$ | 0.11; 0.15 | 250 | +40 | 1.6 | 27, 58 |
| $Sn_{19.3}Cu_{4.7}P_{22}I_8$ | | 0.2 | +600 | | 57 |
| $Sn_{19.3}Cu_{4.7-x}Zn_xP_{22-y-y}I_8$ ($x = 1, 2, 3$) | 0.21–0.29 | 0.4–6.7 | +430 ÷ +620 | | 57 |
| "$Sn_{20.5}Ni_{3.5}P_{22}I_8$"[a] | | | | 1.9 | 57 |
| $Sn_{20.5}As_{22}I_8$ | 0.45 | 0.95 | −180 | 0.5 | 33 |
| $Sn_{24-x-\delta}In_xAs_{22-y}I_8$ | 0.033–0.051 | 135–461 | −322 ÷ −594 | 0.36–0.57 | 29 |
| $Sn_{38}Sb_8I_8$ | 0.4 | ~0.1 | −600 | 0.7 | 11 |
| $Ge_{38}E_8I_8$ (E = P, As) | 0.12–0.16 | | | | 20 |
| $Ge_{38}Sb_8I_8$ | 0.48; 1.16 | | −800 | 1.2 | 11, 20 |
| $Ge_{30}P_{16}Te_8$ | 0.62 | | | 0.9 | 10 |
| $Ge_{30.6}P_{15.4}Se_8$ | 0.41 | ~2 | −200 | | 35 |
| $Ge_{40}Te_{5.3}I_8$ | 0.78 | | | | 23 |
| $Si_{30}P_{16}Te_8$ | 1.24 | 63.3 | 60 | 3.9 | 13 |
| $Si_{30.8}P_{14.1}Te_{7.50}$ | 0.12 | 4750 | | | 40 |
| $Si_{46-x}P_xTe_y$ ($x = 13.3$–14.1; $y = 6.70$–7.25) | | 1000–4000 | ~170 | 3.2 | 40, 52 |
| $Si_{30.3}P_{15.6}Te_{6.6}Se_{1.4}$ | 0.017 | 335 | 240 | 2.0 | 14 |
| $Si_{32.1}P_{13.9}Te_{6.6}Br_{1.0}$ | 0.020 | 2500 | 250 | 5.2 | 14 |
| $Si_{130}P_{42}Te_{21}$ | | 2000 | 170 | 1.5 | 36, 52 |

[a] Composition and structure were not confirmed

very sensitive to the composition. For the above compounds it was calculated to be 0.12 and 1.24 eV, respectively. Another example of a semiconducting clathrate is provided by the type-III representative $Si_{130}P_{42}Te_{21}$, which displays the activated behavior with the room-temperature conductivity of $2 \times 10^3$ S m$^{-1}$ [36].

Interestingly, all investigated Si-based clathrates have holes as dominant charge carriers. They display positive values of the Seebeck coefficient ranging from +60 to +250 µV K$^{-1}$ at room temperature (Table 5.2). With increasing temperature the Seebeck coefficient of type-I and type-III clathrates increases reaching +275 µV K$^{-1}$ at 1,100 K for $Si_{130}P_{42}Te_{21}$, which is the maximal reported value for the Si-based cationic clathrates.

Ge-based cationic clathrates display transport properties of typical semiconductors. They exhibit a wide range of the band gap width, from 0.12 to 1.16 eV and are diamagnets [10, 11, 20, 23, 35]. Unfortunately, not a single representative of these clathrates was thoroughly characterized. Only for two of these compounds the Seebeck coefficient were measured, showing the room temperature values of –200 μV K$^{-1}$ for $Ge_{30.6}P_{15.4}Se_8$ [35] and –800 μV K$^{-1}$ for $Ge_{38}Sb_8I_8$ [11].

Similar to the germanium based compounds, all tin-based cationic clathrates behave as typical semiconductors. They show an activation type dependence of electrical resistivity upon temperature. The absolute values of the electrical conductivity (Table 5.2) cover a wide range of values within three orders of magnitude at room temperature [6, 11, 26, 27, 29, 32, 33, 57, 58]. The calculated values of the band gap range from 0.03 eV for $Sn_{24}P_{19.3}I_8$ [32] to 0.45 eV for $Sn_{20.5}As_{22}I_8$ [33]. It is easy to notice that the band gap is smaller for compounds possessing vacancies in the pnicogen atom positions. The vacancy concentration is the largest in $Sn_{24}P_{19.3}I_8$ and $Sn_{24}P_{19.3}Br_8$, which show narrow band gaps of 0.03 and 0.14 eV [32]. Another factor seemingly affecting the band gap width is the degree of disorder. The most striking example is provided by comparing $Sn_{20.5}As_{22}I_8$ with the In-substituted phases. The latter exhibit lower degree of disorder compared to the parent ternary phase (see Sects. 2.2 and 2.3) and possess a band gap of 0.03–0.05 eV depending on the indium content, which is an order of magnitude smaller than for $Sn_{20.5}As_{22}I_8$ [29, 33]. It is worth noting that in most cases the values of the electrical conductivity measured for samples of the same composition but prepared by different routes do not differ much. The only striking difference is the case of $Sn_{24}P_{19.3}Br_8$ [32, 58]. These samples were synthesized by a similar route but then were pressed by different methods. The sample pressed at 553 K by means of spark-plasma sintering technique showed no admixtures and displayed the electrical conductivity of 333 S m$^{-1}$ at room temperature. Contrastingly, the sample, which was hot-pressed at 723 K showed an admixture of $Sn_4P_3$ of a few weight percent and displayed much higher electrical conductivity of about 6,650 S m$^{-1}$ at 300 K. This raises questions about the nature of such a great discrepancy—is it possible that mixing a clathrate phase with a small admixture of $Sn_4P_3$ having metallic properties can raise the electrical conductivity by more than one order of magnitude?

Investigation of the temperature and frequency dependences of the real and imaginary parts of the complex impedance of ceramic samples was performed between 4.2 and 300 K in the range of 20–10$^6$ Hz for $Sn_{24}P_{19.3}Br_xI_{8-x}$ for a variable Br/I ratio [66]. This study showed that at room temperature and down to about 75 K the electrical resistivity is intrinsic for all compositions. At lower temperatures a substantial contribution to the complex impedance from the grain boundaries appears.

Interestingly, both electrons and holes can be the major charge carriers in the tin-based cationic clathrates. Table 5.2 shows that the Seebeck coefficients have different sign for different compounds. The highest positive value corresponding to hole carriers of about +600 μV K$^{-1}$ is found for clathrates in the Sn–Cu–Zn–P–I system [58], whereas the largest negative value of –600 μV K$^{-1}$ is observed for $Sn_{24-x-\delta}In_xAs_{22-y}I_8$ [29] and $Sn_{38}Sb_8I_8$ [11].

### 5.5.4 Thermoelectric Figure-of-Merit

The dimensionless thermoelectric figure-of-merit defines the performance of a thermoelectric material. It is expressed as $zT = \sigma\, S^2 T\, \kappa^{-1}$, where σ is the electrical conductivity, S the Seebeck coefficient, *T* the absolute temperature and κ the thermal conductivity.

For most Sn-based cationic clathrates the thermal conductivity is sufficiently low and the task of achieving high values of *zT* is associated primarily with the optimization of the charge carriers' transport. There is an interplay between the Seebeck coefficient and electrical conductivity [67]. In most cases, the latter grows as the former decreases. Their optimum values seem to correspond roughly to the optimized concentration of charge carriers around $10^{21}$ $cm^{-1}$. Table 5.2 shows that for most of compounds high values of the Seebeck coefficients are accompanied by low electrical conductivity and vice versa, leading to rather low values of *zT*. Only the $Sn_{24-x-\delta}In_xAs_{22-y}I_8$ clathrates systematically exhibit relatively high *zT* values for varying indium content [29]. The room-temperature values of *zT* for these clathrates reach the maximum value of $zT = 0.04$ for $x = 6.5$ (Fig. 5.10). The temperature dependence of *zT* measured from 77 to 400 K enabled the data interpolation to estimate $zT = 0.4$ at a temperature of 600 K, which is close to the stability limit in open air (623 K).

Silicon-base cationic clathrates are different in that they are chemically and thermally very stable. If appropriately optimized, they can find applications at high temperatures, for instance, as air-stable materials for direct conversion of solar thermal energy into electric energy. At present, the best values achieved are 0.30 at 870 K for the type-I clathrate and 0.36 at 1,100 K for the type-III representative [52] (Fig. 5.10). These values are comparable with the best commercially available air-stable materials; for instance, with the properly doped Si/Ge p-type alloy, which demonstrates zT ≈ 0.6 at 1,000 K [68].

## 5.6 Tin Anionic Clathrates

### 5.6.1 Crystal Structure

Anionic tin clathrates exhibit diverse structural chemistry crystallizing in five structural types: clathrate-I, clathrate-II, clathrate-III, clathrate-VIII, and clathrate-IX. Note that no Si- and Ge-based anionic clathrates-III and Si-based clathrates-VIII are reported. The clathrate structures are built on 3D host frameworks with large polyhedral cages of different topology. 20-vertex pentagonal dodecahedral cages are found in all the clathrate structures, while 24-vertex ones are present in clathrates-I and -III, 26-vertex and 28-vertex cages are unique for clathrate-III and clathrate-II, correspondingly. We refer to other chapters of the current book and recent reviews [7, 54, 55] for the detailed description of the clathrate structure types.

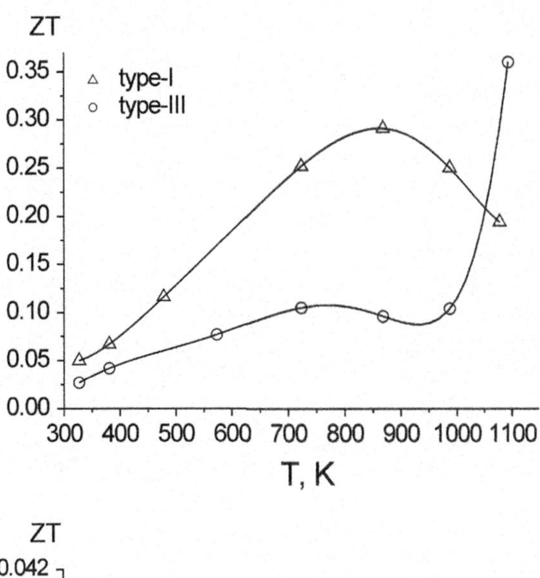


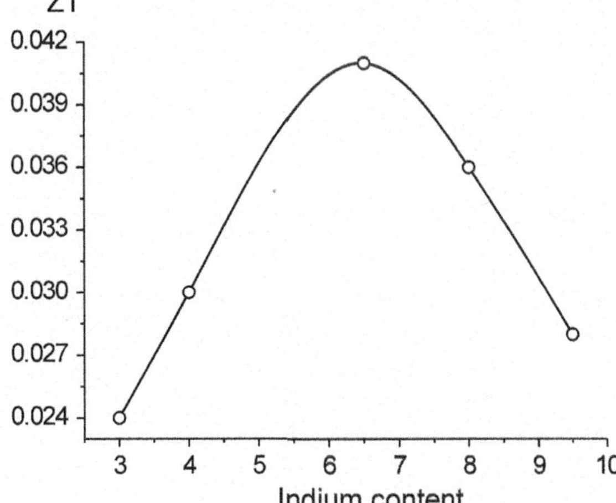


**Fig. 5.10** *Top* Temperature dependence of $zT$ for type-I and type-III clatrhates in the Si–P–Te system; *bottom* $zT$ as a function of indium content $x$ for $Sn_{24-x-\delta}In_xAs_{22-y}I_8$ at 300 K

In the crystal structure of most of the anionic Sn-based clathrates guest positions are completely occupied, whereas vacancies (if any) are formed in the host framework in order to realize electron octet for each atom according to the Zintl rule. The Sn–Sn bond is weaker compared to the Si–Si one (151 kJ $mol^{-1}$ versus 226 kJ $mol^{-1}$) and formation of the framework vacancies is energetically more favorable than filling the conduction band with extra electrons [7]. Determination of the crystal structure of a clathrate containing two different types of atoms and vacancies in the framework is challenging since the presence of two metals and a vacancy in the same crystallographic position cannot be properly refined without extra compositional constrains. Thus, the single crystal X-ray diffraction method alone is often not sufficient for the composition determination. Often scanning electron microscopy with energy (or wave) dispersive X-ray analysis is used for determination of the composition, despite this method provides relative atomic concentration and proper normalization has to be applied. As it was shown for cationic clathrates, $^{119}$Sn Mössbauer spectroscopy is an appropriate method for determination of types and relative amounts of different tin positions. Neutron diffraction in combination with X-ray diffraction is also sufficient for precise determination of the composition of a clathrate compound. Furthermore, transport properties characterization often provides a hint for the composition of the clathrate phases.

Although crystal structures of type-I and type-VIII clathrates are different, these compounds have the same composition, which can be described by a general formula $G_8E_{46}$. Table 5.3 shows that in some cases type-I and type-VIII clathrates can be regarded as polymorphs of the same composition. For instance, low temperature modification of $Ba_8Ga_{16}Sn_{30}$ belongs to the type-VIII clathrate structure. Upon heating it undergoes a reversible transition to the type-I structure [69]. In both cases the composition of the clathrates can be evaluated by a similar approach based on the Zintl electron-count scheme which implies (i) the complete charge transfer between the framework and guest atoms, and (ii) the realization of the electron octet by each atom. In fact, all tin anionic clathrates types, except clathrates-IX, are based on four-bonded atoms, thus electron balanced composition can be calculated for each clathrate type considering that every frameworks atom requires four electrons. For example, in type-I clathrate $Cs_8Sn_{44}$ [70] eight extra electrons from the $Cs^+$ cations are compensated by removing two tin atoms from the hypothetical neutral $Sn_{46}$ framework leaving eight tin atoms with the electron pair, with the formal –1 oxidation state: $(Cs^+)_8(4b - Sn^0)_{38}(3b - Sn^-)_8\square_2$. The heterovalent substitutions are realized for clathrate-I, where part of tin atoms is replaced by Groups 13 (Al, Ga), 12 (Zn, Cd, Hg), and 11 (Cu) elements [7], in such a way that the electron balance is achieved: $(A^{Z+})_8Sn^0_{46-x}M_x^{(4-N_M)-}$, where $x = 8Z/(4-N_M)$, $N_M$ is the number of valence electrons for the metal M, and $Z$ is the valence of the guest cation A.

For a long time only one example of clathrate-II, $Ba_{16}Ga_{32}Sn_{104}$, has been known [71]. This compound was possible to synthesize only upon presence of Na in the reaction mixture, thus incorporation of sodium into the structure cannot be excluded. Recently, two groups reported on new additions to this structure type. Takabatake, et al. reported [72] on clathrate-II in the Ba–K–Ga–Sn system with significant concentration of vacancies in the framework and mixed Ba/K occupancy of the 20- and 28-vertex cages of different size, $K_{10.8}Ba_{13.2}Ga_{36.7}Sn_{89.4}\square_{9.8}$. Total number of vacancies in the framework was fixed to the value determined by energy dispersive x-ray analysis. This composition contradicts the Zintl-Klemm count since 36.7 electrons from the Ga atoms and 39.2 electrons (4 × 9.8) from the $Sn^{1-}$ atoms surrounding vacancies cannot be compensated by 37.2 electrons donated by the guest $K^+/Ba^{2+}$ atoms. As a result of such a discrepancy, the samples of clathrate-II exhibit metallic-like properties. However, in a recent report by Schäfer and Bobev it was pointed out [73] that the large atomic displacement parameter for K/Ba in the 20-vertex cage may indicate that its actual occupancy is

**Table 5.3** Crystal structures of tin anionic clathrates

| Type | Ideal Formula[a] | Examples | Space group, Z |
|---|---|---|---|
| Clathrate-I | $G^{I}_{2}G^{II}_{6}E_{46}$ | $A_xCs_{8-x}Sn_{44}$ A = K, Rb [79, 80]<br>$Cs_8M_4Sn_{42}$, M = Zn, Cd [84, 85]<br>$A_8Hg_4Sn_{42}$, A = K, Rb, Cs [86]<br>$A_8M_8Sn_{38}$, M = Al, Ga, In A = K, Rb, Cs [7]<br>β-$Ba_8Ga_{16}Sn_{30}$ [69, 90] | $Pm\bar{3}n$, Z = 1 |
| | | $A_8Sn_{44}$, A = K, Rb, Cs [49, 78, 80](superstructure) | $Ia\,\bar{3}d$, Z = 8 |
| Clathrate-II | $G^{I}_{16}G^{IV}_{8}E_{136}$ | $Ba_{16}Ga_{32}Sn_{104}$ [71]<br>$K_{8-x}Ba_{16-x}Ga_{40-y}Sn_{96-z-y+z}$ [72]<br>$Cs_8Ba_{16}Ga_{40}Sn_{96}$ [73]<br>$Rb_{9.9}Ba_{13.3}Ga_{36.4}Sn_{99.6}$ [73]<br>$K_2Ba_{14}Ga_{30.4}Sn_{105.6}$ [73, 74] | $Fd\,\bar{3}m$, Z = 1 |
| Clathrate-III | $G^{I}_{10}G^{II}_{16}G^{III}_{4}E_{172}$ | $Cs_{30}Na_{2.5}Sn_{162.6}$ [75]<br>$Cs_{13.8}Rb_{16.2}Na_{2.8}Sn_{162.3}$ [75] | $P4_2/nmn$, Z = 1 |
| Clathrate-VIII | $G^{I}_{8}E^{b}_{46}$ | α-$Ba_8Ga_{16}Sn_{30}$ [69, 92, 93]<br>α-$Ba_8Ga_{16-x}M_xSn_{30}$, M = Cu, Al, Ge, Zn, Sb [91, 94–98] | $I\,\bar{4}3\,m$, Z = 1 |
| Clathrate-IX | $G_{6-x}G^{I}_{2}E^{c}_{25}$ | $K_{6+x}Sn_{25}$ [77, 80]<br>$Rb_5Na_3Sn_{25}$ [81]<br>$K_6Sn_{23}Bi_2$ [77] | $P4_132$, Z = 4 |

[a] G and E denote guest and host atoms, correspondingly. $G^{I}$ —guest atoms situated inside pentagonal dodecahedra, $G^{II}$ , $G^{III}$ , and $G^{IV}$ —guest atoms situated inside larger 24-, 26-, and 28-vertex polyhedra

[b] Distorted polyhedra

[c] Only two guest atoms are located inside regular polyhedra, other G atoms are found inside the channels formed by helical pentagonal polyhedral framework

lower. In the crystal structure of clathrate-II in the Ba–K–Ga–Sn system [73, 74], the bigger 28-vertex cages are vacant, whereas the smaller 20-vertex cages are mixed occupied by $K^{+}/Ba^{2+}$. Careful investigations of the relation between synthetic conditions and compositions of the produced clathrate-II compounds performed by Bobev et al. indicate that clathrate-II indeed exhibits a wide homogeneity range, $K_xBa_{16-x}Ga_{32-x}Sn_{104+x}$, however in all the compounds the larger 28-vertex hexakaidecahera remain empty [74]. The difference in composition for the reported clathrate-II may arise from the different synthetic methods used by these two groups. Further detailed investigation may shed some light on the actual composition of clathrate-II in the K–Ba–Ga–Sn system.

Rare examples of anionic type-III clathrates exhibit an unusual feature of the framework formation [75]. Each of two known examples of these clathrates, $Cs_{30}Na_{(1.33x-10)}Sn_{(172-x)}$ and $Cs_{13.8}Rb_{16.2}Na_{(1.33x-10)}Sn_{(172-x)}$ with x ≈ 9.6(2) contains sodium atoms, which replace $Sn_2$ dumbbells within the framework. No other example of such a substitution is observed in any of the clathrate structure type.

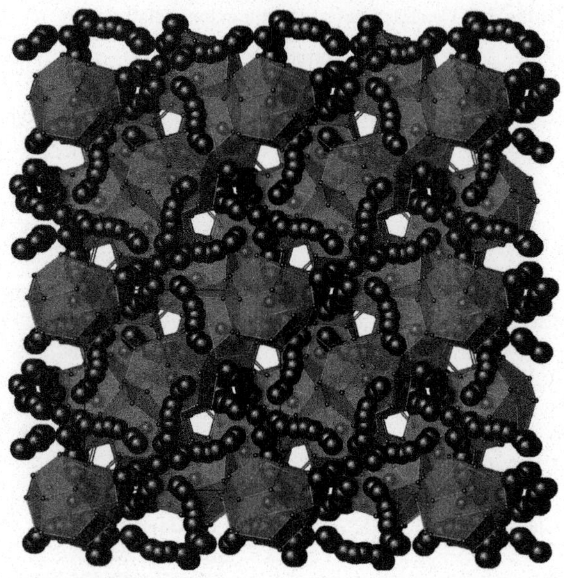

**Fig. 5.11** Crystal structure of clathrate-IX. Two types of guest atoms are shown: first ones are located inside the 20-vertex polyhedral cage, The remaining cations (*black large circles*) are disordered over the open zeolite-like channel system [76]

Clathrate-IX is a special instance by its own since the polyhedral cages in its crystal structure only partially fill the space. The 20-vertex pentagonal dodecahedra centered by guest atoms share pairs of faces forming a helical chain running along the $4_1$ screw axis. There are additional guest positions located in the channels between the helical chains [76]. They have lower coordination numbers and can be occupied by smaller cations such as $Na^+$ (Fig. 5.11). Besides, the type IX clathrate (Table 5.3) is a rare example of clathrate compounds in which the ideal, vacancy-free framework contains three-coordinate atoms. Eight atoms out of the 25 atoms of the framework are three-bonded. If this is a Group 14 atom, it requires one additional electron to complete its electron octet. However the experimentally determined composition of the clathrate-IX $K_6Sn_{25}$ [77] deviates from the composition calculated using the Zintl counting scheme and provides an example of electron-deficient clathrate-IX: $(K^+)_6(4b - Sn^0)_{17}(3b - Sn^-)_8$. Substitution of two Bi atoms for Sn leads to the electron-balanced compound $K_6Sn_{23}Bi_2$ [77], where three-coordinated positions are occupied by two Bi atoms, thus leaving only six three-bonded $Sn^-$ atoms, e.g. $(K^+)_6(4b - Sn^0)_{17}(3b - Sn^-)_6(3b - Bi^0)_2$.

#### 5.6.1.1 Superstructure of Type-I Clathrate

Binary clathrates $A_8Sn_{44}$ (A = Rb, Cs) provide an interesting example of a superstructure formation due to the ordering of vacancies within the clathrate framework [49, 70]. As it was pointed out earlier tin clathrates often exhibit vacancies in the framework for the electron precise composition, e.g. $(A^+)_8(4b - Sn^0)_{36}(3b - Sn^-)_8\square_2$. Two vacancies in the four-bonded framework

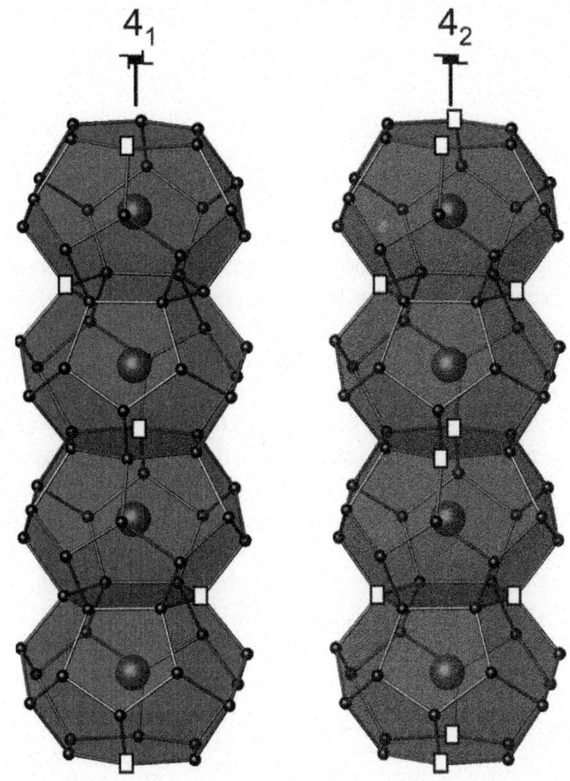


**Fig. 5.12** Columns of 24-vertex cages (tetrakaidecahedra) found in the structures of clathrates-I $A_8Sn_{44}\square_2$ (A = Rb, Cs). In the α-form (superstructure, space group *Ia-3d*), the partially occupied tin site denoted as open square is distributed along $4_1$ axis, while in the β-form (space group $Pm\bar{3}n$) the partially occupied site (*open square*) is distributed along a $4_2$ axis

entail the formation of 8 three-bonded tin atoms with the charge of -1, being compensated by eight electropositive cations $A^+$. The vacancy ordering results in the formation of the 2 × 2 × 2 superstructure. This process is reversible and temperature–driven. At room temperature clathrates $A_8Sn_{44}$ with A = Rb, Cs crystallize in the cubic *Ia–3d* space group with the doubled unit cell parameter $a_s = 2a_0$. Upon heating (up to 80 °C for $Rb_8Sn_{44}$ and 90 °C for $Cs_8Sn_{44}$) the order-disorder phase transition takes place and the structure transforms into the conventional clathrate-I type with the $Pm\bar{3}n$ space group: α-$A_8Sn_{44}$ ⇔ β-$A_8Sn_{44}$. Crystallographically, this phase transition occurs owing to the different modes of distribution of vacancies over the framework positions. In the supercell, the tin position with a partial occupancy of ~0.33 is distributed along the $4_1$ helical axis running within the column of the tetrakaidecahedra (Fig. 5.12), whereas in the high temperature form the partially occupied site with the occupancy of ~0.69 is distributed along the $4_2$ axis. This order-disorder transition is also very subtle and features almost no variation of the local environment of the tin atoms as was shown by $^{119}Sn$ Mössbauer spectroscopy; yet it results in a change in both electrical resistivity and thermopower. The transition is also detectable by differential scanning calorimetry showing an endothermic peak associated with the α-$A_8Sn_{44}$ → β-$A_8Sn_{44}$ transformation [78]. Similar superstructure featuring complete ordering of the vacancies within the framework was reported for the Ge-based anionic clathrate-I $Ba_8Ge_{43}\square_3$ [48].

#### 5.6.1.2 Guest Atoms

Tin-based framework structures are rather flexible and can accommodate cations of different sizes. A variety of guest cations (K, Rb, Cs, Ba) serve as guests in tin anionic clathrates. The importance of the cations size difference is manifested by the crystal structure of clathrate-II. It contains two types of polyhedral cages, 20- and 28-vertex ones, with significantly different volume. Tin clathrates-II can be stabilized by the preferential occupation of the cages by the cations with the pronounced difference in sizes shown by the example of $Cs_8Ba_{16}Ga_{40}Sn_{96}$, in which larger cages are occupied by $Cs^+$, while $Ba^{2+}$ is found to fill smaller cages only [73, 74]. The difference in the cation size for $Rb^+/Ba^{2+}$ appears to be not as significant and the mixed occupation of the 20-vertex cage is observed, whereas the 28-vertex cage is solely occupied by Rb [73, 74]. In the clathrate-I and clathrate-III structures the difference in the cage sizes is not as remarkable (20- and 24-vertex cages in clathrate-I and 20-, 24- and 26-vertex cages in clathrate-III), so all the cages are mixed occupied by $Cs^+/K^+$or $Cs^+/Rb^+$ cations in clathrate-I $A_xCs_{8-x}Sn_{44}$, A = K, Rb [79, 80], and clathrate-III $Cs_{13.8}Rb_{16.2}Na_{2.8}Sn_{162.3}$ [75].

The $Na^+$ cation appears to be too small to fill even the smallest 20-vertex pentagonal dodecahedral $Sn_{20}$ cage, therefore it never appears as a proper guest cation in the tin anionic clathrates. Na-containing clathrate-III and clathrate-IX are the special cases. In the former $Na^+$ cations are incorporated into the framework replacing part of the $Sn_2$ dumbbells [75], whereas in the latter sodium cations occupy positions out of the polyhedral cages [76, 81] (Fig. 5.13). No Sn-based clathrate were reported with $Sr^{2+}$ or $Eu^{2+}$ guest cations, which are larger than $Na^+$ but smaller than $K^+$. This differentiates Sn-based anionic clathrates from Ge-based ones. For the latter, $Sr_8Ga_{16}Ge_{30}$ and $Eu_8Ga_{16}Ge_{30}$ clathrates-I have been reported [82, 83].

#### 5.6.1.3 Comparison with Cationic Tin Clathrates

Comparison of cationic and anionic tin clathrates shows that the former adopt only the clathrate-I structure type, whereas anionic tin clathrates are structurally diverse. A possible explanation accounts for a smaller relative size difference for guest anions compared to guest cations. Clathrates have cages of different sizes and stabilization of one or another clathrate structure type strongly depends on a variation in the guest radius. Cationic species of different sizes are available for the stabilization of anionic clathrates, e.g. ranging from the smallest $Ba^{2+}$ to the largest $Cs^+$ (for coordination number 6: 1.35 to 1.67 Å, 24 % difference). Such a variety in the guest cation's sizes is most likely a reason for the structural diversity of anionic tin clathrates. For the cationic tin clathrates, $I^-$ and $Br^-$ anions are the only

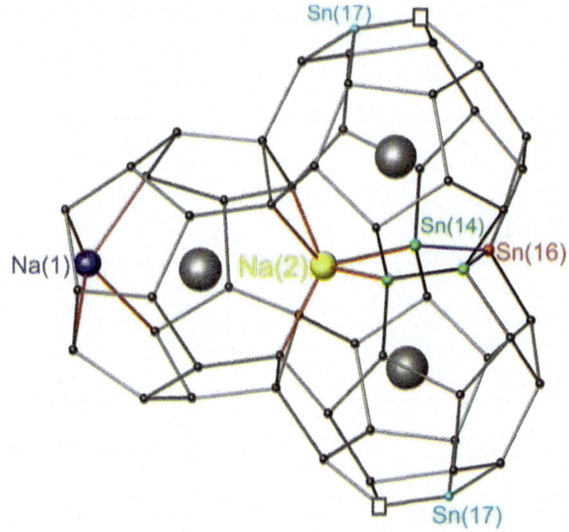


**Fig. 5.13** Location of sodium atoms in the crystal structure of clathrate-III $Cs_{30}Na_{1.33x-10}Sn_{172-x}$. When Na(1) is present the adjacent Sn(17) atoms are missing, when Na(1) is missing only one of the Sn(17) atoms is present. Na(2) is replacing Sn(14)–Sn(16) dumbbells. Atoms numbering is taken from [75]

specimens available, while $Cl^-$ appears to be already too small. Radii for iodine and bromine anions with coordination number 6 are 2.20 and 1.96 Å, only 12 % difference. Both iodine and bromine anions can be accommodated in the 20-vertex and 24-vertex cages, while 26- and 28-vertex cages peculiar for clathrate-III and clathrate-II structures are apparently too large. Anions with larger charge, e.g. $Te^{2-}$ and $Se^{2-}$ are known to get incorporated into clathrate structures as shown by example of silicon clathrate-III $Si_{130}P_{42}Te_{21.2}$ [36] and clathrate-I $Si_{46-x}P_xTe_{8-y}Se_y$ [14], however tin clathrates with Se or Te as guest anions are unknown so far.

## 5.6.2 Thermoelectric Properties

The thermoelectric properties of tin anionic clathrates have been extensively investigated (Table 5.4). Type-I and type-VIII clathrates were studied in detail. The multiple heterovalent substitutions were carried out in order to optimize carrier concentration and maximize power factor $PF = \sigma S^2$, and the greatest success was achieved for the type-VIII clathrates based on $Ba_8Ga_{16}Sn_{30}$. The Ba–Ga–Sn system provides various possibilities of charge carrier concentration tuning by doping different metals into the framework. For ternary clathrate-VIII with the idealized composition $Ba_8Ga_{16}Sn_{30}$, the type of dominate charge carriers can be modified by changing the synthesis method. Crystals grown from Sn-flux are *n*-type semiconductors, whereas using the Ga-flux leads to growing *p*-type semiconductors. Substitution of different elements, M = Al, Cu, Zn, Ge, Sb, into the clathrate framework leads to the drastic enhancement of electric conductivity owing to the optimization of the charge carrier concentrations. However, the

**Table 5.4** Thermoelectric properties of selected tin clathrates at room temperature

| Clathrate | Resistivity, $\rho$, Ω·m | Thermopower, $S$, $\mu VK^{-1}$ | Thermal conductivity, $\kappa$, $Wm^{-1}K^{-1}$ | Figure-of-merit, $zT$ |
|---|---|---|---|---|
| *Clathrates-I* | | | | |
| $Rb_8Sn_{44}$ [78] | $1.6 \times 10^{-5}$ | –7 | 1.7 (200 K) | <0.001 |
| $K_8Ga_8Sn_{38}$ [87] | $10 \times 10^{-5}$ | –220 | 2 | 0.07 |
| $K_8Ga_8Sn_{38}$ [88] | $6 \times 10^{-5}$ | –240 | 1.4 | 0.17 |
| $K_8Ga_8Sn_{38}$ [89] | $8 \times 10^{-5}$ | –250 | 1.4 | 0.16 |
| $K_8Al_8Sn_{38}$ [89] | $16 \times 10^{-5}$ | –235 | 1.2 | 0.09 |
| $K_8In_8Sn_{38}$ [89] | $10 \times 10^{-5}$ | –240 | 1.3 | 0.13 |
| $Cs_8Zn_4Sn_{42}$ [84] | $50 \times 10^{-5}$ | –200 | 1.5 | 0.016 |
| $Cs_8Cd_4Sn_{42}$ [85] | $100 \times 10^{-5}$ | –165 | 1.0 | 0.008 |
| $Ba_8Ga_{16}Sn_{30}$ (*p*-type) [90] | $25 \times 10^{-5}$ | +350 | 0.4 | 0.3 |
| $Ba_8Ga_{16}Sn_{30}$ (*n*-type) [90] | $30 \times 10^{-5}$ | –370 | 0.4 | 0.35 |
| *Clathrates-II* | | | | |
| $K_{10.8}Ba_{13.2}Ga_{36.7}Sn_{89.4-9.8}$ [72] | $1.3 \times 10^{-5}$ | –90 | 1.5 | 0.12 |
| *Clathrates-III* | | | | |
| $Cs_{30}Na_3Sn_{162}$ [75] | Diamagnetic, no thermoelectric properties reported | | | |
| *Clathrates-VIII* | | | | |
| $Ba_8Ga_{16}Sn_{30}$ (*n*-type) [91] | $7 \times 10^{-5}$ | –280 | 0.7 | 0.5 |
| $Ba_8Ga_{16}Sn_{30}$ (*p*-type) [91] | $7 \times 10^{-5}$ | +340 | 0.7 | 0.7 |
| $Ba_8Ga_{16}Sn_{30}$ [92] | $4 \times 10^{-5}$ | –188 | 1.1 (150 K) | 0.2 |
| $Ba_8Ga_{16.6}Sn_{29.4}$ (*n*-type) [93] | $4.2 \times 10^{-5}$ | –108 | 1.2 (150 K) | 0.07 |
| $Ba_8Ga_{16.6}Sn_{29.4}$ (*p*-type) [93] | $6 \times 10^{-5}$ | +150 | 1 (150 K) | 0.1 |
| $Ba_8Ga_{15.65}Cu_{0.125}Sn_{30.25}$ (*n*-type) [94] | $2.5 \times 10^{-5}$ | –180 | 0.7 | 0.55 |
| $Ba_8Ga_{15.89}Cu_{0.003}Sn_{30.11}$ (*p*-type) [94] | $11.5 \times 10^{-5}$ | +350 | 0.7 | 0.45 |
| $Ba_8Ga_{15.7}Zn_{0.07}Sn_{30.3}$ [95] | $4.5 \times 10^{-5}$ | –220 | 0.75 | 0.4 |
| $Ba_8Ga_{10}Al_6Sn_{30}$ [96] | $3.8 \times 10^{-5}$ | –245 | 0.72 | 0.6 |
| $Ba_8Ga_{16.7}Sn_{28.6}Sb_{0.7}$ (*n*-type) [91] | $9 \times 10^{-5}$ | –275 | ~0.7 | 0.4 |
| $Ba_8Ga_{16.7}Sn_{28.6}Sb_{0.7}$ (*p*-type) [91] | $5 \times 10^{-5}$ | +270 | ~0.7 | 0.6 |
| $Ba_8Ga_{16.4}Sn_{25.0}Ge_{4.6}$ [98] | $7 \times 10^{-5}$ | +220 | 1.0 | 0.2 |
| $Ba_8Ga_{16.9}Sn_{19.8}Ge_{9.3}$ [98] | $3 \times 10^{-5}$ | +140 | 1.2 | 0.2 |
| *Clathrates-IX* | | | | |
| $Rb_5Na_3Sn_{25}$ [81] | Diamagnetic, no thermoelectric properties reported | | | |

substitution limit for each M in the $Ba_8Ga_{16}Sn_{30}$ clathrate-VIII is different. Solubility of Al and Ge in the clathrate-VIII was shown to be as high as 10.5 and 9.3 atoms per 46 frameworks atoms, while for Zn and Cu the maximal solubility was found to be only 0.54 and 0.125 atoms, respectively. In the cases of Al, Zn or Sb substitution, the carrier mobility is enhanced without changing the carrier density, which results in higher values of electrical conductivity for the doped samples compared to the undoped ones, overall leading to the enhancement of $zT$. However the effect of the Cu substitution is different depending on the sample charge carries type, *p* or *n*. The Cu doping enhances electric conductivity in the case of the *n*-type sample and has the opposite (negative) impact on the electron transport in the case of the *p*-type material. This effect was attributed to the concomitant change in the composition, e.g. appearance of vacancies in the Ba sites and changing the Ga/Sn ratio. Careful examination of the single-crystal data may shed some light on the observed transport properties as a function of the elemental composition.

Tin anionic clathrates exhibit low values of thermal conductivity in the range of 0.6–2 $Wm^{-1}$ $K^{-1}$. This is lower than the values for anionic Si- and Ge-containing clathrates with similar framework compositions and the same cations. This indicates that clathrate framework atoms play an important role in the scattering of heat-carrying phonons. Thus, the frameworks with heavier Sn atoms scatter phonons more effectively.

The type-VIII tin anionc clathrates show the highest values of the thermoelectric figure-of-merit. Already at room temperature the $zT$ values of 0.4–0.7 can be obtained both for *n*-type and *p*-type materials. The $zT$ for the *n*-type $Ba_8Ga_{16-x}Cu_xSn_{30}$ reaches 1.45 at 500 K, which is the highest value among reported for all clathrates at any temperature [94]. The highest $zT$ value of 1.0 for the *p*-type material is observed for $Ba_8Ga_{16.7}Sn_{28.6}Sb_{0.7}$ at 480 K [97].

## 5.7 Conclusions and Outlook

Clathrates belong to a fancy family of cage compounds and possess aesthetically beautiful highly symmetric structures. They provide the broad field for manipulations with their crystal and electronic structure by conducting multifarious substitutions in a host framework and a guest substructure, by regulating a vacancy concentration, by achieving a desired degree of disorder, and by constructing peculiar superstructures. In response to such operations cationic clathrates and tin anionic clathrates react by varying their transport and thermal properties, reaching high electrical conductivity up to 6,600 S $m^{-1}$, high thermopower of |800| μV $K^{-1}$, and as low thermal conductivity as 0.36 W $m^{-1}$ $K^{-1}$. Were these properties characteristic of a single compound, its dimensionless thermoelectric figure-of-merit would reach $zT = 3.5$, which would greatly exceed a benchmark of $zT = 0.9$ set by $Bi_2Te_3$—a base compound for creating thermoelectric materials for low and mid-temperature applications. In fact, the above record values belong to different compounds, which makes the highest room-temperature ZT value an order of magnitude lower.

Perspective thermoelectric properties ensure recent interest into inorganic clathrates. This is explained not only by expectations of finding a material with outstanding thermoelectric properties, but also by a possibility to go deeper into the underlying physics provided by these compounds. As a result, the number of papers devoted to thermoelectric properties of the clathrates has grown from about ten published annually in the beginning of the new millennium to almost 50 in

2012. Some of these works are devoted to the preparation of new clathrates and investigation of details of their crystal structure. Many more papers describe the ways of optimizing thermoelectric properties of these compounds. Doping and substitutions within the framework are explored to enhance transport of charge carriers, whereas alloying and nanostructuring is used for reducing thermal conductivity. Coming years will witness whether these efforts prove to be successful and if new thermoelectric materials based on the clathrates emerge.

**Acknowledgments** Support of the Russian Foundation for Basic Research, grant # 13-03-00571, is gratefully acknowledged. Clathrate research in Davis is supported by the U.S. Department of Energy, Office of Basic Energy Sciences, Division of Materials Sciences and Engineering under Award DE-SC0008931.

## References

1. J.S. Kasper, P. Hagenmuller, M. Pouchard, C. Cros, Clathrate structure of silicon $Na_8Si_{46}$ and $Na_xSi_{136}$ ($x < 11$). Science **150**, 1713–1714 (1965)
2. C. Cros, M. Pouchard, P. Hagenmuller, Sur 2 Nouvelles phases du systeme silicium-sodium. C. R. Acad. Sci. Paris **260**, 4764 (1965)
3. A. San-Miguel, P. Kéghélian, X. Blase, P. Mélinon, A. Perez, J.P. Itié, A. Polian, E. Reny, C. Cros, M. Pouchard, High pressure behavior of silicon clathrates: a new class of low compressibility materials. Phys. Rev. Lett. **83**, 5290–5293 (1999)
4. A. Ammar, C. Cros, M. Pouchard, N. Jaussaud, J.M. Bassat, G. Villeneuve, M. Duttine, M. Ménétrier, E. Reny, On the clathrate form of elemental silicon, $Si_{136}$: preparation and characterization of $Na_xSi_{136}$ ($x \rightarrow 0$). Solid State Sci. **6**, 393–400 (2004)
5. H.G. von Schnering, H. Menke, $Ge_{38}P_8I_8$ and $Ge_{38}As_8I_8$, a new class of compounds with clathrate structure. Angew. Chem. Int. Ed. Engl. **11**, 43–44 (1972)
6. M.M. Shatruk, K.A. Kovnir, A.V. Shevelkov, I.A. Presniakov, B.A. Popovkin, First tin pnictide halides $Sn_{24}P_{19.3}I_8$ and $Sn_{24}As_{19.3}I_8$: synthesis and the clathrate-I type of the crystal structure. Inorg. Chem. **38**, 3455–3457 (1999)
7. A.V. Shevelkov, K. Kovnir, in Zintl Clathrates, ed. by T.F. Fässler. Structure and Bonding, vol. 139 (Zintl Phases) (Springer, Berlin, 2011), pp. 97–142
8. G.S. Nolas, J.L. Cohn, G.A. Slack, S.B. Schjuman, Semiconducting Ge clathrates: promising candidates for thermoelectric applications. Appl. Phys. Lett. **73**, 178–180 (1998)
9. M.M. Shatruk, K.A. Kovnir, M. Lindsjö, I.A. Presniakov, L.A. Kloo, A.V. Shevelkov, Novel compounds $Sn_{10}In_{14}P_{22}I_8$ and $Sn_{14}In_{10}P_{21.2}I_8$ with clathrate I structure: synthesis and crystal and electronic structure. J. Solid State Chem. **161**, 233–242 (2001)
10. K. Kishimoto, K. Akai, N. Muraoka, T. Koyanagi, M. Matsuura, Synthesis and thermoelectric properties of type-I clathrate $Ge_{30}P_{16}Te_8$. Appl. Phys. Lett. **89**, 172106 (2006)
11. K. Kishimoto, S. Arimura, T. Koyanagi, Preparation and thermoelectric properties of sintered iodine-containing clathrate compounds $Ge_{38}Sb_8I_8$ and $Sn_{38}Sb_8I_8$. Appl. Phys. Lett. **88**, 222115 (2006)
12. E. Reny, S. Yamanaka, C. Cros, M. Pouchard, High pressure synthesis of an iodine doped silicon clathrate compound. Chem. Commun. 2505–2506 (2000)
13. K. Kishimoto, T. Koyanagi, K. Akai, M. Matsuura, Synthesis and thermoelectric properties of type-I clathrate compounds $Si_{46-x}P_xTe_8$. Jpn. J. Appl. Phys. **46**, L746–L748 (2007)
14. N.S. Abramchuk, W. Carrillo-Cabrera, I. Veremchuk, N. Oeschler, A.V. Olenev, Y. Prots, U. Burkhardt, E.V. Dikarev, J. Grin, A.V. Shevelkov, Homo- and heterovalent substitutions in

the new clathrates I $Si_{30}P_{16}Te_{8-x}Se_x$ and $Si_{30+x}P_{16-x}Te_{8-x}Br_x$: synthesis, crystal structure, and thermoelectric properties. Inorg. Chem. **51**, 11396–11405 (2012)

15. R. Nesper, J. Curda, H.G. von Schnering, $Ge_{4.06}I$, an unexpected germanium subiodide—a tetragermanioiodonium(III) iodide with clathrate structure $[Ge_{46-x}I_x]I_8$, x = 8/3. Angew. Chem. Int. Ed. Engl. **25**, 350–352 (1986)
16. K.A. Kovnir, A.N. Uglov, J.V. Zaikina, A.V. Shevelkov, New cationic clathrate: synthesis and structure of $[Si_{40}P_6]I_{6.5}$. Mendeleev Commun. **4**, 135–136 (2004)
17. N. Jaussaud, P. Toulemonde, M. Pouchard, A. San Miguel, P. Gravereau, S. Pechev, G. Goglio, C. Cros, High pressure synthesis and crystal structure of two forms of a new tellurium-silicon clathrate related to the classical type I. Solid State Sci. **6**, 401–411 (2004)
18. N. Jaussaud, M. Pouchard, P. Gravereau, S. Pechev, G. Goglio, C. Cros, A. San Miguel, P. Toulemonde, Structural trends and chemical bonding in Te-doped silicon clathrates. Inorg. Chem. **44**, 2210–2214 (2005)
19. J.V. Zaikina, K.A. Kovnir, U. Schwarz, H. Borrmann, A.V. Shevelkov, Crystal structure of silicon phosphorus telluride, $Si_{46-x}P_xTe_y$ (y = 7.35, 6.98, 6.88; $x \leq 2y$), a cationic clathrate-I. Z. Kristallogr. New Cryst. Struct. **222**, 177–179 (2007)
20. H. Menke, H.G. von Schnering, Die Käfigverbindungen $Ge_{38}A_8X_8$ mit A = P, As, Sb und X = Cl, Br. J. Z. Anorg. Allg. Chem. **395**, 223–238 (1973)
21. H. Menke, H.G. von Schnering, Die partielle Substitution von Ge durch GaAs und GaSb in den Käfigverbindungen $Ge_{38}As_8J_8$ and $Ge_{38}Sb_8J_8$. Z. Anorg. Allg. Chem. **424**, 108–114 (1976)
22. M.A. Kirsanova, L.N. Reshetova, A.V. Olenev, A.V. Shevelkov, On the crystal structure of the germanium-based cationic clathrates $[Ge_{38.3}Sb_{7.7}]I_{7.44}$, $[Ge_{38.1}P_{7.9}]I_8$, and $[Ge_{30.5}Sn_{7.7}P_{7.75}]I_{7.88}$. Russ. J. Coord. Chem. **38**, 192–199 (2012)
23. K.A. Kovnir, N.S. Abramchuk, J.V. Zaikina, M. Baitinger, U. Burkhardt, W. Schnelle, A.V. Olenev, O.I. Lebedev, G. van Tendeloo, E.V. Dikarev, A.V. Shevelkov, $Ge_{40.0}Te_{5.3}I_8$: synthesis, crystal structure, and properties of a new clathrate-I compound. Z. Kristallogr. **221**, 527–532 (2006)
24. M.M. Shatruk, K.A. Kovnir, A.V. Shevelkov, B.A. Popovkin, A new Zintl phase $Sn_{19.3}Cu_{4.7}P_{22}I_8$ with a structure of the clathrate-I type: directional synthesis and structure. Russ. J. Inorg. Chem. **45**, 203–209 (2000)
25. A.V. Shevelkov, L.N. Reshetova, K.A. Kovnir, Unpublished
26. K.A. Kovnir, A.V. Sobolev, I.A. Presniakov, O.I. Lebedev, G. Van Tendeloo, W. Schnelle, Yu. Grin, A.V. Shevelkov, $Sn_{19.3}Cu_{4.7}As_{22}I_8$: a new clathrate-I compound with transition-metal atoms in the cationic framework. Inorg. Chem. **44**, 8786–8793 (2005)
27. K.A. Kovnir, M.M. Shatruk, L.N. Reshetova, I.A. Presniakov, E.V. Dikarev, M. Baitinger, F. Haarmann, W. Schnelle, M. Baenitz, Yu. Grin, A.V. Shevelkov, Novel compounds $Sn_{20}Zn_4P_{22-v}I_8$ (v = 1.2), $Sn_{17}Zn_7P_{22}I_8$, and $Sn_{17}Zn_7P_{22}Br_8$: synthesis, properties, and special features of their clathrate-like crystal structures. Solid State Sci. **7**, 957–968 (2005)
28. M.M. Shatruk, Synthesis and crystal structure of new metal pnictide-halides of host-guest type. PhD Thesis in Inorganic Chemistry, Lomonosov Moscow State University, Moscow, Russia (2000)
29. E.A. Kelm, A.V. Olenev, M.A. Bykov, A.V. Sobolev, I.A. Presniakov, V.A. Kulbachinskii, V.G. Kytin, A.V. Shevelkov, Synthesis, crystal structure, and thermoelectric properties of clathrates in the Sn–In–As–I system. Z. Anorg. Allg. Chem. **637**, 2059–2067 (2011)
30. L.N. Reshetova, A.V. Shevelkov, Synthesis and clathrate-type crystal structure of a solid solution in the Sn–In–P–Br system. Russ. Chem. Bull. **61**, 28–32 (2012)
31. K.A. Kovnir, J.V. Zaikina, L.N. Reshetova, A.V. Olenev, E.V. Dikarev, A.V. Shevelkov, Unusually high chemical compressibility of normally rigid type-I clathrate framework: synthesis and structural study of $Sn_{24}P_{19.3}Br_xI_{8-x}$ solid solution, the prospective thermoelectric material. Inorg. Chem. **43**, 3230–3236 (2004)
32. J.V. Zaikina, W. Schnelle, K.A. Kovnir, A.V. Olenev, Yu. Grin, A.V. Shevelkov, Crystal structure, thermoelectric and magnetic properties of the type-I clathrate solid solutions $Sn_{24}P_{19.3(2)}Br_xI_{8-x}$ ($0 \leq x \leq 8$) and $Sn_{24}P_{19.3(2)}Cl_yI_{8-y}$ ($y \leq 0.8$). Solid State Sci. **9**, 664–671 (2007)

33. J.V. Zaikina, K.A. Kovnir, A.V. Sobolev, I.A. Presniakov, Yu. Prots, M. Baitinger, W. Schnelle, A.V. Olenev, O.I. Lebedev, G. Van Tendeloo, Yu. Grin, A.V. Shevelkov, $Sn_{20.5-3.5}As_{22}I_8$: a largely disordered cationic clathrate with a new type of superstructure and abnormally low thermal conductivity. Chem. Eur. J. **13**, 5090–5099 (2007)
34. M.A. Kirsanova, L.N. Reshetova, A.V. Olenev, A.M. Abakumov, A.V. Shevelkov, Semiclathrates of the Ge–P–Te System: synthesis and crystal structures. Chem. Eur. J. **17**, 5719–5726 (2011)
35. M.A. Kirsanova, T. Mori, S. Maruyama, M. Matveeva, D. Batuk, A.M. Abakumov, A.V. Gerasimenko, A.V. Olenev, Yu. Grin, A.V. Shevelkov, Synthesis, structure, and transport properties of type-I derived clathrate $Ge_{46-x}P_xSe_{8-y}$ (x = 15.4(1); y = 0-2.65) with diverse host-guest bonding. Inorg. Chem. **52**, 577–588 (2013)
36. J.V. Zaikina, K.A. Kovnir, F. Haarmann, W. Schnelle, U. Burkhardt, H. Borrmann, U. Schwarz, Yu. Grin, A.V. Shevelkov, The first silicon-based cationic clathrate III with high thermal stability: $Si_{172-x}P_xTe_y$ ($x = 2y$, $y > 20$) Chem. Eur. J. **14**, 5414–5422 (2008)
37. M. A. Kirsanova, T. Mori, S. Maruyama, A. M. Abakumov, G. Van Tendeloo, A. Olenev, A. V. Shevelkov, Cationic clathrate of type-III $Ge_{172-x}P_xTe_y$ (y ≈ 21.5, x ≈ 2y): synthesis, crystal structure and thermoelectric properties. Inorg. Chem. **52**, 8272−8279 (2013)
38. N. Jaussaud, M. Pouchard, G. Goglio, C. Cros, A. Ammar, F. Weill, P. Gravereau, High pressure synthesis and structure of a novel clathrate-type compound: $Te_{7+x}Si_{20-x}$ (x ~ 2.5). Solid State Sci. **5**, 1193–1200 (2003)
39. M.A. Kirsanova, A.V. Olenev, A.M. Abakumov, M.A. Bykov, A.V. Shevelkov, Extension of the clathrate family: the type X clathrate $Ge_{79}P_{29}S_{18}Te_6$. Angew. Chem. Int. Ed. **50**, 2371–2374 (2011)
40. J.V. Zaikina, K.A. Kovnir, U. Burkhardt, W. Schnelle, F. Haarmann, U. Schwarz, Yu. Grin, A.V. Shevelkov, Cationic clathrate-I $Si_{46-x}P_xTe_y$ ($6.6(1) \leq y \leq 7.5(1)$, $x \leq 2y$): crystal structure, homogeneity range, and physical properties. Inorg. Chem. **48**, 3720–3730 (2009)
41. N. Melnychenko-Koblyuk, A. Grytsiv, P. Rogl, H. Schmid, G. Giester, The clathrate $Ba_8Cu_xGe_{46-x-y}\square_y$: phase equilibria and crystal structure. J. Solid State Chem. **182**, 1754–1760 (2009)
42. M. Christensen, N. Lock, J. Overgaard, B.B. Iversen, Crystal structures of thermoelectric n- and p-type $Ba_8Ga_{16}Ge_{30}$ studied by single crystal, multitemperature, neutron diffraction, conventional X-ray diffraction and resonant synchrotron X-ray diffraction. J. Am. Chem. Soc. **128**, 15657–15665 (2006)
43. S.E. Latturner, J.D. Bryan, N. Blake, H. Metiu, G.D. Stucky, Siting of antimony dopants and gallium in $Ba_8Ga_{16}Ge_{30}$ clathrates grown from gallium flux. Inorg. Chem. **41**, 3956–3961 (2002)
44. J.H. Roudebush, N. Tsujii, A. Hurtando, H. Hope, Yu. Grin, S.M. Kauzlarich, Phase range of the Type-I clathrate $Sr_8Al_xSi_{46-x}$ and crystal structure of $Sr_8Al_{10}Si_{36}$. Inorg. Chem. **51**, 4161–4169 (2012)
45. Ya. Mudryk, P. Rogl, C. Paul, S. Berger, E. Bauer, G. Hilscher, C. Godart, H. Noël, A. Saccone, R. Ferro, Crystal chemistry and thermoelectric properties of clathrates with rare-earth substitution. Phys. B **328**, 44–48 (2003)
46. J. Dünner, A. Mewis, $Ba_8Cu_{16}P_{30}$—a new ternary variant of the clathrate-I type structure Z. Anorg. Allg. Chem. **621**, 195–196 (1995)
47. K. Kovnir, U. Stockert, S. Budnyk, Yu. Prots, M. Baitinger, S. Paschen, A.V. Shevelkov, Yu. Grin, Introducing a magnetic guest to a tetrel-free clathrate: synthesis, structure, and properties of $Eu_xBa_{8-x}Cu_{16}P_{30}$ ($0 \leq x \leq 1.5$). Inorg. Chem. **50**, 10387–10396 (2011)
48. W. Carrillo-Cabrera, S. Budnyk, Yu. Prots, Yu. Grin, $Ba_8Ge_{43}$ revisited: a 2a' × 2a' × 2a' 2a' superstructure of the clathrate-I type with full vacancy ordering. Z. Anorg. Allg. Chem. **630**, 2267–2276 (2004)
49. F. Dubois, T.F. Fässler, Ordering of vacancies in type-I tin clathrate: superstructure of $Rb_8Sn_{44-2}$. J. Amer. Chem. Soc. **127**, 3264–3265 (2005)
50. W. Wang, B.O. Carter, C.L. Brau, A. Steiner, J. Basca, J.T.A. Jones, C. Cropper, Y.Z. Khimyak, D.J. Adams, A.I. Cooper, Reversible methane storage in a polymer-supported

semi-clathrate hydrate at ambient temperature and pressure. Chem. Mater. **21**, 3810–3815 (2009)

51. S. Hashimoto, T. Sugahara, M. Moritoki, H. Sato, K. Ohgaki, Thermodynamic stability of hydrogen plus tetra-n-butyl ammonium bromide mixed gas hydrate in nonstoichiometric aqueous solutions. Chem. Eng. Sci. **63**, 1092–1097 (2008)
52. J.V. Zaikina, T. Mori, K. Kovnir, D. Teschner, A. Senyshyn, U. Schwarz, Yu. Grin, A.V. Shevelkov, Bulk and surface structure and high-temperature thermoelectric properties of inverse Clathrate-III in the Si–P–Te System. Chem. Eur. J. **16**, 12582–12589 (2010)
53. G.A. Jeffrey, in Inclusion Compounds, ed. by I.J.L. Atwood, J.E.D. Davies, D.D. MacNichol, (Academic Press, London, 1983)
54. M. Christensen, S. Johnsen, B.B. Iversen, Thermoelectric clathrates of type I. Dalton Trans. **39**, 978–992 (2010)
55. M. Beekman, G.S. Nolas, Inorganic clathrate-II materials of group 14: synthetic routes and physical properties. J. Mater. Chem. **18**, 842–851 (2008)
56. V.V. Novikov, A.V. Matovnikov, D.V. Avdashchenko, N.V. Mitroshenkov, E. Dikarev, S. Takamizawa, M.A. Kirsanova, A.V. Shevelkov, Low-temperature structure and lattice dynamics of the thermoelectric clathrate $Sn_{24}P_{19.3}I_8$. J. Alloys Compd. **520**, 174–179 (2012)
57. M. Falmbigl, P.F. Rogl, E. Bauer, M. Kriegisch, H. Müller, S. Paschen, On the thermoelectric potential of inverse clathrates. Mater. Res. Soc. Symp. Proc. **1166**, 03–06 (2009)
58. S. Stefanoski, L.N. Reshetova, A.V. Shevelkov, G.S. Nolas, Low-temperature transport properties of $Sn_{24}P_{19.3}Br_8$ and $Sn_{17}Zn_7P_{22}Br_8$. J. Electr. Mater. **38**, 985–989 (2009)
59. M. Falmbigl, G. Rogl, P. Rogl, M. Kriegisch, H. Müller, E. Bauer, M. Reinecker, W. Schranz, Thermal expansion of thermoelectric type-I-clathrates. J. Appl. Phys. **108**, 043529 (2010)
60. U. Aydemir, C. Candolfi, H. Borrmann, M. Baitinger, M. Ormeci, W. Carrillo-Cabrera, C. Chubilleau, B. Lenoir, A. Dauscher, N. Oeschler, F. Steglich, Yu. Grin, Crystal structure and transport properties of $Ba_8Ge_{43-3}$. Dalton Trans. **39**, 1078–1088 (2010)
61. H. Shimizu, T. Kume, T. Kuroda, S. Sasaki, H. Fukuoka, S. Yamanaka, High-pressure Raman study of the iodine-doped silicon clathrate $I_8Si_{44}I_2$. Phys. Rev. B **68**, 212102 (2003)
62. M.A. McGuire, T.K. Reynolds, F.J. DiSalvo, Exploring thallium compounds as thermoelectric materials: seventeen new thallium chalcogenides. Chem. Mater. **17**, 2875–2884 (2005)
63. D.Y. Chung, T.P. Hogan, M. Rocci-Lane, P. Brazis, J.R. Ireland, C.R. Kannewurf, M. Bastea, C. Uher, M.G. Kanatzidis, A new thermoelectric material: $CsBi_4Te_6$. J. Am. Chem. Soc. **126**, 6414–6428 (2004)
64. A.M. White, *Properties of Materials* (Oxford University Press, Oxford, 1999)
65. S. Paschen, V.H. Tran, M. Baenitz, W. Carrillo-Cabrera, Yu. Grin, F. Steglich, Clathrate $Ba_6Ge_{25}$: thermodynamic, magnetic, and transport properties. Phys. Rev. B. **65**, 134435 (2002)
66. A.V. Yakimchuk, J.V. Zaikina, L.N. Reshetova, L.I. Ryabova, D.R. Khokhlov, A.V. Shevelkov, Impedance of $Sn_{24}P_{19.3}Br_xI_{8-x}$ semiconducting clathrates. Low Temp. Phys. **33**, 276–279 (2007)
67. J.F. Snyder, E.S. Toberer, Complex thermoelectric materials. Nature Mat. **7**, 105–114 (2008)
68. C. Wood, Materials for thermoelectric energy conversion. Rep. Prog. Phys. **51**, 459–539 (1988)
69. B. Du, Y. Saiga, K. Kajisa, T. Takabatake, E. Nishibori, H. Sawa, Study of $\alpha\leftrightarrow\beta$ transformation in the dimorphic clathrate $Ba_8Ga_{16}Sn_{30}$. Philos. Mag. **92**, 2541–2552 (2012)
70. K. Kaltzoglou, S.D. Hoffmann, T.F. Fässler, Order-disorder phase transition in type-I clathrate $Cs_8Sn_{44}\Box_2$. Eur. J. Inorg. Chem. **26**, 4162–4167 (2007)
71. R. Kröner, K. Peters, H.G. von Schnering, R. Nesper, Crystal structure of the clathrate-II, $Ba_{16}Ga_{32}Sn_{104}$. Z. Kristallogr. New Cryst. Struct. **213**, 664 (1998)
72. S. Mano, T. Onimaru, S. Yamanaka, T. Takabatake, Off-center rattling and thermoelectric properties of type-II clathrate (K, Ba)$_{24}$(Ga, Sn, $\Box$)$_{136}$ single crystals. Phys. Rev. B **84**, 214101 (2011)

73. M.C. Schäfer, S. Bobev, Tin clathrates with the type II structure. J. Am. Chem. Soc. **135**, 1696–1699 (2013)
74. M.C. Schäfer, S. Bobev, K and Ba distribution in the structures of the clathrate compounds $K_xBa_{16-x}$(Ga, Sn)$_{136}$ (x = 0.8, 4.4, and 12.9) and $K_xBa_{8-x}$(Ga, Sn)$_{46}$ (x = 0.3). Acta Cryst. **69**, 319–323 (2013)
75. S. Bobev, S. Sevov, Clathrate III of group 14 exists after all. J. Am. Chem. Soc. **123**, 3389–3390 (2001)
76. H.G. von Schnering, A. Zürn, J.H. Chang, M. Baitinger, Yu. Grin, The intrinsic shape of the cubic *cP*124 clathrate structure. Z. Anorg. Allg. Chem. **633**, 1147–1153 (2007)
77. T.F. Fässler, C. Kronseder, $K_6Sn_{23}Bi_2$ and $K_6Sn_{25}$—two phases with chiral clathrate structure and their reactivity towards ethylenediamine. Z. Anorg. Allg. Chem. **624**, 561–568 (1998)
78. A. Kaltzoglou, T.F. Fässler, M. Christensen, S. Johnsen, B.B. Iversen, I. Presniakov, A. Sobolev, A.V. Shevelkov, Effects of the order–disorder phase transition on the physical properties of $A_8Sn_{44-2}$ (A = Rb, Cs). J. Mater. Chem. **18**, 5630–5637 (2008)
79. A. Kaltzoglou, T.F. Fässler, C. Gold, E.W. Scheidt, W. Scherer, T. Kumec, H. Shimizu, Investigation of substitution effects and the phase transition in type-I clathrates $Rb_xCs_{8-x}Sn_{44}\square_2$ ($1.3 \leq x \leq 2.1$) using single-crystal X-ray diffraction, Raman spectroscopy, heat capacity and electrical resistivity measurements. J. Solid State Chem. **182**, 2924–2929 (2009)
80. J.T. Zhao, J. Corbett, Zintl Phases in Alkali-Metal-Tin Systems: $K_8Sn_{25}$ with condensed pentagonal dodecahedra of tin. Two $A_8Sn_{44}$ phases with a defect clathrate structure. Inorg. Chem. **33**, 5721–5726 (1994)
81. S. Bobev, S. Sevov, Synthesis and characterization of $A_3Na_{10}Sn_{23}$ (A = Cs, Rb, K) with a new clathrate-like structure and of the chiral clathrate $Rb_5Na_3Sn_{25}$. Inorg. Chem. **39**, 5930–5937 (2000)
82. B. Schujman, G.S. Nolas, R.A. Young, C. Lind, A.P. Wilkinson, G.A. Slack, R. Patschke, M.G. Kanatzidis, M. Ulutagay, S.J. Hwu, Structural analysis of $Sr_8Ga_{16}Ge_{30}$ clathrate compound. J. Appl. Phys. **87**, 1529–1533 (2000)
83. S. Paschen, W. Carrillo Cabrera, A. Bentien, V.H. Tran, M. Baenitz, Yu. Grin, F. Steglich, Structural, transport, magnetic and thermal properties of $Eu_8Ga_{16}Ge_{30}$. Phys. Rev. B. **64**, 2144041 (2001)
84. G.S. Nolas, T.J.R. Weakley, J.L. Cohn, Structural, chemical, and transport properties of a new clathrate compound: $Cs_8Zn_4Sn_{42}$. Chem. Mater. **11**, 2470–2473 (1999)
85. A.P. Wilkinson, C. Lind, R.A. Young, S.D. Shastri, P.L. Lee, G.S. Nolas, Preparation, transport properties, and structure analysis by resonant X-ray scattering of the type I clathrate $Cs_8Cd_4Sn_{42}$. Chem. Mater. **14**, 1300–1305 (2002)
86. A. Kaltzoglou, S. Ponou, T.F. Fässler, Synthesis and crystal structure of mercury-substituted type-I clathrates $A_8Hg_4Sn_{42}$ (A = K, Rb, Cs). Eur. J. Inorg. Chem. **4**, 538–542 (2008)
87. T. Tanaka, T. Onimaru, K. Suekuni, S. Mano, H. Fukuoka, S. Yamanaka, T. Takabatake, Interplay between thermoelectric and structural properties of type-I clathrate $K_8Ga_8Sn_{38}$ single crystals. Phys. Rev. B **81**, 165110 (2010)
88. M. Hayashi, K. Kishimoto, K. Kishio, K. Akai, K. Asada, T. Koyanagi, Preparation and thermoelectric properties of sintered type-I clathrates $K_8Ga_xSn_{46-x}$. Dalton Trans. **39**, 1113–1117 (2010)
89. M. Hayashi, K. Kishimoto, K. Akai, H. Asada, K.T. Kishio, K. Koyanagi, Preparation and thermoelectric properties of sintered n-type $K_8M_8Sn_{38}$ (M = Al, Ga and In) with the type-I clathrate structure. J. Phys. D Appl. Phys. **45**, 455308 (2012)
90. Y. Saiga, K. Suekuni, B. Dua, T. Takabatake, Thermoelectric properties and structural instability of type-I clathrate $Ba_8Ga_{16}Sn_{30}$ at high temperatures. Solid State Commun. **152**, 1902–1905 (2012)
91. Y. Saiga, K. Suekuni, S.K. Deng, T. Yamamoto, Y. Kono, N. Ohya, T. Takabatake, Optimization of thermoelectric properties of type-VIII clathrate $Ba_8Ga_{16}Sn_{30}$ by carrier tuning. J. Alloys Compd. **507**, 1–5 (2010)

92. D. Huo, T. Sakata, T. Sasakawa, M.A. Avila, M. Tsubota, F. Iga, H. Fukuoka, S. Yamanaka, S. Aoyagi, T. Takabatake, Structural, transport, and thermal properties of the single-crystalline type-VIII clathrate $Ba_8Ga_{16}Sn_{30}$. Phys. Rev. B **71**, 075113 (2005)
93. K. Suekuni, K. Avila, M.A. Umeo, K. Fukuoka, H. Yamanaka, S. Nakagawa, T.T. Takabatake, Simultaneous structure and carrier tuning of dimorphic clathrate $Ba_8Ga_{16}Sn_{30}$. Phys. Rev. B **77**, 235119 (2008)
94. Y. Saiga, B. Dua, S.K. Deng, K. Kajisa, T. Takabatake, Thermoelectric properties of type-VIII clathrate $Ba_8Ga_{16}Sn_{30}$ doped with Cu. J. Alloys Compd. **537**, 303–307 (2012)
95. B. Du, Y. Saiga, K. Kajisa, T. Takabatake, Thermoelectric performance of Zn-substituted type-VIII clathrate $Ba_8Ga_{16}Sn_{30}$ single crystals. J. Appl. Phys. **111**, 013707 (2012)
96. S.K. Deng, Y. Saiga, K. Suekuni, T. Takabatake, Effect of Al substitution on the thermoelectric properties of the type VIII clathrate $Ba_8Ga_{16}Sn_{30}$. J. Electr. Mater. **40**, 1124–1128 (2011)
97. Y. Kono, N. Ohya, Y. Saiga, K. Suekuni, T. Takabatake, K. Akai, S. Yamamoto, Carrier doping in the type VIII clathrate $Ba_8Ga_{16}Sn_{30}$ through Sb substitution. J. Electr. Mater. **40**, 845–850 (2011)
98. K.K. Kishimoto, H. Yamamoto, K. Akai, T. Koyanagi, Effect of Ge substitution on carrier mobilities and thermoelectric properties of sintered p-type $Ba_8Ga_{16+x}Sn_{30-x-y}Ge_y$ with the type-VIII clathrate structure. J. Phys. D Appl. Phys. **45**, 445306 (2012)

# Chapter 6
# Inorganic Clathrates for Thermoelectric Applications

**Stevce Stefanoski, Matt Beekman and George S. Nolas**

**Abstract** The unique crystal structures clathrates possess, in which atoms reside inside polyhedral cavities typically formed by framework group-14 elements, directly results in unconventional physical properties, including low thermal conductivity for these crystalline solids and their potential for thermoelectric heat-to-electrical energy conversion. A unique "tunablility" exists in these materials by altering the type of guest and host atoms, as well as the guest content. Poor thermal conduction, akin to amorphous solids, and electronic properties typical for crystalline materials define the so called phonon glass-electron crystal (PGEC) concept. This concept, first proposed by Glen Slack, identified clathrates as promising thermoelectric materials. In order to manufacture efficient thermoelectric devices the required thermoelectric performance must be realized in mechanically robust materials comprised of earth-abundant non-toxic elements, and low-cost manufacturing processes. Considering the recent advances in thermoelectric energy conversion technologies and the current interest for research in materials with excellent thermoelectric properties, it is likely that clathrates will play a role in thermoelectric power generation applications. In this chapter we provide an overview of the interesting transport properties these materials possess, then present some of the recent developments in the field of thermoelectric clathrates

S. Stefanoski (✉)
Geophysical Laboratory R-G18, Carnegie Institution of Washington,
5251 Broad Branch Rd. NW, Washington, DC 20015-1305, USA
e-mail: sstefanoski@carnegiescience.edu

M. Beekman (✉)
Department of Natural Sciences, Oregon Institute of Technology,
3201 Campus Drive, Klamath Falls, OR 97601, USA
e-mail: matt.beekman@oit.edu

G. S. Nolas
Department of Physics, University of South Florida, 4202 E. Fowler Ave.,
Tampa, FL 33620, USA
e-mail: gnolas@usf.edu

G. S. Nolas (ed.), *The Physics and Chemistry of Inorganic Clathrates*,
Springer Series in Materials Science 199, DOI: 10.1007/978-94-017-9127-4_6,

while highlighting potential strategies for further improvement of their thermoelectric properties.

## 6.1 Introduction

The development of technologies that convert waste heat into useful electrical energy provides one pathway to reduce our dependence on fossil fuels while decreasing greenhouse gas emissions [1]. One such technology utilizes thermoelectric (TE) materials for solid-state heat-to-electrical energy conversion. The thermoelectric effect was reported in 1823 by Thomas Seebeck, in the proceedings of the Prussian Academy of Sciences entitled *Magnetic polarization of metals and ores produced by a temperature difference*, who indicated that electrical currents arise in closed circuits made of different conductors held at different junction temperatures [2]. This effect is also known as the Seebeck effect (Fig. 6.1a).

The ratio of the voltage, $\Delta V$, developed as a result of the temperature difference, $\Delta T$, across the material is related to an intrinsic property of the material, known as the Seebeck coefficient or thermopower, $S$, where $S = \Delta V/\Delta T$. A few years later Peltier [3] observed a related phenomenon where heat was either absorbed or released depending on the direction of the current through a junction of two dissimilar metals (Fig. 6.1b). It was observed that the heat absorbed or released, $Q$, is proportional to the current, $I$, produced across the junction, $Q = \Pi I$, where $\Pi$ is the Peltier coefficient. The Seebeck and Peltier effects are related to each other by the relation $\Pi = ST$ [4]. An efficient TE material would produce a high voltage at a given temperature difference across the junction, therefore high $S$ is desirable for practical TE applications. The performance of any TE material at a temperature $T$ is assessed by its dimensionless figure of merit, $ZT$, defined as:

$$ZT = \frac{S^2\sigma}{\kappa}T \tag{6.1}$$

where $\sigma$ is the electrical conductivity and $\kappa$ the total thermal conductivity which includes contributions from the crystal lattice, $\kappa_L$, and electrons, $\kappa_e$ ($\kappa = \kappa_L + \kappa_e$).

Multiple TE couples, such as those shown in Fig. 6.1a and b, arranged electrically in series and thermally in parallel form a TE module (Fig. 6.1c). The device is essentially a solid-state heat engine or refrigerator [5]. TE devices are compact, quiet (no moving parts), and provide localized heating or cooling as solid state refrigerators or heat pumps or can generate electrical power [1]. The efficiency, $\eta$, of a TE couple (power generation mode) is given by the ratio between the power delivered to the load, $W$, and the heat flow rate through the couple, $Q_H$ [6]:

$$\eta = \frac{W}{Q_H} = \frac{T_H - T_C}{T_H}\left[\frac{\sqrt{1+ZT}-1}{\sqrt{1+ZT}+T_H/T_C}\right] \tag{6.2}$$

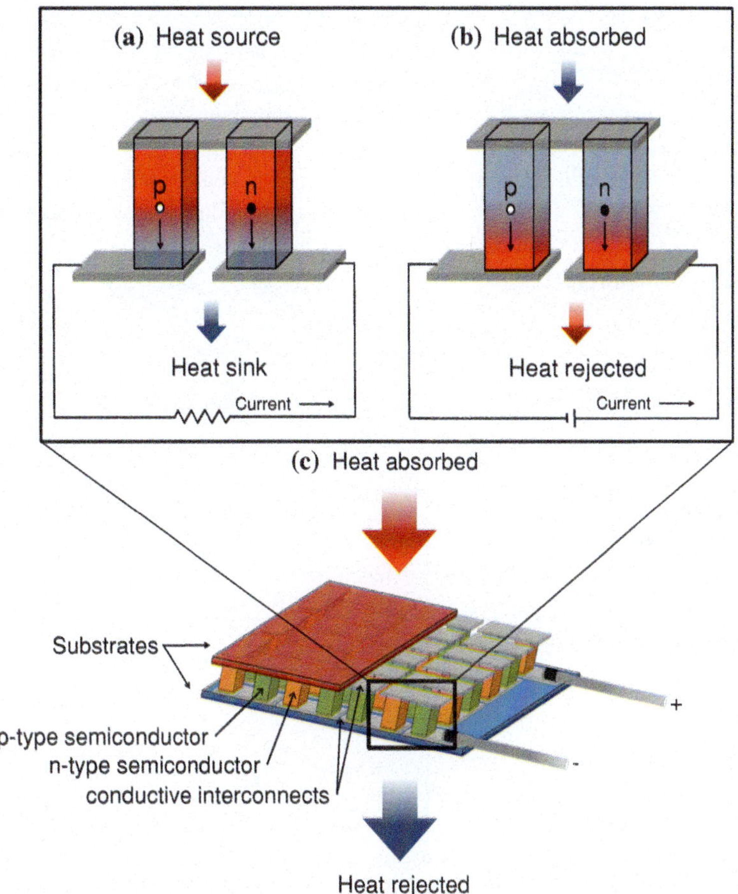


**Fig. 6.1** **a** The Seebeck effect: an electrical potential difference is induced across a thermoelectric material when a temperature difference is imposed generating an electrical current, i.e. solid state conversion of heat into electrical energy. **b** The Peltier effect: a *p*–*n* junction absorbs or releases heat when current passes through the junction, in the respective directions. **c** An example of a thermoelectric module consisting of an array of *p*–*n* couples connected electrically in series and thermally in parallel

with $T_H$ and $T_C$ being the temperatures of the hot and cold side junctions, respectively, and Z ($=S^2\sigma/\kappa$) is the figure of merit for the couple. From Eq. 6.2, as $ZT$ approaches infinity $\eta$ approaches the Carnot efficiency (the term in front of the brackets), i.e. the upper theoretical limit to the efficiency of an ideal thermodynamic cycle. The quantity in brackets encompasses the irreversible transport processes that tend to reduce the efficiency.

For almost half a century $ZT$ has been limited to $\sim$1. However, the expanded use of TE technology in practical applications will require higher $ZT$ values [7]. It is therefore imperative to search for new materials with improved TE properties.

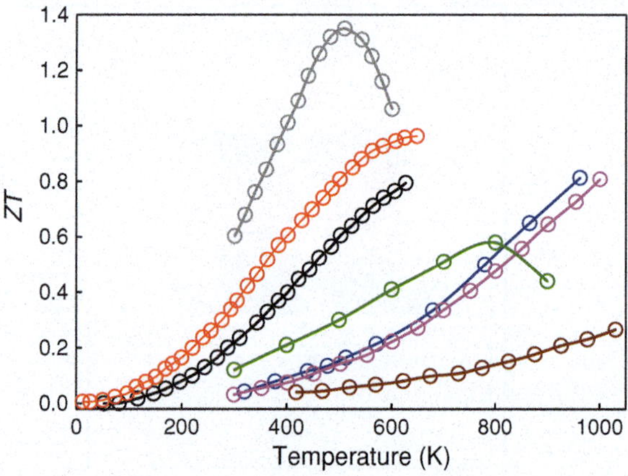


**Fig. 6.2** *ZT* as a function of temperature for representative clathrates compositions: $Ba_8Ga_{15.8}Cu_{0.02}Sn_{30.1}$ (*gray*) [9], $Sr_8Ga_{16}Ge_{30}$ (*red*) [10], $Eu_8Ga_{16}Ge_{30}$ (*black*) [10], $Sr_8A_{16.3}Ga_{10.3}Ge_{29.4}$ (*green*) [11], $Ba_8Ga_{16}Ge_{30}$ (*blue*, *pink*) [12, 13], and $Ba_8Al_{14}Si_{31}$ (*dark red*) [14]

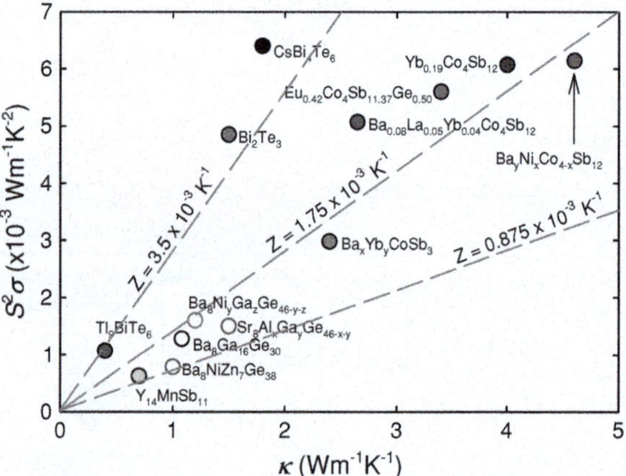


**Fig. 6.3** Power factor versus $\kappa$ for selected bulk thermoelectric materials [9, 11–13, 16–23]. Clathrates are shown with *open symbols*. *Dashed lines* are curves of constant *Z* values as indicated

Clathrates have attracted interest as potential TE materials over the last two decades. *ZT* as a function of temperature for selected state-of-art clathrates is shown in Fig. 6.2. The relatively high *ZT* values make clathrates relevant for TE power conversion applications, competitive with other state-of-art materials in terms of their TE performance [8]. The flexibility in synthesizing clathrates with different compositions potentially allows for peak efficiencies over a broad range of temperatures of interest for waste heat recovery applications [1, 5, 7].

From Eq. 6.1 optimizing *ZT* requires "maximizing" the numerator, i.e. good electronic properties, and "minimizing" the denominator, i.e. poor thermal conduction. The major challenge in the search for materials with desirable TE properties is that these properties are interrelated; changing one affects the rest. This motivated the "phonon-glass electron-crystal" (PGEC) concept introduced by Slack [15], discussed later in this chapter. The power factor as function of $\kappa$ at the temperatures for which *ZT* is maximum for several clathrates compositions is given in Fig. 6.3. Lines of constant slope in such a plot represent curves of constant $Z$ allowing for a comparison of the contributions to $ZT$ from the power factor and $\kappa$, respectively. As illustrated in Fig. 6.3, high $ZT$ for clathrates primarily result from low $\kappa$ values. Their $\kappa$ values are lower than $Bi_2Te_3$ alloys, one of the best TE materials for cooling applications, and skutterudites. We also note the clathrates investigated so far typically exhibit values for the power factor that are significantly lower than most other new or established TE materials (Fig. 6.3).

## 6.2 Phonon Glass-Electron Crystal

While thermal energy can be transported through a solid *via* a variety of different mechanisms [24], the two most important for TE applications are diffusive transport of energy by the mobile charge carriers (electronic thermal conductivity, $\kappa_e$) and phonons (lattice thermal conductivity, $\kappa_L$). Since it is a relatively good approximation to treat $\kappa_e$ and $\kappa_L$ as independent for many solids, significant emphasis has been placed on the development of TE materials with low $\kappa_L$ values [6]. Amorphous solids are typically characterized by low carrier mobilities and thus low $\sigma$ values, however they also exhibit some of the lowest known thermal conductivities (excluding porous materials) and serve as useful benchmark materials for TE materials research.

Slack [25] and Cahill et al. [26] explored the theoretical limits on $\kappa$ for solids within a phonon model of heat transport. Their work utilized the concept of the minimum thermal conductivity, $\kappa_{min}$. At this minimum value the mean free path for all heat carrying phonons in a material approaches the phonon wavelengths [25]. In this limit, the material behaves as an Einstein solid in which energy transport occurs *via* a random walk of energy transfer between localized vibrations in the solid. Experimentally, $\kappa_{min}$ is often comparable to the value in the amorphous state of the same composition. In principle $\kappa_{min}$ can be achieved by the introduction of one or more phonon scattering mechanisms that reduce the phonon mean free path to its minimum value over a broad range of frequencies, and therefore reduces $\kappa_L$ over a broad range of temperatures. In practice, there are relatively few crystalline compounds for which this limit is approached.

The conceptual framework that motivated the investigation of open-structured semiconductors for TE applications was originally introduced in 1995 by Slack in his seminal contribution to the CRC Handbook of Thermoelectrics [15]. This volume, and Slack's contribution in particular, continue to remain highly relevant references for TE materials research. In addition to providing research guidelines to aid the search for compound semiconductors that have "good" electrical properties for TE cooling applications, Slack emphasized that the reduction of $\kappa_L$ to the minimum possible value would result in a significant improvement in TE performance [15]. The challenge lies in accomplishing this without detrimentally affecting the electrical transport, i.e. charge carrier mobility should ideally retain values characteristic of good single-crystal semiconductors.

Slack noted several crystalline compounds known to have $\kappa_L$ values that approach $\kappa_{min}$. These included $YB_{68}$, $NH_4Cl$, and AgI, however of particular relevance are the clathrate hydrates because of their low thermal conductivities with glass-like temperature dependences [27, 28]. The common characteristic of these materials is the presence of one or more constituents that are ordered in a crystalline lattice, forming the framework, while the other constituent(s) possesses disorder resulting in strong phonon scattering . The design of materials with structural features similar to crystalline materials that have thermal conductivities near the amorphous limit was one of the key features of his approach, and helped guide the search for new TE materials. If the electrical properties are not substantially compromised by this structural “disorder” high TE performance may be achieved, with the ideal model system being a “phonon glass, electron crystal” (PGEC). As the name suggests, the vibrational states and/or phonon scattering in a PGEC material would ideally mimic those present in glasses [15]. While many of the known crystals having thermal conductivities that approach $\kappa_{\min}$ do not have favorable electrical properties for thermoelectrics [15], Slack proposed [15, 29, 30] two open-structured material systems for which the TE properties had not yet been fully investigated: filled skutterudites antimonides and inorganic clathrates. The realization of exceptional TE performance in both of these material systems and their continued intensive development is an indication of the predictive power of the PGEC concept and how it has proven to be one of the more influential guiding principles in TE materials research.

## 6.3 Thermal Conductivity of Inorganic Clathrates

Due to remarkable flexibility in composition intermetallic clathrates display a wide variety of electrical behavior, from very good metals to semiconductors (as will be discussed in Sect. 6.4). Thus, in some compositions the charge carriers dominate $\kappa$ [31], while in others they play a relatively minor role. Compositions with pronounced metallic conduction are not typically of interest for TE applications, thus we will not consider the thermal conductivity of these compositions in detail. In contrast, many clathrates can be described as doped semiconductors. In such materials the relative importance of the electronic contribution depends on the value of $\kappa_L$ which, as discussed below, can be very low for intermetallic clathrates. $\kappa_e$ is typically estimated using the measured $\sigma$ and the Wiedemann-Franz relation [32] thus allowing for an estimate of $\kappa_L$ from the measured $\kappa$ values.

The first studies of thermal conduction in intermetallic clathrates were carried out by Nolas et al. who measured the transport properties of a number of different Si, Ge, and Sn-based clathrate compositions [33–35]. $Sr_8Ga_{16}Ge_{30}$ and $Eu_8Ga_{16}Ge_{30}$ compounds have low $\kappa_L$ values with temperature dependences that are remarkably similar to the “universal” behavior of amorphous dielectric solids [35] (see Fig. 6.4). Employing a phenomenological Debye–Callaway model (kinetic theory) in which several phonon scattering mechanisms are accounted for in a linear combination of scattering rates, very good agreement with the experimental data was obtained in successfully modelling this glass-like temperature

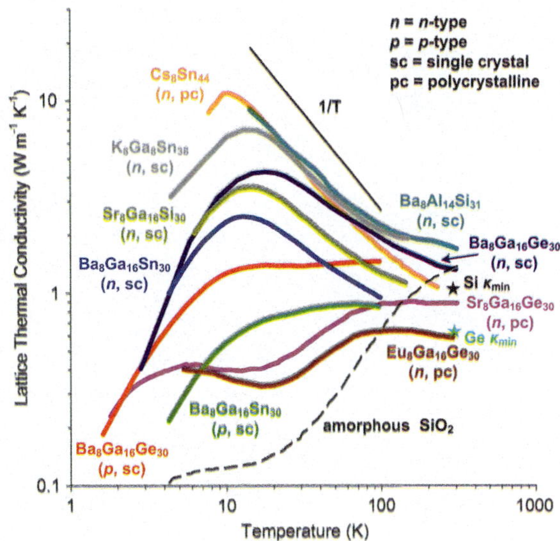


**Fig. 6.4** Low temperature $\kappa_L$ for various type I clathrates [35, 40–44], as well as data for amorphous $SiO_2$ (*dashed curve*) and room temperature $\kappa_{min}$ for elemental silicon and germanium (*stars*). The *solid black line* indicates a 1/*T* temperature dependence characteristic of phonon scattering dominated by Umklapp processes in crystalline solids

dependence [35, 36]. This analysis suggested that the thermal transport at low temperatures (<100 K) is affected by the dynamic disorder from Sr and Eu that strongly interact with the heat carrying phonons. The presence of atomic tunneling of the guest was later described by a variety of independent experimental and computational techniques [37–39]. Subsequent studies have similarly used variations based on the Debye–Callaway model to describe the guest-host interaction, and this approach has been successful in describing the thermal transport in many clathrate compositions, including Sn-based type I clathrates [40].

Lattice thermal conductivity data below 300 K for representative clathrates are summarized in Fig. 6.4 [35, 40–44]. These data, for a representative number of compositions collected from both polycrystalline and single crystalline specimens, allow for a comparison of the effect of guest and framework composition on $\kappa_L$. While some compositions have lattice thermal conductivities that are characteristic of glasses, others have $\kappa_L$ values that more closely resemble the typical temperature dependence of defect-free crystalline solids in which Umklapp processes [24] produce a monotonically decreasing $\kappa_L$ with increasing temperature above ~10 K.

The unusual thermal conductivity observed in several intermetallic clathrates at low temperatures ($T < 300$ K, see Fig. 6.4) has motivated nearly two decades of scientific investigations aimed at understanding the lattice dynamics and phonon scattering processes in these materials. These investigations which include thermophysical [31, 40, 42, 45], crystallographic [36, 40, 41, 46, 47], Raman scattering [48–50], neutron and nuclear inelastic scattering [51–53], X-ray absorption [54], ultrasound [37, 55, 56], nuclear magnetic resonance [57, 58], and computational/theoretical studies [59–62], for example, have produced a wealth of information concerning structure-property correlations in clathrates. Much of the literature has been summarized in a number of reviews [10, 63–69] (see Fig. 6.5

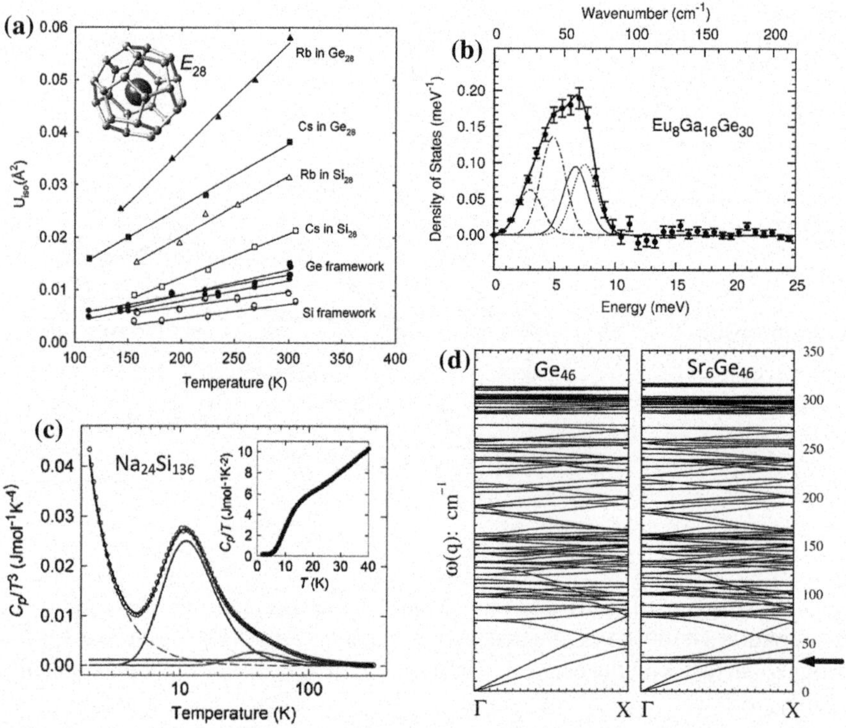


**Fig. 6.5** Selected experimental and theoretical data exemplifying the presence of low-energy vibrations of the guest atoms in intermetallic clathrates. **a** Isotropic atomic displacement parameters ($U_{iso}$) obtained from single-crystal X-ray diffraction refinements ($A_8Na_{16}X_{136}$, A = Rb or Cs, X = Si or Ge) represent the mean square displacement of an atom from its "equilibrium" crystallographic site. Guest atoms exhibit significantly larger $U_{iso}$ than the corresponding Si and Ge framework, with a more pronounced temperature dependence [63]. **b** Partial phonon density of states derived from nuclear inelastic scattering (NIS) experiments reveals the low energy Eu guest vibrations in $Eu_8Ga_{16}Ge_{30}$ at 20 K. The selective nature of NIS demonstrates that Eu does not contribute any vibrational modes above 9 meV at this temperature [51]. **c** Experimental heat capacity ($C_p$, data points) of $Na_{24}Si_{136}$ plotted as $C_p/T^3$ versus $T$ (Inset: $C_p/T$ vs. $T$ for $T < 40$ K). A Debye-Callaway model (*black curve*) incorporating contributions from electrons (*dashed curve*), Einstein oscillators (*red* and *blue curves* for Na in the $Si_{20}$ and $Si_{28}$ polyhedra, respectively), and Si framework (*green curve*) accurately describes the data over a wide temperature range and indicates the presence of low energy Na vibrations with Einstein temperatures of 55 and 166 K, respectively [31]. **d** Ab initio phonon dispersion curves for empty $Ge_{46}$ and $Sr_6Ge_{46}$ calculated from density functional theory, assuming a Leonard-Jones potential for the guest-host interaction [59]. The low energy modes and avoided crossing produced by incorporation of Sr into the $Ge_{46}$ framework are indicated by the arrow. **a** Reproduced from [63] by permission the Royal Society of Chemistry. **b** Reprinted with permission from [51]. Copyright (2005), American Physical Society. **c** Reprinted with permission from [31]. Copyright (2010), American Physical Society. **d** Reprinted with permission from [59]. Copyright (2001), American Physical Society

for selected experimental and computational results). Because of the complexity of the specific thermal transport processes, the development of a detailed and complete understanding of the thermal transport mechanisms in complex materials can be challenging,[1] requiring a breadth of complementary techniques to help discriminate between plausible models and their limits of applicability.

From the extensive collection of available data produced by the abovementioned studies, among others, emerges the understanding that the guest atoms in intermetallic clathrates can have a profound effect on the lattice dynamics thermal properties, and provides support for the PGEC concept. Low energy vibrations or "rattling" associated with the guest atoms were investigated by a variety of complementary experimental and theoretical techniques (Fig. 6.5), for a range of different guest species and framework compositions, indicating that the guests in many cases can undergo large amplitude, low frequency vibrations that are often in the energy range of the framework acoustic phonons responsible for heat transport. The majority of available data are consistent with a model in which the guest atoms participate in resonant scattering of the framework acoustic phonons resulting in very low $\kappa_L$ values below 300 K for several compositions. This general model is the one most often invoked in describing $\kappa_L$ of clathrates in the literature. The importance of "off-center" dynamic, tunneling, and static disorder due to the guest shifting away from the center of its cage has been elucidated in a number of theoretical and experimental studies [35–38, 47, 50, 60].

Coupling between guest and framework in clathrate structures (both hydrate and intermetallic clathrates) may induce a symmetry avoided crossing in the phonon dispersion relation [28, 71]. In such cases significant phonon scattering can be expected. In addition, the avoided crossing can also produce a reduction in the group velocity of the framework phonons, which also reduces the heat transfer rate. The relative strength of these two distinct effects is in part determined by the strength of the guest-host coupling [71]. With negligible guest-host coupling, the guest atom vibrations would be truly independent and no modification of the phonon dispersion (nor scattering) occurs. In this case the guests behave as true Einstein oscillators. Shortly after the initial observations of glass-like thermal conductivity in intermetallic clathrates, computational studies by Dong et al. [59] demonstrated the avoided crossing in the calculated phonon dispersion curves of the hypothetical clathrate "$Sr_6Ge_{46}$" (see Fig. 6.5), a feature that was more recently observed experimentally using single-crystal inelastic neutron scattering on *n*-type $Ba_8Ga_{16}Ge_{30}$ by Lee et al. [53] and Christensen et al. [52]. Similar studies on "glass-like" $Sr_8Ga_{16}Ge_{30}$, $Eu_8Ga_{16}Ge_{30}$, or *p*-type $Ba_8Ga_{16}Ge_{30}$ (see Fig. 6.4) have not been reported as yet. The relative influence of the avoided crossing on scattering of the framework phonons is determined by the strength of the guest-framework coupling. This continues to be an important scientific

[1] Consider the compound semiconductor and TE material PbTe with the relatively "simple" NaCl crystal structure. A complete understanding of the mechanisms that result in low $\kappa$ for PbTe continues to be investigated [70].

question that is typically separately investigated for each composition since the underlying mechanisms that determine the thermal conductivity are very sensitive to composition, as illustrated in Fig. 6.4. Current research efforts are directed toward understanding the applicability and limitations of the various models for thermal transport in clathrates that have been proposed to explain the magnitude and temperature dependence of the thermal conductivity in the various temperature regimes of interest [35–38, 40, 47, 50, 52, 61, 71–73].

In their original interpretation of the thermal conductivity data, Nolas et al. attributed the differences in thermal transport behavior with composition (Fig. 6.4) to the relative size of the guest and cage [35, 36] which in turn influences the amplitude of the guest atom vibrations. In essence, a larger difference in size between guest and cage is expected to result in more pronounced "rattling" thus lower $\kappa_L$. The radius, mass, and chemistry of the guest atom (relative to the framework) all influence both the frequency of vibration and coupling between the guest and framework, which in turn determines the influence on the phonon dispersion and scattering. By examining literature data for a large number of clathrate compositions, Suekuni et al. [74] highlighted the general inverse correlation between $\kappa_L$ and the free space of the guest atom. The systematic transition from crystal-like to glass-like $\kappa$ with increasing $x$ in $Sr_8Ga_{16}Si_{30-x}Ge_x$ was also attributed to the increasing available space for the Sr guest and corresponding influence of "rattling" on the thermal transport [42]. The qualitative effect of the free space on $\kappa$ was directly observed in the selective filling of the two different sized cages with Na in $Na_xSi_{136}$ clathrates [75]. The latter constitutes a simple model system lacking complications of varying framework and guest composition, allowing direct comparison in a binary material system.

While an increase in available space in the cages of the clathrate structure can be directly correlated to an increase in "rattling" thus reducing $\kappa_L$ at low temperatures, other factors influence $\kappa_L$. Remarkably different thermal transport has been observed in $Ba_8Ga_{16-x}Ge_{30+x}$ [40, 72] and $Ba_8Ga_{16-x}Sn_{30+x}$ [40] type I clathrates as the Ga content is varied. $n$-type and $p$-type compositions exhibit crystalline and glass-like $\kappa$, respectively (Fig. 6.4) indicating the dependence on the framework composition and/or carrier type [40]. These observations, as well as those discussed in the above paragraphs, raise important questions about the fundamental physics of the unusual low temperature transport in these materials, aspects that continue to be actively investigated [35–38, 40, 47, 50, 52, 61, 71–73]. However, based on the data to date (Fig. 6.2) intermetallic clathrates are prospective TE materials at temperatures above room temperature, thus the question naturally arises: What influence does the "rattling" behavior have on high temperature $\kappa$ and resulting TE performance? In spite of the significant variation in low temperature $\kappa_L$ with composition, by 300 K the overwhelming majority of clathrate compositions have been reported to have $\kappa_L$ near or below 1 W/m-K, comparable to that of amorphous solids and approaching $\kappa_{min}$ of elemental Si and Ge. Above room temperature, low $\kappa_L$ appears to be a universal property for nearly all intermetallic clathrates investigated to date [35, 40–44].

A common characteristic of all intermetallic clathrates, independent of composition, is a relatively large, complex unit cell. For type I clathrates, the number of atoms per primitive unit cell, $N$, is 54. The effect of $N$ on $\kappa_L$ for crystalline solids at high temperature was analyzed, using relatively simple models, by Slack [25]. A well-known inverse correlation between $N$ and $\kappa_L$ at the Debye temperature $\Theta_D$ for a crystal is described by the relation [25]

$$\kappa_L = \frac{B\bar{M}V^{1/3}\Theta_D^2}{N^{2/3}\gamma^2} \tag{6.3}$$

where $B \approx 3.04 \times 10^{-8}$ for a large number of materials, $\bar{M}$ is the mean atomic weight, $V$ is the average volume per atom in the crystal, and $\gamma$ is the Grüneisen parameter. Equation 6.3 is derived by considering face-centered cubic crystals, and is based on the assumption that only acoustic modes carry the heat current and that phonon- phonon scattering is the only scattering mechanism [25]. At temperatures near or above $\Theta_D$ a significant fraction of acoustic phonons in the crystal are excited thus (acoustic) phonon-phonon scattering often becomes a dominant scattering mechanism [24]. The distribution of thermal energy amongst the $3(N - 1)$ optic branches (not accounted for in the model represented by Eq. 6.3), which have small group velocities, can also result in a low $\kappa_L$ [41]. Dong et al. demonstrated the effect of $N$ on $\kappa_L$ for clathrates by molecular dynamics-based calculations, specifically for $Ge_{46}$ (no guest atoms) and diamond-structured Ge. This study predicted $Ge_{46}$ should have a factor of 10 reduction in $\kappa_L$ as compared with diamond-structured Ge [59]. Thermal conductivity measurements later demonstrated low $\kappa$ values for "guest-free" $Si_{136}$ [76]. Inspection of the data summarized in Fig. 6.4 indicates that irrespective of composition, carrier type, or the low-temperature behavior, the high temperature $\kappa_L$ of intermetallic clathrates appears to be much less dependent on composition and/or phonon scattering mechanism(s). This suggests that the common low $\kappa_L$ above 300 K may be attributed in part to the large $N$ common to all compositions. Moreover, $\kappa_L$ for clathrates often approaches $\kappa_{min}$ at elevated temperatures. From this, one can conclude that any significant improvements in the high temperature TE properties of intermetallic clathrates will likely come from improvements in the power factor, since $\kappa_L$ will not likely be reduced much further.[2] Conversely, any intermetallic clathrate composition having good electrical transport properties will likely have low $\kappa$ in the temperature regime of interest for TE applications, since $\kappa_L$ approaches $\kappa_{min}$ at high $T$ due to the complex unit cell.

[2] Recently, the lowest thermal conductivities were reported in $Ba_8Au_6P_{40}$, attributed to phonon scattering from a high density of twin and intergrowth interfaces in the material [77]. It will be interesting to see if this phenomenon can be extended to other clathrate compositions.

## 6.4 Electronic Transport in Intermetallic Clathrates

The electronic properties of inorganic clathrates are often discussed in terms of a rigid band model [63]. According to this approximation electropositive guest atoms donate their valence electrons to the $sp^3$ bonded framework atoms. Compositions where electronegative guest elements contribute holes to the framework orbitals are known as "inverse" [64] (cf. Chap. 5). In order to consider clathrates for TE applications it is necessary to optimize their electronic properties thereby maximizing the power factor in Eq. 6.1. From the relationship between the electrical resistivity, $\rho$, and $\sigma$ [32]

$$1/\rho = \sigma = nq\mu \tag{6.4}$$

where $n$ is the carrier concentration, $q$ is the carrier charge, and $\mu$ is the carrier mobility, high values for $\sigma$ require high concentrations of high-mobility carriers. Furthermore, the mobility of electrons, $\mu_e$, depends on the particular carrier scattering mechanisms and the electronic structure of the material. Near room temperature $\mu_e \propto T^\alpha$ where $\alpha = -3/2$ and 3/2 for electron-phonon interactions and electron-ionized impurity interactions, respectively [6]. The Seebeck coefficient can be correlated to the electronic structure of the material by the Mott expression [78]:

$$S = \frac{\pi^2 k_B}{3q} k_B T \left[\frac{d(\ln \sigma(E))}{dE}\right]_{E=E_F} \quad \text{or} \quad S \propto \frac{1}{DOS(E_F)} \left[\frac{d(DOS(E))}{dE}\right]_{E=E_F} \tag{6.5}$$

where $E_F$ is the Fermi energy and DOS is the electronic density of states. Taking into account the energy-dependent electrical conductivity $\sigma(E) = n(E)q\mu(E)$, Eq. 6.5 can be rewritten as

$$S = \frac{\pi^2 k_B}{3q} k_B T \left[\frac{1}{n}\frac{dn(E)}{dE} + \frac{1}{\mu}\frac{d\mu(E)}{dE}\right]_{E=E_F} \tag{6.6}$$

so that $S$ can increase either by increasing the energy-dependence of $n(E)$, for instance by a local increase in DOS, or by an increased energy-dependence of $\mu(E)$, for example by scattering mechanisms that strongly depend on the energy of the charge carriers [79]. $S$ can also be thought of as the energy of the charge carriers relative to the Fermi energy [4]. For metals $S$ becomes [4]

$$S = \frac{\pi^2 k_B}{e} \frac{k_B T}{E_F} \tag{6.7}$$

and for semiconductors Eq. 6.5 becomes [4]

$$S = \frac{k_B}{2e} \frac{E_g}{k_B T} \tag{6.8}$$

where $E_g$ is the band gap. At room temperature ($k_B T/E_F \propto 0.002$) $S$ is on the order of a few μV/K for metals and >100 μV/K for semiconductors. For metals or degenerate semiconductors, assuming parabolic bands and energy-independent scattering [80], $S$ is given by [81]:

$$S = \frac{8\pi^2 k_B}{3eh^2} m^* k_B T \left(\frac{\pi}{3n}\right)^{2/3} \tag{6.9}$$

where $h$ is the Planck's constant and $m^*$ is the effective mass of the charge carriers. Large $m^*$ result in high $S$, but low $\sigma$, since the inertial effective mass is also related to $m^*$ and heavier carriers will have lower velocities, and therefore lower mobilities (Eq. 6.4). High $m^*$ and low $\mu$ is typically found in ionic compounds, whereas low $m^*$ and high $\mu$ are typical for covalent materials and materials made of elements with small electronegativity differences [81]. A weighted mobility, $U = \mu(m^*/m_0)^{3/2}$ where $m_0$ is the free electron mass, of 1,800 $cm^2/V{\cdot}s$ is required in undoped compositions with low average electronegativity differences of the constitute elements in order to achieve PGEC [15]. From the relation $S \propto n^{-2/3}$ (Eqs. 6.4 and 6.9) it is clear that in order to obtain high $ZT$ values in clathrates, a compromise between high $S$ and high $\sigma$ is necessary, as high $n$ results in high $\sigma$ but low $S$. Maximum power factors can, in principle, be obtained in semiconductor compositions with carrier concentrations on the order of $10^{19}$ carriers/$cm^3$ [81]. For doping levels less than $10^{19}$ carriers/$cm^3$ charge impurity scattering that may lower $\mu$ will be screened out presumably by a higher dielectric constant [15]. "Engineering" the electronic structure may thus be achieved by (i) varying the type and concentration of guest atoms, (ii) appropriate framework framework substitution or (iii) both (i) and (ii) simultaneously. Since desirable TE properties for clathrates should be explored for semiconducting compositions, it is imperative to obtain Zintl compositions.

A conspicuous aspect of clathrate-II compositions is the ability to vary the concentration of guest atoms and therefore alter the electronic structure and/or carrier concentration via approach (i) [63]. The type of atoms that can reside within polyhedra formed by the framework is, to some extent, determined by the relative size of the polyhedral cavities as compared to the ionic radii of the guest elements [82]. A variety of compositions with the clathrate-II structure type and general formula $A_8B_{16}M_xE_{136-x}$ (A, B = Na, K, Ba, Rb, Cs; M = Ga, Ni, Cu, Ag, Au, In; E = Si, Ge, Sn) have been synthesized and their electronic properties investigated [82–93]. One example in which the first approach (i) has been demonstrated is the clathrate-II $Na_xSi_{136}$ ($0 \leq x \leq 24$) [31, 83, 94–97], the first clathrate-II composition on which synthesis and physical properties have been reported [96–98]. Alteration of the electronic properties by varying the Na content

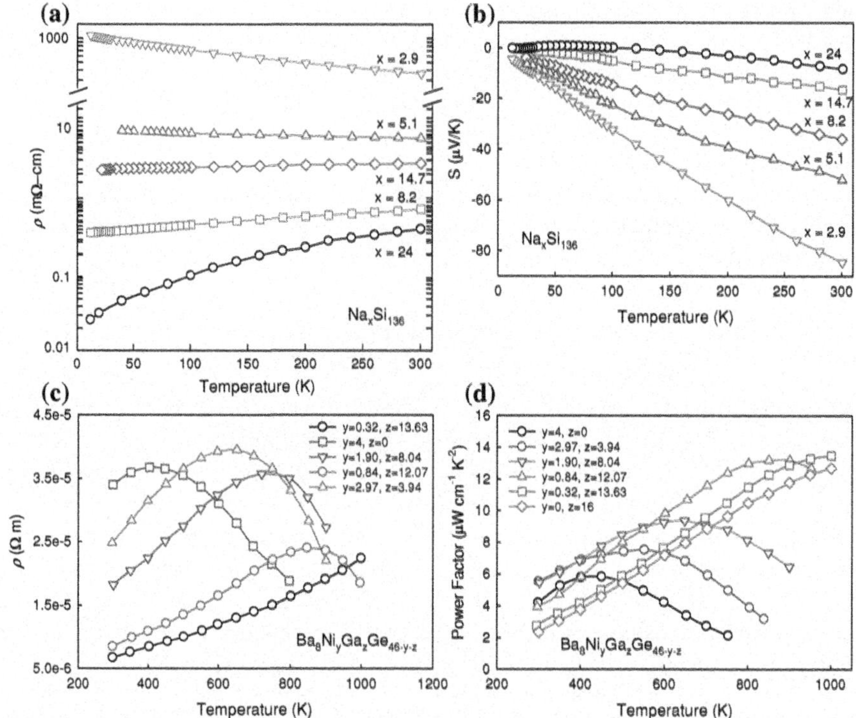


**Fig. 6.6** Electrical properties of selected clathrates: **a** $\rho$ and **b** $S$ of clathrate-II $Na_xSi_{136}$ [83]; **c** $\rho$ and **d** power factor of clathrate-I $Ba_8Ni_yGa_zGe_{46-y-z}$ [13]. **a** Reprinted with permission from [83]. Copyright (2012) American Chemical Society. **b** Reprinted with permission from [13]. Copyright © 2010 WILEY-VCH Verlag GmbH & Co. KGaA, Weinheim

is shown in Fig. 6.6 [83]. There is a clear difference in the magnitude and temperature dependence of $\rho$ as a function of the Na content, $x$ (Fig. 6.6a). The temperature dependence of $\rho$ indicates an apparent metal-to-semiconductor transition while $\rho$ increases with increasing temperature for $x > 8$, typical of metals, and decreases with increasing temperature for $x < 8$, typical for semimetallic or semiconducting materials. The modulus of $S$ increases with temperature, as expected in metals and heavily doped semiconductors with negligible phonon drag (Fig. 6.6b). The relatively low magnitude of $S$ for $Na_{24}Si_{136}$ is typical for metals (Eq. 6.6) in which the location of the Fermi level is well inside the conduction band. As the Na content decreases, the modulus of $S$ increases. These experimental observations are corroborated by theoretical studies based on the similarity between the density of states profiles for Na (both 3s and 3p) and the total density of states indicating a significant hybridization between metallic Na and the Si framework wave functions [99, 100]. Although there have been relatively few reports on varying the guest content in clathrate-I compositions, in contrast to clathrate-II structures, various $A_2B_6M_xE_{46-x}$ (A, B = Na, K, Cs, Rb, Sr, Eu;

M = Sr, Ni, Zn, Hg, Ag, Au, Cu; E = Si, Ge, Sn) compositions have been experimentally synthesized and their electronic properties investigated [101–111]. Compositions with the clathrate-I structure $Na_xBa_ySi_{46}$ are known to exhibit superconductivity [102, 110] (cf. Chap. 7).

Approach (ii) was observed first in clathrate-I $Sr_8Ga_{16+x}Ge_{30-x}$ [33]. The doping level, $x$, was varied by changing the Ga/Ge ratio while maintaining a fixed Sr concentration. By decreasing the carrier concentration the temperature dependence of the electrical properties changed, from metallic to semiconducting. Varying the Ga/Ge ratio has also been reported for clathrate-I and VIII $Eu_8Ga_{16-x}Ge_{30+x}$ [112]. These compositions were also found to exhibit excellent magnetic properties making them relevant for magnetocaloric applications (cf. Chap. 9).

In order to achieve charge balance in clathrate-I compositions with the general stoichiometry $A_{8+x}(TM_1)_{y1}(TM_2)_{y2}\ldots(TM_n)_{yn}M_zX_{46-y1-y2-\ldots yn-z}$, where A is the guest atom residing inside the polyhedra, TM is the transition metal substituted on the framework sites, and M and X are framework elements, the following Zintl-Klemm composition rule [63] should hold: $z = (8 + x)\cdot q_A - |\Delta q_1|y_1 - |\Delta q_2|y_2 - \ldots - |\Delta q_n|y_n$, where $q_A$ is the charge of A and $\Delta q_n$ the nominal charge of the $n$th TM atom (the difference between its valence electron count and that of a group 14 element). For example, for the ternary composition $Ba_8Ga_{16}Ge_{30}$ $x = 0$, $q_A = 2$, and $\Delta q_1 = -1$, therefore z = 0' i.e. using this analysis charge balance is achieved and $Ba_8Ga_{16}Ge_{30}$ is expected to be a semiconductor. Substitution on the framework was also achieved in clathrate-I $Ba_8Ni_yGa_zGe_{46-y-z}$ (Fig. 6.6c, d) [13]. Ionized impurities and lattice defects were introduced in these compositions resulting in an increase in ionized impurity scattering of carriers and point defect scattering of lattice phonons leading to improved TE properties as compared to $Ba_8Ga_{16}Ge_{30}$ (Fig. 6.3d) [12, 13, 113]. With increasing Ni concentration the relative ionized impurity scattering increases as well, resulting in an intrinsic semiconducting behavior for high Ni concentrations at elevated temperatures (Fig. 6.6c) [13]. Framework substitution of Cu for Ga in the clathrate-VIII $Ba_8Ga_{16-x}Cu_xSn_{30}$ resulted in a factor-of-two increase in carrier mobility, consequently resulting in one of the highest peak *ZT* values for clathrates thus far reported (Fig. 6.2) [9].

A combination of (i) and (ii) has been reported in clathrate-II $K_{8+x}Ba_{16-x}Ga_{40-y}Sn_{96-z}\square_{y+z}$ [86], an example of clathrate-II compositions following the Zintl–Klemm concept. By varying the K/Ba ratio and framework alloying with Ga, a metal to semiconductor transition was observed [86], with the highest $S$ values observed for the composition with highest $\rho$ values, as expected. Further insight into these clathrate-II compositions is warranted, while new compositions continue to be reported [92, 93].

Another important parameter that needs to be taken into consideration for TE applications is $E_g$. Theoretical analyses by Mahan [114] indicated that the best semiconductor band gap for large *ZT* is 10 $k_BT_{op}$, where $T_{op}$ is the operating temperature of the TE device. At room temperature this value is approximately 0.25 eV and is comparable to that of the well known $Bi_2Te_3$ alloy materials [8]. Values of $E_g$ for selected clathrates obtained either by theoretical considerations or experimental data are listed in Table 6.1. As indicated above, the band gap of

**Table 6.1** Estimated experimental and theoretical band-gaps ($E_g$) for selected clathrates

| Composition | Clathrate-type | $E_g$ (eV) | Reference |
|---|---|---|---|
| $Ba_8Ni_6Ge_{40}$ | I | 0.11 | 13 |
| $Ba_8Cu_{5.33}Ge_{40.67}$ | I | 0.10 | 13 |
| $Ba_8Zn_xGe_{46-x}$, $x = 6, 8, 10$ | I | 0.35–0.83 | 115 |
| $Ba_8Ga_{16}Ge_{30}$ | I | 0.89 | 116 |
| $Ba_8Ga_{16}Sn_{30}$ | I | 0.60 | 116 |
| $Sr_8Ga_{16}Ge_{30}$ | I | 0.50 | 117 |
| $Ba_8Al_{16}Ge_{30}$ | I | 0.43 | 118 |
| $Ba_8Al_{16}Si_{30}$ | I | 0.37 | 118 |
| $Sn_{136}$ | II | 0.46 | 119 |
| $Si_{136}$ | II | 1.9 | 120 |
| $Ge_{136}$ | II | 0.7 | 121 |
| $Ba_8Ga_{16}GeSn_{30}$ | VIII | 0.32 | 122 |
| $Sr_8Al_xGa_yGe_{46-x-y}$ | VIII | 0.56 | 123 |

clathrates can be tuned in a wide range, from metals to semiconductors, by choosing the appropriate combination of host and guest elements.

The nature and size of the band gap can therefore be tuned either by selecting the appropriate structure type, by varying the type and concentration of guest atoms, or by framework alloying, as shown in Fig. 6.7. Figure 6.7a shows the electronic DOS for clathrate-I and VIII $Ba_8Ga_{16}Sn_{30}$ based on first principle calculations [124]. Clathrate-I $Ba_8Ga_{16}Sn_{30}$ has a direct band gap of 0.6 eV, consistent with the Zintl–Klemm concept, whereas clathrate-VIII $Ba_8Ga_{16}Sn_{30}$ exhibits a more complicated band structure with a DOS peak arising in the fundamental gap [124]. Density Functional Theory (DFT) calculations of the electronic DOS for $Si_{136}$ and $A_8Ga_{16}Si_xGe_{30-x}$, (A = Ba, Sr; $x = 0$, 5) are shown in Fig. 6.7b, and for $Ba_8Al_{16}Si_{30}$ and $Ba_8Al_{16}Ge_{30}$ in Fig. 6.7c, d [118]. $Si_{136}$ is a semiconductor with a 1.9 eV band gap [120], although it was reported that the band gap of $(Si_{1-y}Ge_y)_{136}$ clathrate alloys can be tuned in the range of 0.6 to 1.9 eV [125].

Clathrate-II $Si_{136}$ was first theoretically predicted [120] and latter experimentally confirmed to be a wide band gap semiconductor [127]. The band gap of $Si_{136}$ is expanded by approximately 0.7 eV relative to that for diamond-Si [63]. The opening of this band gap can be understood in term of a slight distortion from the ideal tetrahedral bonding in diamond-Si, as well as the high density of 5-membered rings in the $Si_{136}$ structure [120]. Theoretical predictions on $Ge_xSi_{136-x}$ clathrate alloys indicate that these clathrates possess a direct band gap for a particular range of $x$ values [125]. The band gap in these alloys can be tuned from 0.6 to 1.9 eV, in the visible range of the electromagnetic spectrum, making them of interest for potential use in optoelectronic and photovoltaic applications, in addition to TE applications [63, 125, 128]. The calculated band gap[3] for

[3] The true band gaps for these materials are expected to be larger than those calculated as Local Density Approximation (LDA) typically underestimates the band gap [121].

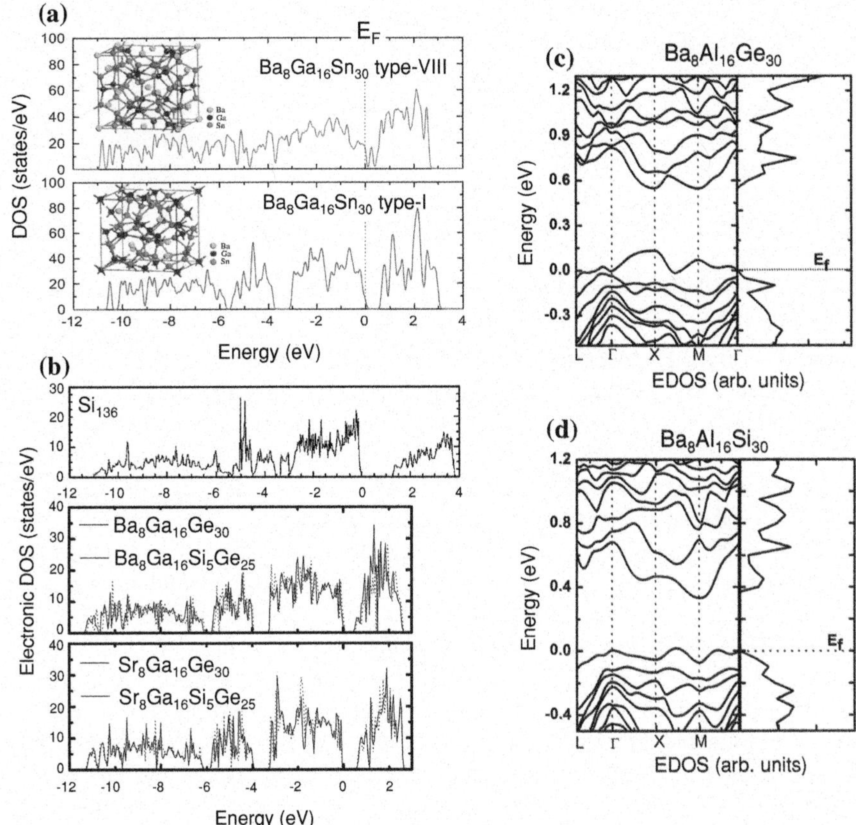


**Fig. 6.7** Electronic properties of selected clathrates: **a** DOS for clathrate-I and VIII $Ba_8Ga_{16}Sn_{30}$ [124]; **b** DOS for clathrate-II $Si_{136}$ [84] and clathrates $A_8Ga_{16}Si_xGe_{30-x}$, (A = Ba, Sr; $x = 0, 5$) [126]; **c** DOS and band gap for $Ba_8Al_{16}Ge_{30}$ [118]; **d** DOS and band gap for $Ba_8Al_{16}Si_{30}$ [118]. **a** Reprinted with permission from [124]. Copyright © 2008, Elsevier. **b** Reprinted with permission from [84]. Copyright (2006), American Physical Society. Reprinted with permission from [126]. Copyright (2006), American Physical Society. **c**, **d** Reprinted with permission from [118]. Copyright (2008), Institute of Physics

$Ba_8Al_{16}Ge_{30}$ (0.43 eV) was found to be direct (Fig. 6.7c), whereas the band gap of $Ba_8Al_{16}Si_{30}$ (0.37 eV), which is energetically more stable than the Ge counterpart by −0.38 eV per atom, is expected to be indirect (Fig. 6.7d) [118].

Based on the preceding discussion, several criteria need to be fulfilled in order to optimize the physical properties of clathrates for TE applications. Achieving the optimal charge carrier concentration, high $\mu$ (and therefore $\sigma$), optimal band gap, and high $S$ values are, however, insufficient for designing a good TE clathrate. It is necessary to synthesize both $n$ and $p$-type compositions with equally good TE performance in order to build TE devices (Fig. 6.1). The majority of the known clathrates are $n$-type, and a few $p$-type compositions have been synthesized to date

[65, 115, 129–136]. In addition the quality, crystallinity, and microstructure for a particular composition plays an important role. For example, single-crystal $Na_8Si_{46}$ has nearly two orders of magnitude lower $\rho$ than a polycrystalline specimen with the same composition [107, 137]. It is therefore equally important to investigate new synthetic tools in order to assess the transport properties of different compositions (cf. Chap. 3).

## 6.5 Concluding Remarks

Since the original suggestion by Slack [30] and subsequent demonstration by Nolas et al. [33] that intermetallic clathrates are promising candidates for TE energy conversion, significant effort has advanced the understanding of the mechanisms underlying the unusual thermal transport and lattice dynamics in these materials, which continues to be a topic of active research [35–38, 40, 47, 50, 52, 61, 71–73]. While differences in composition and structural features have been shown to have a significant influence on their low temperature thermal properties, resulting in glass-like $\kappa_L$ in some cases and more typical crystalline behavior in others, low $\kappa_L$ values at elevated temperatures appears to be a universal property of intermetallic clathrates. This can be attributed in part to the structural features common to all clathrates, namely a relatively large number of atoms in the primitive unit cell in relatively complex crystal structures. Based on the compositions investigated thus far, it can be conjectured that very low $\kappa_L$ can be expected for any new composition or variant. Since the power factors of intermetallic clathrates are somewhat lower than that observed for other high-temperature TE materials, such as skutterudites for example, efforts aimed at improving the electrical transport properties could result in significant improvement in their TE performance. This will likely be required for these materials to transition from the research and development stage to practical use. If the promising results of recent efforts to improve the TE properties of well-established TE materials, e.g. lead chalcogenides [138, 139], can be taken as an indication, efforts towards TE enhancement of clathrate compounds *via* various approaches and modification of their electronic structure are of interest since $\kappa_L$ for all clathrate compositions are expected to be low. The enhancement in $S$ recently reported for Ce containing clathrate-I compositions [140] as well as methods for power factor enhancement by pressure tuning [141] and strain effects [142] represent encouraging steps toward the goal of improving the thermoelectric properties of intermetallic clathrates.

Clathrate-I compositions remain the most extensively studied of the intermetallic clathrate structure types, and several compositions exhibiting potential for TE applications. While the number of known compositions in other structure types continues to increase, the TE properties of these other structure types remains relatively unexplored, despite the expectation of good TE performance in clathrate-II compositions [143] and the ability to simultaneously modify the framework composition and guest content. In this respect, there are significant research opportunities to further expand on the understanding of these materials as well as develop better TE materials.

Good TE performance is a necessary but not sufficient criterion for taking a material system to the device development stage. For comercial viability, the material must be comprised of relatively inexpensive, earth-abundant, and non-toxic chemical elements, a constraint that presents a significant challenge for further development of some candidate TE materials, including Ge clathrates. The mechanical properties of TE materials are also of great relevance, and are the subject of Chap. 10 of this volume. The compositional flexibility afforded by the different clathrate structure types enables a variety of viable compositions, some of which are currently under investigation. Achieving the requisite TE performance is worthy of future effort as it would likely have a significant impact on TE energy conversion technologies.

**Acknowledgements** The authors gratefully acknowledge the support of the U.S. Department of Energy, Basic Energy Sciences, Division of Materials Science and Engineering, under Award No. DE-FG02-04ER46145 at the University of South Florida.

## References

1. T. Tritt, M.A. Subramanian, MRS Bull. **31**, 188 (2006)
2. T.J. Seebeck, Abh. K. Akad. Wiss., 265 (1823)
3. J.C. Peltier, Ann. Chem. **LVI**, 371 (1834)
4. A.F. Ioffe, *Semiconductor Thermoelements and Thermoelectric Cooling* (Infosearch, London, 1957)
5. L.E. Bell, Science **321**, 1457 (2008)
6. G.S. Nolas, J. Sharp, H.J. Goldsmid, in *Thermoelectrics, Basic Principles and New Materials Developments* (Springer, Berlin, 2001)
7. J. Yang, T. Caillat, MRS Bull. **31**, 224 (2006)
8. J.R. Sootsman, D.Y. Chung, M.G. Kanatzidis, Angew. Chem. Intl. Ed. **48**, 8616 (2009)
9. S. Deng, X. Tang, P. Li, Q. Zhang, J. Appl. Phys. **109**, 103704 (2011)
10. G.S. Nolas, in *Thermoelectrics Handbook: Macro- to Nano-Structured Materials*, ed. by D.M. Rowe (2006) pp. 33/1–8
11. Y. Sasaki, K. Kishimoto, T. Koyanagi, H. Asada, K. Akai, J. Appl. Phys. **105**, 073702 (2009)
12. J. Martin, H. Wang, G.S. Nolas, Appl. Phys. Lett. **92**, 222110 (2008)
13. B.X. Shi, J. Yang, S. Bai, J. Yang, H. Wang, M. Chi, J.R. Salvador, W. Zhang, L. Chen, W. Wong-Ng, Adv. Funct. Mater. **20**, 755 (2010)
14. C.L. Condron, S.M. Kauzlarich, F. Gascoin, J. Snyder, Chem. Mater. **18**, 4939 (2006)
15. G. Slack, in *CRC Handbook of Thermoelectrics*, ed. by D.M. Rowe (CRC Press, Boca Raton, 1995)
16. M. Flambigl, N. Nasir, A. Grytsiv, P. Rogl, S. Seichter, A. Zavarsky, E. Royanian, E. Bauer, J. Phys. D Appl. Phys. **45**, 215308 (2012)
17. X. Tang, Q. Zhang, L. Chen, T. Goto, T. Hirai, J. Appl. Phys. **97**, 093712 (2005)
18. Q. Shen, L. Chen, T. Goto, T. Hirai, J. Yang, Appl. Phys. Lett. **79**, 4165 (2001)
19. X. Tang, W. Xie, H. Li, W. Zhao, Q. Zhang, Appl. Phys. Lett. **90**, 012102 (2007)
20. B. Wölfing, C. Kloc, J. Yeuber, E. Bucher, Phys. Rev. Lett. **86**, 4350 (2001)

21. D.Y. Chung, T. Hogan, P. Brazis, M.R. Lane, C. Kannewurf, M. Bastea, C. Uher, M. Kanatzidis, Science **287**, 1024 (2000)
22. X. Shi, J. Yang, J.R. Salvador, M. Chi, J.Y. Cho, H. Wang, S. Bai, J. Yang, W. Zhang, L. Chen, J. Am. Chem. Soc. **133**, 7837 (2011)
23. B. Du, Y. Saiga, K. Kajisa, T. Takabatake, J. Appl. Phys. **111**, 013707 (2012)
24. T.M. Tritt, *2004–Thermal Conductivity: Theory, Practice, and Applications* (Kluwer Academic, New York, NY, 2004)
25. G.A. Slack, in *Solid State Physics*, vol. 34, ed. by H. Ehrenreich, F. Seitz, D. Turnbull (Academic Press, New York, 1979), p. 1
26. D.G. Cahill, S.K. Watson, R.O. Pohl, Phys. Rev. B **46**, 6131 (1992)
27. R.G. Ross, P. Andersson, G. Backstrom, Nature **290**, 322 (1981)
28. J.S. Tse, M.A. White, J. Phys. Chem. **92**, 5006 (1988)
29. G.A. Slack, V.G. Tsoukala, J. Appl. Phys. **76**, 1665 (1994)
30. G.A. Slack, Mater. Res. Soc. Symp. Proc. **478**, 47 (1997)
31. M. Beekman, W. Schnelle, H. Borrmann, M. Baitinger, Yu. Grin, G.S. Nolas, Phys. Rev. Lett. **104**, 018301 (2010)
32. N.W. Ashcroft, N.D. Mermin, *Solid State Physics* (Saunders College, Philadelphia, 1976)
33. G.S. Nolas, J.L. Cohn, G.A. Slack, S.B. Schujmann, Appl. Phys. Lett. **73**, 178 (1998)
34. G.S. Nolas, J.L. Cohn, J.S. Dyck, C. Uher, J. Yang, Phys. Rev. B **65**, 165201 (2002)
35. J.L. Cohn, G.S. Nolas, V. Fessatidis, T.H. Metcalf, G.A. Slack, Phys. Rev. Lett. **82**, 779 (1999)
36. G.S. Nolas, T.J.R. Weakley, J.L. Cohn, R. Sharma, Phys. Rev. B **61**, 3845 (2000)
37. V. Keppens, B.C. Sales, D. Mandrus, B.C. Chakoumakos, C. Laermans, Philos. Mag. Lett. **80**, 807 (2000)
38. I. Zerec, V. Keppens, M.A. McGuire, D. Mandrus, B.C. Sales, P. Thalmeier, Phys. Rev. Lett. **92**, 185502 (2004)
39. R.P. Hermann, V. Keppens, P. Bonville, G.S. Nolas, F. Grandjean, G.J. Long, H.M. Christen, B.C. Chakoumakos, B.C. Sales, D. Mandrus, Phys. Rev. Lett. **97**, 017401 (2006)
40. M.A. Avila, K. Suekuni, K. Umeo, H. Fukuoka, S. Yamanaka, T. Takabatake, Phys. Rev. B **74**, 125109 (2006)
41. B.C. Sales, B.C. Chakoumakos, R. Jin, J.R. Thompson, D. Mandrus, Phys. Rev. B **63**, 245113 (2001)
42. K. Suekuni, M.A. Avila, K. Umeo, T. Takabatake, Phys. Rev. B **75**, 195210 (2007)
43. T. Tanaka, T. Onimaru, K. Suekuni, S. Mano, H. Fukuoka, S. Yamanaka, T. Takabatake, Phys. Rev. B **81**, 165110 (2010)
44. C.L. Condron, J. Martin, G.S. Nolas, P.M.B. Piccoli, A.J. Schultz, S.M. Kauzlarich, Inorg. Chem. **45**, 9381 (2006)
45. X. Zheng, S.Y. Rodriguez, L. Saribaev, J.H. Ross Jr, Phys. Rev. B **85**, 214304 (2012)
46. B.C. Chakoumakos, B.C. Sales, D.G. Mandrus, G.S. Nolas, J. Alloys Comp. **296**, 80 (2000)
47. M. Christensen, N. Lock, J. Overgaard, B.B. Iversen, J. Am. Chem. Soc. **128**, 15657 (2006)
48. G.S. Nolas, C.A. Kendziora, Phys. Rev. B **62**, 7157 (2000)
49. Y. Takasu, T. Hasegawa, N. Ogita, M. Udagawa, M.A. Avila, K. Suekuni, I. Ishii, T. Suzuki, T. Takabatake, Phys. Rev. B **74**, 174303 (2006)
50. Y. Takasu, T. Hasegawa, N. Ogita, M. Udagawa, M.A. Avila, K. Suekuni, T. Takabatake, Phys. Rev. Lett. **100**, 165503 (2008)
51. R.P. Hermann, W. Schweika, O. Leupold, R. Rüffer, G.S. Nolas, F. Grandjean, G.J. Long, Phys. Rev. B **72**, 174301 (2005)
52. M. Christensen, A.B. Abrahamsen, N.B. Christensen, F. Juranyi, N.H. Andersen, K. Lefmann, J. Andreasson, C.R.H. Bahl, B.B. Iversen, Nature Mater. **7**, 811 (2008)
53. C.H. Lee, H. Yoshizawa, M.A. Avila, I. Hase, K. Kihou, T. Takabatake, J. Phys. Conf. Ser. **92**, 012169 (2007)
54. R. Baumbach, F. Bridges, L. Downward, D. Cao, P. Chesler, B. Sales, Phys. Rev. B **71**, 024202 (2005)

55. I. Ishii, H. Higaki, S. Morita, M.A. Avila, T. Sakata, T. Takabatake, T. Suzuki, Phys. B **383**, 130 (2006)
56. V. Keppens, M.A. McGuire, A. Teklu, C. Laermans, B.C. Sales, D. Mandrus, B.C. Chakoumakos, Phys. B **95**, 316 (2002)
57. X. Zheng, S.Y. Rodriguez, J.H. Ross Jr, Phys. Rev. B. **84**, 024303 (2011)
58. W. Gou, Y. Li, J. Chi, J.H. Ross Jr, M. Beekman, G.S. Nolas, Phys. Rev. B **71**, 174307 (2005)
59. J. Dong, O.F. Sankey, C.W. Myles, Phys. Rev. Lett. **86**, 2361 (2001)
60. F. Bridges, L. Downward, Phys. Rev. B **70**, 140201(R) (2004)
61. J. Dong, O.F. Sankey, G.K. Ramachandran, P.F. McMillan, J. Appl. Phys. **87**, 7726 (2000)
62. G.K.H. Madsen, G. Santi, Phys. Rev. B **72**, 220301(R) (2005)
63. M. Beekman, G.S. Nolas, J. Mater. Chem. **18**, 842 (2008)
64. M. Flambigl, P.F. Rogl, in *Modules, Systems, and Applications in Thermoelectrics*, ed. by D.M. Rowe (CRC Press, Boca Raton, 2012)
65. G.S. Nolas, G.A. Slack, S.B. Schujman, in *Semiconductors and Semimetals*, vol. 69, ed. by T.M. Tritt (Academic Press, San Diego, 2000), p. 255
66. G.S. Nolas, in *Chemistry, Physics, and Materials Science of Thermoelectric Material: Beyond Bismith Telluride*, ed. by M.G. Kanatzidis, S.D. Manhanti, T.P. Hogan (Kluwer Academic/Plenum Publishers, New York, 2003), pp. 107–120
67. M. Christensen, S. Johnsen, B.B. Iversen, Dalton Trans. **39**, 978 (2010)
68. T. Takabatake, K. Suekuni, T. Nakayama, E. Kaneshita, Rev. Mod. Phys. arXiv:1402. 5756v1 (2014)
69. F. DiSalvo, Science **285**, 703 (1999)
70. O. Delaire, J. Ma, K. Marty, A.F. May, M.A. McGuire, M-H. Du, D.J. Singh, A. Podlesnyak, G. Ehlers, M.D. Lumsden, B.C. Sales, Nature Mater. **10**, 614 (2011)
71. E.S. Toberer, A. Zevalkink, G.J. Snyder, J. Mater. Chem. **21**, 15843 (2011)
72. A. Bentien, M. Christensen, J.D. Bryan, A. Sanchez, S. Paschen, F. Steglich, G.D. Stucky, B.B. Iversen, Phys. Rev. B **69**, 045107 (2004)
73. H. Euchner, S. Pailhès, L.T.K. Nguyen, W. Assmus, F. Ritter, A. Haghighirad, Y. Grin, S. Paschen, M. de Boissieu, Phys. Rev. B **86**, 224303 (2012)
74. K. Suekuni, M.A. Avila, K. Umeo, H. Fukuoka, S. Yamanaka, T. Nakagawa, T. Takabatake, Phys. Rev. B **77**, 235199 (2008)
75. A.D. Ritchie, M.B. Johnson, J.F. Niven, M. Beekman, G.S. Nolas, J. Gryko, M.A. White, J. Phys.: Condens. Matter **25**, 435401 (2013)
76. G.S. Nolas, M. Beekman, J. Gryko, G.A. Lamberton Jr, T.M. Tritt, P.F. McMillan, Appl. Phys. Lett. **82**, 910 (2003)
77. J. Fulmer, O.I. Lebedev, V.V. Roddatis, D.C. Kaseman, S. Sen, J.-A. Dolyniuk, K. Lee, A.V. Olenev, K. Kovnir, J. Am. Chem. Soc. **135**, 12313 (2013)
78. N.F. Mott, H. Jones, *The Theory and Properties of Metals and Alloys* (Dover Publications, New York, 1958), p. 310
79. J.P. Heremans, V. Jovovic, E. Toberer, A. Saramat, K. Kurosaki, A. Charoenphakdee, S. Yamanaka, G.J. Snyder, Science **321**, 554 (2008)
80. M. Cutler, J.F. Leavy, R.L. Fitzpatrick, Phys. Rev. **133**, A1143 (1964)
81. G.J. Snyder, E. Toberer, Nature Mater. **7**, 105 (2008)
82. S. Bobev, S. Sevov, J. Solid State Chem. **153**, 92 (2000)
83. S. Stefanoski, C.D. Malliakas, M.G. Kanatzidis, G.S. Nolas, Inorg. Chem. **51**, 8686 (2012)
84. K. Biswas, C.W. Myles, Phys. Rev. B **74**, 115113 (2006)
85. S. Latturner, B.B. Iversen, J. Sepa, V. Srdanov, G. Stucky, Phys. Rev. B **63**, 125403 (2001)
86. S. Mano, T. Onimaru, S. Yamanaka, T. Takabatake, Phys. Rev. B **84**, 214101 (2011)
87. G.S. Nolas, D.G. Vanderveer, A.P. Wilkinson, J.L. Cohn, J. Appl. Phys. **91**, 8970 (2002)
88. A.M. Guloy, Z. Tang, R. Ramlau, B. Böhme, M. Baitinger, Y. Grin, Eur. J. Inorg. Chem. **17**, 2455 (2009)
89. M. Beekman, W. Wong-Ng, J.A. Kaduk, A. Shapiro, G.S. Nolas, J. Solid State Chem. **180**, 1076 (2007)

90. M. Beekman, J.A. Kaduk, J. Gryko, W. Wong-Ng, A. Shapiro, G.S. Nolas, J. Alloys Comp. **470**, 365 (2009)
91. M. Beekman, G.S. Nolas, Mater. Res. Soc. Proc. **1044**, 1044-U05-05 (2007)
92. M.C. Shäfer, S. Bobev, Acta Cryst. C **69**, 319 (2013)
93. M.C. Shäfer, S. Bobev, J. Am. Chem. Soc. **135**, 1696 (2013)
94. M. Beekman, G.S. Nolas, Phys. B **383**, 111 (2006)
95. H. Horie, T. Kikudome, K. Teramura, S. Yamanaka, J. Solid State Chem. **182**, 129 (2009)
96. C. Cros, M. Pouchard, P. Hagenmuller, J. Solid State Chem. **2**, 570 (1970)
97. A.D. Ritchie, M.A. MacDonald, P. Zhang, M.A. White, M. Beekman, J. Gryko, G.S. Nolas, Phys. Rev. B **82**, 155207 (2010)
98. J.S. Kasper, P. Hagenmuller, M. Pouchard, C. Cross, Science **150**, 1713 (1965)
99. V.L. Smelyanski, J.S. Tse, Chem. Phys. Lett. **264**, 459 (1997)
100. J.C. Conesa, C. Tablero, P. Wahon, J. Chem. Phys. **120**, 6142 (2004)
101. S. Stefanoski, G.S. Nolas, Cryst. Growth Des. **11**, 4533 (2011)
102. H. Kawaji, H.-O. Horie, S. Yamanaka, M. Ishikawa, Phys. Rev. Lett. **74**, 1427 (1995)
103. M. Hayashi, K. Kishimoto, K. Kishio, K. Akai, H. Asada, T. Koyanagi, Dalton Trans. **39**, 1113 (2010)
104. A. Kaltzoglou, T. Fässler, C. Gold, E.-W. Scheidt, W. Scherer, T. Kume, H. Shimizu, J. Solid State Chem. **182**, 2924 (2009)
105. L.T.K. Nguyen, U. Aydemir, M. Baitinger, E. Bauer, H. Borrmann, U. Burkhardt, J. Custers, A. Haghighirad, R. Höfler, K.D. Luther, F. Ritter, W. Assmus, Yu. Grin, S. Paschen, Dalton Trans. **39**, 1071 (2010)
106. J. Martin, G.S. Nolas, H. Wang, J. Yang, J. Appl. Phys. **102**, 103719 (2007)
107. S. Stefanoski, J. Martin, G.S. Nolas, J. Phys. Condens. Matter **22**, 485404 (2010)
108. G.K. Ramachandran, P.F. McMillan, J. Dong, J. Gryko, O.F. Sankey, Mat. Res. Soc. Symp. **626**, Z13.2.1 (2000)
109. B.C. Sales, B.C. Chakoumakos, R. Jin, J.R. Thompson, D. Mandrus, Phys. Rev. B **63**, 245113 (2001)
110. S. Yamanaka, E. Enishi, H. Fukuoka, M. Yasukawa, Inorg. Chem. **39**, 56 (2000)
111. H. Zhang, H. Bormann, N. Oeschler, C. Candolfi, W. Schnelle, M. Schmidt, U. Burkhardt, M. Baitinger, J.-T. Zhao, Yu. Grin, Inorg. Chem. **50**, 1250 (2011)
112. A. Bentien, V. Pascheco, S. Paschen, Y. Grin, F. Steglich, Phys. Rev. B **71**, 165206 (2005)
113. H. Anno, M. Hokazono, M. Kawamura, K. Matsubara, in *Proceedings of 22nd International Conference on Thermoelectrics* (IEEE, Piscataway, 2003), p. 102
114. G.D. Mahan, J. Appl. Phys. **65**, 1578 (1989)
115. T. Eto, K. Kishimoto, K. Koga, K. Akai, T. Koyanagi, H. Anno, T. Tanaka, H. Kurisu, S. Yamamoto, M. Matsuura, Mater. Trans. **50**, 631 (2009)
116. V.L. Kuznetsov, L.A. Kuznetsova, A.E. Kaliazin, D.M. Rowe, J. Appl. Phys. **87**, 7871 (2000)
117. I. Fujita, K. Kishimoto, M. Sato, H. Anno, T. Koyanagi, J. Appl. Phys. **99**, 093707 (2006)
118. E. Nenghabi, C.W. Myles, J. Phys. Condens. Matter **20**, 415214 (2008)
119. C.W. Myles, J. Dong, O.F. Sankey, Phys. Rev. B **64**, 165202 (2001)
120. G.B. Adams, M. O'Keeffe, A.A. Demkov, Phys. Rev. B **49**, 8048 (1994)
121. J. Dong, O.F. Sankey, J. Phys. Condens. Matter **11**, 6129 (1999)
122. Y. Kono, N. Ohya, T. Taguchi, K. Suekuni, T. Takabatake, S. Yamamoto, K. Akai, J. Appl. Phys. **107**, 123720 (2010)
123. Y. Sasaki, K. Kishimoto, T. Koyanagi, H. Asada, K. Akai, J. Appl. Phys. **105**, 073702 (2009)
124. Y. Li, J. Gao, N. Chen, Y. Liu, Z.P. Luo, R.H. Zhang, X.Q. Ma, G.H. Cao, Phys. B **403**, 1140 (2008)
125. K. Moriguchi, S. Munetoh, A. Shintani, Phys. Rev. B **62**, 7138 (2000)
126. E.N. Nenghabi, C.W. Myles, Phys. Rev. B **77**, 205203 (2008)
127. J. Gryko, P.F. McMillan, R.F. Marzke, G. Ramachandran, D. Patton, S.K. Deb, O.F. Sankey, Phys. Rev. B **62**, R7707 (2000)

128. A.D. Martinez, L. Krishna, L.L. Baranowski, M.T. Lusk, E.S. Toberer, A.C. Tamboli, IEEE J. Photovoltaics **3**, 1305 (2013)
129. G.J. Miller, in *Chemistry, Structure, and Bonding of Zintl Phases and Ions*, ed. by S.M. Kauzlarich (Wiley, New York, 1996), p. 1
130. S. Stefanoski, Y. Dong, G.S. Nolas, J. Solid State Chem. **204**, 166 (2013)
131. D. Cederkrantz, A. Saramat, G.J. Snyder, E.C. Palmqvist, J. Appl. Phys. **106**, 074509 (2009)
132. K. Kishimoto, T. Koyanagi, K. Akai, M. Matsuura, Jpn. J. Appl. Phys. **46**, L746 (2007)
133. M. Saisho, L. Bin, Y. Nagatomo, Y. Nakakohara, R. Teranishi, S. Munetoh, J. Phys. Conf. Series **379**, 012009 (2012)
134. S. Latturner, X. Bu, N. Blake, H. Metiu, G. Stucky, J. Solid State Chem. **151**, 61 (2000)
135. S. Laumann, M. Ikeda, H. Sassik, A. Prokofiev, S. Paschen, J. Mater. Res. **26**, 1861 (2011)
136. Y. Kono, N. Ohya, Y. Saiga, K. Suekuni, T. Takabatake, K. Akai, S. Yamamoto, J. Electron. Mater. **40**, 845 (2011)
137. G.S. Nolas, J.M. Ward, J. Gryko, L. Qiu, M. White, Phys. Rev. B **64**, 153201 (2001)
138. J.P. Heremans, V. Jovovic, E.S. Toberer, A. Saramat, K. Kurosaki, A. Charoenphakdee, S. Yamanaka, G.J. Snyder, Science **321**, 554 (2008)
139. Y. Pei, X. Shi, A. LaLonde, H. Wang, L. Chen, G.J. Snyder, Nature **473**, 66 (2011)
140. A Prokofiev, A. Sidorenko, K. Hradil, M. Ikeda, R. Svagera, M. Waas, H. Winkler, K. Neumaier, S. Paxhen, Nature Mater. **12**, 1096 (2013)
141. J.F. Meng, N.V. Chandra Shekar, J.V. Badding, G.S. Nolas, J. Appl. Phys. **89**, 1730 (2001)
142. J. Martin, G.S. Nolas, H. Wang, J. Yang, J. Appl. Phys. **102**, 103719 (2007)
143. A. Karttunen, T.F. Fassler, Chem. Phys. Chem. **14**, 1807 (2013)

# Chapter 7
# High Pressure Synthesis of Superconducting Silicon Clathrates and Related Compounds

**Shoji Yamanaka**

**Abstract** The high pressure and high temperature (HPHT) synthesis method is useful in developing metal intercalated silicon clathrate compounds. The barium containing binary type I clathrate compound $Ba_8Si_{46}$ can be prepared under HPHT conditions of 3 GPa at 800 °C, and becomes a superconductor with a transition temperature $T_c$ of 8.0 K. A variety of clathrate related compounds with covalently bonded nano-cage structures have been prepared using HPHT conditions with most showing superconductivity. An attempt to prepare carbon analogs of silicon clathrate compounds by the HPHT 3D polymerization of $C_{60}$ is also introduced.

## 7.1 Introduction

The discovery of fullerene $C_{60}$ with a soccer-ball-shaped carbon cage in 1985 came as a surprise [1]. Soon after another big surprise followed, alkali metal intercalated $C_{60}$ in the interstices showed superconductivity with high transition temperatures, $K_3C_{60}$ ($T_c = 19.5$ K), and $RbCs_2C_{60}$ ($T_c = 33$ K) [2, 3]. Carbon and silicon chemistry and physics are often comparatively discussed. Another exciting nano-cage compound with silicon had already been discovered as clathrate compounds by Cros et al. in 1965 [4, 5], much earlier than the discovery of $C_{60}$. People expected superconductivity in silicon clathrate compounds having expanded frameworks with alkali metal atoms in the nano-cages [6], although the bonding schemes are very different between the two kinds of cages; silicon forms extended three dimensional framework with $sp^3$ covalent bonds whereas carbon forms isolated molecules with $sp^2$ bonds in fullerenes. Alkali metal intercalated silicon clathrate compounds so far known at that time did not show superconductivity. In

S. Yamanaka (✉)
Department of Applied Chemistry, Graduate School of Engineering, Hiroshima University, Higashi-Hiroshima 739-8527, Japan
e-mail: syamana@hiroshima-u.ac.jp

G. S. Nolas (ed.), *The Physics and Chemistry of Inorganic Clathrates*, Springer Series in Materials Science 199, DOI: 10.1007/978-94-017-9127-4_7,

1995 the barium doped type I silicon clathrate $Na_2Ba_6Si_{46}$ was prepared for the first time by conventional thermal decomposition of the Zintl compound $Na_2BaSi_4$ [7]. Surprisingly, it turned out to be a superconductor with $T_c \sim 4$ K [8]. This is the first known superconducting clathrate compound.

Later, as will be described in this chapter, high pressure and high temperature (HPHT) conditions were applied to develop new clathrate compounds in binary and ternary silicide systems, and a number of clathrates and related compounds with new types of nano-cage structures have been discovered. Interestingly most show superconducting behavior. The traditional clathrate compounds obey the classification of isomorphous gas hydrates, type I, type II, and type III [9]. In this chapter, clathrate compounds are more widely characterized as the compounds composed of covalently bonded metal-endohedral polyhedral cages, which are linked to each other by face sharing polyhedra similar to gas hydrates. Fullerene $C_{60}$ molecules rich in C = C double bonds can be three dimensionally polymerized under HPHT conditions [10]. An attempt to develop carbon analogs of silicon clathrate compounds from the polymerization of $C_{60}$ will also be introduced [11].

## 7.2 First Superconductor in Silicon Clathrate Compounds

Alkali metal intercalated silicon clathrate compounds can be obtained by thermal decomposition of alkali metal monosilicides ASi (A = Na, K, Rb) [4, 5, 12], which contain $[Si_4]^{4-}$ Zintl anions with alkali metal cations. In thermal decomposition in vacuum or in inert gas atmosphere, alkali metals are partially removed, forming a Si $sp^3$ clathrate cage network with alkali metal atoms encaged. In an attempt to prepare a silicon clathrate compounds containing divalent metals, $BaSi_2$ should be the first candidate as a starting material among the Zintl alkaline earth disilicides $AESi_2$ (AE = Ca, Sr, Ba) since it consists of $[Si_4]^{4-}$ Zintl anions, similar to alkali metal monosilicides. $BaSi_2$ was subjected to thermal decomposition to obtain $Ba_8Si_{46}$. $BaSi_2$ was found to be very stable upon heating; therefore, a solid solution between NaSi and $BaSi_2$ was prepared [7]. The crystal structure of the solid solution $Na_2BaSi_4$ was analyzed later by Huang and Corbett [13]. They reported that the compound consists of $[Si_4]^{4-}$ Zintl anions with Ba and Na, as shown in Fig. 7.1. The compound was thermally decomposed under vacuum at about 500 °C into a mixture of $BaSi_2$ and a Ba containing silicon clathrate phase by the removal of excess Na;

$$17\ Na_2BaSi_4 \rightarrow Na_2Ba_6Si_{46} + 11\ BaSi_2 + 32\ Na \uparrow$$

The byproduct $BaSi_2$ is soluble in water, and a type I clathrate compound with an approximate composition $Na_2Ba_6Si_{46}$ was isolated after stirring the decomposed product in water [7]. It should be noted that silicon clathrate compounds are stable in alkaline solutions as well as in acid solutions except hydrofluoric acid. The clathrate is even more stable than elemental Si in alkaline solutions, and a small amount of Si contamination can be removed by prolonged stirring in an alkaline solution.

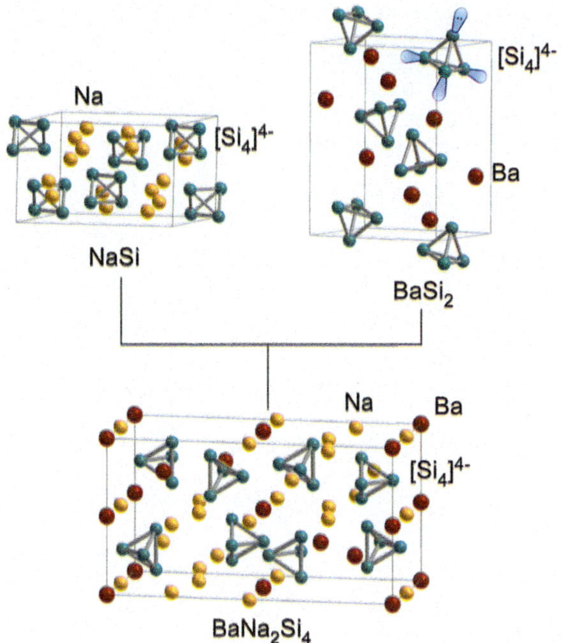


**Fig. 7.1** Preparation of the solid solution $BaNa_2Si_4$ from NaSi and $BaSi_2$. All of the compounds consist of $[Si_4]^{4-}$ Zintl anions. Lone pair orbitals are shown in one of the Zintl anions of $BaSi_2$

The structure of the barium containing type I silicon clathrate compound $Na_2Ba_6Si_{46}$ was analysed by the X-ray Rietveld refinement method [7]. The Na atoms mainly occupy the 2*a* sites in @$Si_{20}$ dodecahedral cages, and the Ba atoms occupy the 6*d* sites in the larger @$Si_{24}$ tetradecahedral cages. The compound became superconducting at $T_c \sim 4$ K, as shown in Fig. 7.2 [8]. This is the first superconductor found in clathrate compounds. As shown later, the composition of the clathrate $Na_xBa_ySi_{46}$ varies depending on the decomposition temperature of $Na_2BaSi_4$, and the $T_c$ varies in a range of 2.6–5 K [14]. A potassium substituted clathrate compound $K_2Ba_6Si_{46}$ was similarly prepared [15], and also showed superconductivity with a similar $T_c$ (Fig. 7.2). The lattice parameter of $Na_2Ba_6Si_{46}$ ($a = 10.26$ Å) is slightly larger than that of $Na_8Si_{46}$ ($a = 10.20$ Å) due to the substitution of Na atoms with larger Ba atoms.

When we focus on the two different kinds of polyhedral cages of the type I clathrate compound $Na_2Ba_6Si_{46}$, $[Na@Si_{20}]_2[Ba@Si_{24}]_6$, it is isostructural with the $W_3O$ type or the A15 structure. The intermetallic superconductors $Nb_3Ge$ and $Nb_3Sn$ are the prototype of this class with space group $Pm\bar{3}n$ the same as that of type I clathrate compounds. Figure 7.3 shows a schematic structural model of $Na_2Ba_6Si_{46}$ in comparison with $Nb_3Sn$, where Na atoms like Sn are located at the corner and the body center positions and Ba atoms like Nb are in pairs on the faces of the cube. The Ba@$Si_{24}$ cages are aligned along the axes by sharing hexagonal faces to make infinite rods, which run along the three axes without crossing each other.

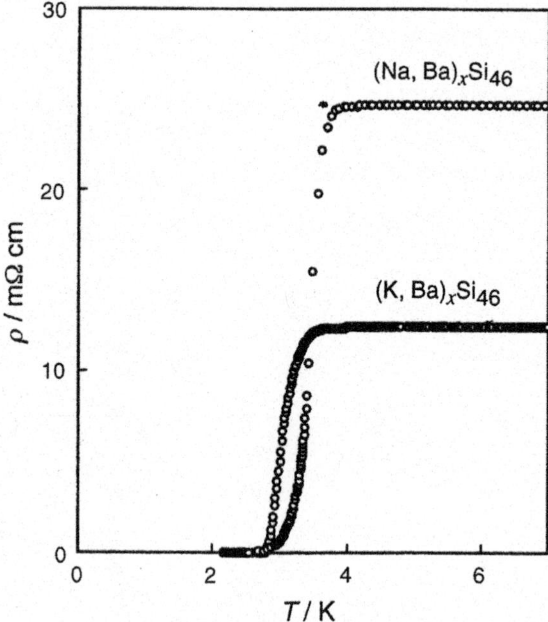


**Fig. 7.2** Temperature dependence of the resistivity of the compressed powder sample of (Na, Ba)$_x$Si$_{46}$ and (K, Ba)$_x$Si$_{46}$ near the superconducting transition temperature

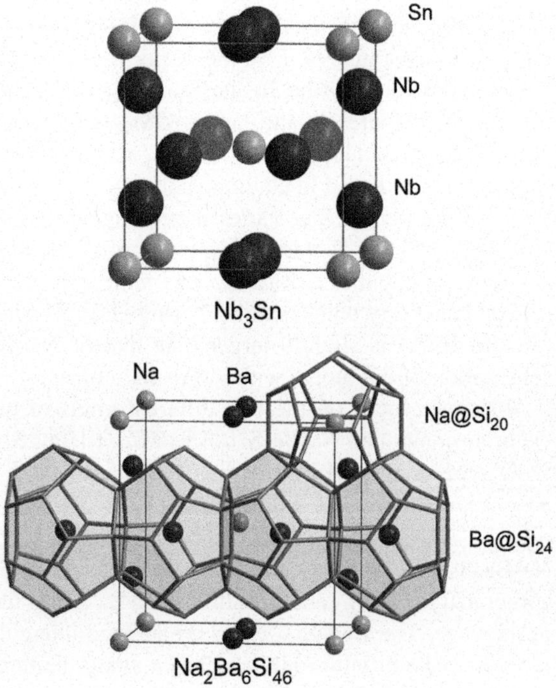


**Fig. 7.3** Schematic structural model of $Na_2Ba_6Si_{46}$ (*bottom*), showing the relation with the structure of $Nb_3Sn$ (*top*) of the A15 structure

The Ba containing type II silicon clathrate compound $Na_{16}Ba_8Si_{136}$ was prepared by Rachi et al. [16] by a similar thermal decomposition of the Zintl compound $Na_2BaSi_4$ in vacuum at a temperature of 350 °C, a lower temperature than that used for the preparation of the type I clathrate compound $Na_2Ba_6Si_{46}$. The lattice parameter was determined to be 14.69 Å, the Na atoms occupying the smaller dodecahedral cages and Ba atoms in the larger hexakaidecahedral (Ba@$Si_{28}$) cages. The type II compound does not become a superconductor down to 2 K.

Before the discovery of the superconductivity in $Na_2Ba_6Si_{46}$, Ba containing type I ternary silicon clathrate compounds $Ba_8Si_{40}T_6$ (T = Au, Ag, Ni, Pt, Pd, Cu) had been prepared by a simple melting of the constituent element mixtures by Cordier and Woll [17]. These are substituted type I silicon clathrate compounds with the Si atoms at 6*c* sites substituted with transition metals (T), as shown in Fig. 7.4. When superconductivity was found in $Na_2Ba_6Si_{46}$ it was expected that the Ba containing type I clathrate compounds $Ba_8Si_{40}T_6$ should also show superconductivity in a similar way. These compounds are metallic but not superconductors. A possible reason for this non-superconductivity may be found in the substitution of the silicon atoms at the 6*c* sites in the framework with transition metals whose electronegativities are larger than silicon, and thus may trap electrons for superconductivity. We tried to prepare the binary type I clathrate $Ba_8Si_{46}$ without substitution in the framework by arc melting of Ba and Si. This produced only a mixture of $BaSi_2$ and Si. In the end, the binary clathrate compound $Ba_8Si_{46}$ was prepared using HPHT conditions.

## 7.3 High Pressure Synthesis of Ba Containing Silicon Clathrate Compounds

### *7.3.1 High Pressure Apparatus for HPHT Synthesis*

In HPHT synthesis, a small pressure medium including furnace (heater), sample cell, and thermocouple is compressed in a high pressure apparatus. There are two types of high pressure systems used for the synthesis of clathrate and intermetallic compounds, a belt-type apparatus and a multi-anvil apparatus. These apparatuses are also widely used for geological high pressure studies and in situ X-ray diffraction studies using synchrotron radiation under HPHT conditions [18].

The belt-type is a modified piston-cylinder apparatus, consisting of a cylindrical belt with a tungsten carbide toroidal core for a tapered sample cavity (die) and two punches, or anvils, with concave surfaces, as shown in Fig. 7.5. The pressure medium has a cylindrical shape. To prevent outflow of the pressure medium between anvils pyrophyllite gaskets are used. A large scale belt-type apparatus is widely used for the production of diamond and c-BN.

There are two types of multianvil compression systems, a one-stage cubic anvil system and a two-stage multianvil system. In the former system, a cubic pressure medium, as shown in Fig. 7.6b, is compressed by six anvils (Fig. 7.6a) [19]. The

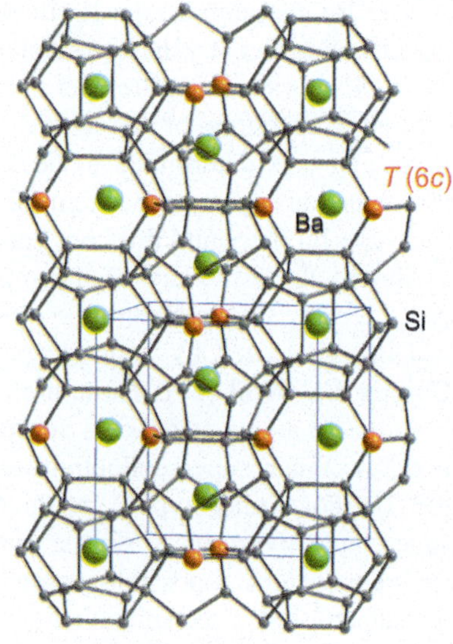


**Fig. 7.4** The crystal structure of $Ba_8T_6Si_{40}$ (T = Ni, Pd, Pt, Cu, Ag, Au)

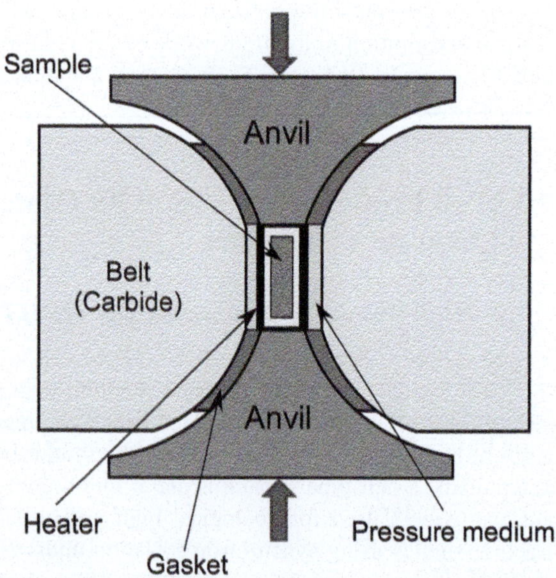


**Fig. 7.5** Schematic drawing of a belt-type high pressure system

one-stage press is mainly used for a pressures below 5 GPa. A typical two-stage system is the Kawai type (Fig. 7.7), [20] referred to as a 6–8 compression system, where an octahedral pressure medium (Fig. 7.7e) is placed in the center of the "nest" surrounded by eight truncated cubic anvils (Fig. 7.7c) and squeezed by advancement of the eight inner second-stage cubic anvils which are driven with six outer first-stage anvils (Fig. 7.7a, b). In order to separate multianvils more than a dozen gaskets are needed for a single run. However the shapes are simple;

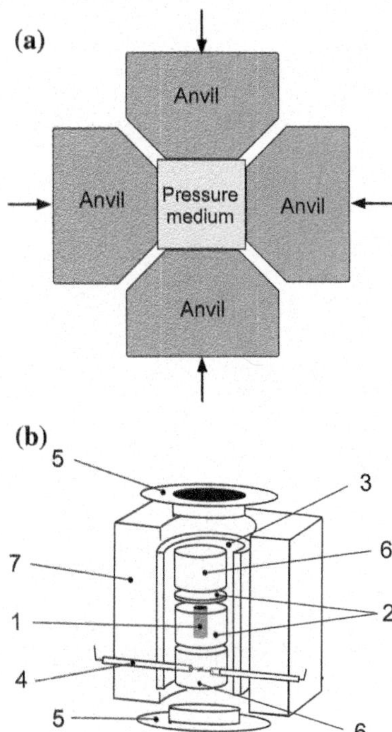


**Fig. 7.6** Schematic drawing of **a** a cubic multi-anvil high pressure system composed of six anvils, and **b** the cross-section of a cubic high pressure medium containing sample (1) in an h-BN sample cell (2), carbon ring heater (3), thermocouple (4), copper electrodes for heater (5), pyrophyllite pressure medium (6,7) [19]

multianvil systems are versatile for HPHT synthesis up to 15 GPa and a temperature up to 1,500 °C. To generate higher pressure cubic multianvils with a smaller truncation edge length (TEL) must be used with the space available for samples being extremely smaller.

Figure 7.7e shows a typical sample assembly in an octahedral pressure medium used in the Kawai type high pressure apparatus. The sample is contained in an h-BN cell which is surrounded by a graphite tube heater. The heater is in turn surrounded by a thermal insulator of stabilized zirconia or calcined pyrophyllite which is in turn placed in a pierced MgO octahedron as a pressure medium. The MgO octahedron is doped with Cr or Co oxide to reduce thermal conductivity. In case of a two-stage compression system, the truncation edge length (TEL) of anvils is 4–6 mm to generate a pressure ~15 GPa, and the inner diameter of an h-BN cell for samples is about 1.5–3 mm. A thin metal heater made of Mo, Ta, or Pt can be used for carbon. The metal film heater can provide a larger space for the samples; however, it is often reactive with the samples making silicides through an h-BN cell. A graphite heater is stable. However, it cannot be used above 13 GPa and above 1,000 °C since it changes into a diamond-like insulator at these conditions. The temperature can be controlled by a monitoring thermocouple or by control of the electric power. The reaction can normally be completed within an hour. The temperature can be cooled down to room temperature in a few minutes by turning off the power, or cooled down slowly to grow single crystals.

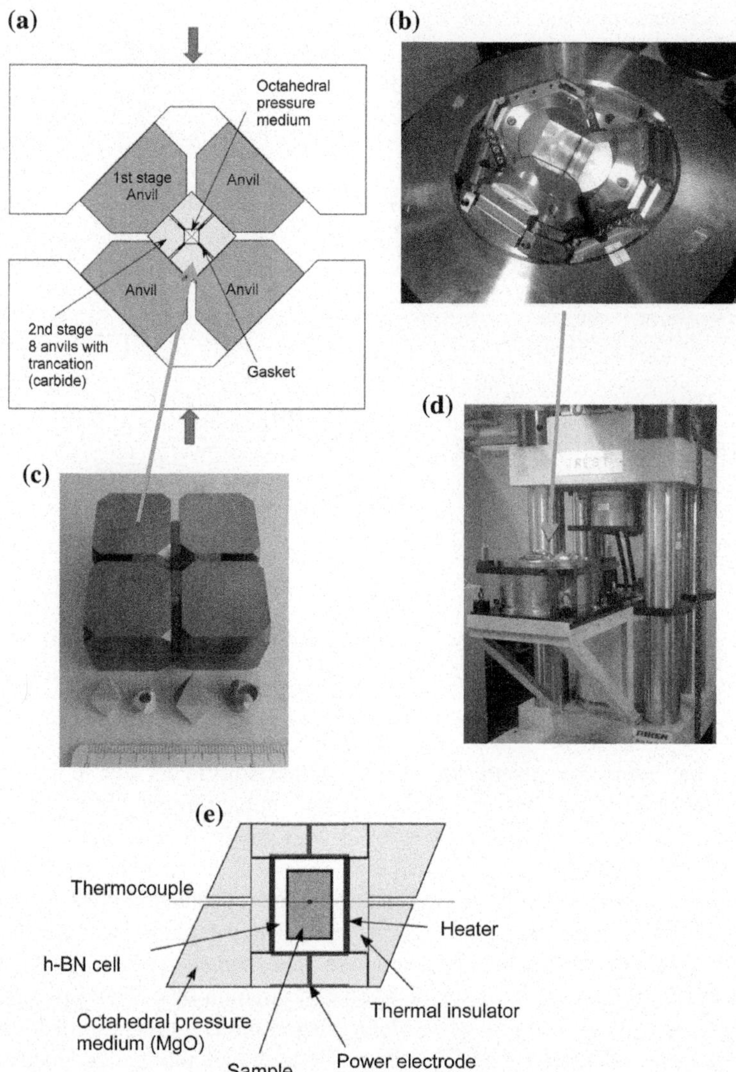


**Fig. 7.7** Schematic drawing of **a** Kawai-type two-stage multianvil high pressure system, and **e** octahedral pressure assembly for sample. The arrangement of the anvils in the drawing is an image showing the two-stage compression. The second stage anvils surrounding the octahedral pressure medium are compressed from the [111] or [100] directions of cubic multianvils. *Photos* **b** first stage multianvils for [111] press, **c** second stage cubic anvils with truncation edge lengths of 4 and 6 mm, and Co doped MgO octahedra, and **d** 1,000-t press

### 7.3.2 Silicides Under High Pressure

Alkaline-earth metal and lathanoid disilicides ($MSi_2$) adopt a variety of silicon sub-networks depending on the size of the metals. Schematic structural models of $MSi_2$ are summarized in Fig. 7.8 [11]. As shown in the figure, only $BaSi_2$ consists of isolated $[Si_4]^{4-}$ anions like alkali metal monosilicides. This is the reason why $BaSi_2$ was first used in an attempt to prepare a clathrate compound containing alkaline earth metal. The other disilicides have extended networks of three coordinated $Si^{-1}$. The length (*l*) of the cubic root of the cell volume per unit formula $MSi_2$ is plotted as a function of the metallic radii of M in Fig. 7.9 [21]. The length *l* decreases in the following order, $BaSi_2 > SrSi_2 > CaSi_2 > \alpha$-$ThSi_2$ structure compounds ($EuSi_2$, $LaSi_2$, $CeSi_2$, $PrSi_2$, $ThSi_2$, $NdSi_2$, $GdSi_2$, $SmSi_2$, $YSi_2$, $DySi_2$) > $AlB_2$ structure compounds ($YbSi_2$, $ErSi_2$, $TmSi_2$, $LuSi_2$). A pioneer study on the application of HPHT conditions for the study of Zintl silicides was made by Evers et al. [22]. They showed that the silicon subnetworks of Zintl metal disilicides $MSi_2$ were flexible under HPHT conditions. The crystals are transformed to the next Zintl disilicide structure with the smaller size metal atoms; $SrSi_2$ and $CaSi_2$ are transformed to the $\alpha$-$ThSi_2$ structure [21, 22]. The orthorhombic $BaSi_2$ undergoes a phase transition to the layer structured $EuGe_2$ (a polytype of the $CaSi_2$ structure) at 5.2 GPa and 400 °C, and to the $SrSi_2$ structure at 5.2 GPa and 600 °C [23, 24]. The high pressure form of $BaSi_2$, with the $EuGe_2$ layered structure, is a superconductor with $T_c = 6.8$ K [25]. $CaSi_2$ is transformed to the $AlB_2$ structure containing graphite-like Si layers at 15 GPa with $T_c$ as high as 14 K under pressure [26]. High-pressure crystal chemistry of binary Zintl phases has been reviewed in a recent paper by Demchyna et al. [27]

### 7.3.3 HPHT Synthesis of Superconducting Type I Clathrate $Ba_8Si_{46}$

A molar mixture of 8 $BaSi_2$ + 30 Si forms the type I silicon clathrate compound $Ba_8Si_{46}$ by HPHT treatment under a pressure of 3 GPa at 800 °C [28]. The crystal structure was refined by Rietveld analysis. The lattice parameter ($a = 10.328$ Å) is larger than that of $Ba_6Na_2Si_{46}$ ($a = 10.26$ Å), indicating that all of the metal sites are occupied by large Ba atoms, as shown in Fig. 7.10a. The compound shows superconductivity with $T_c = 8.0$ K (Fig. 7.10b). It seems unusual to use high pressure conditions for the synthesis of a compound with a cage-like structure. However, if we calculate the volume change accompanied by the reaction it is evident that the reaction is accompanied by a large decrease in volume ($\Delta V$), and the use of HPHT is favourable for the reaction:

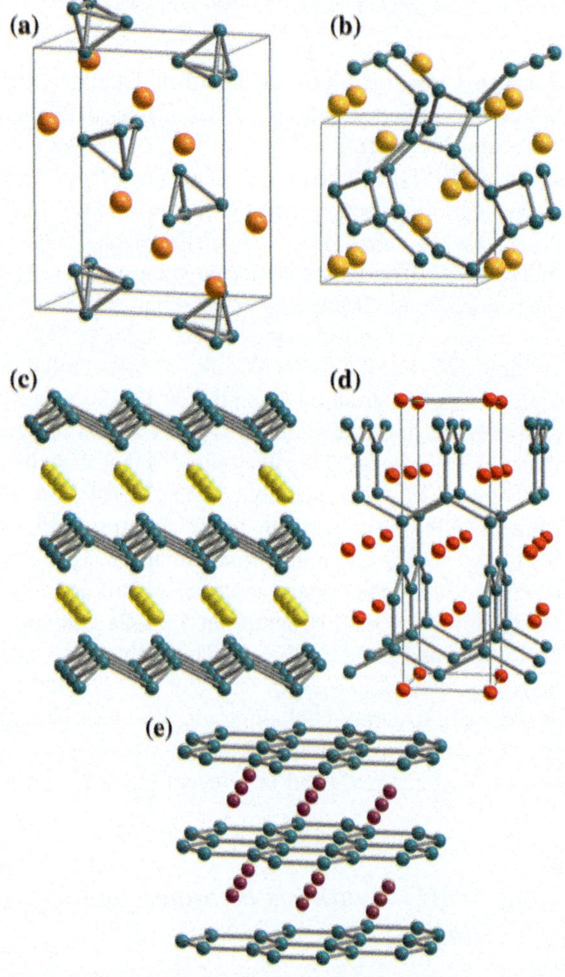


**Fig. 7.8** Schematic illustration of the structures of Zintl disilicides with different kinds of alkaline-earth and lantanoid metals: **a** $BaSi_2$, **b** $SrSi_2$, **c** $CaSi_2$, **d** $LaSi_2$, and **e** $YbSi_2$. The *green–blue* color small balls are silicon atoms [11]

$$4\ Ba_2Si_4 + 30\ Si \rightarrow Ba_8Si_{46} \quad \Delta V = -207\ \mathrm{cm^3/mol}\ (-15\%).$$

The Zintl anion $[Si_4]^{4-}$ in $BaSi_2$ is voluminous with four lone electron pairs directing outward, as shown in Fig. 7.1. In the clathrate compound, all of the lone pairs are changed into Si–Si σ bonds. This leads to a significant decrease in the molecular volume in the system. HPHT syntheses obey the Le Chatelier's principle; the decrease in the molecular volume of the reaction system is an important driving force for synthesis by HPHT reaction.

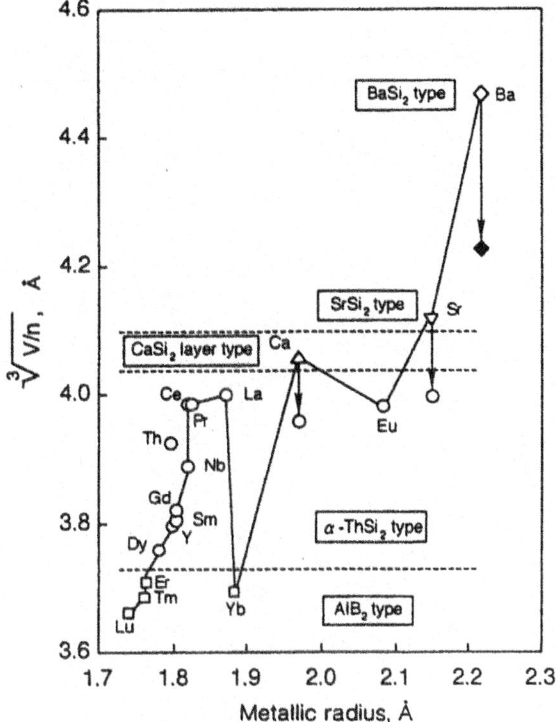


**Fig. 7.9** Formation region of various types of Zintl phases of disilicides in a plot of $\sqrt[3]{V/n}$ (the cubic root of the cell volume per unit formula $MSi_2$) versus the metallic radii for 12 coordination [21]

## 7.3.4 *Compositional Dependence of the $T_c$ in Ba Containing Silicon Clathrate Compounds*

### 7.3.4.1 Single Crystal Study of the Ba Deficient Compounds $Ba_{8-x}Si_{46}$

Single crystals of $Ba_8Si_{46}$ were obtained by using different temperature protocols under a pressure of 3 GPa [19]. A molar mixture of 8 $BaSi_2$ + 30 Si was heated at a higher temperature of 1,300 °C and 3 GPa for 2 min, and slowly cooled down to 700 °C in 2 h followed by rapid cooling to room temperature. Single crystals of $Ba_8Si_{46}$ suitable for X-ray structural study were obtained. The structural analysis results are shown in Table 7.1. It was revealed that the crystal has a Ba deficiency at the 2*a* site of the smaller cages with a composition of $Ba_{7.76}Si_{46}$ or $(Ba_6)^{6d}(Ba_{1.76})^{2a}Si_{46}$ [19]. $T_c$ was 9.0 K, higher than that observed for polycrystalline samples obtained at a lower temperature of 800 °C and 3 GPa (8.0 K). Upon evacuation at 527°C, Ba atoms at the 2*a* sites, in @$Si_{20}$ cages, are deintercalated to a composition of $(Ba_6)^{6d}(Ba_{0.63})^{2a}Si_{46}$, and the lattice parameter decreases from 10.3141(7) to 10.2652(8) Å. Note that only the Ba atoms in the smaller cages are removed. In the case of the type II silicon clathrate $Na_xSi_{136}$, the Na atoms in the smaller cages (@$Si_{20}$) are preferentially removed in a similar way on evacuation at

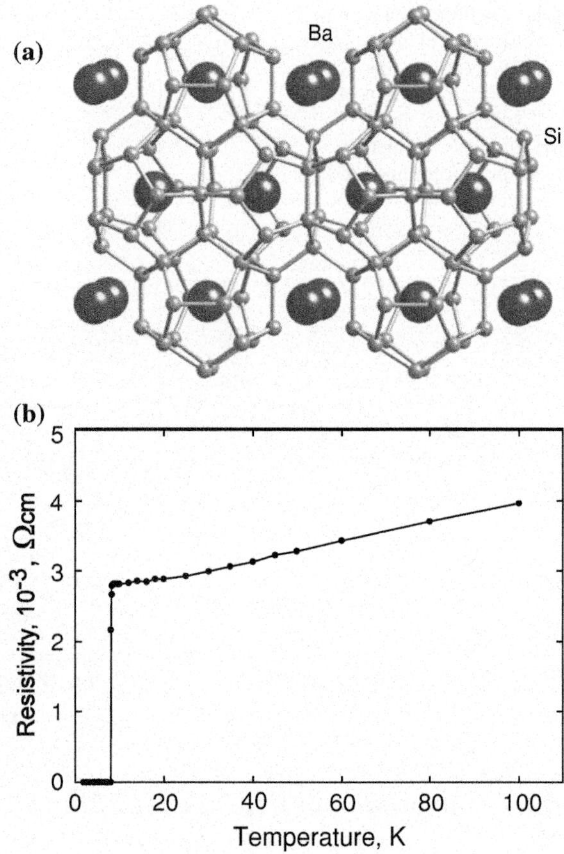


**Fig. 7.10** Schmatic structure of $Ba_8Si_{46}$ (**a**), and the temperature dependence of the electrical resistivity (**b**), showing a superconducting transition at 8.0 K

**Table 7.1** Atomic parameters for the structure of $Ba_8Si_{46}$ [19]

| Formula: $Ba_{7.76}Si_{46}$ | | Space group: $Pm\bar{3}n$ (No. 223) | | | | |
|---|---|---|---|---|---|---|
| Atom | Site | *x* | *y* | *z* | $B_{eq}$ (Å$^2$) | *ocp* |
| Ba1 | 2*a* | 0 | 0 | 0 | 1.16(2) | 0.881(5) |
| Ba2 | 6*d* | 1/4 | 1/2 | 0 | 1.70(1) | 1 |
| Si1 | 6*c* | 1/4 | 0 | 1/2 | 1.14(4) | 1 |
| Si2 | 16*i* | 0.1854(1) | 0.1854(1) | 0.1854(1) | 1.16(1) | 1 |
| Si3 | 24*k* | 0 | 0.3061(1) | 0.1213(1) | 1.26(3) | 1 |

elevated temperatures [12]. After almost all of the Na atoms in the smaller cages are removed, Na atoms in the larger cages start to be removed, forming the almost guest-free compound $Na_xSi_{136}$ ($x < 1$). The type I clathrate $Na_8Si_{46}$ can have only a stoichiometric composition, and decomposes to Si with the diamond structure at temperatures above 550 °C.

The superconducting transition temperature $T_c$ decreases from 9.0 to 6.0 K as the Ba content decreases from $Ba_{7.76}Si_{46}$ to $Ba_{6.63}Si_{46}$, as shown in Fig. 7.11. The first HPHT synthesis of $Ba_8Si_{46}$ showed $T_c = 8.0$ K. It is very likely that the powder sample might also have a Ba deficient composition corresponding to $T_c = 8.0$ K.

#### 7.3.4.2 $Na_xBa_6Si_{46}$

The first Ba containing silicon clathrate compound was prepared by the thermal decomposition of the Zintl compound $Na_2BaSi_4$. When the decomposition temperature was changed, the compounds with different Na contents were obtained [14]. The decomposition temperature, composition and $T_c$ of the decomposed product are listed in Table 7.2. The $T_c$ increased with decreasing Na content. The decomposition at 650 °C gave a mixture of $BaSi_2$ and Si, suggesting that the type I clathrate compound $Na_xBa_6Si_{46}$ is not stable above this temperature. The sample with a composition $Na_{0.2}Ba_{6.0}Si_{46}$ can be post-doped with Na atoms by treating with Na metal vapour at 400 °C in a vacuum-sealed glass tube. This treatment increased the Na content to $Na_{0.7}Ba_{6.0}Si_{46}$, and $T_c$ decreased from 4.8 to 3.0 K. The compositional dependence on $T_c$ is shown in Fig. 7.11 as a function of the Ba content in $Ba_xBa_6Si_{46}$, and the Na content in $Na_xBa_6Si_{46}$. It should be noted that $T_c$ of $Na_xBa_6Si_{46}$ decreases with the increase of the Na content, whereas $T_c$ of $Ba_x$-$Ba_6Si_{46}$ increases with the increase of the Ba content. The two curves for $Ba_x$-$Ba_6Si_{46}$ and $Na_xBa_6Si_{46}$ appear to meet together at $T_c \sim 5$ K with $x$ approaching 0. $T_c$ is dependent on the Ba as well as the Na contents. These findings may suggest that the electronic band structure of the type I clathrate compounds cannot be understood in terms of a simple rigid band model. Ba and Na atoms are only partially ionized and the orbitals interact with the silicon framework.

#### 7.3.4.3 Alkaline-Earth Metal Doped Type I Clathrate Compounds

Other alkaline-earth metal (AE) doped type I silicon clathrate compounds, such as $Ca_8Si_{46}$ and $Sr_8Si_{46}$, have not been prepared but only solid solutions with Ba are formed, such as $Ba_{8-x}Ca_xSi_{46}$ ($x \leq 4$) and $Ba_{8-x}Sr_xSi_{46}$ ($x \leq 6$) [29, 30]. The solid solutions also show superconductivity with $T_c$ decreasing with increasing substitution with Sr or Ca.

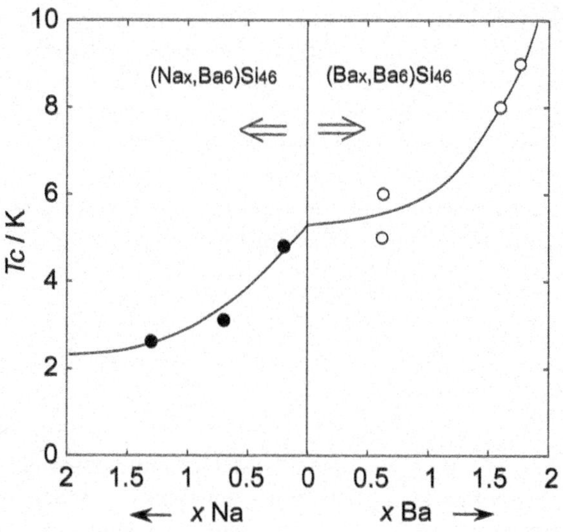


**Fig. 7.11** The $T_c$ values of $Ba_xBa_6Si_{46}$ and $Na_xBa_6Si_{46}$ silicon clathrates plotted as a function of $x$ [19]

**Table 7.2** Compositions and superconducting transition temperature ($T_c$) of silicon clathrate compounds obtained by decomposition of $Na_2BaSi_4$ at different temperatures [14]

| Decomposition temperature °C | Composition | $T_c$ (K) |
|---|---|---|
| 500 | $Na_{1.5}Ba_{6.0}Si_{46}$ | 2.6 |
| 550 | $Na_{0.3}Ba_{6.2}Si_{46}$ | 4.0 |
| 600 | $Na_{0.2}Ba_{6.0}Si_{46}$ | 4.8 |
| 650 | Si + $BaSi_2$ | – |
| Na post-doped | $Na_{0.7}Ba_{6.0}Si_{46}$ | 3.0 |

#### 7.3.4.4 Transition Metal Substituted Clathrates $Ba_8Si_{46-x}T_x$ (T = Ag, Au, Ni, Cu; $0 \leq x \leq 6$)

The solid solutions $Ba_8Si_{46-x}T_x$ (T = Ag, Au, Ni, Cu; $0 \leq x \leq 6$) can be prepared as single phases using similar HPHT conditions as those used for the preparation of $Ba_8Si_{46}$. The transition metal atoms randomly replaces Si on the 6*c* site. The solid solutions with $x < 3$ are superconductors, and $T_c$ decreased from 8.0 K with increasing $x$ [31]. Optical reflection spectra were measured on $Ba_8Ag_xSi_{46-x}$ ($0 \leq x \leq 6$) at room temperature. A systematic decrease in the plasmon energy was found with increasing $x$ indicating that the carrier concentration decreases with increasing $x$.

#### 7.3.4.5 HPHT Synthesis and Superconductivity of $Ba_8Si_{46-x}Ge_x$

The type I clathrate $Ba_8Ge_{43}$ has vacancies at the 6c sites of the germanium framework [32, 33]. It does not show superconductivity. The stoichiometric compound $Ba_8Ge_{46}$ and solid solutions $Ba_8Ge_{46-x}$ ($x < 3$) have not been prepared.

Ge substituted silicon clathrate compounds $Ba_8Si_{46-x}Ge_x$ ($0 \leq x \leq 23$) were prepared using HPHT conditions of 3 GPa at 800 °C [34]. All the solid solutions with $x \leq 23$ showed superconductivity with $T_c$ decreasing from 8.0 to 2.0 as the Ge content increased from $x = 0$ to 23. The decrease in $T_c$ is associated with the expansion of the lattice parameter from 10.33 to 10.50 Å. The larger Ge–Si average distance may result in a weaker electron-phonon interaction, leading to lower $T_c$ in the Ge substituted compounds. Single crystals with different compositions were obtained. The crystal structure and the distribution of the Ge and Si atoms were determined by single crystal X-ray diffraction. A slight Ba deficiency was also observed at 2*a* site for all the clathrates. The Ge atoms replace the Si atoms at the 24*k* site for the composition range $x < 8$, and then at the 16*i* sites.

### 7.3.5 Characterization of the Superconductivity of $Ba_8Si_{46}$

The band structure of Ba intercalated silicon clathrate compounds $Na_2Ba_6Si_{46}$ and $Ba_8Si_{46}$ were calculated in comparison with $Na_8Si_{46}$ and the hypothetical metal free type I clathrate $Si_{46}$ [35, 36], as shown in Fig. 7.12. The band structure of $Na_2Ba_6Si_{46}$ is very similar to that of $Ba_8Si_{46}$, and not shown here. The band structure of $Na_8Si_{46}$ can be understood by a rigid band model; the electrons doped from Na atoms fill the conduction band of $Si_{46}$ and the compound becomes metallic. It has been reported that the Na metal bands are located at the bottom of the Si conduction band in the electron density of states (DOS) [37]. In the case of $Ba_6Na_2Si_{46}$ and $Ba_8Si_{46}$, the Ba 5d orbitals are strongly hybridized with Si orbitals forming a new hybridized band at the Fermi level. The electrons doped from the metal atoms fill the hybridized bands and have a high DOS at the Fermi level. This appears to be favourable for superconductivity.

It is interesting to note that the hypothetical metal free clathrate compound $Si_{46}$ is a new silicon allotrope. The band calculation predicts that the compound has a wide band gap of 1.8 eV compared with 1.2 eV for the diamond structure. A wide optical band gap of 1.9 eV has been observed on the almost metal free type II clathrate compound $Na_xSi_{136}$ [38].

NMR measurements were carried out on the superconducting clathrate compound $Na_xBa_ySi_{46}$ by Shimizu et al. [39]. The spin-lattice relaxation time $T_1$ on the three kinds of nuclei, $^{29}Si$, $^{137}Ba$, and $^{23}Na$, showed the Korringa relation ($T_1T$ = const.) down to 4 K, implying that the compound has a metallic electronic structure and the conduction electrons are on Si, Na, and Ba atoms. The temperature dependence of the peak positions of the resonance lines of the $^{23}Na$ and $^{137}Ba$ signals suggests a Knight shift contribution indicating that the Na and Ba atoms are not fully ionized and the orbitals are coupled with the $Si_{46}$ orbitals. This is the reason that the superconductivity depends on the composition of the clathrates, and Na and Ba contents show significant effects on $T_c$ (Fig. 7.11). As shown in Fig. 7.13, three separate NMR peaks (S1, S2, and S3) were observed for $^{29}Si$, corresponding to three non-equivalent Si sites, 16*i*, 6*c*, and 24*k* sites, respectively.

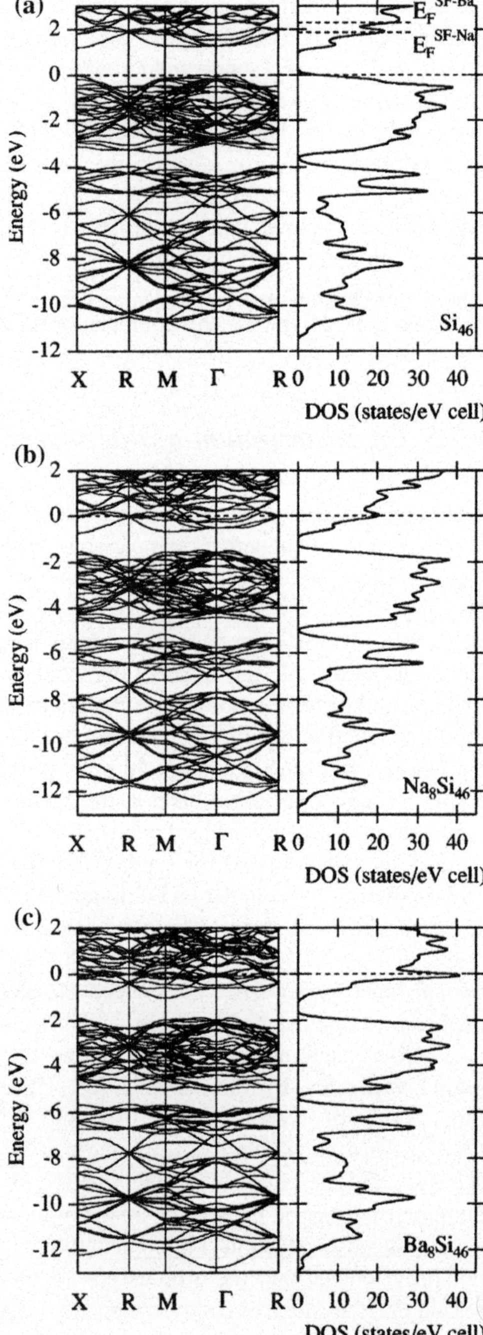


**Fig. 7.12** Band structure and density of states for **a** $Si_{46}$, **b** $Na_8Si_{46}$, and **c** $Ba_8Si_{46}$. The Fermi level is denoted by *horizontal broken lines*. Copyright American Institute of Physics [36]

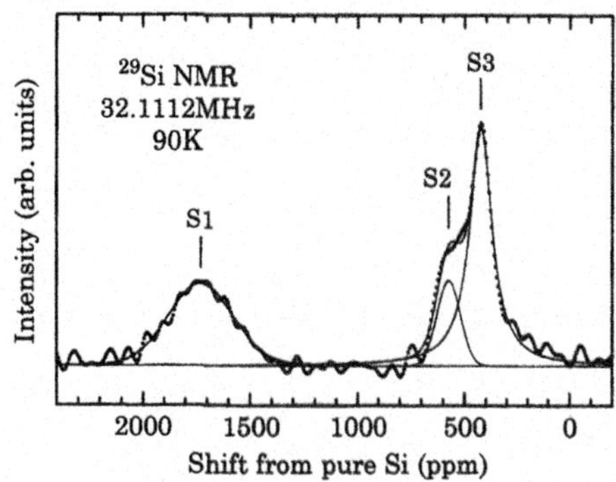


**Fig. 7.13** $^{29}$Si NMR spectrum of $Na_2Ba_6Si_{46}$ at 90 K. The NMR shift is measured from the resonance in the semiconducting pure Si. The peak areas for S1, S2, and S3 are in the ratio of 16:6:24. Copyright American Institute of Physics [39]

The peak areas are in the ratio of the number of Si atoms at these sites. It was pointed out that the peak S1 for Si in the 16*i* site shows a remarkably large shift. The $^{29}$Si-NMR in non-superconducting silicon clathrate compounds is observed around the frequencies found for the S2 and S3 peaks for Si in the 6*c* and 24*k* sites. The remarkably large shift for the S1 site may be caused by hybridization with the Ba orbitals. Such a large shift in $Na_xBa_ySi_{46}$ might have some relevance to superconductivity.

The contribution of the Ba 5d orbitals was also suggested by a Compton scattering study of $Ba_8Si_{46}$ in comparison with the scattering of the metal free clathrate compound $Si_{136}$ [40]. The superconducting gap was directly measured by high resolution XPS by Yokoya et al. [41] as shown in Fig. 7.14, and determined to be $\Delta E_{gap}/k_BT_c = 4.38$, slightly larger than a typical value (3.52) for a BCS superconductor. Tanigaki et al. [42] studied the isotope effect on the superconductivity of $Ba_8^{28}Si_{46}$ and $Ba_8^{30}Si_{46}$. They also studied the specific heat and Raman scattering effect. The Debye temperature was estimated to be 370 K. The superconducting gaps measured by the XPS and the specific heat imply that $Ba_8Si_{46}$ is a typical phonon-mediated BCS superconductor in the medium coupling regime.

### *7.3.6 Ba–Si Binary System Under High Pressure*

A new silicon clathrate compound $Ba_{24}Si_{100}$ isotypic with $K_8Sn_{25}$ was obtained under HPHT conditions with a slightly reduced pressure of 1.5 GPa as compared to that (3 GPa) used for the preparation of $Ba_8Si_{46}$ [43]. As shown in Fig. 7.15a, the new compound has a structure consisting of Ba dodecahedral silicon cages Ba@$Si_{20}$ linked by shared pentagonal faces forming a chiral (space group $P4_132$) zeolite-like framework. The rest of Ba atoms occupy the resulting interstices outside the linked Ba@$Si_{20}$ cages. The silicon framework of $Ba_{24}Si_{100}$ can be derived from that of the type I clathrate structure by subtraction of 4 Si atoms from the @$Si_{24}$ cage, as illustrated in Figs. 7.15b, c. The new clathrate $Ba_{24}Si_{100}$ can be classified as a derivative from type I. It shows superconductivity at $T_c = 1.4$ K

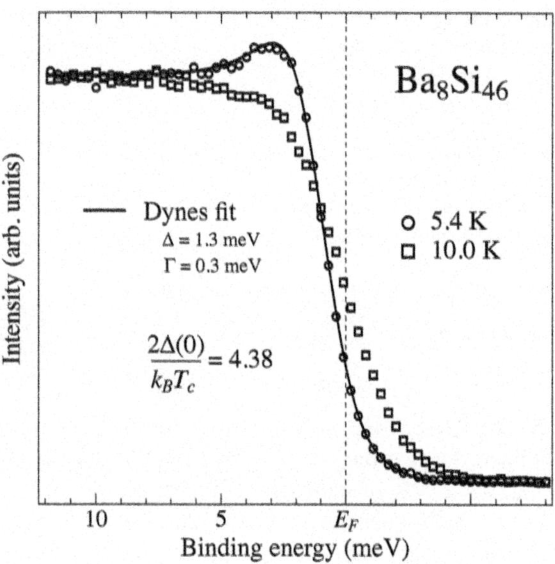


**Fig. 7.14** Ultrahigh-resolution of photoelectron spectra of $Ba_8Si_{46}$ measured at 5.4 K (superconducting state, *open circles*) and 10.0 K (normal state, *open squares*) using an energy resolution of 2.4 meV. Copyright American Institute of Physics [41]

[44]. Viennois et al. reported a $T_c$ = 1.55 K for $Ba_{24}Si_{100}$ [45]. $Ba_{24}Ge_{100}$ is isomorphous with $Ba_{24}Si_{100}$ [46]. $Ba_{24}Ge_{100}$ can be prepared under ambient pressure. The structure and properties of $Ba_{24}Ge_{100}$ have been well characterized. It shows superconductivity at $T_c$ = 0.24 K [47, 48]. The superconductivity of $Ba_{24}Ge_{100}$ showed a strong pressure dependence and $T_c$ increased to 3.8 K at 2.7 GPa.

Under a much higher pressure, >10 GPa, a new silicon-rich phase $BaSi_6$ was obtained in the Ba–Si binary system [49]. The structure is shown in Fig. 7.16. The barium atoms are placed in the channel running along the *a* axis. The Ba atoms are surrounded by 18 silicon atoms with an average distance of 3.435 Å, forming an irregularly shaped cage ($@Si_{18}$). The molecular volume change accompanied by the following reaction is −16 %:

$$BaSi_2 + 4\ Si \rightarrow BaSi_6 \qquad \Delta V = -83.9\ cm^3/mol, (-16\%),$$

larger than that for $Ba_8Si_{46}$ (−15 %). The Ba–Si binary compounds prepared under HPHT conditions are summarized in Fig. 7.17 as a function of pressure [49]. As the pressure increases, the silicon content, or Si/Ba ratio of the product, increases with the increase of the molecular volume change. $BaSi_6$ is isomorphous with $EuGa_2Ge_4$. $SrSi_6$ and $CaSi_6$ are also isomorphous with $BaSi_6$, and have also been prepared by similar HPHT conditions [50, 51]. These are not superconductors. In a recent HPHT study on the binary Na–Si system, a new compound $NaSi_6$, isomorphous with $BaSi_6$, has been obtained [52].

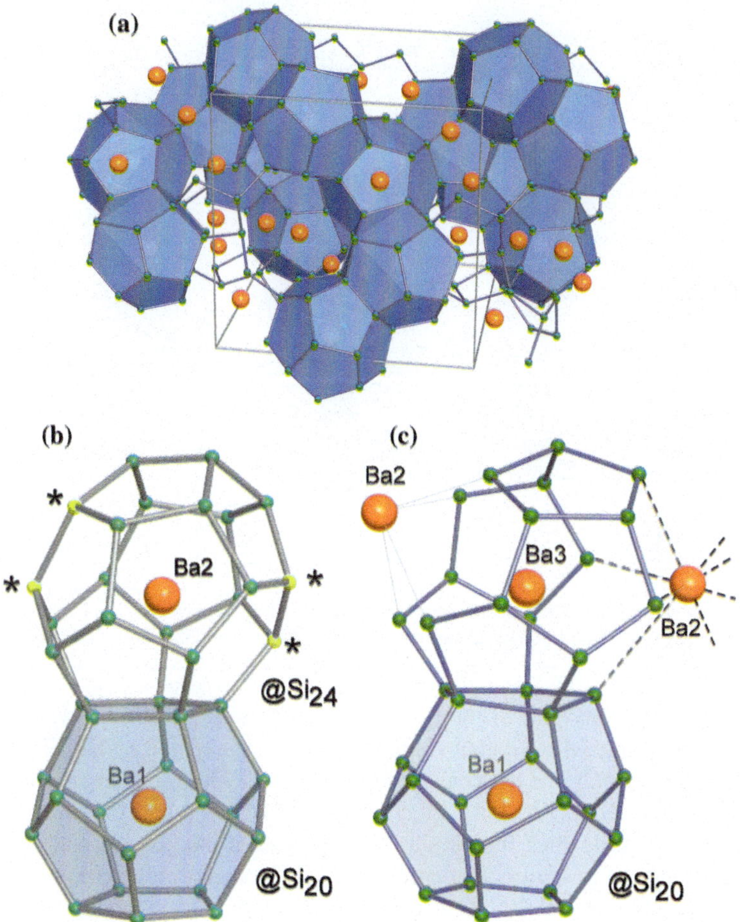


**Fig. 7.15** Schematic illustration of the structure of **a** $Ba_{24}Si_{100}$ composed of Ba containing polyhedra (Ba@$Si_{20}$) by face sharing. The polyhedra form a zeolite-like 3D framework, and the rest of Ba atoms (*red color*) are located in the space between the polyhedra. There are three kinds of Ba sites in $Ba_{24}Si_{100}$. The part of the structure of $Ba_{24}Si_{100}$ (**c**) can be derived from type I clathrate $Ba_8Si_{46}$ (**b**) by removing four Si atoms (marked by *) from the Ba@$Si_{24}$ polyhedron [11, 46]

## 7.4 HPHT Synthesis of New Clathrate Compounds

### *7.4.1 Iodine–Si Binary System*

The first type I clathrate compound $I_8Ge_{43.3}I_{2.7}$ (lattice parameter $a$ = 10.814 Å) containing electronegative element iodine was prepared by a rapid thermal decomposition of $GeI_2$ [53]. The iodine atoms occupy the centers of all the cages

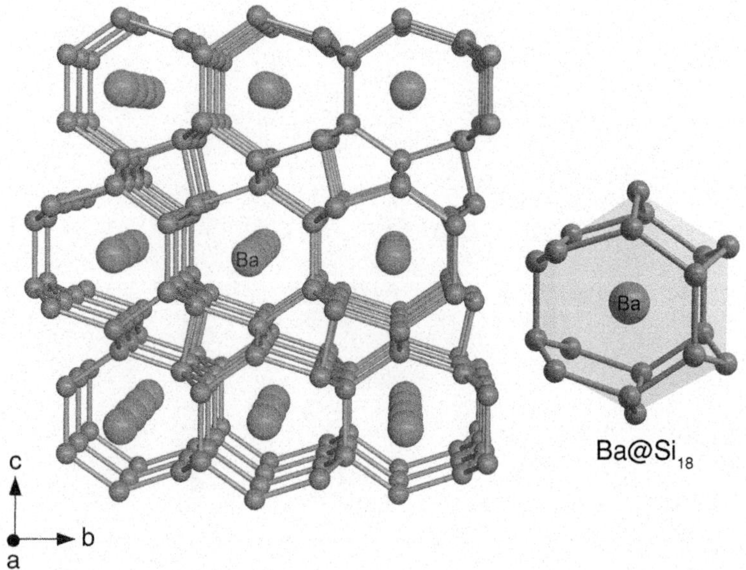


**Fig. 7.16** Schematic illustration of the structure of $BaSi_6$ composed of irregular shaped polyhedral Ba@$Si_{18}$ (*right*). Ba atoms are placed in the tunnels running along the *a* axis [11, 49]

and replace randomly part of the framework Ge 6*c* sites. A similar type I silicon clathrate compound with a composition $I_8(I_2Si_{44})$ was prepared by HPHT synthesis at 5 GPa and 700 °C. The lattice parameter $a = 10.4195(7)$ Å is larger than that of $Ba_8Si_{46}$ ($a = 10.328(2)$ Å) [54]. Rietveld analysis showed that iodine replaces Si on the 24*k* site instead of the 6*c* site. We expected a hole doped clathrate compound, but it is an insulator. The type I iodine silicon clathrate compound $I_{10}Si_{44}$ shows a remarkable stability with pressure up to 35 GPa before an irreversible amorphization is observed [30].

### *7.4.2 La–Si Binary System*

The La–Si binary system is very attractive in relation with elemental carbon chemistry. $LaSi_2$ adopts the α-$ThSi_2$ structure as shown in Fig. 7.8d and the coordination of Si in $LaSi_2$ is a planner triangle. The structure is composed of $sp^2$-bonded silicon. $LaSi_2$ is metallic and shows superconductivity with $T_c = 2.5$ K [55]. $LaSi_2$ is a Zintl phase with delocalized electrons. If this sub-network is realized with all-$sp^2$ carbon without metal atoms, a new 3D carbon allotrope will be obtained. The hypothetical 3D all-$sp^2$ carbon phase is called "hyper graphite" [56, 57]. Metallic conduction is expected from theoretical band structure calculations of the hypothetical structure.

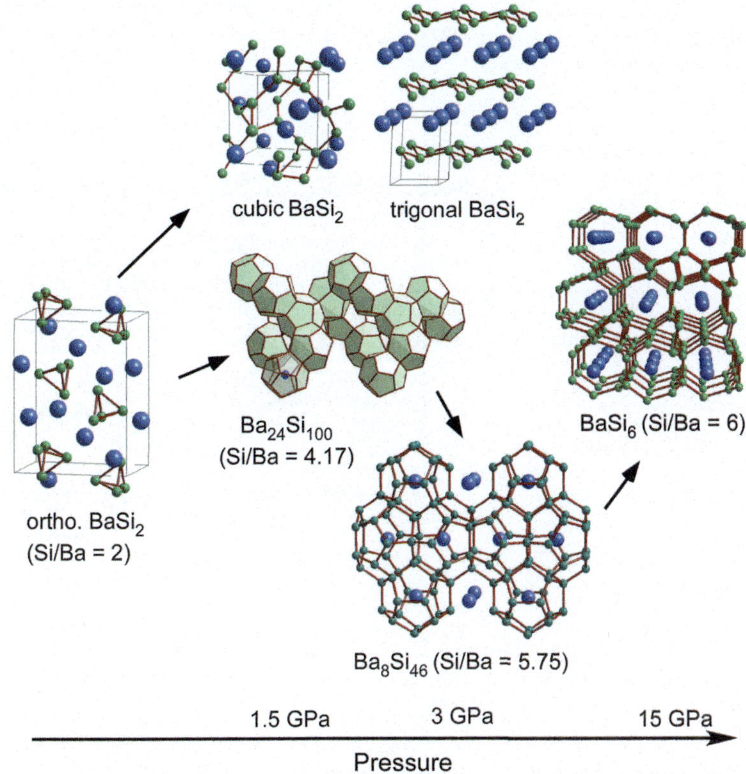


**Fig. 7.17** Schematic illustration of various crystal structures in the binary system Ba–Si found by HPHT treatments; orthorhombic $BaSi_2$ is transformed to trigonal and cubic $BaSi_2$ [49]

In the La–Si binary system, the compound with the highest Si/La ratio is $LaSi_2$ at ambient pressure. $LaSi_2$ is not transformed to the other structure even up to a pressure of 10 GPa; however, by using a much higher pressure, 13.5 GPa, new Si rich compounds $LaSi_5$ and $LaSi_{10}$ are obtained [58]. Figure 7.18 shows the La–Si binary phase diagram in the Si-rich region under a pressure of 13.5 GPa. $LaSi_5$ and $LaSi_{10}$ peritectically decompose at 980 and 750 °C, respectively, at 13.5 GPa. $LaSi_5$ has two polymorphs, α- (high temperature) and β- (low temperature) forms. The α-form has the averaged structure of the disordered β-form, and becomes a superconductor with $T_c = 11.5$ K. In the α-form, the $sp^2$ hybridized Si atoms form coplanar six membered rings which share the edges side-by-side to make sila-polyacene ribbons along the *b*-axis, as shown in Fig. 7.19. The unhybridized *p* orbitals of Si atoms would be delocalized over the sila-polyacene ribbons for superconductivity.

Figure 7.20 shows the schematic structure of $LaSi_{10}$, which is a superconductor with $T_c = 6.4$ K. The structure is composed of La containing silicon polyhedra $La@Si_{18}$ with hexagonal beer barrel shapes. The barrels are stacked along the

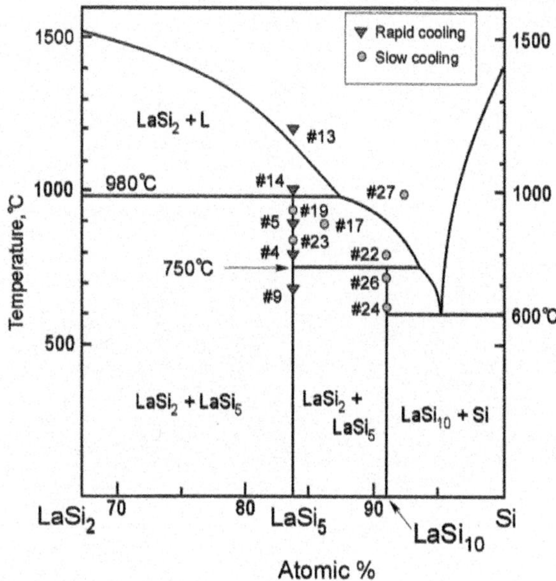


**Fig. 7.18** Phase diagram of the La–Si binary system in the Si-rich region under a pressure of 13.5 GPa [58]

*c*-axis to form straight columns by sharing the flat hexagonal faces. Such one-dimensional columns are bundled side-by-side by sharing side edges. In the center of the bundle is a straight Si linear chain; each Si atom in the chain has trigonal bipyramid fivefold coordination. The beer barrel-type polyhedron $La@Si_{18}$ can be compared with the irregular shaped $Ba@Si_{18}$ polyhedron found in $BaSi_6$ (Fig. 7.16). The two kinds of polyhedra resemble each other as shown in Fig. 7.20b–c. In $Ba@Si_{18}$ one Si–Si distance (3.62 Å) in the hexagonal face is longer than the others (~2.4 Å), and thus the bond is open, while the one Si–Si distance (2.42 Å) on the side of the polyhedron is much shorter than the others, having a bond like a waist band of the polyhedron.

### *7.4.3 Ternary Silicides*

#### 7.4.3.1 Ca–Al–Si

$CaAl_2Si_2$ is a typical ternary Zintl compound with the α-$ThCr_2Si_2$ structure. A number of intermetallic compounds belong to this family. It melts congruently at about 940 °C under ambient pressure. However, under a pressure of 5 GPa it peritectically decomposes into a mixture of $Ca_2Al_3Si_4$ and aluminium metal at temperatures above 600 °C. The structure of $Ca_2Al_3Si_4$ is composed of an $Al_3Si_4$ framework and a layer structured $Ca_2$ sub-network interpenetrating with each other as shown in Fig. 7.21. The layer structured sub-network $Ca_2$ is isomorphous with black phosphorus [59]. $Ca_2Al_3Si_4$ is a superconductor with $T_c = 6.4$ K.

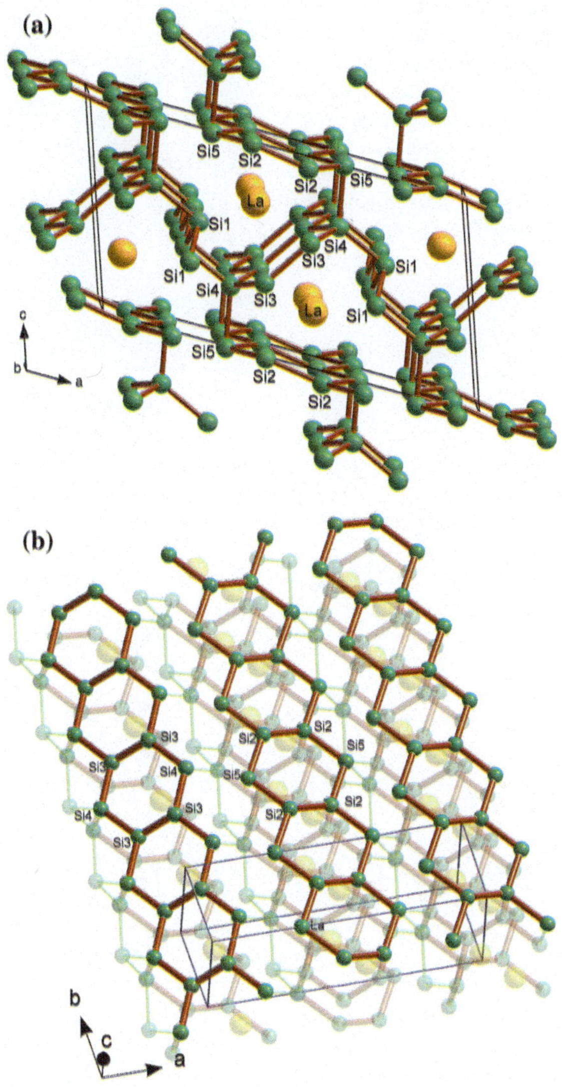


**Fig. 7.19** Schematic illustration of the structure of **a** α-$LaSi_5$, and **b** a view along the *c*-axis, showing the presence of poly sila-acene ribbons running along the *b*-axis [58]

Under a much higher pressure of 15 GPa, $CaAl_2Si_2$ decomposed to $Ca(Al,Si)_2$ with the Laves phase structure [60]. The Laves phase compounds with transition metals are well known for superconductors with the C15 structure, which also belongs to space group $Fd\bar{3}m$ isotypic with the type II silicon clathrate compounds. Solid solutions $Ca(Al_{1-x}Si_x)_2$ ($0.35 \leq x \leq 0.75$) can be directly prepared under HPHT conditions of 1,000 °C and 13 GPa. The stoichiometric Laves phase compound CaAlSi shows superconductivity at $T_c = 2.6$ K. This is the first Laves phase compound composed of only commonly found elements. The compound can

**Fig. 7.20** Schematic illustration of the structure of $LaSi_{10}$ **a** composed of La@$Si_{18}$ clathrate columns **b** [58]. Similar column structures can be seen in $BaSi_6$. La@$Si_{18}$ **b** of $LaSi_{10}$ and Ba@$Si_{18}$ **c** of $BaSi_6$ are compared

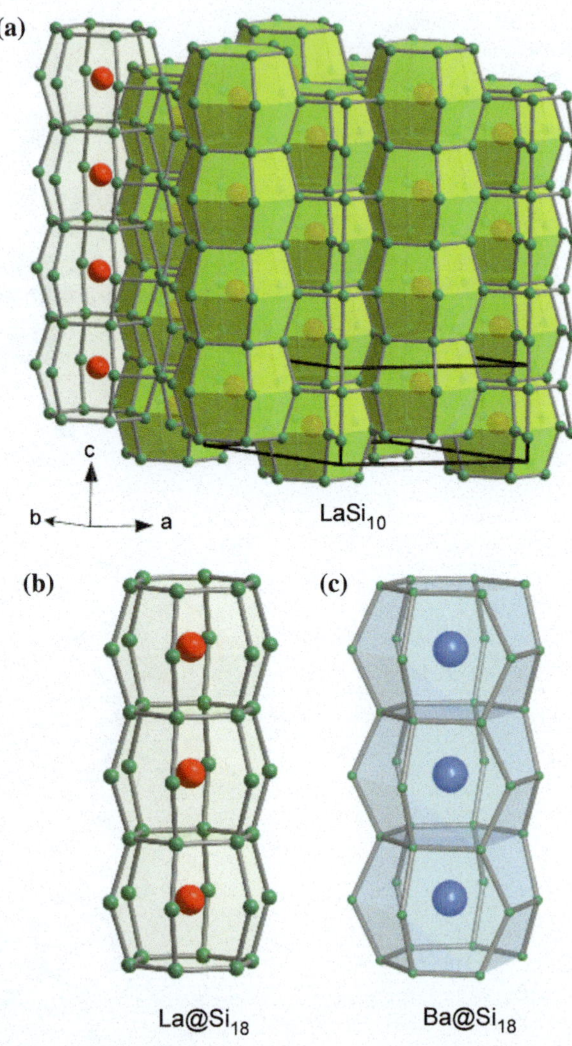


be seen as a type of clathrate compound composed of face sharing truncated tetrahedral cages as shown in Fig. 7.22. The Ca atoms are encapsulated in @$Al_6Si_6$ polyhedra.

#### 7.4.3.2 Ba–Al–Si

The ternary compound $BaAl_2Si_2$ has two polymorphs, the α-form, ambient pressure phase isostructural with the α-$BaCu_2S_2$ structure (space group *Pnma*), and the β-form, high pressure phase isostructural with the α-$ThCr_2Si_2$ structure (space group *I4/mmm*) [61]. The structures are compared in Fig. 7.23. When the β-form

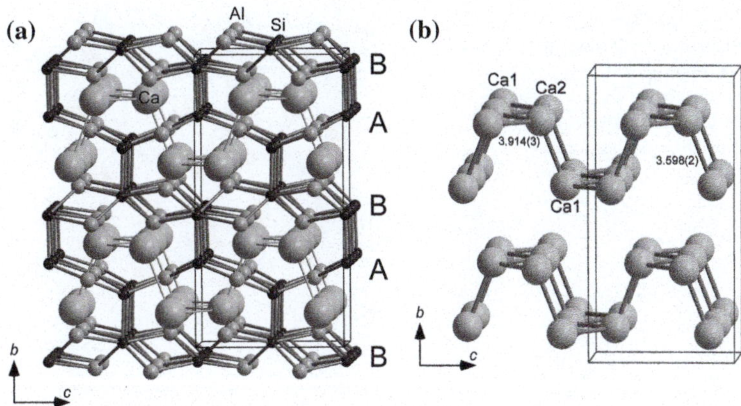


**Fig. 7.21** Schematic structural illustration of $Ca_2Al_3Si_4$, containing layer structured [$Ca_2$] sub-network isomorphous with black phosphorus [59]

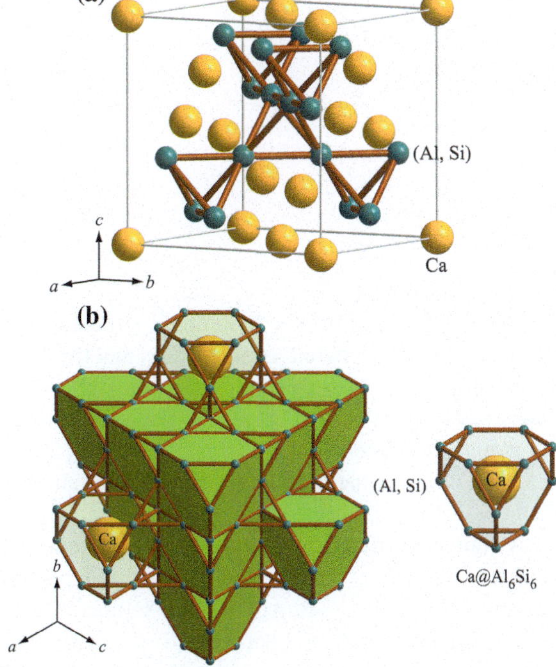


**Fig. 7.22** Schematic structure of **a** the Laves phase compound $Ca(Al,Si)_2$, and **b** a structural view composed of truncated tetrahedral cages with Ca atoms in the center, $Ca@Al_6Si_6$ [60]

prepared under HPHT conditions (5 GPa, 1000 °C) is annealed at 600 °C under ambient pressure, it transformes to the α-form. The Ba atoms in the two polymorphs are placed in large cavities formed by [$Al_2Si_2$] framework. X-ray diffraction study cannot distinguish between Al and Si, space group *Cmcm* was once

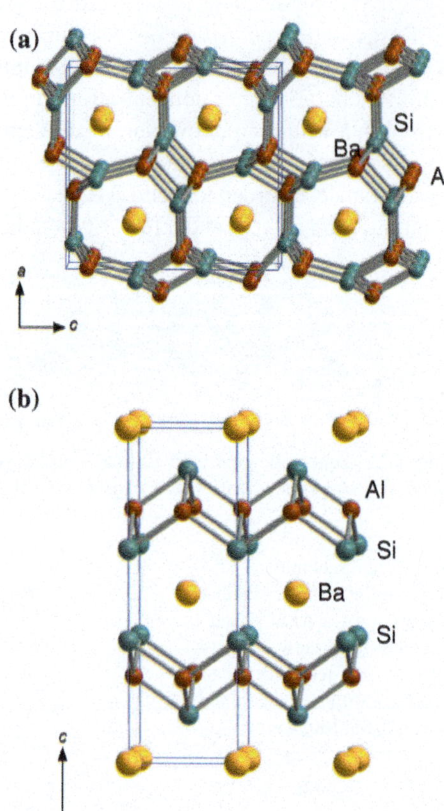


**Fig. 7.23** Polymorphs of $BaAl_2Si_2$; **a** α-(low pressure) form isostructural with α-$BaCu_2S_2$, and **b** β-(high pressure) form isostructural with the α-$ThCr_2Si_2$ structure [61]

tentatively assigned for the disordered structure of the α-form. Single crystals of the α-form were prepared using Al flux, and the Al and Si sites were distinguished by neutron diffraction by Condron et al. [62]. The crystal structure space group *Pnma* is strictly isostructural with α-$BaCu_2S_2$ as shown in Fig. 7.23. In the α-form, Ba atoms are 14-coordinated with 7 Si + 7 Al atoms with distances in a range of 3.32–3.67 Å, comparable with the Ba–Si distances 3.35–3.65 Å in the large cage $Ba@Si_{24}$ of the clathrate compound $Ba_8Si_{46}$.

A Raman study of the α-form under pressure suggests that the compound is transformed to the β-form irreversibly at a pressure of 20 GPa [63]. The isostructural Ge analog α-$BaAl_2Ge_2$ (low temperature form) is transformed thermally to the β-form (high temperature form) with the α-$ThCr_2Si_2$ structure under ambient pressure [64]. In contrast, $BaAl_2Si_2$ has the pressure induced polymorph. In the high pressure form β-$BaAl_2Si_2$ (Fig. 7.23b) the Si–Si distance is 3.433(5) Å, much larger than the normal Si–Si distance 2.4–2.7 Å of clathrate compounds. The β-form should have a layered character. Both polymorphs of $BaAl_2Si_2$ do not show superconductivity.

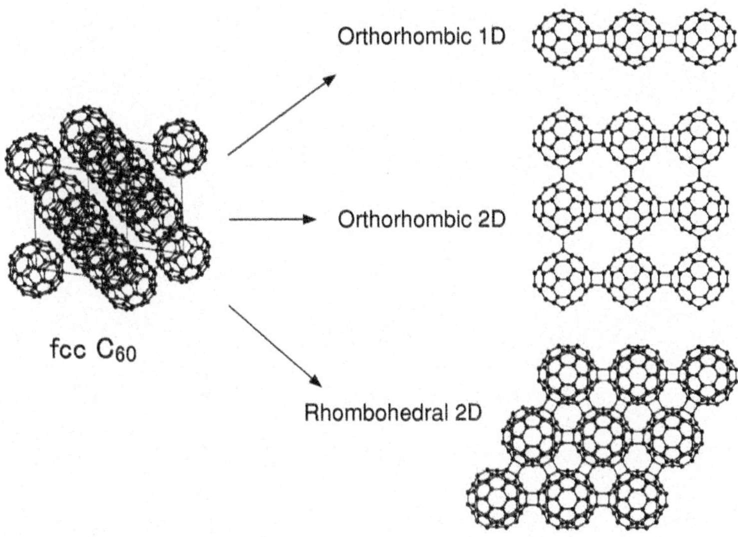


**Fig. 7.24** Polymerization of fcc packed $C_{60}$ molecules via [2 + 2] cycloaddition. The orthorhombic 2D polymer is obtained by topochemical polymerization of $C_{60}$ within the {100} planes. The rhombohedral 2D polymer is formed by the polymerization within the {111} planes

## 7.5 Carbon Clathrate Compounds

$C_{60}$ molecules rich in C=C double bonds are easily polymerized with the aid of pressure via [2 + 2] cycloaddition of the hexagon-hexagon double bonds on adjacent molecules, as shown in Fig. 7.24. If the carbon cages of $C_{60}$ can be polymerized three dimensionally in an ordered manner, the resulting 3D polymer can be an analog of silicon clathrate compounds. This is also a new carbon allotrope [11].

$C_{60}$ monomer single crystals were embedded in hexagonal BN (h-BN) powder, then compressed under HPHT conditions. $C_{60}$ polymer single crystals were obtained, keeping the starting outer shapes of the monomer crystals [65, 66]. Orthorhombic two-dimensional (O-2D), and rhombohedral 2D (R-2D) polymers are obtained at 2.5 GPa, and 5 GPa at 500 °C, respectively. The single crystal structures were determined as shown in Fig. 7.25 [65, 66]. The bond distances and bond angles of the polymer crystals were precisely determined for the first time by the single crystal X-ray analysis. The 2D polymerization proceeds topochemically within the {100} and {111} planes of the fcc monomer crystals for the O-2D and R-2D polymers, respectively.

In order to obtaine 3D polymer single crystals, the O-2D $C_{60}$ polymer crystals were compressed in the second step using a much higher pressure of 15 GPa at 500–600 °C. Single crystals of $C_{60}$ monomer (fcc) were also similarly treated in a one-step compression to 15 GPa to have different types of three-dimensional polymers [67].

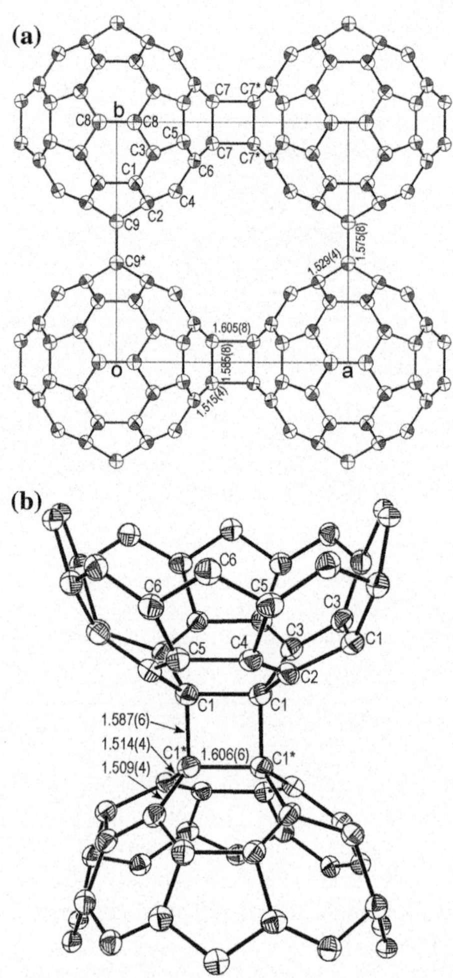


**Fig. 7.25** Structure details of the linking of $C_{60}$ moleclues in **a** orthorhombic 2D and **b** rhombohedral 2D $C_{60}$ polymers determined by single crystal studies [65, 66]

### *7.5.1 Structure of Three Dimensional (3D) Polymers*

The $C_{60}$ polymers obtained in different HPHT polymerization procedures are summarized in Table 7.3. Note that the crystal symmetries of the 3D polymers were unchanged from those of the corresponding lower dimensional precursors, although the lattices greatly shrink after the HPHT treatments. This suggests that the conversion to the 3D polymers should be topochemical reactions like the formation of 2D polymer from the monomer crystals. Single crystals of orthorhombic 3D (O-3D) $C_{60}$ polymer (*Immm*) were obtained from the O-2D $C_{60}$ crystals (*Immm*). The X-ray structural analysis results are shown in Fig. 7.26 [68]. Spherical $C_{60}$ molecules are substantially deformed to rectangular parallelepiped

**Table 7.3** Polymerization procedures for the different types of 2D and 3D $C_{60}$ polymers. Space group (S. G.) and crystallographic parameters are compared

| Monomer $C_{60}$ | |
|---|---|
| S. G. | Lattice parameter (Å) and density ($d$, g/cm$^3$) |
| *Fm*-3 (*Pa*-3) | $a = 14.17$ $d_{calcd.} = 1.69$ |

| Conditions (from monomer) | 2D $C_{60}$ polymer S. G. | Lattice parameter (Å) and density ($d$, g/cm$^3$) | Conditions | 3D $C_{60}$ polymer S. G. | Lattice parameter (Å) and density ($d$, g/cm$^3$) |
|---|---|---|---|---|---|
| 2.5 GPa, 500 °C | *Immm* | $a = 9.026(2)$ $b = 9.083(2)$ $c = 15.077(3)$ $d_{calcd.} = 1.94$ | 15 GPa, 600 °C | *Immm* | $a = 7.86(2)$ $b = 8.59(3)$ $c = 12.73(4)$ $d_{calcd.} = 2.78$ |
| 5.0 GPa, 500 °C | *R*-3*m* | $a = 9.175(1)$ $c = 24.568(3)$ $d_{calcd.} = 2.00$ | | | |
| 15 GPa, 550 °C | | | | *Fm*-3 (*R*-3) | $a = 11.93$ $d_{calcd.} = 2.82$ |

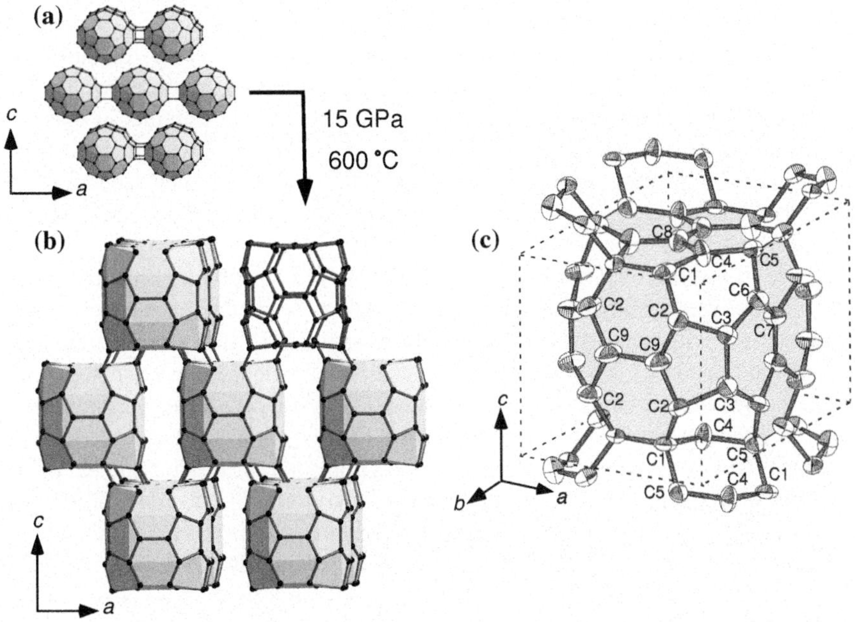


**Fig. 7.26** Crystal structure of the 3D $C_{60}$ polymer (**b**), in comparison with the starting 2D $C_{60}$ polymer (**a**), and an ORTEP plot of the cuboidal $C_{60}$ unit (**c**), which is connected to the eight neighboring units at the corners. Copyright American Institute of Physics [68]

(cuboidal) shapes in the 3D polymer crystals. Each cuboid is bonded to eight neighbouring units at the corners via [3 + 3] cycloaddition to form the body centred orthorhombic lattice (space group *Immm*) [68, 69].

Another type of 3D $C_{60}$ polymer was obtained from the polymerization of $C_{60}$ monomer crystals in one-step compression to 15 GPa at 600 °C [67]. It crystallized in the fcc symmetry like monomer crystals. The single crystal analysis coupled with a molecular dynamic calculation suggested a rhombohedral 3D (space group $R\bar{3}$) structure. Each $C_{60}$ unit is bonded to 12 neighbouring $C_{60}$ units; six units are bonded via [3 + 3] cycloaddition using the pentagons between the neighbouring units, and the rest of 6 units are linked via [2 + 2] cycloaddition using the pentagon and hexagon single bond edges.

### *7.5.2 Properties of Three Dimensional Polymers*

The band structure and the DOS of the orthorhombic 3D $C_{60}$ (O-3D) polymer were calculated on the basis of the coordinates obtained by the X-ray refinement and theoretically optimized structures [68, 70, 71]. All of the calculations suggested the metallic conductivity of the O-3D polymer with a large density of states at the Fermi level. The calculation by Yang et al. [70] suggested that the electronic band structure has steep and flat bands near the Fermi level, and possible occurrence of interband nesting which may enhance electron-phonon coupling for superconductivity. With these features it is possible that the O-3D $C_{60}$ polymer would be a potential superconductor even without doping. The O-3D polymer is electron conductive with a conductivity of $10^{-1}$–$10^{-2}$ $Scm^{-1}$ at room temperature [68]. However, it is not superconducting down to 2 K. The temperature dependence of the conductivity suggests 3D Mott's variable range hopping (VRH). This is probably due to a lack of long-range order in the O-3D $C_{60}$ polymer, and the presence of disorder and dangling bonds. The R-3D $C_{60}$ polymer obtained by the direct 3D polymerization of the monomer crystals was found to be semiconducting in agreement with band structure calculation results [67]. It is interesting to prepare metal doped $C_{60}$ 3D polymers like metal doped silicon clathrate compounds. It is expected that carbon clathrate compounds will yield high $T_c$ superconductivity with an effective electron-phonon interaction [72].

Some 3D $C_{60}$ polymers have been reported to be extremely hard and have bulk moduli even larger than those of diamond. Blank et al. [73–75] reported that the graphitized $C_{60}$ at high pressures and temperatures >700 °C could scratch the (111) crystallographic plane of diamond, and called it "ultrahard fulleride." The micro-Vickers hardness ($Hv$, $kg/mm^2$) measured on various $C_{60}$ polymers are shown in Fig. 7.27. The monomer $C_{60}$ crystal is as soft as $Hv = 15$, and the 2D polymers have the values of the order of $Hv = 200$–$300$. The 3D polymer crystals having a density of 2.6–2.8 $g/cm^3$ are in a range of $Hv = 3{,}500$–$4{,}500$, comparable to c-BN values. The $R\bar{3}$ 3D polymer having 12 coordination units is the hardest polymer.

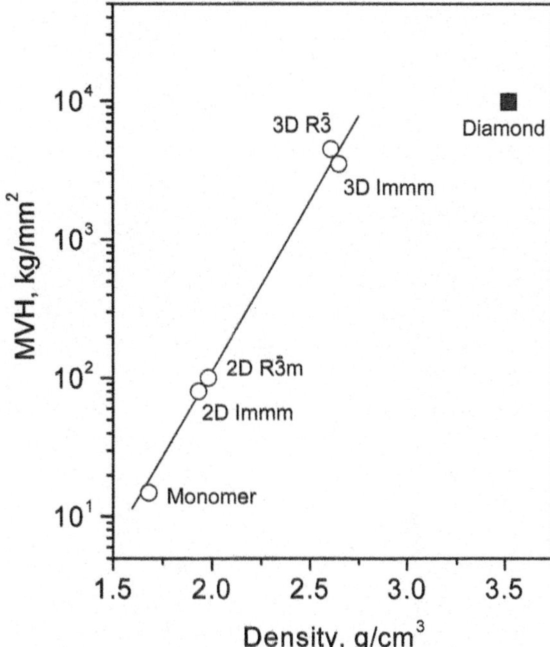


**Fig. 7.27** Micro Vickers hardness of various $C_{60}$ phases as a function of the observed density [67]

**Table 7.4** Superconductivity of silicon clathrates and related compounds

| Compounds | Preparation | Structure | $T_c$, K | Ref |
|---|---|---|---|---|
| $Ba_8Si_{46}$ | HPHT | Type I clathrate | 8.0 | 28 |
| $Ba_xBa_6Si_{46}$ | HPHT | Type I clathrate | 6–9 | 19 |
| $Na_2Ba_6Si_{46}$ | From $Na_2BaSi_4$ | Type I clathrate | 4 | 8 |
| $Na_xBa_6Si_{46}$ | From $Na_2BaSi_4$ | Type I clathrate | 2.6–4.8 | 14 |
| $K_2Ba_6Si_{46}$ | From $K_2BaSi_4$ | Type I clathrate | 4 | 15 |
| $Ba_8Si_{46-x}Ag_x$ | HPHT | Type I clathrate | <8 | 31 |
| $Ba_8Si_{46-x}Ge_x$ | HPHT | Type I clathrate | <8 | 34 |
| $Ba_{8-x}Ca_xSi_{46}$ | HPHT | Type I clathrate | <8 | 29 |
| $Ba_{8-x}Sr_xSi_{46}$ | HPHT | Type I clathrate | <8 | 30 |
| $Ba_{24}Si_{100}$ | HPHT | | 1.4 | 44 |
| $Ba_{24}Ge_{100}$ | HT | | 0.24 | 48 |
| $LaSi_2$ | HT | $\alpha$-$ThSi_2$ | 2.5 | 55 |
| $LaSi_5$ | HPHT | | 11.5 | 58 |
| $LaSi_{10}$ | HPHT | | 6.7 | 58 |
| $Ca_2Al_3Si_4$ | HPHT | | 6.4 | 59 |
| CaAlSi | HPHT | Laves phase | 2.6 | 60 |
| $CaSi_2$ | HP > 16 GPa | $AlB_2$ | 14 | 26 |
| $CaSi_2$ | HPHT | $\alpha$-$ThSi_2$ | 1.58 | 21 |
| $BaSi_2$ | HPHT | $EuGe_2$ | 6.8 | 25 |

## 7.6 Summary

In binary silicon clathrate compounds such as $Na_8Si_{46}$, $Na_xSi_{136}$, and $Ba_8Si_{46}$, the Si atoms in the framework need not use the electrons from the doped metals to complete their octets. The extra electrons from the guest metal atoms are doped to the antibonding bands of the framework. The binary compounds can be classified into 'intercalation' clathrate compounds. Most of clathrate compounds are prepared by substituting the framework Si atoms with lower valent elements such as Al and Ga, and the electrons doped from the guest metal atoms are used to complete the octets of the lower valent elements to form four fold $sp^3$ covalent bonds with silicon. The substituted compounds can be classified into 'Zintl' clathrate compounds. Unlike Zintl type clathrate compounds, 'intercalation' type clathrate compounds cannot be prepared by a simple melting or heating of the constituent elements, but by using special techniques, such as thermal decomposition of NaSi and $Na_2BaSi_4$. HPHT synthesis is also a useful method for the preparation of intercalation type clathrates and silicon-rich intermetallic compounds. The driving force of making covalent bonded cage structures in HPHT synthesis is the Le Chaterie's principle on pressure. Covalent bonded clathrates and related structures with excess electrons are good candidates for new superconductors. The superconductors introduced in this chapter are summarized in Table 7.4.

## References


1. H.W. Kroto, J.R. Heath, S.C. O'Brien, R.F. Curl, R.F. Smalley, Nature **318**, 162 (1985)
2. R.M. Fleming, M.J. Rosseinsky, A.P. Ramirez, D.W. Murphy, J.C. Tully, R.C. Haddon, T. Siegrist, R. Tycko, S.H. Glarum, P. Marsh, G. Dabbagh, S.M. Zahurak, A.V. Makhija, C. Hampton, Nature **352**, 701 (1991)
3. K. Tanigaki, T.W. Ebbesen, S. Saito, J. Mizuki, J.S. Tsai, Y. Shimakawa, Y. Kubo, S. Kuroshima, Nature **352**, 222 (1991)
4. J.S. Kasper, P. Hagenmuller, M. Pouchard, C. Cros, Science **150**, 1713 (1965)
5. C. Cros, M. Pouchard, P. Hagenmuller, C. R. Acad. Sci. **260**, 4764 (1965)
6. B. Roy, K.E. Sim, A.D. Caplin, Phil. Mag. B **65**, 1445 (1992)
7. S. Yamanaka, H. Horie, H. Nakano, M. Ishikawa, Fullerene Sci. Tech. **3**, 21 (1995)
8. H. Kawaji, H. Horie, S. Yamanaka, M. Ishikawa, Phys. Rev. Lett. **74**, 1427 (1995)
9. Ya. Mudryk, P. Rogl, C. Paul, S. Berger, E. Bauer, G. Hilscher, C. Godart, H. Noël, A. Saccone, R. Ferro, Phys. B **328**, 44 (2003)
10. B. Sundqvist, in *Fullerenes*, ed. by K.M. Kadish, R.S. Rudorff (Wiley, New York, 2000), Chap. 15
11. S. Yamanaka, Dalton Trans. **39**, 1885 (2010)
12. H. Horie, T. Kikudome, K. Teramura, S. Yamanaka, J. Solid State Chem. **182**, 129 (2009)
13. B. Huang, J.D. Corbett, Solid State Sci. **1**, 555 (1999)
14. H. Kawaji, K. Iwai, S. Yamanaka, M. Ishikawa, Solid State Commun. **100**, 393 (1996)
15. S. Yamanaka, H. Horie, H. Kawaji, M. Ishikawa, Eur. J. Solid State Inorg. Chem. **32**, 799 (1995)

16. T. Rachi, K. Tanigaki, R. Kumashiro, J. Winter, H. Kuzmany, Chem. Phys. Lett. **409**, 48 (2005)
17. G. Cordier, P. Woll, J. Less-Common Metals **169**, 291 (1991)
18. M.H. Manghnini, S. Akimono, Y. Shono eds., *High Pressure Research in Mineral Physics*, Geophysical monograph 39 (Tera Scientific Publishing Company, American Geophysical Union, Tokyo 1987)
19. H. Fukuoka, J. Kiyoto, S. Yamanaka, Inorg. Chem. **42**, 2933 (2003)
20. N. Kawai, S. Endo, Rev. Sci. Instrum. **41**, 1178 (1970)
21. H. Nakano, S. Yamanaka, J. Solid State Chem. **108**, 260 (1994)
22. J. Evers, G. Oehlinger, A. Weiss, J. Solid State Chem. **20**, 173 (1977)
23. M. Imai, T. Hirano, Phys. Rev. B **58**, 11922 (1998)
24. M. Imai, T. Kikegawa, Chem. Mater. **15**, 2543 (2003)
25. M. Imai, K. Hirata, T. Hirano, Phys. C **245**, 12 (1995)
26. P. Bordet, M. Affronte, S. Sanfilippo, M. Núñez-Regueiro, O. Laborde, G.L. Olcese, A. Palenzona, S. LeFloch, D. Levy, M. Hanfland, Phys. Rev. B **62**, 11392 (2000)
27. R. Demchyna, S. Leoni, H. Rosner, U. Schwarz, U. Schwarz, Z. Kristallogr. **221**, 420 (2006)
28. S. Yamanaka, E. Enishi, H. Fukuoka, M. Yasukawa, Inorg. Chem. **39**, 56 (2000)
29. P. Toulmonde, Ch. Adessi, X. Blasé, S. Miguel, Phys. Rev. B **71**, 094504 (2005)
30. A. San-Miguel, P. Toulemond, High Pressure Res. **25**, 159 (2005)
31. Y. Nozue, G. Hosaka, E. Enishi, S. Yamanaka, Mol. Cryst. Liq. Cryst. **341**, 1313 (2000)
32. W. Carrillo-Cabrera, J. Curda, K. Peters, S. Paschen, M. Baenitz, Y. Grin, H.G. von Schnering, Z. Kristallogr. **215**, 321 (2000)
33. W. Carrillo-Cabrera, S. Budnyk, Y. Prots, Y. Grin, Z. Anorg. Allg. Chem. **630**, 2267 (2004)
34. H. Fukuoka, J. Kiyoto, S. Yamanaka, J. Solid State Chem. **175**, 237 (2003)
35. S. Saito, A. Oshiyama, Phys. Rev. B **51**, 2628 (1995)
36. K. Moriguchi, M. Yonemura, A. Shintani, S. Yamanaka, Phys. Rev. B **61**, 9859 (2000)
37. J. Gryko, P.F. McMillan, R.F. Marzke, A.P. Dodokin, A.A. Demkov, O.F. Sankey, Phys. Rev. B **57**, 4172 (1998)
38. J. Gryko, P.F. McMillan, R.F. Marzke, G.K. Ramachandran, D. Patton, S.K. Deb, O.F. Sankey, Phys. Rev. B **62**, R7707 (2000)
39. F. Shimizu, Y. Maniwa, K. Kume, H. Kawaji, S. Yamanaka, M. Ishikawa, Phys. Rev. B **54**, 13242 (1996)
40. M. Itou, Y. Sakurai, M. Usuda, C. Cros, H. Fukuoka, S. Yamanaka, Phys. Rev. B **71**, 125125 (2005)
41. T. Yokoya, A. Fukushima, T. Kiss, K. Kobayashi, S. Shin, K. Moriguchi, A. Shintani, H. Fukuoka, S. Yamanaka, Phys. Rev. B **6417**, 172504 (2001)
42. K. Tanigaki, T. Shimizu, K.M. Itoh, J. Teraoka, Y. Moritomo, S. Yamanaka, Nat. Mater. **2**, 653 (2003)
43. H. Fukuoka, K. Ueno, S. Yamanaka, J. Organomet. Chem. **611**, 543 (2000)
44. T. Rachi, H. Yoshino, R. Kumashir et al., Phys. Rev. B **72**, 144504 (2005)
45. R. Viennois, P. Toulemonde, C. Paulsen, S. Miguel, J. Phys. Condens. Matter **17**, L311 (2005)
46. H. Fukuoka, K. Iwai, S. Yamanaka, H. Abe, K. Yoza, L. Haming, J. Solid State Chem. **151**, 117 (2000)
47. W. Carrillo-Cabrera, J. Curda, H.G. von Schnering, S. Paschen, Y. Grin, Z. Kristallogr. **215**, 207 (2000)
48. F.M. Grosche, H.Q. Yuan, W. Carrillo-Cabrera, S. Paschen, C. Langhammer, F. Kromer, G. Sparn, M. Baenitz, Y. Grin, F. Steglich, Phys. Rev. Let **87**, 247003 (2001)
49. S. Yamanaka, S. Maekawa, Z. Naturforsch. B **61**, 1493 (2006)
50. A. Wosylus, Y. Prots, U. Burkhard, W. Schnelle, U. Schwarz, Y. Grin, Z. Naturforsch. **61b**, 1485 (2006)
51. A. Wosylus, Yu. Prots, U. Burkhard, W. Schnelle, U. Schwarz, Sci. Tech. Adv. Mater. **8**, 383 (2007)

52. O.O. Kurakevych, T.A. Strobel, D.Y. Kim, T. Muramatsu, V.V. Struzhkin, Cryst. Growth Des. **13**, 303 (2013)
53. R. Nesper, J. Curda, H.G. von Schnering, Angew. Chem. Int. Ed. Engl. **25**, 350 (1986)
54. E. Reny, S. Yamanaka, C. Cros, M. Pouchard, Chem. Commun. 505 (2000)
55. W.E. Henry, C. Betz, H. Muir, Bull. Am. Phys. Soc. **7**, 474 (1962)
56. Y. Takagi, M. Fujita, K. Kusakabe, Mol. Cryst. Liq. Cryst. **340**, 379 (2000)
57. G.-M. Rignanese, J.-C. Charlier, Phys. Rev. B **78**, 125415 (2008)
58. S. Yamanaka, S. Izumi, S. Maekawa, K. Umemoto, J. Solid State Chem. **2009**, 182 (1991)
59. M. Tanaka, S. Zhang, Y. Tanaka, K. Inumaru, S. Yamanaka, J. Solid State Chem. **198**, 445 (2013)
60. M. Tanaka, S. Zhang, K. Inumaru, S. Yamanaka, Inorg. Chem. **52**, 6039 (2013)
61. S. Yamanaka, M. Kajiyama, S.N. Sivakumar, H. Fukuoka, High Pressure Res. **24**, 481 (2004)
62. C.L. Condron, H. Hope, P.M.B. Piccoli, A.J. Schultz, S.M. Kauzlarich, Inorg. Chem. **46**, 4523 (2007)
63. H. Shimizu, T. Kume, Y. Narita, S. Sasaki, T. Kikudome, S. Yamanaka, Phys. Stat. Sol. B **244**, 357 (2007)
64. S. Leoni, W. Carrilo-Cabrera, W. Schnelle, Y. Grin, Solid State Sci. **5**, 139 (2003)
65. X. Chen, S. Yamanaka, Chem. Phys. Lett. **360**, 501 (2002)
66. X. Chen, S. Yamanaka, K. Sako, K.Y. Inoue, M. Yasukawa, Chem. Phys. Lett. **356**, 291 (2002)
67. S. Yamanaka, N.S. Kini, A. Kubo, S. Jida, H. Kuramoto, J. Am. Chem. Soc. **130**, 4303 (2008)
68. S. Yamanaka, A. Kubo, K. Inumaru, K. Komaguchi, N.S. Kini, T. Inoue, T. Irifune, Phys. Rev. Lett. **96**, 076602 (2006)
69. S. Yamanaka, A. Kubo, N.S. Kini, K. Inumaru, Phys. B **383**, 59 (2006)
70. J. Yang, J.S. Tse, Y. Yao, T. Iitaka, Angew. Chem. Int. Ed. **46**, 6275 (2007)
71. Y. Yamagami, S. Saito, Phys. Rev. B **79**, 045425 (2009)
72. D. Connetable, V. Timoshevskii, B. Masenelli, J. Beille, J. Marcus, B. Barbara, A.M. Saitta, G.M. Rignanese, P. Melinon, S. Yamanaka, X. Blase, Phys. Rev. Lett. **91**, 247001 (2003)
73. V.D. Blank, S.G. Buga, N.R. Serebryanaya, V.N. Denisov, G.A. Dubitsky, A.N. Ivlev, B.N. Mavrin, MYu. Popov, Phys. Lett. A **205**, 208 (1995)
74. V. Blank, M. Popov, G. Pivovarov, K. Lvova, K. Gogolinsky, V. Reshetov, Diam. Related. Mater. **7**, 427 (1998)
75. V.D. Blank, S.G. Buga, G.A. Dubitsky, N.R. Serebryanaya, MYu. Popov, B. Sundqvist, Carbon **36**, 319 (1998)

# Chapter 8
# Light Element Group 13–14 Clathrate Phases

**Susan M. Kauzlarich and Fan Sui**

**Abstract** There has been a renewed interest in clathrate compounds composed of light elements as promising thermoelectric materials due to their potential for chemical tuning. Clathrate structures are ideal frameworks for investigating the phonon glass electron crystal (PGEC) model for efficient band engineering. In this model, the guest atom provides for phonon scattering (phonon glass) to reduce thermal conductivity while tuning the chemical composition of the framework allows for control over electronic transport (electron crystal). This chapter provides an overview of the synthesis, structure, and properties of light element group 13-Si compounds with the clathrate structure. The primary focus will be on alkali and alkaline earth metal containing clathrates, $A_8E_xSi_{46-x}$ (A = Sr, Ba, Eu, Na, K; E = B, Al, Ga). Additionally, hydrogen capacity in Si clathrate structures will be presented. By reviewing the current status of the field, we will demonstrate the potential of these materials for electronic and thermoelectric applications and new avenues for research.

## 8.1 Introduction

Inorganic clathrate-structured compounds are of interest as the complex unit cells allow for the tuning of the bulk materials properties by doping elements and/or through subtle synthetic adjustments to their ratios [1–4]. Several reviews of clathrates provide additional details on the wide range of compositions, structures, and properties of this fascinating class of compounds [5–8]. Compounds

S. M. Kauzlarich (✉) · F. Sui
Department of Chemistry, University of California, One Shields Ave, Davis, CA, USA
e-mail: smkauzlarich@ucdavis.edu

F. Sui
e-mail: fsui@ucdavis.edu

G. S. Nolas (ed.), *The Physics and Chemistry of Inorganic Clathrates*, Springer Series in Materials Science 199, DOI: 10.1007/978-94-017-9127-4_8, 

crystallizing in the clathrate structure types (referred to as type I, II, III, etc.) have the general form of a covalently bonded framework of polyhedra in which an atom is encapsulated. There currently exists a rich literature of reports on thermoelectric properties of ternary type-I clathrates with the general formula (1 or $2)_8(13)_x(14)_{46-x}$, where the numbers in parentheses indicate the appropriate group from the Periodic Table and the subscripts indicate stoichiometry. Other applications that might be envisioned include hydrogen and energy storage where the group 1 or 2 element can be replaced with hydrogen or lithium [9–11]. Finally, the Si-based phases are potentially wide band gap semiconductors [12] and are attractive for solar cell applications either as a single junction cell or a top cell material for a tandem Si-based solar cell [13]. This chapter will focus on the light elements containing clathrates where group 13 is limited to B, Al, and Ga and group 14 is limited to Si. A summary of recent work on hydrogen encapsulated Si clathrate structures will also be presented.

Group 13 and Si-containing clathrate phases have so far been found to crystallize in the type I or type VIII clathrate structure with the formula, (1 or $2)_8(13)_x(14)_{46-x}$. Figure 8.1 is a schematic of the type I clathrate, space group $Pm\bar{3}n$. Group 13 and 14 elements occupy the three crystallographic sites (designated by the Wyckoff notation: 6*c*, 16*i* and 24*k*) of the framework while group 2 elements occupy the 2*a* and 6*d* sites in the center of the two different polyhedra. The two types of polyhedral cages are described as pentagonal dodecahedra made up of 20 atoms and tetrakaidecahedra made up of 24 atoms. Type VIII is related to type I, but shows significant deformation of the polyhedral faces as shown in Fig. 8.2 [3]. Type VIII clathrate crystallizes in the noncentrosymmetrical space group, $I\bar{4}3m$, and only has one type of polyhedron. Group 13 and 14 elements occupy 4 Wyckoff positions of the framework (2*a*, 8*c*, 12*d*, and 24*g*) with group 2 elements occupying the 8*c* site. The type VIII clathrate has been reported only for the Sr-containing phase, $Sr_8Al_xGa_{16-x}Si_{30}$ ($x = 8 - 13$) [14, 15]. The structures can be achieved by controlling the relative amounts of Al and Ga according to the formula, $Sr_8Al_x$-$Ga_{16-x}Si_{30}$ (type I for $x = 0 - 7$, and type VIII for $x = 8 - 13$) [14, 15].

## 8.2 Synthesis

The literature provides an array of syntheses, producing polycrystalline and single crystal samples. In general, for the compositions, (1 or $2)_8(13)_xSi_{46-x}$, the alkaline earth cations have been limited to Sr and Ba and alkali metal limited to K, Rb and the group 13 element has been limited to B, Al, Ga. It appears that in the case of $Ba_8Al_xSi_{46-x}$, x can be varied from 8 to almost 16; [16] however, for $Sr_8Al_xSi_{46-x}$ only a narrow phase width is achieved with $x \sim 10$ [17]. $A_8Ga_xSi_{46-x}$ has been synthesized with A = Ba, Sr and the focus of research has been on $x \sim 16$ [18–21]. There are examples of the solid solutions of Ba or Sr containing type I clathrate with Eu, [3, 22–24] but the Eu only containing phase has not yet been reported. The alkali metal cations have been limited to K and Rb, thus far [25–28].

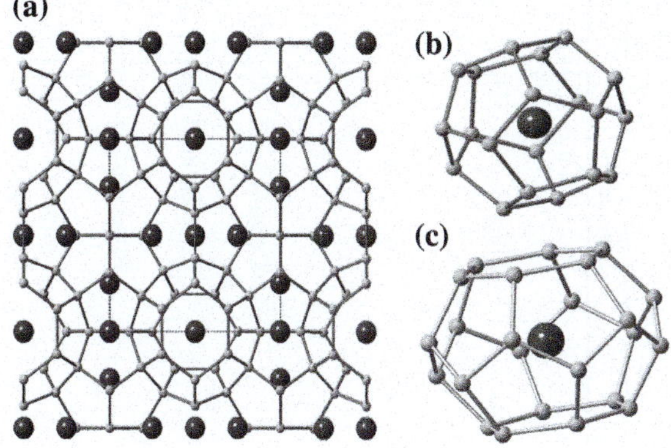


**Fig. 8.1** A view of the Type I clathrate structure type showing a view of the (**a**) unit cell down the *a* axis, (**b**) the 20 vertex cage and (**c**) the 24 vertex cage. The large (*green*) spheres are the guest ion and the small (*yellow*) spheres are the framework atoms

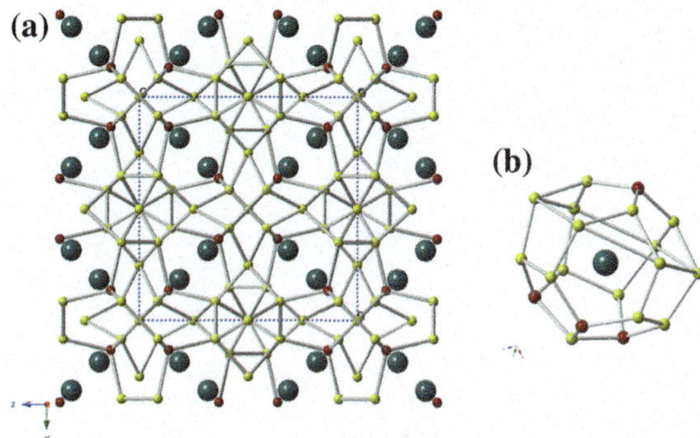


**Fig. 8.2** A view of the Type VIII clathrate structure type showing a view of the (**a**) unit cell down the *a* axis, and (**b**) the 20 vertex cage. The large (*green*) spheres are the guest ion and the small (*yellow* and *red*) spheres are the framework atoms

There are only two examples of B either as a substitutional element or as the primary group 13 element [29, 30]. Table 8.1 provides a listing of various compositions reported to date with their lattice parameter and the synthetic route.

*Synthetic routes of polycrystalline samples.* Polycrystalline samples can be synthesized from stoichiometric reactions of the elements in sealed tubes. Pure elements in the appropriate ratios can be reacted in glassy carbon tube, sealed in a steel autoclave under argon atmosphere and heated in a tube furnace at temperatures of to 1,325 K for several hours. The elements can also be contained in tantalum tubes, sealed in evacuated (or with some argon pressure) fused silica ampules and heated at temperatures between 1,000 and 1,325 K for as little as several hours or as long as several days. Typically, niobium tubes are avoided for silicon containing phases because of the reaction with the metal to produce $NbSi_2$. The first borosilicide with type I clathrate structure, $K_7B_7Si_{39}$, was synthesized from a stoichiometric reaction [30]. The elements were sealed in tantalum ampoules, encapsulated in fused silica and heated to about 1,100 K. The reaction was not complete, and the impurities were removed by reaction with sodium hydroxide and concentrated acid. Similarly, $K_8Ga_8Si_{38}$ single crystal was grown from a stoichiometric reaction of the elements at 1,220–970 K [25, 27]. $Ba_8Al_x$-$Si_{46-x}$ and $Eu_yBa_{8-y}Al_xSi_{46-x}$ were prepared by heating the elements in capped

**Table 8.1** Base structure, composition, lattice parameter and synthesis of (1 or 2)$_8$(13)$_x$Si$_{46-x}$ compounds

| Base structure | Composition | Lattice parameter, *a* (Å)[a] | Synthesis | References |
|---|---|---|---|---|
| $Ba_8Al_{16}Si_{30}$ Type I | $Ba_8Al_{16}Si_{30}$ | 10.6068 (30) | Stoichiometric | Eisenmann et al. [31] |
| | $Ba_8Al_xSi_{46-x}$ x = 8, 12, 16 | (x = 8) 10.4890(1) (x = 12) 10.5654(1) (x = 16) 10.6285(1) | Arc-melted and annealed | Mudryk et al. [2] |
| | $Ba_8Al_xSi_{46-x}$ $8 \leq x < 16$ | (x = 8) 10.4868 (1) (x = 14) 10.5984 (1) (x = 15) 10.6322 (1) (x ~ 16) 10.6319 (1) | Arc-melted and annealed | Tsujii et al. [16] |
| | $Ba_{7.5}Al_{13}Si_{29}$ ($Ba_{7.3}Al_{14}Si_{31}$) | 10.6373 (11) | Al flux | Condron et al. [22, 24] |
| | $Ba_8B_xAl_ySi_{31}$ ($x = 0.17$–$0.32$, $y = 14,15$) | (x = 0.32; y = 14) 10.6140 (4) | Al flux | Condron et al. [29] |
| | $Sr_yBa_{8-y}Al_{14}Si_{32}$ ($0.6 \leq y \leq 1.3$) | 10.6136 (1) | Al flux | Roudebush et al. [33] |
| | $Sr_{0.7}Ba_{7.3}Al_{14}Si_{31}$ | 10.61710(5) | Al flux | Condron et al. [24] |
| | $Eu_2Ba_6Al_8Si_{36}$ | 10.4951(2) | Arc-melt | Mudryk et al. [3] |
| | $Eu_{0.27}Ba_{7.22}Al_{13}Si_{29}$ ($Eu_{0.3}Ba_{7.7}Al_{14}Si_{31}$) | 10.631 (10) | Al flux | Condron et al. [22, 24] |
| | $Ba_8B_xAl_{14}Si_{31}$ $0.17 \leq x \leq 0.32$ | (x = 0.17) 10.6263 (5) (x = 0.32) 10.6140 (4) | Al flux | Condron et al. [29] |
| $Ba_8Ga_{16}Si_{30}$ Type I | $Ba_8Ga_{16}Si_{30}$ | 10.5300 (9) 10.5532(4) | Stoichiometric | Refs. [18, 20, 31] |
| | $Ba_8Ga_{10}Si_{36}$ and $Ba_8Ga_{16}Si_{30}$ | | | |
| | $Eu_2Ba_6Ga_8Si_{36}$ | 10.4373(2) | Arc-melt | Mudryk et al. [3] |

(continued)

**Table 8.1** (continued)

| Base structure | Composition | Lattice parameter, $a$ (Å)[a] | Synthesis | References |
|---|---|---|---|---|
| $Sr_8Al_{10}Si_{36}$ Type I | $Sr_8Al_{16}Si_{36}$ | 10.4767(8) | Stoichiometric | Eisenmann et al. [31] |
| | $Sr_8Al_xSi_{46-x}$ x ~ 10 | 10.4606(11)-10.4661(9) | Arc-melt and anneal | Roudebush et al. [17] |
| | $EuSr_7Al_{10}Si_{36}$ | 10.4541(1) | Arc-melt | Mudryk et al. [3] |
| $Sr_8Ga_{16}Si_{30}$ Type I | $Sr_8Ga_{16}Si_{30}$ | 10.4595(40) | Stoichiometric | Eisenmann et al. [31] |
| | $Sr_8Ga_{16-x}Si_{30+x}$ $x = 2.18, 4.05$ | (x = 2.18) 10.4553(2)<br>(x = 4.05) 10.4234(3) | Induction melt and anneal | Ref. [19] |
| $Sr_8Ga_xSi_{46-x}$ Type I | $Sr_8Al_xGa_{16-x}Si_{30}$ x = 1-7 | (x = 5) 10.479<br>(x = 6) 10.481<br>(x = 7) 10.482 | | Ref. [14] |
| $Sr_8Ga_{16}Si_{30}$ Type VIII | $Sr_8Al_xGa_{16-x}Si_{30}$ x = 8-12 | (x = 8) 10.4416<br>(x = 9) 10.4422<br>(x = 10) 10.4437<br>(x = 11) 10.4491<br>(x = 12) 10.4517 | | Ref. [14] |
| $Ba_8Al_xSi_{46-x}$ Type I | $K_7B_7Si_{39}$ | 9.952(1)- 9.971(1) | Stoichiometric | Ref. [30] |
| | $K_8Ga_8Si_{38}$ | 10.427 (1)<br>10.4261(2) | Stoichiometric | Refs. [25, 27] |
| | $Rb_8Ga_8Si_{38}$ | 10.469 (2) | Stoichiometric | von Schnering et al. [26] |

[a] Room temperature powder diffraction

alumina crucibles sealed in fused silica under argon at 1,073 K for 7 days. The polycrystalline samples were ground and reheated several times to get fairly homogeneous material with the compositions x $\sim$ 14 and y $\sim$ 1 [24]. In general, type I clathrate with (1 or $2)_8(13)_xSi_{46-x}$ composition are air and acid stable, so most impurities can be removed by washing the sample with acid or base, depending upon the solubility of the impurity.

Compounds with the composition $(2)_8(13)_xSi_{46-x}$ ($8 \leq x \leq 15$) can also be prepared by arc melting the elements under an argon atmosphere close to stoichiometry [typically the alkaline earth element is provided in slight excess (1–5 %)]. As-cast samples are typically further annealed in alumina crucibles which are sealed in fused silica tubes for various lengths of time (2 days to 1 week) at various temperatures, from 1,020 to 1,270 K. After annealing, the annealed samples can be further etched or mechanically polished to remove any oxidized surface.

*Single crystal growth: Self-flux reaction.* Many type I clathrates have been reported to be grown from self-flux reaction [34]. Most examples are for the $Ba_8(13)_xGe_{46-x}$ phases as compounds containing Ge show some of the best thermoelectric properties. However, the procedures for adjusting composition for group 13 containing Si clathrates should be similar. An example is the use of Al flux for the synthesis of $Ba_8Al_{14}Si_{31}$ [22]. The flux ratio of 2:70:30 (Ba:Al:Si) was used to produce $Ba_8Al_{14}Si_{31}$. The excess Al was removed by centrifugation of the reaction vessel at moderate temperatures. The residual flux can be further removed by acid or base treatment. A $Ba_8Al_{14}Si_{31}$ crystal from an Al flux synthesis is shown in Fig. 8.3.

## 8.3 Structure

Good quality single crystals with type I clathrate structure type can be grown from aluminum flux reaction. The crystal structure and thermal motion of the atoms as a function of temperature have been widely investigated by X-ray crystallography.

*$Ba_8Al_xSi_{46-x}$.* Condron et al. synthesized $Ba_{7.5}Al_{13}Si_{29}$ and $Eu_{0.27}Ba_{7.22}Al_{13}Si_{29}$ from Al flux reaction and characterized the samples' crystal structures [22]. In both cases, vacancies are presented in guest atom positions and framework positions. Eu atoms go to 2*a* sites, which is the guest atom site in the smaller pentagonal dodecahedron ($5^{12}$) cages, as Eu has smaller ionic radius than Ba and prefers occupying the smaller cages. The framework has three different Wyckoff positions, 6*c*, 16*i* and 24*k*, among which the aluminum substitution preference can affect the properties to a great extent. Since the electron densities of Al and Si are similar, laboratory X-ray source failed to distinguish the two and the refinement gave similar bond lengths of the different sites [22].

To investigate the occupancies of Al and Si, Condron et al. studied the structure by neutron diffraction and $^{27}$Al NMR on $Ba_8Al_{14}Si_{31}$ [36]. Neutron diffraction has better resolution when distinguishing Al and Si as they have relatively larger difference in scattering lengths. In the refinement of the neutron diffraction data, the Ba sites were considered fully occupied and the scattering amplitude was

**Fig. 8.3** A $Ba_8Al_{14}Si_{31}$ crystal prepared from an Al flux. The grid is a mm scale bar

restricted to the total Al/Si ratio as determined from electron microprobe. This gave a slight deficiency in framework, 0.965 occupied by Al and Si, and higher Al ratio at 6*c* sites than 16*i* and 24*k* sites. They concluded that Al preferentially fills the 6*c* site before filling the 16*i* and 24*k* sites [36]. The framework deficiency model better explained the metallic behavior of $Ba_8Al_{14}Si_{31}$ than the cation deficiency model. The Al distribution was further studied with $^{27}$Al NMR, shown in Fig. 8.4. The symmetric 6*c* site was assigned to the 1,600 ppm peak while the broad quadrupole peak at 500 ppm should come from Al at 16*i* and 24 *k* sites. The integrated ratio of the signals gave the result that there are 2.6 Al at the 6*c* site and 11.4 Al in the combined 24*k* and 16*i* sites, which again confirmed Al's preference in occupying the 6*c* site [36]. This result was further corrected when studying the boron-doped version of this structure and the sharp peak at 1,600 ppm was assigned to Al–Si residual flux that was on the surface of the crystal [29]. Solid state $^{27}$Al NMR on $Ba_8Al_xGe_{46-x}$ provided further evidence that the sharp peak at 1,600 ppm is attributed to residual Al–Si flux [35].

An effort to incorporate boron into the Al–Si clathrate structure was moderately successful. Single crystals of the nominal composition of $Ba_8B_xAl_{14}Si_{31}$ ($0.17 \le x \le 0.32$) were prepared [29]. In this example, boron was verified by chemical analysis and solid state $^{11}$B-$^{27}$Al spin echo double resonance (SEDOR) NMR spectroscopy. The $^{27}$Al MAS and multiple-quantum MAS (MQMAS) NMR signal in the frequency range from 150 to 650 ppm was investigated in detail for one sample, $x = 0.17$ (Fig. 8.5). Based on a number of NMR experiments, a model was proposed describing the signal in this region as a superposition of signals of chemically similar nuclei. Three regions can be identified corresponding to the number of nonequivalent crystallographic sites in the clathrate framework. The distribution of frequencies indicated a large amount of slightly different local environments and their small differences were attributed to a nonregular connectivity of the atoms in the framework. Additionally, the distributions of shifts are also influenced by small shifts of the cations from the center of the cages; therefore an absolute assignment of the signal regions to sites in the clathrate structure could not be given. $^{11}$B NMR was also used to investigate the structure. The NMR spectra from all samples indicated two contributions to the signal of varying intensity ratios, interpreted as boron located on at least two sites or that due to

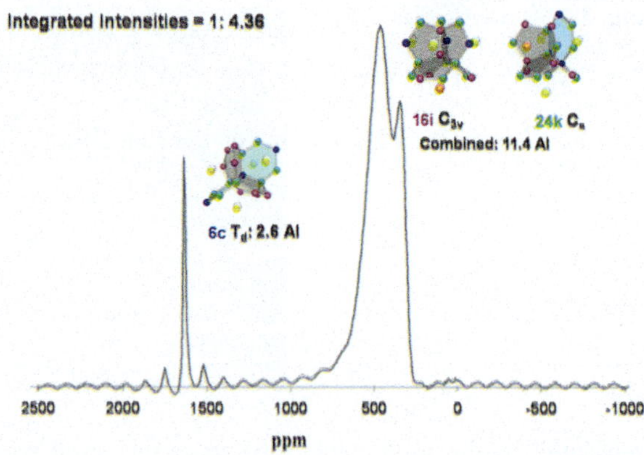


**Fig. 8.4** $^{27}Al$ MAS NMR spectrum for $Ba_8Al_{14}Si_{31}$. The figures show the local site symmetry of the framework atoms with the 6*c*, 16*i*, and 24 *k* indicated by *blue*, *purple* and *green*, respectively. The cation atoms are indicated by *orange* and *yellow* for the 2*a* and 6*d* sites, respectively (Reprinted with permission from [36]. Copyright 2006, American Chemical Society.)

local bonding two chemically nonequivalent positions exist. This is similar to what has been observed for the boron NMR in $K_7B_7Si_{39}$ [30].

The phase range of $Ba_8Al_xSi_{46-x}$ and structural details were investigated by Tsujii et al. and Roudebush et al. [16, 37]. Tsujii studied the phase range of arc-melted and annealed samples by SEM, electron microprobe, and powder X-ray diffraction. The clathrate type I structure of $Ba_8Al_xSi_{46-x}$ has a significant phase width with the nominal Al content varying from 8 and 15, as shown in Fig. 8.6. Powder X-ray diffraction also confirmed the lattice parameters increase with increasing Al content. Roudebush et al. studied the structure of the $Ba_8Al_xSi_{46-x}$ ($x = 8, 10, 12, 14, 15$) samples prepared in a similar manner by means of neutron diffraction [37]. This study provided details on the Al site occupancy preferences as the Al content increased. The Al site occupancies refinement was performed based on a bond difference model.

As the Al content increased, both 16*i* and 24*k* site occupancies showed an increasing trend to contain more Al atoms (Fig. 8.7). There is a complex dependence on site occupancy and composition, x, that could be described as two trends which were associated with low levels of Al and higher levels of Al. As the amount of Al increases in the structure, the occupancy in the 6*c*, 16*i*, and 24*k* sites decrease, increase and plateau, and increase, respectively. The Al site occupancies are in good agreement with the rules proposed for $Ba_8Al_{16}Ge_{30}$ when $x \leq 14$, as the main premise of the rules are to avoid group 13–13 bonding [38]. Understanding the site occupancy of the Al along with the anisotropic atomic displacement of the Ba atom provides some insight into the low thermal conductivity in this system. The anisotropic atomic displacement in Ba 6*d* sites may play an important role in decreasing thermal conductivity, as the rattling in the larger cage generates more scattering as phonon vibrations. Roudebush et al. studied the ratio of the primary displacement parameters ($U_{11}U_{33}/U_{22}$) and found that the value increased with Al content, indicating that increasing cage volume increases displacement in the direction normal to the six-member ring.

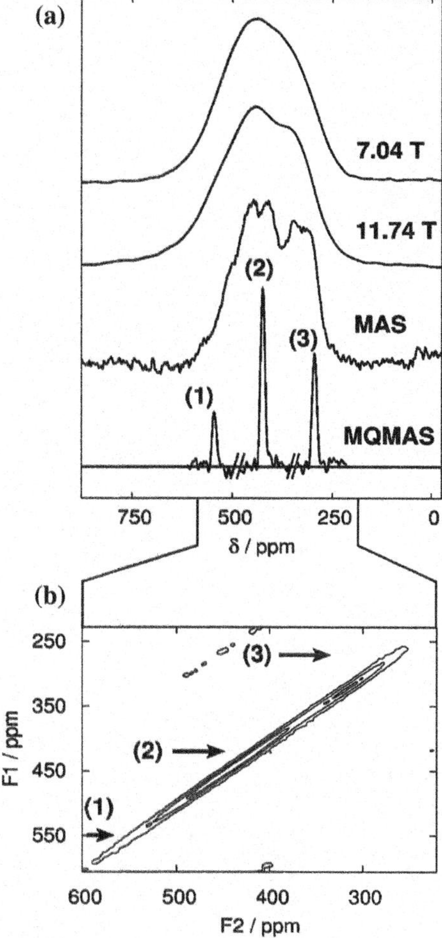


**Fig. 8.5** $^{27}Al$ NMR signals from $Ba_8B_{0.17}Al_{14}Si_{31}$. Wide-line experiments measured in the designated magnetic fields, $B_0$, and the MAS spectrum with 35 kHz rotation frequency and $B_0 = 11.74$ T. Slices extracted from the MQMAS experiment. **b** MQMAS spectrum obtained with 35 kHz rotation frequency. The F1 frequencies corresponding to the extracted slices in (**a**) are marked by *arrows*. Reprinted with permission from [29]. Copyright (2008) American Chemical Society

*$Sr_8Al_xSi_{46-x}$*. Unlike the $Ba_8Al_xSi_{46-x}$ solid solutions, $Sr_8Al_xSi_{46-x}$ has a narrow composition range, $9.54(6) \leq x \leq 10.30(8)$. The structural details of $Sr_8Al_xSi_{46-x}$ have been recently investigated [17]. The only single phase composition was found for x ~ 10. It was proposed that the framework structure must distort to accommodate the smaller Sr cation and this makes the clathrate type I structure with the charge compensated composition of $Sr_8Al_{16}Si_{30}$ unfavorable.

*$Ba_8Ga_{16}Si_{30}$*. Qiu et al. synthesized $Ba_8Ga_{16}Si_{30}$ via a stoichiometric route and characterized its structure [20]. The lattice parameter showed smooth expansion with increasing temperature. And the refinement of site mixed occupancies showed that Ga prefers to occupy the 6*c* site.

*$Eu_2Ba_6Ga_{16}Si_{30}$*. Eu substituted $Ba_8Ga_{16}Si_{30}$, $Eu_2Ba_6Ga_{16}Si_{30}$, has also been synthesized and characterized [3]. The lattice parameter is 10.4373(2) Å, smaller

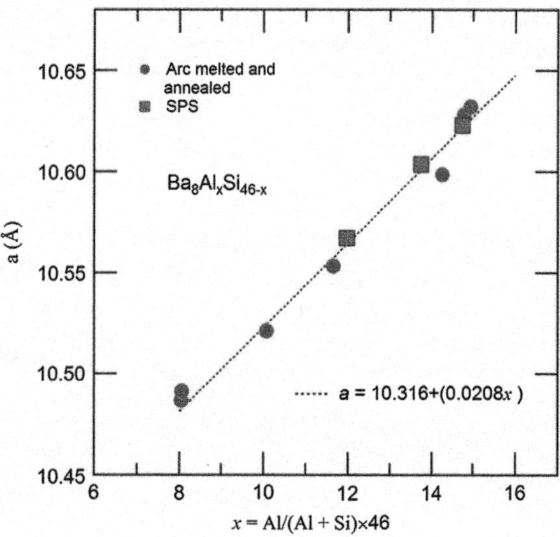


**Fig. 8.6** Lattice constant, *a*, of annealed samples of $Ba_8Al_xSi_{46-x}$, as a function of x, where the value of x was determined by electron microprobe analysis. The *dotted line* provides the fit to the data points. (Reprinted from [16] Copyright 1990, with permission from Elsevier.)

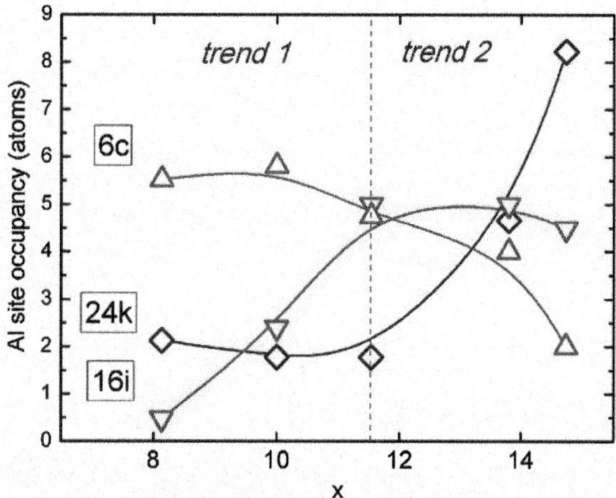


**Fig. 8.7** Al site occupancies versus total Al content, x, from electron microprobe for $Ba_8Al_xSi_{46-x}$. The *dashed line* divides the plot into two regions. The *left* where the occupancy of the 16*i* site increases and the *right* where the occupancy of the 24 *k* site increases. Reprinted with permission from [37]. Copyright (2012) American Chemical Society

than the $Ba_8Ga_{16}Si_{30}$ phase, as expected. Single crystal X-ray data revealed full atomic order among Eu and Ba atoms with the smaller rare earth ions occupying the smaller polyhedral (2*a*) sites [3].

*Sr-Ga-Si.* Crystals of $Sr_8Ga_{16}Si_{30}$ were originally synthesized by Eisenmann et al. [31]. Suekuni et al. synthesized Ge doped $Sr_8Ga_{16}Si_{30-x}Ge_x$ in full range of $0 \leq x \leq 30$ [32]. With the tuning of the cage volume, the cation vibrational parameters could be correlated with the thermal conductivity. With increasing germanium, the thermal conductivity decreased and showed glasslike behavior when $x = 30$. They studied the structure of the samples with laboratory powder X-ray diffraction and found that the lattice parameters increased with increasing Ge.

*$Sr_8Al_xGa_{16-x}Si_{30}$.* Kishimoto et al. studied the Al-Ga solid solution phase of $Sr_8(Al,Ga)_{16}Si_{30}$ and found that there was a phase change from type I clathrate to type VIII clathrate depending on the Al-Ga ratio [14]. The phase information was determined from powder XRD and Rietveld refinement. According to the formula, $Sr_8Al_xGa_{16-x}Si_{30}$, for $0 < x < 4$, the samples had the type-I clathrate structure with $Sr(Al,Ga)_4$ as a secondary phase. For $5 < x < 7$, the samples exhibited a single phase of the type-I clathrate structure. For $8 < x < 13$, the samples exhibited the type-VIII clathrate structure with some type-I and $SrAl_2Si_2$. For $14 < x < 16$, the samples showed mixtures of the type-I clathrate structure and $SrAl_2Si_2$ [14]. They concluded that the phase change is related to the formation energies of the two types of clathrates.

*Alkali metal contained type I clathrate.* A new alkali metal contained type I clathrate phase, $K_7B_7Si_{39}$ was synthesized by the Grin group [30]. The composition and structure were mainly studied by single-crystal X-ray diffraction. There was enough boron in the structure to verify the composition with single-crystal X-ray diffraction. Further evidence was provided by $^{11}B$ solid state NMR spectroscopy. The 2*a* site in the structure was only half occupied, resulting a vacancy in the unit cell and the electron precise description of the composition is $K_7\square B_7Si_{39}$ (where $\square$ = defect). The boron substitutes in the framework at 16*i* site while 24*k* and 6*c* sites are fully occupied by silicon. This is different with what has been found for the group 13, 14 clathrates before, whose vacancies and substitutions of the group 13 element mainly occur at 6*c* site. This was attributed to the fact that boron has a smaller radius than aluminum and substitution at the six member ring, which is comprised of 24*k* and 6*c* sites, would be expected to increase stress or strain in the structure.

In addition to the $K_7\square B_7Si_{39}$ compound, the electron precise compounds $K_8Ga_8Si_{38}$ and $Rb_8Ga_8Si_{38}$ have been synthesized [25–28]. There is apparently no phase width to these compounds and they have not been well studied to date.

## 8.4 Properties

Clathrates with an Al–Si framework have lower densities and higher melting points than their Ga-Ge counterparts, which make them candidates for high temperature and weight sensitive applications. The $Ba_8Al_{16}Si_{30}$ type-I clathrate system is isostructural to the $Ba_8Ga_{16}Ge_{30}$ system, but exhibits relatively higher total thermal conductivity (~20 vs. 10 mW/cm·K), and lower Seebeck values (~−30 vs. −60 μV $K^{-1}$at RT). Light elements in general are expected to provide higher lattice thermal conductivity values since their strong covalently-bound atoms along with their low mass lead to high phonon group velocity. However, the lattice thermal conductivities of both structures are comparable.

To date, $Ba_8Al_xSi_{46-x}$ type-I clathrate samples have been aluminum deficient (Si rich), *n*-type, and exhibit metallic conductivity. High-temperature thermoelectric properties were measured on $Ba_8Al_xSi_{46-x}$ with $x = 16$ synthesized by direct methods; the high temperature zT is promising, with a value of 0.35 for the composition, $Ba_8Al_{14}Si_{31}$, at T = 1,000 °C [23]. The phase composition was determined for these samples by microprobe analysis, assuming the cation site to be fully occupied, as powder X-ray diffraction suggested a single-phase clathrate I structure. Further research measuring properties of arc-melted samples for the solid solution series, $Ba_8Al_xSi_{46-x}$ showed that the valence precise phase (x = 16) could not be achieved by this synthetic method and $x_{max}$ ~15 was determined by structure and electron microprobe analysis [16]. The important correlation of the lattice constant with increasing Al content was verified. The melting point for the phases x = 12, 14, 15 were above 1,350 K (Fig. 8.8), making these good materials for high temperature application. The Seebeck coefficients are negative with increasing absolute values with x. The phase with x = 15 provided a carrier concentration of $3 \times 10^{21}$ $cm^{-3}$ and the highest zT of 0.24 at 1,000 K. If it is possible to electronically tune this system further in order to achieve and optimal carrier concentration of $4 \times 10^{20}$ $cm^{-3}$ then a zT of ~0.7 is expected [16]. Optimization of the sintering process, along with partial substitution of Al by Zn or Ba by K, might provide a way to enhance zT further. The x = 15 phase has also been prepared by combining arc melting and spark plasma sintering methods [39]. The carrier concentration is consistent with the results of Tsujii et al.: $1 \times 10^{21}$ $cm^{-3}$. They were able to show for a sample with carrier concentration of $9.7 \times 10^{20}$ $cm^{-3}$ a zT of ~0.4 at 900 K, which is consistent with the prediction that high zT's may be discovered in this system with further optimization [39].

The various $Ba_{8-x}A_xAl_ySi_{46-y}$ have so far been limited to A = Sr, Ba, Eu. This is expected because of the size of the polyhedra limits the size of the cation guest species. One might expect the optimal composition to be $Ba_6A_2Al_ySi_{46-y}$ where the large cation, $Ba^{2+}$ occupies the large polyhedron and either $Sr^{2+}$ or $Eu^{2+}$ occupy the smaller polyhedron in type I structure, as proposed by Mudryk et al. [2]. Single crystals of $Ba_6Eu_2Al_8Si_{36}$, presumed to have a lattice defect and therefore the general formula, $Ba_6Eu_2Al_xSi_{42-3/4x}\square_{4-1/4x}$ where □ indicates a lattice defect, were prepared by arc-melting and annealing. The Eu cation preferentially occupies the smaller polyhedron (*2a* site) and no defects on the framework site could be located by single crystal X-ray diffraction [2]. In order to charge balance, each defect site takes 4 electrons, so the $Ba_6Eu_2Al_8Si_{36}$ phase should be a semiconductor. The shape of the electrical resistivity is consistent with metallic behavior and there is an additional reduction of resistivity at a ferromagnetic ordering temperature of about 32 K. The ferromagnetic ordering was confirmed by magnetic susceptibility and specific heat data. The magnetic properties were investigated in detail and Eu was determined to be $Eu^{2+}$ over the entire temperature range (2–300 K) by $L_{III}$ absorption edge measurements. The Seebeck coefficient is small and negative indicating a large number of electrons as the predominant carriers [2]. The data taken together suggests that there are no localized defects and that this phase is metallic.

The reported phase, $SrBa_7Al_{16-x}Si_{30+y}$, was investigated by room temperature Seebeck, resistivity, and thermal conductivity measurements [24] because it was expected to have a promising zT at high temperatures. Further work on this system and measurements of powders that were compacted via spark plasma sintering

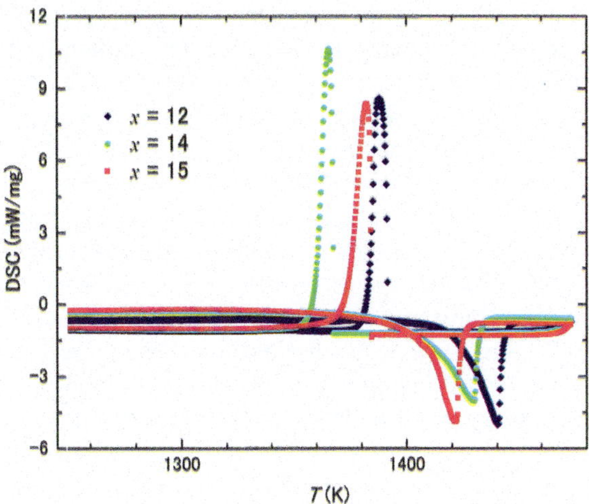


**Fig. 8.8** DSC Traces as a function of temperature for $Ba_8Al_xSi_{46-x}$ (Reprinted with permission from [16]. Copyright 1990, with permission from Elsevier.)

showed higher thermal conductivities, demonstrating that there is some phase width for this system [31]. The phase width was determined to be $Sr_yBa_{8-y}Al_{14}Si_{32}$ ($0.6 \leq y \leq 1.3$) and since this phase was prepared by Al flux, a single crystal structure was determined. Sr was found only on the 2*a* site, as expected because it is the smaller than Ba. Single crystal X-ray diffraction was collected at 90 and 12 K and indicated that Al substitutes all of the framework sites to some extent, but in decreasing amounts on the sites according to $6c > 24k > 16i$. No framework or cation vacancies were detected from a combination of single crystal X-ray diffraction and electron microprobe analysis. Microprobe analysis indicated that both Sr content and Al content varied but were not correlated in any obvious manner. Again, this phase is fairly metallic and this is attributed to the fact that it is far from the charge balanced state. The addition of Sr to the structure does not appreciably lower the thermal conductivity. This is attributed to the fact that Sr only occupies the 2*a* site which does not provide for large displacement parameters, because the size of the cation appears to be optimal for the size of the cage. However, the thermal conductivity is still quite low, suggesting that if the carrier concentration could be sufficiently optimized, this system might be a worthwhile target for thermoelectric applications. A single parabolic band model utilizing the experimental parameters provided evidence for a five-fold increase in zT at 800 K, if the carrier concentration could be lowered appropriately.

$Sr_8Al_xSi_{46-x}$ has been prepared by arc-melting and annealing and crystalize in the type I clathrate structure [37]. Unlike $Ba_8Al_xSi_{46-x}$, where x can vary from ~8 to 15, the composition of $Sr_8Al_xSi_{46-x}$ is rather narrow with x ~ 10. $Ba_8Al_{10}Si_{36}$ and $Sr_8Al_{10}Si_{36}$ are high melting point compounds, with melting points of ~1,150 and 1,000 °C, respectively. Electronic structure calculations indicated that the idealized $Sr_8Al_6Si_{40}$ composition with all *6c* sites occupied by Al would have a very small band gap that closes upon providing additional Al on the Si sites.

$Ba_8Ga_xSi_{46-x}$ has also been investigated. The phases, x = 10 and 16, were reported and their structure, high temperature transport and thermal properties reported [40]. It was reported that $Ba_8Ga_{16}Si_{30}$ showed metallic whereas $Ba_8Ga_{10}Si_{36}$ showed semiconductive transport properties. They proposed that the semiconductive nature of $Ba_8Ga_{10}Si_{36}$ was due to vacancies in the framework sites, verified by Rietveld refinement of the powder diffraction data. Compositions close to 16 (x = 15.72, 16.61, and 17.13) have been investigated [41]. For these compositions, the optimal thermoelectric properties were for x ~ 17. The reasons for the inconsistencies between composition and properties for this system are unclear. Most recently this solid solution was re-investigated for x = 14 – 18 [42]. Similar to what has been reported for $Ba_8Al_xSi_{46-x}$, [16] this research demonstrated that the maximum x ~ 15 rather than Zintl formula of x = 16. The zT was shown to be 0.55 at 900 K for the Ga content of 14.51.

Based on the variety of results described above for $A_8(13)_xSi_{46-x}$, it is highly likely that x = 16 cannot be achieved for the Type I clathrate with group 13 element = Al, Ga. It is not clear why the x = 16 composition cannot be prepared, but all of the reports suggest that this is the case. Detailed characterization of the phase prepared and hot pressed pellet is important to provide any validity to the presented transport and thermal properties. Estimation of how much zT can be improved requires control of carrier concentration.

There has been recent interest in the phases, $A_8(13)_8Si_{38}$, where A = alkali metal such as K, Rb, Cs, because there is the better possibility of controlling carrier concentration and thereby the transport properties of these type I clathrate structures [13, 25–28]. $K_8Ga_8Si_{38}$ has been shown to be a semiconductor by optical absorption and electrical resistivity measurements [27]. By plotting $(\alpha hv)^{0.5}$ and $(\alpha hv)^2$, where $\alpha$ is the optical absorption coefficient and $hv$ is the photon energy, $K_8Ga_8Si_{38}$ was determined to be a semiconductor with an indirect band gap, $E_g$(indirect) of approximately 0.10 eV. The semiconducting behavior was verified by the fact that the electrical resistivity increased by a factor of $10^4$ with decreasing temperature. The electrical resistivity is about two times higher than that observed for $K_7\square B_7Si_{39}$ whose temperature dependent resistivity is also consistent with semiconducting behavior [30] (Table 8.2).

## 8.5 Theoretical Calculations

Quantum chemical calculations have been applied to understand the atomic interaction and the transporting properties. There are a few examples of published density of state calculations of the phases described above and this section will give a review of the theoretical studies.

*Al–Si phases.* Roudebush et al. studied the type I clathrate phase $Sr_8Al_{10}Si_{36}$ and the ordered model they use for calculation is an idealized phase with composition $Sr_8Al_6Si_{40}$ which has only *6c* sites occupied by aluminum atoms [17]. The electron density of states (DOS) reveals a small band gap below the Fermi level

**Table 8.2** Room temperature transport properties reported for (1 or $2)_8(13)_xSi_{46-x}$ compounds

| Clathrate | S | ρ | κ | $κ_l$ | n | μ | zT | References |
|---|---|---|---|---|---|---|---|---|
| $Ba_8Al_{15}Si_{31}$ | −50 | 0.9 | 2.2 | 1.2 | | 7.4 | 0.1 | Anno et al. [39] |
| $Ba_8Al_{14}Si_{31}$<br>T = 400 K | −68 | 0.9 | 2.7 | 0.5 | $1 \times 10^{21}$ | | 0.12 | Condron et al. [23] |
| $EuBa_8Al_{13}Si_{33}$ | −45 | 1.4 | 2.2 | 0.8 | $1 \times 10^{21}$ | | 0.05 | Condron et al. [23] |
| $SrBa_7Al_{16}Si_{30}$ | −30 | 1 | 2.3 | 0.5 | $1 \times 10^{21}$ | | 0.12 | Roudebush et al. [33] |
| $Ba_8Ga_{16}Si_{30}$ | −66 | 2 | 1.55 | 1.2 | | | | Qiu et al. [20] |
| $Ba_8Ga_{16}Si_{30}$ | −47 | 0.85 | | | $8.7 \times 10^{20}$ | 8.4 | | Kuznetsov et al. [18] |
| $Ba_8Ga_{16}Si_{30}$ | | | | 1.0 | $1.4 \times 10^{21}$ | | | Nataraj et al. [40] |
| $Ba_8Ga_{10}Si_{30}$ | | | | 1.0 | $3.0 \times 10^{19}$ | | | Nataraj et al. [40] |
| $Ba_{7.81}Ga_{15.72}Si_{29.83}$ | −35 | 0.58 | 1.9 | | $6.3 \times 10^{21}$ | 1.67 | | Deng et al. [41] |
| $Ba_{8.01}Ga_{16.61}Si_{28.93}$ | −40 | 0.54 | 2.0 | | $3.2 \times 10^{21}$ | 3.55 | | Deng et al. [41] |
| $Ba_{7.93}Ga_{17.13}Si_{28.72}$ | −50 | 0.73 | 1.95 | | $1.1 \times 10^{21}$ | 7.32 | | Deng et al. [41] |
| $Ba_8Ga_{14}Si_{31}$ | −50 | | | 1.1 | $2.3 \times 10^{21}$ | 6.1 | | Anno et al. [42] |
| $Ba_8Ga_{15}Si_{31}$ | −65 | | | 1.2 | $5.6 \times 10^{20}$ | 10.4 | | Anno et al. [42] |
| $Sr_8Al_xGa_{16-x}Si_{30}$, x = 6, Type I | −10 | 0.24 | | 8.0 | $4.6 \times 10^{21}$ | 6.7 | | Kishimoto et al. [14] |
| $Sr_8Al_xGa_{16-x}Si_{30}$, x = 9, Type VIII | −33 | 0.4 | | 2.0 | $0.7 \times 10^{21}$ | 24 | | Kishimoto et al. [14] |

*Units* S = μV/K; ρ = mΩ·cm; κ = W/m·K; $κ_l$ = W/m·K; μ = $cm^2$ /V·s

with antibonding states partially occupied (Fig. 8.9). The DOS is formed mainly by *s* and *p* states of Si and partially of *s* and *p* states of Al, with overlap of *s* and *d* states of Sr, which are present in a wide range of energies. Calculations of the idealized phase $Ba_8Al_6Si_{40}$ was performed as comparison and gave very similar result but with a larger band gap size. For the calculation on and idealized $Sr_8Al_{16}Si_{30}$ phase, Al was placed fully in the *6c* and partially in the *16i* sites. In this model, the Fermi level was shifted to a lower energy level and the band gap was closed. This is consistent with the experimental results on $Sr_8Al_{10}Si_{36}$ showing metallic transport properties. Analysis of distribution of the electron localizability indicator confirms the direct covalent bonding within the framework and ionic interactions between the cations and the framework [17].

*Ga-Si phases.* In the study of $K_8Ga_8Si_{38}$, first principle calculation was performed on a guest atom free $Si_{46}$ structure, a type I clathrate $K_8Si_{46}$ phase and the $K_8Ga_8Si_{38}$ phase [13, 27, 28]. The DOS of $Si_{46}$ confirms it is a semiconductor, with $E_g$ approximately 0.67 eV. When adding guest atom K into the system, the shape of DOS is similar to that obtained for $Si_{46}$, except that the Fermi level shifted to higher energy level because of the electron filling from K. $K_8Si_{46}$ is correctly predicted to be metallic. In the case of $K_8Ga_8Si_{38}$, the Ga, being one electron deficient, compensates the eight electrons from K, and DOS shows that it once again has a gap with filled valence and empty conduction bands. The calculations indicate that $K_8Ga_8Si_{38}$ should be semiconducting. The calculated $E_g$ of $K_8Ga_8Si_{38}$ is 0.52 eV, smaller than that calculated for $Si_{46}$, but 0.15 eV larger than that of diamond-phase Si (Fig. 8.10).

## 8.6 Hydrogen Containing Clathrates

While developing the synthetic methodology to prepare macroscopic amount of hydrogen capped silicon nanoparticles as a possible hydrogen storage material, [43] we discovered that a slight variation of this route resulted in silicon type I clathrate with hydrogen in the cages [9, 10]. The general reaction with the Zintl salts, ASi (A = Na, K), is shown below:

$$ASi + NH_4X \rightarrow A_x(H_2)_ySi_{46} + AX + NH_3 (A = Na, K; X = Cl, Br, I)$$

The ASi used in this route can be prepared by either conventional high temperature routes or via low temperature reaction of the element, Si, with the metal hydrides [44]. The reaction to prepare clathrates with hydrogen encapsulation was performed either in solution (dioctyl ether) at 250 °C or as a solid state reaction at 200 °C under vacuum. The solid state reaction is prepared by mechanical mixing (or ball-milling) of the two solids, ASi and $NH_4X$, pressing them into a pellet which is then placed into a furnace and heated under vacuum to 200 °C. The solution reaction uses the same precursors, but they are dispersed in a solvent and

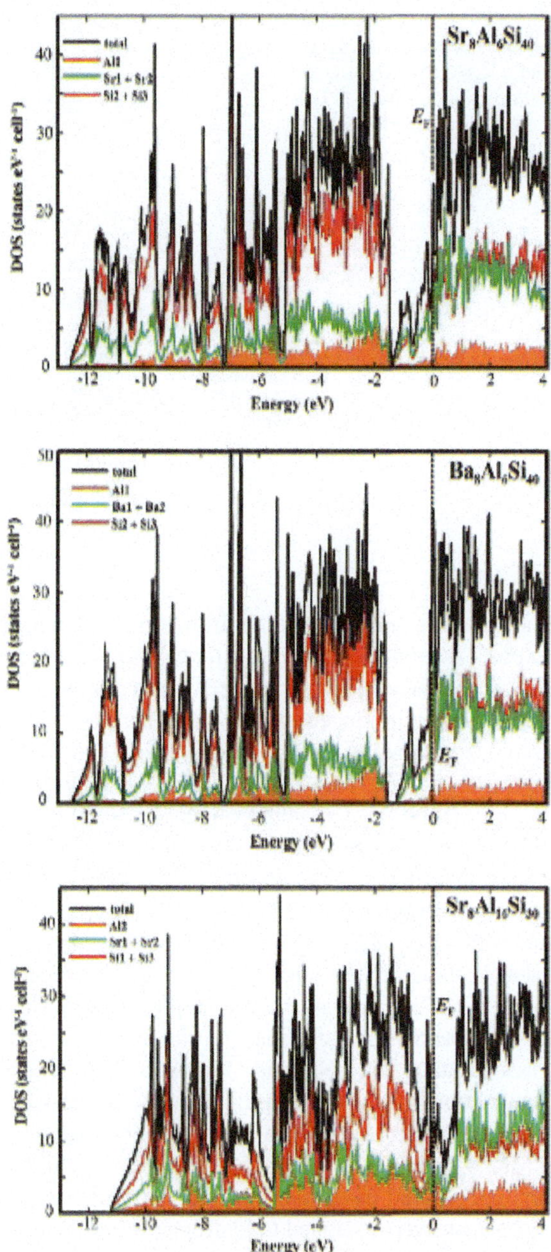


**Fig. 8.9** Total electronic density of states along with the contribution of different atoms for the model systems $Sr_8Al_6Si_{40}$ (Al occupies the *6c* site), $Ba_8Al_6Si_{30}$ (Al occupies the *6c* site) and $Sr_8Al_{16}Si_{30}$ (Al fully occupies the *6c* site and partially occupies the *16i* site). Reprinted with permission from [17]. Copyright (2012) American Chemical Society

the reaction takes place in a round bottom flask equipped with a reflux column under flowing argon at the reflux temperature of dioctyl ether, ~250 °C. These are significantly lower temperatures than what have been reported in the literature for the transformation of NaSi to the clathrate structure [45]. At these

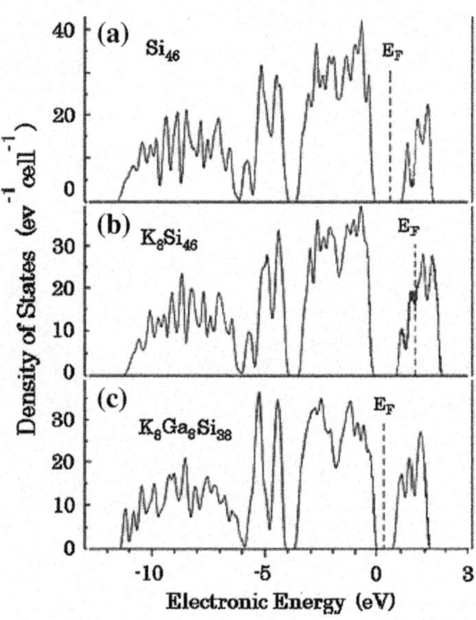


**Fig. 8.10** Calculated density of states of **a** $Si_{46}$ **b** $K_8Si_{46}$ and **c** $K_8Ga_8Si_{38}$. Reprinted with permission from [27]. Copyright (2011) Royal Chemical Society

temperatures, both the product from the solution route and the solid state route lead to phase pure type I clathrate structure, as can be observed from the powder X-ray diffraction pattern shown in Fig. 8.11 for the Na version of the structure. The solid state reaction produced micron-sized particles and the solution reaction produced nano-sized particles (40 nm in diameter). There is a good fit of the clathrate type I structure to the experimental data. The cubic cell is only slightly larger ($a$ = 10.1999(4) Å), than the values reported for a fully sodium occupied clathrate, $Na_8Si_{46}$, $a$ = 10.19648(2) Å [46]. However, the amount of Na in each cage could not be determined based on this refinement. This material has been structurally characterized with HRTEM and EDS. The agreement between the experimental and simulated HRTEM for the clathrate type I structure is excellent. The Na content was determined by energy dispersive spectroscopy (EDS) and Na:Si = 4.7:46 was obtained or approximately $Na_5Si_{46}$. Chemical analysis (C, H, N, Na, and Si) indicates that the stoichiometry is $Na_6(H_2)_2Si_{46}$.

Solid state $^1H$, $^{23}Na$, and $^{29}Si$ MAS NMR has been used for the further characterization of this compound; the spectra are shown in Fig. 8.12. The solid state $^1H$ NMR spectrum is shown in Fig. 8.4a. There are four distinct resonances in the $^1H$ NMR: 4.2 ppm, 3.2 ppm 1.2 ppm and 0.79 ppm. FTIR data were also obtained on this sample and there is no evidence for a Si–H stretching mode; therefore, these resonances were assigned to hydrogen trapped in the cavities. Solid state NMR has been reported on a type II THF·$H_2O$ clathrate, which under hydrogen pressure can store up to 2.1 wt% $H_2$ [47]. They propose that the hydrogen is in the small cages with the resonance at ~4.2 ppm and that hydrogen in the large cages gives rise to a resonance at about 0 ppm [47]. The spectrum for the $Na_6(H_2)_2Si_{46}$ shows a sharp resonance at about 4.2 ppm consistent with molecular hydrogen that can freely rotate in the large cavities and the other resonances consistent with hydrogen that

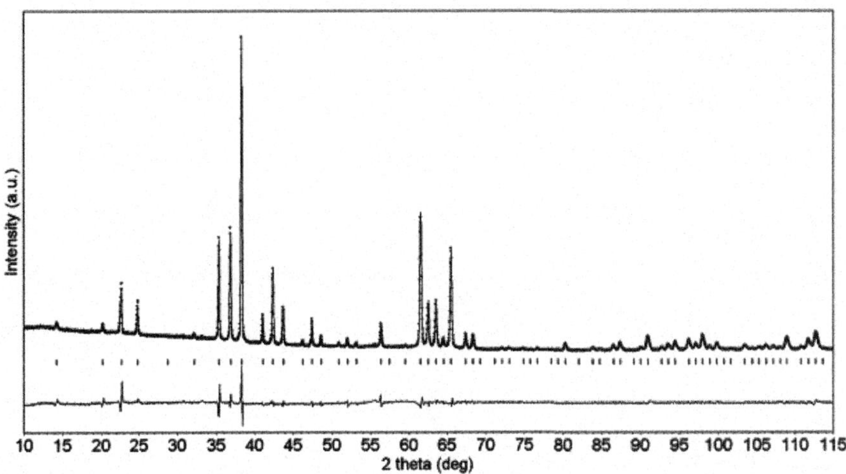


**Fig. 8.11** Rietveld profile fit for $Na_x(H_2)_ySi_{46}$. Experimental data points and the theoretical fit are shown. The data were refined for the clathrate type I structure (*tick marks*) and the difference between the observed and theoretical patterns is shown. Reprinted with permission from [9]. Copyright (2007) American Chemical Society

might be interacting with the cages. The solid state $^{23}$Na spectrum, Fig. 8.4b, shows the two characteristic Knight shifts for the Na atom in a $Si_{20}$ cage (2a crystallographic site) at ~1,730 ppm and for the Na in the $Si_{24}$ cage (6d crystallographic site) at ~2,018 ppm [45]. The integrated intensities is 0.998:1 for the 2a:6d crystallographic sites, suggesting that the large cage is Na deficient. Further modeling of the $^1$H NMR is necessary to provide a consistent picture of where the $H_2$ molecule resides. The Na deficiency would allow for the hydrogen to fill the remainder of the cavities in this compound. Solid state $^{29}$Si spectrum, Fig. 8.4c, is consistent with clathrate type I structure. There are three crystallographic silicon sites (16*i*, 6*c*, and 24*k*), with three distinct resonances, at ~617 ppm for the silicon in 16*i* site, at ~653 ppm for the Si in 6*c*, and ~842 ppm for the Si in the 24*k* site [48]. The resonances are labeled according to their crystallographic site symmetry. The integrated intensities are consistent with the stoichiometry of $Si_{46}$ with all sites completely filled. Taken together, the solid state NMR and FTIR evidence points to hydrogen in the cages and the clathrate being sodium deficient, consistent with results from EDS and chemical analysis.

Further evidence for hydrogen encapsulation comes from an additional publication on $K_{8-x}(H_2)_ySi_{46}$. In this example, additional results from cross-polarization (CP) MAS provided further evidence for no Si–H covalent bonding, suggesting that hydrogen in these materials is present as $H_2$ in interstitial sites. In addition, high annular dark-field (HAADF) STEM experimental and simulated images indicated that the K is deficient in both the *2a* and *6d* sites. Figure 8.13 shows the results of thermal gravimetry (TG) mass spectrometry (MS) data, confirming the loss of hydrogen at ~400 °C.

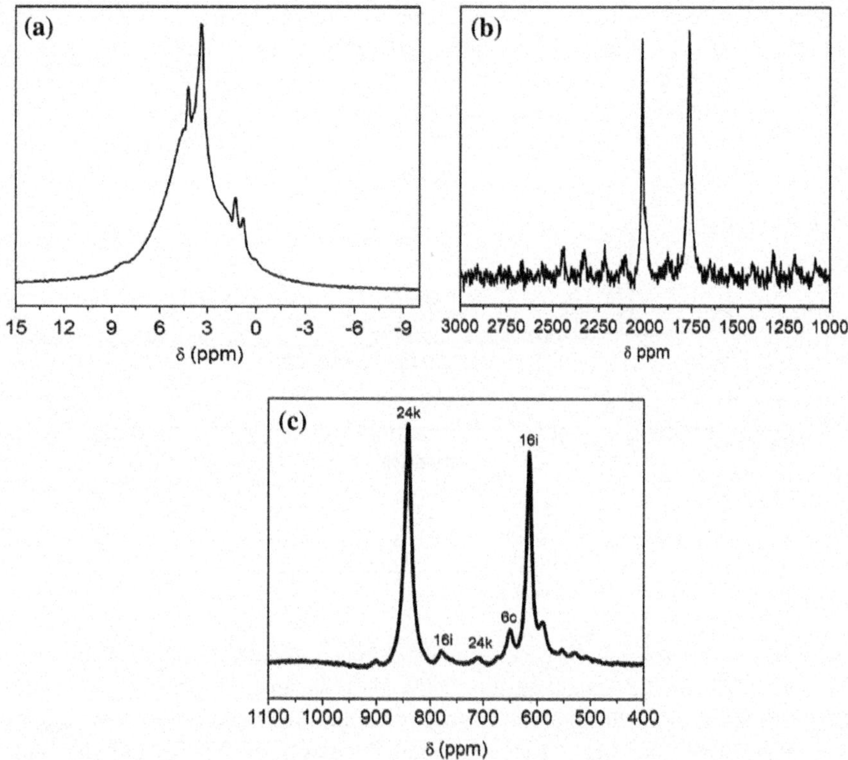


**Fig. 8.12** **a** $^1H$, **b** $^{23}Na$, and **c** $^{29}Si$ solid state MAS NMR spectra for $Na_x(H_2)_ySi_{46}$. The assignment for the Si sites are provided in (**c**). Reprinted with permission from [9]. Copyright (2007) American Chemical Society

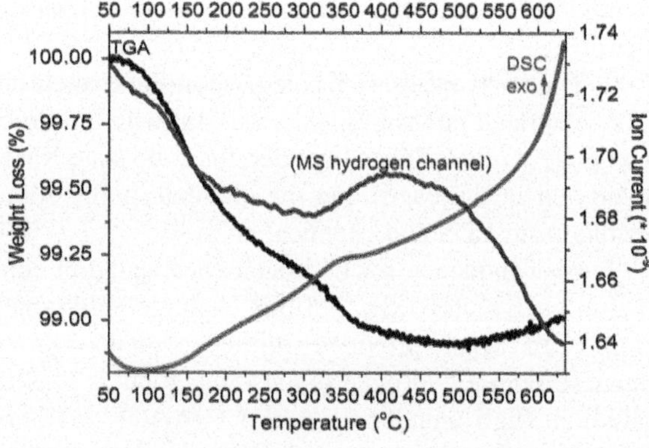


**Fig. 8.13** TGA/DSC/MS data for $K_{8-x}(H_2)_ySi_{46}$. TG, DSC, and hydrogen MS traces are indicated as *black*, *red*, and *blue*, respectively. Reprinted with permission from [10]. Copyright (2010) American Chemical Society

Some clathrate structures have cages that are accessible from the surface to the interior, providing access for hydrogen migration through the structure. The results on type I clathrates are promising and provide the motivation for further investigation of other inorganic clathrates as possible hydrogen storage materials.

## 8.7 Summary

Light element containing clathrate structured solids have potential for many applications, from high temperature materials, and thermoelectrics to hydrogen storage materials. The barium-containing phases have high melting points (above 1,000 °C). The alkaline earth-containing clathrate I compounds cannot reach the electron precise Zintl formulism and therefore exhibit metallic conduction. Further control of the carrier concentration through the addition of various framework atoms or mixed occupancy of the cation sites might provide high zT at high temperatures for these compounds. As more compositions of light elements are explored, it can be expected that additional important properties and functions of these materials will be discovered.

## References


1. B.B. Iversen, A.E.C. Palmqvist, D.E. Cox, G.S. Nolas, G.D. Stucky, N.P. Blake, H. Metiu, J. Solid State Chem. **149**(2), 455–458 (2000)
2. Y. Mudryk, P. Rogl, C. Paul, S. Berger, E. Bauer, G. Hilscher, C. Godart, H. Noel, J. Phys. Condens. Matter **14**(34), 7991–8004 (2002)
3. Y. Mudryk, P. Rogl, C. Paul, S. Berger, E. Bauer, G. Hilscher, C. Godart, H. Noel, A. Saccone, R. Ferro, Phys. B-Condens. Matter **328**(1–2), 44–48 (2003)
4. M. Christensen, S. Johnsen, B.B. Iversen, Dalton Trans. **39**(4), 978–992 (2010)
5. T.F. Fassler, S. Hoffmann, Zeitschrift Fur Kristallographie **214**(11), 722–734 (1999)
6. M. Beekman, G.S. Nolas, J. Mater. Chem. **18**(8), 842–851 (2008)
7. S. Yamanaka, Dalton Trans. **39**(8), 1901–1915 (2010)
8. A.V. Shevelkov, K. Kovnir, Z. Clathrates, in *Zintl Phases: Principles and Recent Developments*, vol. 139, ed. by T.F. Fassler, Structure and Bonding (Springer, Berlin, 2011), pp. 97–142
9. D. Neiner, N.L. Okamoto, C.L. Condron, Q.M. Ramasse, P. Yu, N.D. Browning, S.M. Kauzlarich, J. Am. Chem. Soc. **129**(45), 13857–13862 (2007)
10. D. Neiner, N.L. Okamoto, P. Yu, S. Leonard, C.L. Condron, M.F. Toney, Q.M. Ramasse, N.D. Browning, S.M. Kauzlarich, Inorg. Chem. **49**(3), 815–822 (2010)
11. T. Langer, S. Dupke, H. Trill, S. Passerini, H. Eckert, R. Poettgen, M. Winter, J. Electrochem. Soc. **159**(8), A1318–A1322 (2012)
12. G.B. Adams, M. Okeeffe, A.A. Demkov, O.F. Sankey, Y.M. Huang, Phys. Rev. B **49**(12), 8048–8053 (1994)
13. Y. Imai, M. Imai, J. Alloy. Compd. **509**(9), 3924–3930 (2011)
14. K. Kishimoto, N. Ikeda, K. Akail, T. Koyanagi, Appl. Phys. Exp. **1**(3), 031201 (2008)
15. H. Shimizu, Y. Takeuchi, T. Kume, S. Sasaki, K. Kishimoto, N. Ikeda, T. Koyanagi, J. Alloy. Compd. **487**(1–2), 47–51 (2009)

16. N. Tsujii, J.H. Roudebush, A. Zevalkink, C.A. Cox-Uvarov, G.J. Snyder, S.M. Kauzlarich, J. Solid State Chem. **184**(5), 1293–1303 (2011)
17. J.H. Roudebush, N. Tsujii, A. Hurtando, H. Hope, Y. Grin, S.M. Kauzlarich, Inorg. Chem. **51**(7), 4161–4169 (2012)
18. V.L. Kuznetsov, L.A. Kuznetsova, A.E. Kaliazin, D.M. Rowe, J. Appl. Phys. **87**(11), 7871–7875 (2000)
19. W. Carrillo-Cabrera, R.C. Gil, Y. Grin, Z. Kristallogr. New Crystal Struct. **217**(2), 179–180 (2002)
20. L.Y. Qiu, I.P. Swainson, G.S. Nolas, M.A. White, Phys. Rev. B **70**(3), 035208 (2004)
21. A. Bentien, B.B. Iversen, J.D. Bryan, G.D. Stucky, A.E.C. Palmqvist, A.J. Schultz, R.W. Henning, J. Appl. Phys. **91**(9), 5694–5699 (2002)
22. C.L. Condron, R. Porter, T. Guo, S.M. Kauzlarich, Inorg. Chem. **44**(25), 9185–9191 (2005)
23. C.L. Condron, S.M. Kauzlarich, F. Gascoin, G.J. Snyder, Chem. Mater. **18**(20), 4939–4945 (2006)
24. C.L. Condron, S.M. Kauzlarich, Inorg. Chem. **46**(7), 2556–2562 (2007)
25. R. Kroner, K. Peters, H.G. von Schnering, R. Nesper, Z. Kristallogr.-New Crystal Struct. **213**(4), 667–668 (1998)
26. H.G. von Schnering, R. Kroner, H. Menke, K. Peters, R. Nesper, Z. Kristallogr.-New Crystal Struct. **213**(4), 677–678 (1998)
27. M. Imai, A. Sato, H. Udono, Y. Imai, H. Tajima, Dalton Trans. **40**(16), 4045–4047 (2011)
28. K. Nakamura, S. Yamada, T. Ohnuma,Mat. Trans. **54**(3):276–285 (2013)
29. C.L. Condron, S.M. Kauzlarich, T. Ikeda, G.J. Snyder, F. Haarmann, P. Jeglic, Inorg. Chem. **47**(18), 8204–8212 (2008)
30. W. Jung, J. Loerincz, R. Ramlau, H. Borrmann, Y. Prots, F. Haarmann, W. Schnelle, U. Burkhardt, M. Baitinger, Y. Grin, Angew. Chem. Int. Ed. **46**(35), 6725–6728 (2007)
31. B. Eisenmann, H. Schafer, R. Zagler, J. Less-Common Metals **118**(1), 43–55 (1986)
32. K. Suekuni, M.A. Avila, K. Umeo, T. Takabatake, Phys. Rev. B **75**, 195210 (2007)
33. J.H. Roudebush, E.S. Toberer, H. Hope, G.J. Snyder, S.M. Kauzlarich, J. Solid State Chem. **184**(5), 1176–1185 (2011)
34. R.A. Ribeiro, M.A. Avila, Philos. Mag. **92**(19–21), 2492–2507 (2012)
35. W. Gou, S.Y. Rodriguez, Y. Li, J. Ross, H. Joseph, Phys. Rev. B **80**, 144108 (2009)
36. C.L. Condron, J. Martin, G.S. Nolas, P.M.B. Piccoli, A.J. Schultz, S.M. Kauzlarich, Inorg. Chem. **45**(23), 9381–9386 (2006)
37. J.H. Roudebush, C. de la Cruz, B.C. Chakoumakos, S.M. Kauzlarich, Inorg. Chem. **51**(3), 1805–1812 (2012)
38. M. Christensen, B.B. Iversen, Chem. Mater. **19**(20), 4896–4905 (2007)
39. H. Anno, M. Hokazono, R. Shirataki, Y. Nagami, J. Mater. Sci. **48**(7), 2846–2854 (2013)
40. D. Nataraj, J. Nagao, M. Ferhat, T. Ebinuma, J. Appl. Phys. **93**(5), 2423–2428 (2003)
41. S.-K. Deng, X.-F. Tang, R.-S. Tang, Chin. Phys. B **18**(7), 3084–3086 (2009)
42. H. Anno, H. Yamada, T. Nakabayashi, M. Hokazono, R. Shirataki, J. Solid State Chem. **193**, 94–104 (2012)
43. D. Neiner, S.M. Kauzlarich, Chem. Mater. **22**(2), 487–493 (2010)
44. X. Ma, F. Xu, T.M. Atkins, A.M. Goforth, D. Neiner, A. Navrotsky, S.M. Kauzlarich, Dalton Trans. (46), 10250–10255
45. M. Pouchard, C. Cros, P. Hagenmuller, E. Reny, A. Ammar, M. Menetrier, J.M. Bassat, Solid State Sci. **4**(5), 723–729 (2002)
46. G.K. Ramachandran, J.J. Dong, J. Diefenbacher, J. Gryko, R.F. Marzke, O.F. Sankey, P.F. McMillan, J. Solid State Chem. **145**(2), 716–730 (1999)
47. H. Lee, J.W. Lee, D.Y. Kim, J. Park, Y.T. Seo, H. Zeng, I.L. Moudrakovski, C.I. Ratcliffe, J.A. Ripmeester, Nature **434**(7034), 743–746 (2005)
48. G.K. Ramachandran, P.F. McMillan, J. Diefenbacher, J. Gryko, J.J. Dong, O.F. Sankey, Phys. Rev. B **60**(17), 12294–12298 (1999)

# Chapter 9
# Structural and Physical Properties of Rare-Earth Clathrates

**Silke Paschen, Matthias Ikeda, Stevce Stefanoski and George S. Nolas**

**Abstract** Clathrates that contain rare-earth elements as guest atoms have been of active interest since the discovery of intermetallic clathrates. A large body of work focussed on thermoelectric properties of Eu-containing clathrates. The very low lattice thermal conductivities that are reached in Eu-containing type-I clathrates are generally attributed to the pronounced rattling of Eu in oversized host cages and to the occurrence of split sites in the larger of the two cages of the structure. The potential of Eu-containing clathrates for magnetic refrigeration has been recognized more recently. Here, key features are the large magnetic moment of Eu, together with the second order character of the paramagnetic to ferromagnetic phase transition. The incorporation of other magnetic rare-earth elements into the clathrate cages has long remained elusive. Only very recently the successful synthesis of a cerium containing type-I clathrate was reported. Interestingly, a sizable enhancement of the thermopower is observed and attributed to a rattling enhanced Kondo interaction. This discovery may trigger a wealth of future investigations.

S. Paschen (✉) · M. Ikeda (✉)
Institute of Solid State Physics, Vienna University of Technology,
Wiedner Hauptstr. 8-10, 1040 Vienna, Austria
e-mail: paschen@ifp.tuwien.ac.at

M. Ikeda
e-mail: ikeda@ifp.tuwien.ac.at

S. Stefanoski (✉)
Geophysical Laboratory, Carnegie Institution of Washington, 5251 Broad Branch Road NW,
Washington, DC 20015, USA
e-mail: sstefanoski@carnegiescience.edu

G. S. Nolas (✉)
Department of Physics, University of South Florida, 4202 E. Fowler Ave.,
Tampa, FL 33620, USA
e-mail: gnolas@usf.edu

G. S. Nolas (ed.), *The Physics and Chemistry of Inorganic Clathrates*,
Springer Series in Materials Science 199, DOI: 10.1007/978-94-017-9127-4_9,

## 9.1 Introduction

Intermetallic clathrates, first described in the 1960s [1], are inclusion complexes in which atoms of one substance (guest) are completely enclosed in cavities formed by the crystal lattice of another substance (host or framework). Their structure is closely related to clathrate hydrates formed by water and certain molecules at high pressures and low temperatures, which have first been observed in the arctic ice [2]. Today, clathrates are classified into nine different structure types [3, 4]. Most investigated are type-I clathrates of which many intermetallic representatives are known. The structural units of clathrates are different kinds of cages which build up the structure in a space filling arrangement. A type-I clathrate for instance contains two pentagondodecahedra and six tetrakaidecahedra per unit cell. Unlike in the case of zeolites the shared faces of the polyhedra have too small apertures to let the enclosed guest species pass from one cage to the other.

The bonding situation in clathrates may, in a first approximation, be understood in terms of the Zintl concept [5]: The more electropositive guest atoms (cations) donate electrons to the more electronegative host atoms (anions) such that the latter complete their valence requirements (octet rule) when forming a covalently bonded ($sp^3$ like) space-filling framework (in inverse clathrates, the roles of cations and anions are exchanged, see Chap. 5). For the type-I clathrate $Sr_8Ga_{16}Ge_{30}$, for instance, this concept can be expressed by the notation $(Sr^{2+})_8(Ga^{1-})_{16}(Ge^0)_{30}$ where the superscripts are the formal charges of the atoms: +2 for Sr (acts as a cation), $-1$ for Ga (needs in addition to its own three valence electrons one extra electron to form the four framework bonds), and 0 for Ge (its four valence electrons are used in the four framework bonds).

This chapter focuses on clathrates that contain rare-earth elements. The earliest representative of this family of compounds was $Eu_8Ga_{16}Ge_{30}$ [6]. The possibility to obtain this compound followed naturally from the existence of $Sr_8Ga_{16}Ge_{30}$ [7], together with the fact that Eu frequently assumes a two-valent oxidation state and that the ionic radius of $Eu^{2+}$ is similar to the one of $Sr^{2+}$ in related compounds [7]. Only later, the added value of the larger mass and smaller ionic radius of Eu compared to Sr, as well as the importance of the spin degree of freedom of Eu were recognized. The vast majority of experiments on rare-earth-containing clathrates was done on $Eu_8Ga_{16}Ge_{30}$. Therefore, a separate section (Sect. 9.2) reviews these results. In Sect. 9.3, results on other rare-earth-containing clathrates are summarized.

## 9.2 $Eu_8Ga_{16}Ge_{30}$

The transport properties of the clathrate $Eu_8Ga_{16}Ge_{30}$ were first investigated by Cohn et al. [6]. The lattice thermal conductivity was shown to be extremely low and "glass-like" [6], as was the case for $Sr_8Ga_{16}Ge_{30}$ [7], much like the type-I clathrate hydrate [8]. This discovery triggered immense interest in the thermoelectrics community and initiated a wealth of investigations. This section reviews, in six subsections, different aspects of the research on $Eu_8Ga_{16}Ge_{30}$.

### 9.2.1 Crystallographic and Electronic Structure

The material $Eu_8Ga_{16}Ge_{30}$ investigated by Cohn et al. [6] had the type-I crystal structure (space group $Pm\bar{3}n$, Pearson symbol $cP54$, Fig. 9.1), just as the other clathrates ($Cs_8Sn_{44}$, $Ba_8Ga_{16}Si_{30}$, $Sr_8Ga_{16}Ge_{30}$, and $Sr_4Eu_4Ga_{16}Ge_{30}$) studied in that pioneering work. Shortly later, a second modification of $Eu_8Ga_{16}Ge_{30}$ with type-VIII clathrate structure (space group $I\bar{4}3m$, Pearson symbol $cI54$) [9] was discovered [10]. Type-I $Eu_8Ga_{16}Ge_{30}$ ($\beta$-phase) is the high-temperature phase. It exists only between 696 and 699 °C, the congruent melting temperature of this phase. Below 696 °C, the low-temperature phase type-VIII $Eu_8Ga_{16}Ge_{30}$ ($\alpha$-phase) is stable [10].

Both structures are characterized by covalent $E_{46}$ networks ($E$ = Ga, Ge) of fourfold bonded $E$ atoms, forming polyhedral cages occupied by Eu atoms. The type-I clathrate structure is built up of two different polyhedral cages: $E_{20}$ pentagonal dodecahedra centered by Eu1 at the site $2a$ (2 per formula unit) and $E_{24}$ tetrakaidecahedra centered by Eu2 at the site $6d$ (six per formula unit, Fig. 9.1). The type-VIII clathrate structure consists of only one type of cage, a distorted $E_{20+3}$ polyhedron, centered by Eu at the $8c$ site (eight per formula unit) [10]. The structural transition between the two modifications was modeled using periodic minimal surface approximations; this revealed a topological relationship between the two four-connected networks [12].

The atomic coordinates, important interatomic distances, and the anisotropic atomic displacement parameters of both type-I and type-VIII $Eu_8Ga_{16}Ge_{30}$ are tabulated [10]. The displacement ellipsoid of the Eu2 atoms in type-I $Eu_8Ga_{16}Ge_{30}$ is plate-like. A better description can be obtained by replacing the Eu2 ($6d$) site with a Eu2' ($24k$) site with 1/4 occupation [10, 13, 14]. In type-VIII $Eu_8Ga_{16}Ge_{30}$ the displacement ellipsoid of the Eu atoms is much less flattened and there is no need to introduce this split site model. Neutron diffraction measurements on single crystalline type-I $Eu_8Ga_{16}Ge_{30}$ revealed large, but only weakly temperature dependent isotropic atomic displacement parameters for Eu2 indicating either a static displacement from the $6d$ position or a dynamic one with very low activation energy [15]. The Eu2 nuclear density distribution clearly reveals this displacement (Fig. 9.2). The anomalous behavior of Eu2 is generally considered as an important ingredient for the very low lattice thermal conductivity of type-I $Eu_8Ga_{16}Ge_{30}$, as will be further discussed below.

The electronic structure of both $Eu_8Ga_{16}Ge_{30}$ modifications was investigated by density functional theory. For the undoped compounds band gaps of 0.6–0.9 eV were found [16]. The conduction bands hybridize with the unoccupied $d$-states of the guest atoms. When the material is $n$-doped these bands are populated.

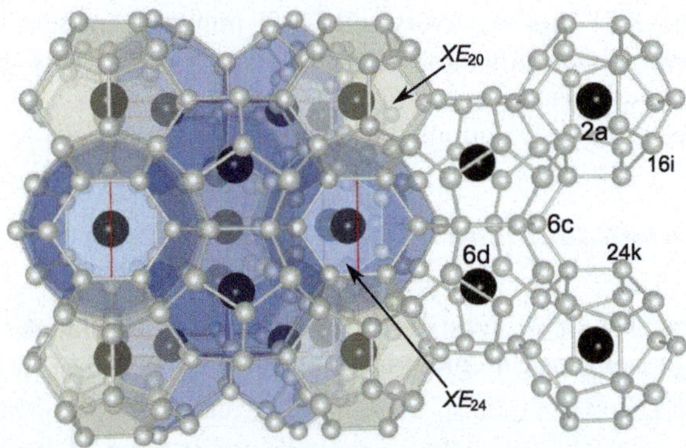


**Fig. 9.1** Crystal structure of type-I $Eu_8Ga_{16}Ge_{30}$. Eu: *large black spheres*; Ga, Ge: *small grey spheres*. The different crystallographic sites are labeled. The *red lines* represent one unit cell. The structure was drawn with the software VESTA 3 [11]

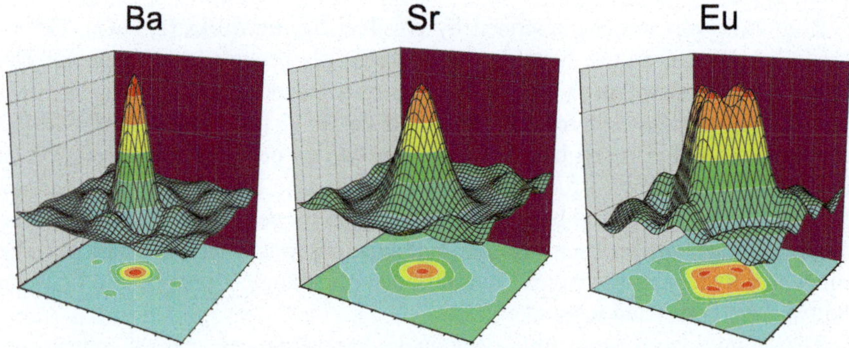


**Fig. 9.2** Nuclear density at the center of the large cage of type-I $M_8Ga_{16}Ge_{30}$ (M = Ba, Sr, Eu) at 40 K, determined as difference Fourier map from single-crystal neutron diffraction. Reprinted with permission from Sales et al. [15]. Copyright (2013) by the American Physical Society

Therefore, the electronic properties of *n*-doped clathrates are expected to depend more strongly on the guest atom than *p*-doped clathrates (where the *d*-states remain empty) [16].

### 9.2.2 Stoichiometry

Bandstructure calculations predict semiconducting behavior for $Eu_8Ga_{16}Ge_{30}$ [16] but experiments typically show properties characteristic of heavily doped, degenerate semiconductors. Additionally, these properties vary strongly from one sample to another. Therefore, a thorough investigation of the composition–property relationship was undertaken [17, 18]. It was shown that the exact composition, in particular the value of $x$ in $Eu_8Ga_{16-x}Ge_{30+x}$, varies for polycrystalline samples that underwent different thermal treatments. Samples with different $x$ had different charge carrier concentrations $n$ [17]. This is well understood from the Zintl counting scheme introduced in Sect. 9.1, where $x$ in $(Eu^{2+})_8(Ga^{1-})_{16-x}(Ge^0)_{30+x}$ corresponds directly to the number of excess electrons per formula unit (a negative $x$ corresponds to excess holes). Since all thermoelectric quantities depend sensitively on $n$ [18] (see Sect. 9.2.3), $x$ has to be determined and controlled for reproducible and predictable behavior. The homogeneity range was determined as $0.49 \leq x \leq 1.01$ for the type-I and $0.28 \leq x \leq 0.48$ for the type-VIII modification. Within these homogeneity ranges the composition of the clathrate phases could be influenced by the annealing temperature. All samples were $n$-type conductors [17]. The exact 8:16:30 stoichiometry could not be reached.

Smaller $x$ values could be obtained by a new synthesis technique for clathrates [19], rapid quenching by melt spinning [20]. A melt-spun sample with nominally $x = 0$ showed only minute amounts of foreign phases in high-resolution transmission electron microscopy. Thus, the sample's composition must be very close to the ideal stoichiomety. Its electrical conductivity indeed had semiconductor-like character.

### 9.2.3 Thermoelectric Properties

By far most published work on $Eu_8Ga_{16}Ge_{30}$ deals with thermoelectric aspects. In this section the thermoelectric properties of type-I and type-VIII $Eu_8Ga_{16}Ge_{30}$ are reviewed. The three quantities entering the dimensionless thermoelectric figure of merit

$$ZT = \frac{S^2 T}{\rho \kappa} \tag{9.1}$$

are the thermopower $S$, the electrical resistivity $\rho$, and the thermal conductivity $\kappa$, which are all functions of the temperature $T$. The thermal conductivity is composed of an electronic contribution $\kappa_e$ and a lattice contribution $\kappa_l$ as

$$\kappa = \kappa_e + \kappa_l. \tag{9.2}$$

$\kappa_e(T)$ is generally estimated from the temperature-dependent electrical resistivity using the Wiedemann-Franz law

$$\kappa_e(T) = \frac{\pi^2 k_B^2}{3e^2} \cdot \frac{T}{\rho(T)}. \tag{9.3}$$

$S$, $\rho$ and $\kappa_e$ all depend sensitively on the charge carrier concentration $n$ which, in the simplest case (single parabolic band), is related to $R_H$ via

$$R_H(T) = \frac{1}{n(T)e}. \tag{9.4}$$

In rare-earth clathrates an anomalous contribution to the Hall coefficient [21] may complicate the analysis, in particular if the system orders ferromagnetically [22]. As both type-I and type-VIII $Eu_8Ga_{16}Ge_{30}$ are ferromagnetic at low temperatures (see Sect. 9.2.6), this has to be taken into account [10, 17, 20].

For $Eu_8Ga_{16-x}Ge_{30+x}$, the dependence of $\rho$ and $S$ on $n$ was studied in detail to obtain information on the dominant scattering mechanism of the charge carriers [18]. According to semiclassical transport theory [23] the electrical conductivity $\sigma$ depends on $n$ via

$$\sigma = \sigma_0 n^{p_\rho} \tag{9.5}$$

where the exponent $p_\rho$ is determined by the dominating scattering mechanism. For instance, $p = 2/3$ is the exponent for scattering of electrons from acoustic phonons and $p = 2$ is for impurity scattering. The analogous relation

$$S = S_0 n^{p_S} \tag{9.6}$$

holds for $S$ only if scattering of the charge carriers is dominated by a single mechanism. The exponent should be $p_S = -2/3$, independent of the scattering mechanism. Between 2 and 400 K, Eqs. (9.5) and (9.6) were used to approximate the data of type-I and type-VIII $Eu_8Ga_{16-x}Ge_{30+x}$. Figure 9.3 shows $\sigma(n)$ and $S(n)$ data and the corresponding fits at two selected temperatures.

The temperature dependencies of $p_\rho$ and $p_S$ (not shown) indicate that impurity scattering plays an important role in type-VIII $Eu_8Ga_{16-x}Ge_{30+x}$ at the lowest temperatures, whereas scattering from acoustic phonons dominates for both modifications at high temperatures. The low-temperature $p_\rho$ values of type-I $Eu_8Ga_{16-x}Ge_{30+x}$ are surprisingly low (close to 2/3) but $p_S$ is clearly larger than −2/3. It was speculated that this is due to deviations from the rigid-band behavior [18].

The lattice thermal conductivity of $Eu_8Ga_{16}Ge_{30}$ was studied by several groups below room temperature [6, 13, 15, 18, 20, 24, 25]. In all cases it was estimated via Eqs. (9.2) and (9.3) using $\rho(T)$ data. This naturally puts a larger error bar on $\kappa_l$ than on the electronic transport quantities $\rho$ and $S$. In addition, in the generally employed steady-state heat flow technique, measurements of $\kappa$ above about 100 K are encumbered by radiation losses. Thus, differences in absolute values between results from different groups should be discussed with caution. Here we focus on the overall trends and relative changes within the sample series measured and analyzed in the same way.

The most striking and robust observation is that $Eu_8Ga_{16-x}Ge_{30+x}$ shows qualitatively different $\kappa_l(T)$ characteristics for the type-I and type-VIII modifications.

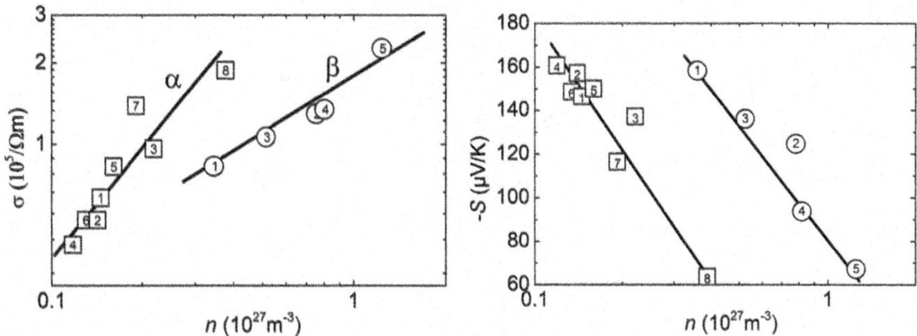


**Fig. 9.3** Electrical conductivity $\sigma$ versus charge carrier concentration $n$ isotherm at 110 K (*left*) and thermopower $S$ versus $n$ at 333 K (*right*) of type-I ($\beta$, *circles*) and type-VIII ($\alpha$, *squares*) $Eu_8Ga_{16}Ge_{30}$. Reprinted with permission from Bentien et al. [18]

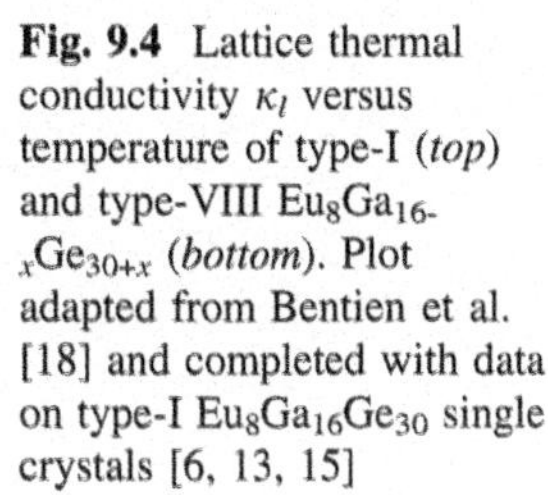

**Fig. 9.4** Lattice thermal conductivity $\kappa_l$ versus temperature of type-I (*top*) and type-VIII $Eu_8Ga_{16-x}Ge_{30+x}$ (*bottom*). Plot adapted from Bentien et al. [18] and completed with data on type-I $Eu_8Ga_{16}Ge_{30}$ single crystals [6, 13, 15]

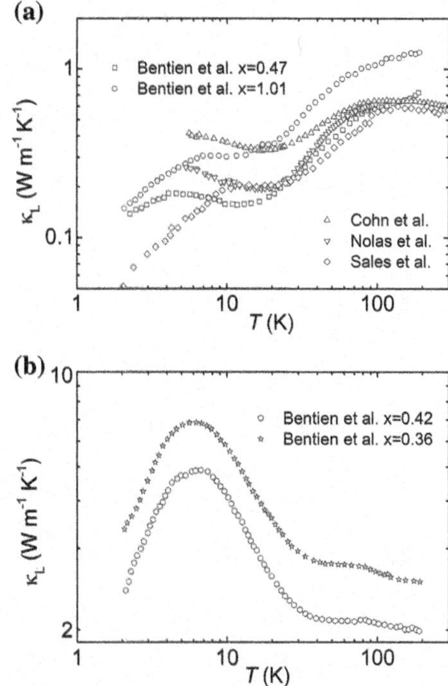


$\kappa_l$ of the type-VIII phase increases with decreasing temperature, passes through a pronounced maximum somewhat below 10 K and then decreases [18, 24]. By contrast, $\kappa_l$ of the type-I phase shows an overall decrease with decreasing temperature [6, 13, 15, 18, 24], with two shoulders (or very shallow maxima) somewhat below 100 and 10 K (Fig. 9.4). It was frequently pointed out that $\kappa_l(T)$ of the type-VIII phase is typical of crystalline materials whereas $\kappa_l(T)$ of the type-I phase resembles more the behavior of amorphous materials or glasses.

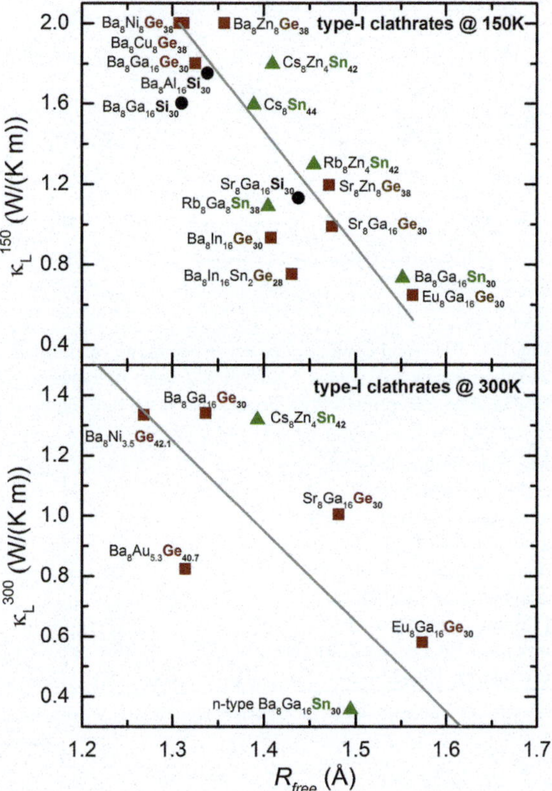


**Fig. 9.5** Lattice thermal conductivity $\kappa_l$ at 150 K (*top*) and 300 K (*bottom*) versus guest free space $R_{\text{free}}$. Figure adapted from [30] and completed with data from [15, 31–34]. The $\kappa_l$ data at 300 K are, except for $Cs_8Zn_4Sn_{42}$, all for single crystalline samples

A number of mechanisms were suggested to explain the anomalous reduction of the low-temperature $\kappa_l$ of the type-I phase.

One important experimental observation was that $\kappa_l(T)$ is essentially the same for single and polycrystalline type-I $Eu_8Ga_{16-x}Ge_{30+x}$ (Fig. 9.4). Thus, scattering from grain boundaries cannot be the dominating scattering process in the studied temperature range. Instead, the low and glass-like $\kappa_l$ appears to be an intrinsic property of the material.

Similar behavior has been observed early on in type-I clathrate hydrates (ice clathrates) [8]. Here, structural investigations together with lattice dynamics calculations have established that the interaction of localized guest vibrations (of the noble gas molecules in the oversized water cages) with the host acoustic phonon branches leads to an avoided crossing of dispersing phonon branches and nondispersing guest vibrations of the same symmetry, and yields a resonant damping of heat-carrying phonons at certain points in the Brillouin zone [26]. In analogy with this interpretation, "rattling" was also put forward to explain the low thermal conductivity of type-I $Eu_8Ga_{16}Ge_{30}$ [6]. Direct evidence for the validity of this picture in intermetallic type-I clathrates had to await inelastic neutron scattering investigations [27, 28] as well as lattice dynamics calculations [28] (see also Sect. 9.2.4).

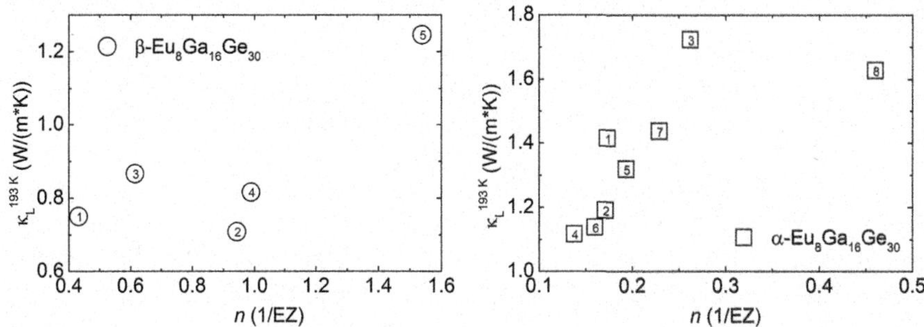


**Fig. 9.6** Lattice thermal conductivity $\kappa_l$ versus charge carrier concentration $n$ isotherm at 193 K of type-I ($\beta$, *left*) and type-VIII $Eu_8Ga_{16}Ge_{30}$ ($\alpha$, *right*). Plot based on data from Bentien et al. [18]

Type-I clathrate hydrates are electrical insulators. By contrast, intermetallic type-I clathrates are typically degenerate semiconductors (see above). That the charge carrier type impacts strongly on the low-temperature thermal conductivity of intermetallic clathrates was first pointed out by Bentien et al. [25]: $n$-type $Ba_8Ga_{16}Ge_{30}$ shows crystal-like $\kappa_l(T)$ but $p$-type $Ba_8Ga_{16}Ge_{30}$ was found to have glass-like character, similar to type-I $Eu_8Ga_{16}Ge_{30}$. This result came as a surprise since before the size of the atomic displacement parameter of the guest atoms [15] was considered as a direct measure of the rattling strength and the related suppression of the lattice thermal conductivity. Apparently, the situation is more complex and phonon-electron interactions cannot be neglected.

Figures 9.5 and 9.6 show that indeed both structural and electronic properties influence the lattice thermal conductivity of intermetallic clathrates. In Fig. 9.5, $\kappa_l$ is plotted for different type-I clathrates versus the guest free space, an analysis done for many type-I clathrates by Suekuni et al. [29, 30]. For calculating the guest free space

$$R_{\text{free}} = R_{\text{cage}} - R_{\text{guest atom}} - R_{\text{host atom}} \tag{9.7}$$

we used for $R_{\text{cage}}$ the distance between the 6$c$ guest and the 24k host site of the $E_{24}$ cage, for $R_{\text{guest atom}}$ the covalent radius of the guest atom, and for $R_{\text{host atom}}$ the mean covalent radius of the different host atoms

$$R_{\text{host atom}} = \frac{\sum_{i=1}^{46} R_i}{46}. \tag{9.8}$$

Suekuni et al. used a slightly different definition where $R_{\text{host atom}}$ is the covalent radius of the most abundant host atom [29, 30]. It is clear that the trend of lower $\kappa_l$ for lower $R_{\text{free}}$ is more pronounced at 150 K (Fig. 9.5, top) than at 300 K (Fig. 9.5, bottom).

In Fig. 9.6, $\kappa_l$(193 K) of type-I (left) and type-VIII $Eu_8Ga_{16}Ge_{30}$ (right) is plotted versus the charge carrier concentration [18]. For both phases, $\kappa_l$ increases with increasing $n$. As suggested in [18], a larger charge carrier concentration might lead to a stronger screening of the rattling cation and thus to a reduction of the coupling between the acoustic phonons and the rattling modes. Interestingly, in the series $Sr_8Ga_{16}Si_{30-x}Ge_x$ the guest free space increase with $x$ and the charge carrier concentration decrease with $x$ [29]. Therefore, it remains ambiguous which of the two effects plays the major role in reducing $\kappa_l$ at low temperatures.

Even though the discussion on the mechanism responsible for reducing the lattice thermal conductivity at low temperatures is interesting from the fundamental point of view, it is not clear to which extent it is relevant for thermoelectric applications at temperatures well above room temperature, where clathrates show promising *ZT* values. In Suekuni et al. [29], for example, it was shown that for temperatures above 80 K, the two end compounds ($x = 0$ and $x = 30$) of the series $Sr_8Ga_{16}Si_{30-x}Ge_x$ have higher $\kappa_l$ than all intermediate samples ($0 < x < 30$), pointing to the dominating role of Si/Ge site disorder at these temperatures. Clearly, further experiments are needed to clarify this important point.

### 9.2.4 Spectroscopy

Various spectroscopic techniques have been used to study $Eu_8Ga_{16}Ge_{30}$. Focus in these investigations was on the rattling modes of the guest atoms. Unfortunately, one of the most powerful techniques, inelastic neutron scattering (INS), is encumbered by the large neutron absorption cross section of $^{151}Eu$. For neutron diffraction (e.g., Fig. 9.2), $^{153}Eu$ enriched $Eu_8Ga_{16}Ge_{30}$ was used [14, 15] but INS results are not yet available.

For other, Eu-free type-I clathrates, INS on large single crystalline samples has in recent years considerably advanced our understanding of the dynamical properties. For $Ba_8Ga_{16}Ge_{30}$, the phonon dispersion was mapped out and an avoided crossing of the guest atom rattler modes and the host atom acoustic phonon branches was observed [27], much like the early findings on ice clathrates [26] discussed above. With even higher resolution experiments on $Ba_8Ni_{3.5}Ge_{42.1}$ a hybridization between the rattler modes and the acoustic phonon branches could be shown (Fig. 9.7, left) and was interpreted to result from a series of anticrossings, that acts like a low-pass filter for acoustic phonons [28]. Importantly, both studies deduce a phonon lifetime that is much too long to explain the low lattice thermal conductivities by Umklapp scattering processes [27, 28]. Thus, not a reduced lifetime but a reduced average phonon group velocity and the suppression of weight of the acoustic branches (Fig. 9.7, right) appear to be the dominating effects. Obviously similar studies on type-I $Eu_8Ga_{16}Ge_{30}$, the system with the most extreme enhancement of the atomic displacement parameters (Fig. 9.2), would be of great interest.

In the absence of INS results, rattling in $Eu_8Ga_{16}Ge_{30}$ has been explored with other spectroscopic techniques. A brief summary of the most important results is given below.

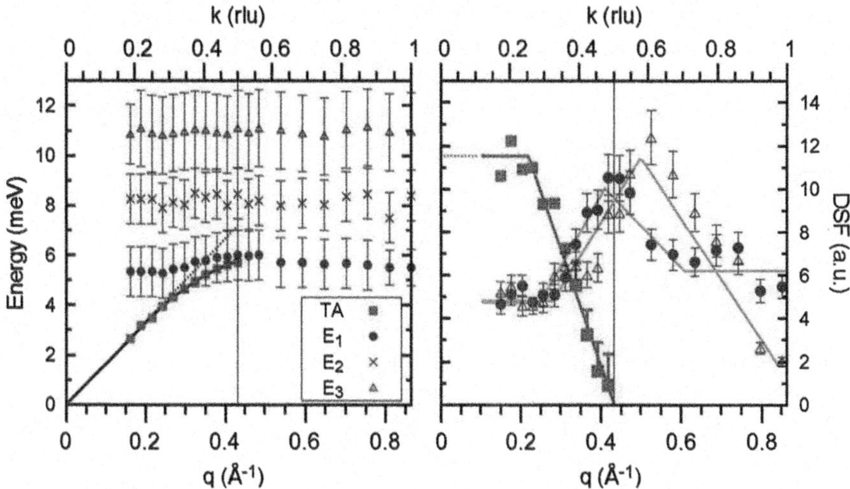


**Fig. 9.7** (*left*) Energy dispersions of the transverse acoustic (TA) and the Einstein-like ($E_1$, $E_2$, $E_3$) phonons. The error bars represent the FWHM of the phonon linewidth. (*right*) Dynamical structure factor (DSF) versus wave vector $q$ in the direction (0*kk*). Figure adapted from [28]

$^{151}$Eu is a Mössbauer active nuclide. This allows to determine the Eu partial phonon density of states, averaged over the Brillouin zone, by nuclear inelastic scattering. For type-I $Eu_8Ga_{16}Ge_{30}$, a broad feature was observed at 25 K, and modelled with one Gaussian contribution for Eu at the 2*a* site (in the $E_{20}$ cage) and three Gaussian contributions for Eu at the 6*d* site (in the $E_{24}$ cage), all with energies below 8 meV [35]. Thus, there is no sign of participation of the Eu vibrational modes in higher-energy optical modes of the framework at this temperature. The fact that more than two contributions are needed to model the data was taken as evidence for the anisotropy of the potential in the $E_{24}$ cage.

This anisotropy was also studied by Mössbauer spectroscopy at temperatures down to 32 mK. The absence of magnetic hyperfine splitting in the Eu 6*d* derived signal was attributed to incoherent tunelling of Eu between the split sites in the $E_{24}$ cage [36]. The characteristic tunneling frequency of about 450 MHz was also confirmed by microwave absorption measurements [36].

Raman-scattering spectroscopy of type-I $Eu_8Ga_{16}Ge_{30}$ can only be used to explore the rattling modes of Eu at the 6*d* site because the vibrational modes of Eu at the 2*a* site are Raman inactive. The Eu 6*d* rattling mode was first detected in polycrystalline samples at room temperature at about 21 $cm^{-1}$ (2.6 meV) [37]. Later, temperature-dependent studies on single crystals, together with first-principle calculations, allowed for a full assignment of all Raman active modes of type-I $Eu_8Ga_{16}Ge_{30}$. Also, two additional low-energy modes were detected and argued to arise from the split positions and the associated symmetry lowering at the 6*d* site [38].

Further evidence for off-center positions of Eu in the $E_{24}$ cage comes from extended X-ray absorption fine structure (EXAFS) experiments. They reveal that

Eu is on-center in the $E_{20}$ cage, but off-center by about 0.45 Å in the $E_{24}$ cage [39]. The Einstein temperatures estimated from these investigations are 80 K for Eu in the $E_{20}$ cage and 95 K for the shortest bond in the $E_{24}$ cage. The pair distribution functions of the longer Eu-cage atom bonds in the $E_{24}$ cage are strongly broadened and only weakly temperature dependent, indicating strong static disorder for these distances [39].

### *9.2.5 Magnetic Properties*

Both modifications of $Eu_8Ga_{16}Ge_{30}$ have been shown to order ferromagnetically early on [15, 24]. The Curie temperature $T_C$ was reported to be about 35 K (33 K in [15], 36 K in [24]) for type-I $Eu_8Ga_{16}Ge_{30}$, and 10.5 K for type-VIII $Eu_8Ga_{16}Ge_{30}$ [24]. An analysis of the entropy below the respective transition temperatures revealed both transitions to be bulk effects [24]. Single-crystal neutron structure refinements revealed that, at 12 K and in zero applied magnetic field, the magnetization is directed along the principle axis of the cubic crystal [15]. For both modifications, no appreciable hysteresis was found in the magnetic state, pointing to the soft nature of the ferromagnetism, with very small coercive fields of the order of 1–10 G [15, 24]. At low temperatures and in fields of a few T, the saturation magnetization reaches the free-ion value of $7\mu_B$/Eu ion [15, 24]. At high temperatures and in small fields, Curie-Weiss behavior with an effective moment close to the free ion value of $7.93\mu_B$/Eu is observed [15, 24].

The mechanism for the magnetic ordering was suggested to be the indirect exchange interaction between the magnetic moments of the $Eu^{2+}$ ions mediated by the conduction electron spin polarization (Ruderman-Kittel-Kasuya-Yosida or RKKY interaction $J_{RKKY}$) [15, 24]. This interaction oscillates as a function of the distance between the nearest neighbor moments, leading to a ferromagnetic coupling at small distances, but to an antiferromagnetic coupling at larger distances. The period of the oscillation also depends on the charge carrier concentration $n$. Nevertheless, both type-I and type-VIII $Eu_8Ga_{16}Ge_{30}$ samples were found to have Curie temperatures that were almost independent of $n$ [17]. To understand this behavior, the $J_{RKKY}(n)$ dependence was calculated for type-I and type-VIII $Eu_8Ga_{16}Ge_{30}$, using a simple free-electron model and taking more than 1,000 nearest Eu atoms into account. It was found that $J_{RKKY}(n)$ is very similar for both modifications [17]. The similar Curie temperatures within the type-I and type-VIII $Eu_8Ga_{16}Ge_{30}$ sample series were attributed to the fact that all samples lie in the vicinity of the extremum of the first oscillation of $J_{RKKY}(n)$ [17]. The fact that $T_C$ is about three times larger for type-I than for type-VIII $Eu_8Ga_{16}Ge_{30}$, on the other hand, had to be attributed [17] to the effective mass ($m^*$) of the charge carriers being about three times higher in type-I than in type-VIII $Eu_8Ga_{16}Ge_{30}$ [18]. Since $J_{RKKY} \propto m^*$, higher $m^*$ will lead to higher $T_C$.

The ferromagnetic order persists even if Eu is partially substituted by non-magnetic atoms. This is illustrated in Table 9.1. The observed reduction of $T_C$ in

**Table 9.1** Curie temperatures $T_C$ of Eu containing clathrates with the corresponding references

| Composition | $T_C$ (K) | Reference |
|---|---|---|
| type-I $Eu_8Ga_{16}Ge_{30}$ | ≈ 35 | Sales et al. [15], Paschen et al. [24] |
| type-VIII $Eu_8Ga_{16}Ge_{30}$ | 10.5 | Paschen et al. [24] |
| type-VIII $Eu_8Ga_{16}Ge_{30}$ | 13 | Phan et al. [40] |
| type-I $Eu_6Sr_2Ga_{16}Ge_{30}$ | 20 | Woods et al. [41] |
| type-I $Eu_4Sr_4Ga_{16}Ge_{30}$ | 15 | Woods et al. [41] |
| type-I $Eu_2K_6Cd_5Ge_{41}$ | 9.3 | Paschen et al. [42] |
| type-I $Eu_2K_6Zn_5Ge_{41}$ | 4.0 | Paschen et al. [42] |
| type-I $Eu_2K_6Ga_{10}Ge_{36}$ | 3.8 | Paschen et al. [42] |
| type-I $Eu_2Ba_6Al_8Si_{36}$ | 32 | Mudryk et al. [43] |
| type-I $Eu_2Ba_6Cu_4Si_{42}$ | 5 | Mudryk et al. [43] |
| type-I $Eu_2Ba_6Cu_4Ga_4Si_{38}$ | 4 | Mudryk et al. [43] |

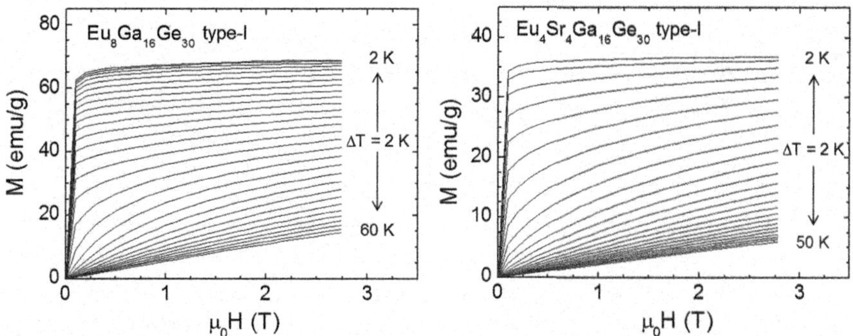


**Fig. 9.8** Magnetization versus field isotherms, $M(\mu_0 H)$, measured between (*left*) 2 and 60 K for the type-I clathrate $Eu_8Ga_{16}Ge_{30}$ and (*right*) 2 and 50 K for type-I $Eu_4Sr_4Ga_{16}Ge_{30}$. Reprinted with permission from Phan et al. [44]. Copyright (2011) American Institute of Physics

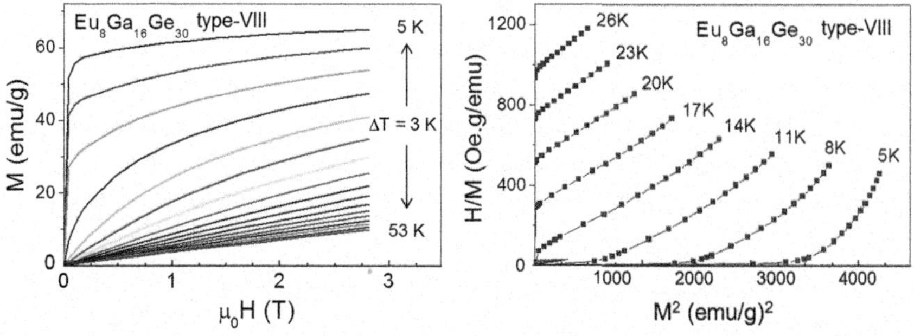


**Fig. 9.9** (*left*) Magnetization isotherms measured between 5 and 53 K for the type-VIII clathrate $Eu_8Ga_{16}Ge_{30}$. (*right*) Arrott plots for representative temperatures around $T_C$ for type-VIII $Eu_8Ga_{16}Ge_{30}$. Reprinted with permission from Phan et al. [40]. Copyright (2011) American Institute of Physics

**Fig. 9.10** Magnetization versus temperature, *M*(*T*), curves taken at 0.01 mT for type-I $Eu_8Ga_{16}Ge_{30}$ and $Eu_4Sr_4Ga_{16}Ge_{30}$ and at 10 mT for type-VIII $Eu_8Ga_{16}Ge_{30}$. No difference can be discerned between heating and cooling curves (same symbols). Data on the type-I clathrates are from Phan et al. [44]

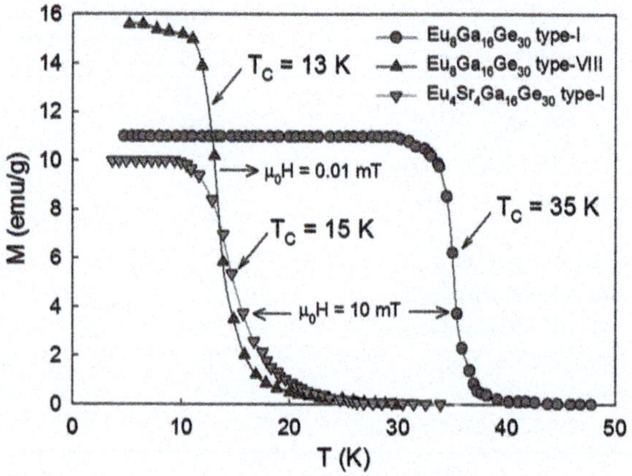

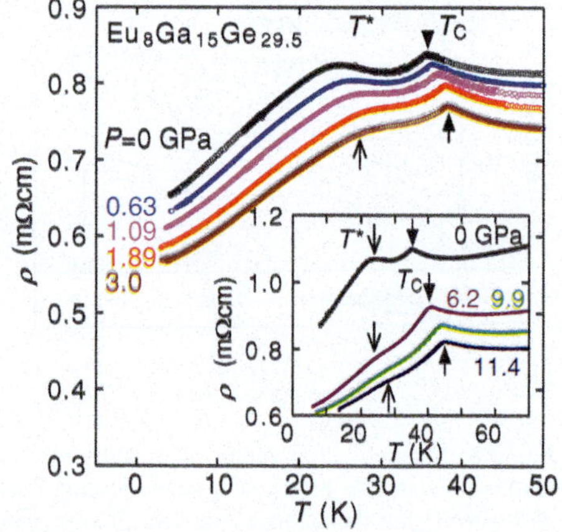


**Fig. 9.11** Temperature dependence of the electrical resistivity of $Eu_8Ga_{15}Ge_{29.5}$ under pressure. Reprinted with permission from [48]

diluted systems is likely due to an enhanced mean Eu–Eu distance, in combination with a variation of $n$ and $m^*$.

Examples of magnetization ($M$) versus magnetic field ($H$) isotherms are shown in Fig. 9.8 for type-I $Eu_8Ga_{16}Ge_{30}$ (left) and type-I $Eu_4Sr_4Ga_{16}Ge_{30}$ (right) [44], and in Fig. 9.9 (left) for type-VIII $Eu_8Ga_{16}Ge_{30}$ [40]. The above discussed soft magnetic behavior is clearly revealed.

The nature of the magnetic phase transition of type-VIII $Eu_8Ga_{16}Ge_{30}$ has been analyzed with the help of $H/M$ versus $M^2$ plots, known as Arrott plots (Fig. 9.9, right) [40]. According to the Banerjee criterion [45], a magnetic transition is of second order if all $H/M$ versus $M^2$ curves have a positive slope. By contrast, if some of these curves show a negative slope at some point, the transition is of first order [45, 46]. Based on this criterion, the magnetic transition in type-VIII $Eu_8Ga_{16}Ge_{30}$ is of second order. The same analysis was also done for type-I $Eu_8Ga_{16}Ge_{30}$ and $Eu_4Sr_4Ga_{16}Ge_{30}$ and revealed second order paramagnetic to ferromagnetic transitions also here [47]. The second-order nature of these magnetic transitions is consistent with the absence of a thermal hysteresis in *M* versus *T* curves measured upon heating and cooling, as shown in Fig. 9.10, as well as with the observed scaling behavior [47].

The effect of pressure on the magnetism of Eu containing clathrates has been studied by means of electrical resistivity and Hall effect measurements [43, 48]. From pressure-dependent electrical resistivity measurements on type-I $Eu_2Ba_6Al_8Si_{36}$, $T_C$ was shown to increase from 32.8 K at ambient pressure to 36.5 K at 16 kbar; this was interpreted as evidence for local-moment magnetism [43]. In the RKKY picture this behavior suggests that pressure moves the system towards the extremum of the first (ferromagnetic) oscillation of $J_{RKKY}$.

The type-I clathrate $Eu_8Ga_{15}Ge_{29.5}$ was investigated up to 11.4 GPa [48]. At ambient pressure, its temperature-dependent resistivity shows a peak at $T_C = 36$ K and a broader feature at $T^* = 23$ K (Fig. 9.11). Upon pressurizing, both the features at $T_C$ and $T^*$ are shifted to higher temperatures. The peak at $T_C$ remains sharp whereas the feature at $T^*$ gets successively washed out up to the highest pressure of 11.4 GPa.

Also other studies of type-I $Eu_8Ga_{16}Ge_{30}$ have provided evidence for a second feature at a characteristic temperature below $T_C$. For instance, the magnetic specific heat, in addition to the lambda-type anomaly at $T_C$, shows a broad shoulder at about 10 K [10]. This anomaly is more clearly revealed by the temperature dependence of the magnetic entropy change [47] that will be further discussed in the context of magnetic refrigeration in the next section.

### 9.2.6 Magnetocaloric Properties

Materials that exhibit large magnetic entropy changes over a wide temperature range possess a large refrigeration capacity, and are therefore of interest for magnetocaloric applications. Type-I and type-VIII $Eu_8Ga_{16}Ge_{30}$ clathrates are two such materials, as will be shown in this section, after a brief introduction to the magnetic refrigeration process.

Magnetic refrigeration based on the magnetocaloric effect (MCE) is receiving increased attention as an alternative to conventional techniques. Whereas conventional gas-compression refrigerators operate at about 40 % of the Carnot efficiency [49, 50], the efficiency of magnetic refrigerators containing Gd is 60 % [49]. MCE is an environmentally friendly technology that does not use hazardous chemicals, green house gasses, or ozone depleting materials [51]. MCE, also known as magnetic cooling or adiabatic demagnetization, is a magneto-thermodynamic phenomenon in which an externally applied changing magnetic field results in a reversible change in temperature of a given material [51]. It is instructive to think of such materials as having two heat reservoirs: phonon excitations associated with the lattice vibrations, and magnetic excitations related to the spin degrees of freedom. The four stages of a typical MCE cycle [49] are depicted in Fig. 9.12.

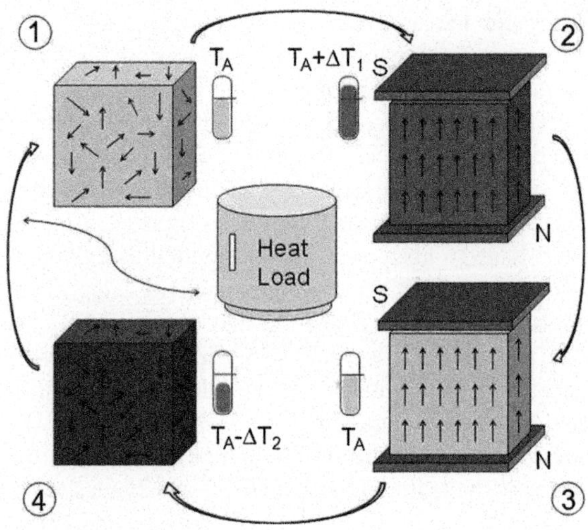


**Fig. 9.12** Schematic depiction of the four stages of the MCE

In the first stage (1 → 2) a magnetocaloric substance with randomly oriented magnetic moments at ambient temperature $T_A$ is subjected to an external magnetic field. In this stage the material undergoes an adiabatic magnetization, that is, the initially randomly oriented magnetic moments are aligned by the magnetic field with no heat exchanged through the process. Since the entropy change, $\Delta S_M$, in a system at temperature $T$ absorbing an infinitesimal amount of heat $\delta Q$ is given by $\Delta S_M = \delta Q/T$, the net result is heating up of the specimen to a temperature of $T_A + \Delta T_1$. In the second stage (2 → 3), known as isomagnetic enthalpic transfer, heat is removed from the system by fluid or gas while the magnetic field is unchanged, bringing the system back to $T_A$. Adiabatic demagnetization is performed in the third stage (3 → 4) where the direction of the magnetic moments is randomized by removing the magnetic field, which results in cooling the specimen at a temperature below the ambient, $T_A - \Delta T_2$. In the final stage (4 → 1) the magnetocaloric substance is placed in thermal contact with the environment being cooled and heat migrates to the working material.

With their large magnetic moments and large change in magnetization at relatively low magnetic fields type-I and type-VIII $Eu_8Ga_{16}Ge_{30}$ are of interest for magnetocaloric applications at modest fields.

The magnitude of the MCE and its dependence on temperature and magnetic field is intimately related to the nature of the corresponding magnetic phase transition [50]. As discussed in Sect. 9.2.5, both type-I and type-VIII $Eu_8Ga_{16}Ge_{30}$ show second order paramagnetic to ferromagnetic phase transitions. In particular, the magnetization curves do not display any thermal hysteresis. This is beneficial for active magnetic refrigeration since there is no heat loss in the processes 1 → 2 and 3 → 4 in the MCE cycle shown in Fig. 9.12.

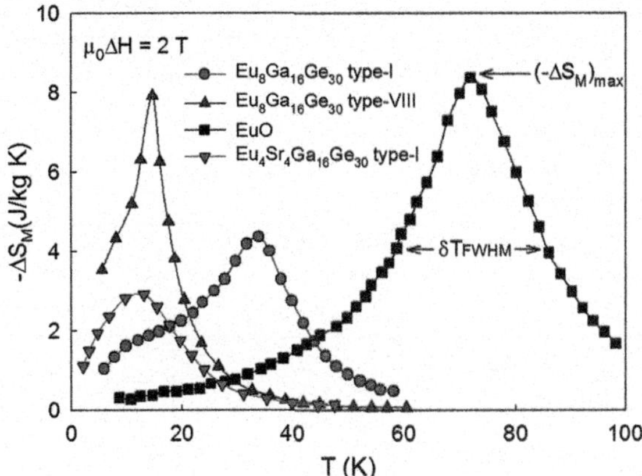


**Fig. 9.13** Magnetic entropy change ($\Delta S_M$) as a function of temperature ($T$) for $Eu_8Ga_{16}Ge_{30}$ and $Eu_4Sr_4Ga_{16}Ge_{30}$ extracted from *M–H–T* curves via the Maxwell Eq. 9.9

The entropy change ($\Delta S_M$) of a given material can be calculated from the family of *M-H* isotherms using the Maxwell's equation [52]

$$\Delta S_M = \mu_0 \int_0^{H_{max}} \left(\frac{\delta M}{\delta T}\right)_H dH \tag{9.9}$$

where $\mu_0$ is the magnetic permeability in vacuum and $H_{max}$ is the maximum magnetic field when the external magnetic field is switched on (Fig. 9.12) [51, 52].

Figure 9.13 shows the temperature dependence of $-\Delta S_M$ for $\mu_0 \Delta H = 2$ T for type-I and type-VIII $Eu_8Ga_{16}Ge_{30}$ and for type-I $Eu_4Sr_4Ga_{16}Ge_{30}$, calculated from the *M*(*H*) isotherms in Figs. 9.8 and 9.9 (left), using Eq. 9.9. For comparison, $\Delta S_M$ of EuO, one of the best candidate materials for magnetic refrigeration at 70 K [53], is also given in Fig. 9.13. The four materials exhibit peaks at their respective Curie temperatures (Fig. 9.13).

For type-I $Eu_8Ga_{16}Ge_{30}$ an additional shoulder is visible at about 10 K. This feature was suggested to be due to a second magnetic transition resulting from interactions between Eu-atoms occupying the 2*a* and 24 *k* (split) site [47].

The refrigeration capacity (*RC*) of a material can be calculated as [52, 54]

$$RC = -\int_{T_1}^{T_2} \Delta S_M(T) dT. \tag{9.10}$$

*RC* is a measure of the heat that is transferred from the cold end to the hot end of a refrigerator in an ideal thermodynamic cycle. *RC* is typically approximated as [54, 55]

$$RC \approx [-\Delta S_M(T)]_{\max} \times \delta T_{FWHM} \tag{9.11}$$

**Table 9.2** $-\Delta S_M$ and $RC$ for type-I and type-VIII Eu-clathrates and EuO

| Composition | $T_{max}$ (K) | $-\Delta S_M$(J/kg K) | $RC$ (J/kg) |
|---|---|---|---|
| EuO | 73 | 8.5 | 120 |
| Type-I $Eu_8Ga_{16}Ge_{30}$ | 13 | 4.5 | 80 |
| Type-VIII $Eu_8Ga_{16}Ge_{30}$ | 35 | 8 | 60 |
| Type-I $Eu_4Sr_4Ga_{16}Ge_{30}$ | 15 | 3 | 49 |

$T_{max}$ is the temperature at $(-\Delta S_M)_{max}$

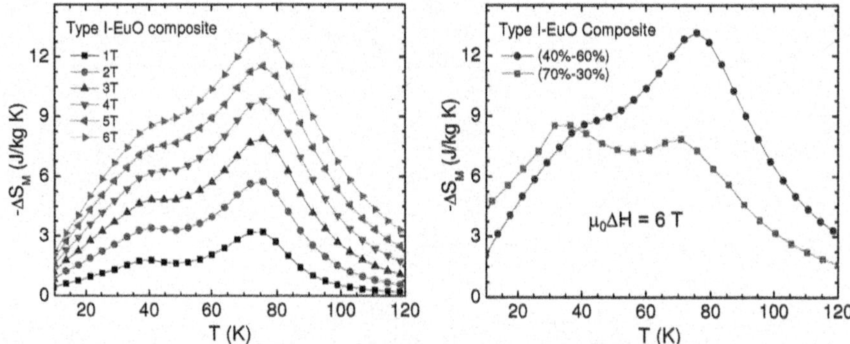


**Fig. 9.14** Temperature dependence of $-\Delta S_M$ at different fields for the (*left*) 40–60 % $Eu_8Ga_{16}Ge_{30}$ type-I clathrate-EuO composite and for the (*right*) 40–60 % and 70–30 % composites. Reprinted with permission from Chaturvedi et al. [55]. Copyright (2011) American Institute of Physics

where $[\Delta S_M(T)]_{max}$ is the maximum entropy change and $\delta T_{FWHM}$ is the full width at half maximum (FWHM) of the $\Delta S_M(T)$ curve (Fig. 9.13) [52, 54–56]. The $-\Delta S_M$ and $RC$ values for the Eu-clathrates and EuO, calculated from the $\Delta S_M(T)$ curves in Fig. 9.13, are given in Table 9.2.

The type-I clathrate possesses a larger $RC$ than the type-VIII clathrate, even though the magnitude of $\Delta S_M$ of the former is about half that of the latter. The reason for this is that the type-I clathrate undergoes a second magnetic transition at 10 K in addition to the ferromagnetic transition at 35 K (Fig. 9.13), resulting in a broadened $\Delta S_M(T)$ curve at low temperatures and consequently an enhanced $RC$. Since $RC$ is effectively the surface area underneath the $\Delta S_M(T)$ curve and the type-I $Eu_8Ga_{16}Ge_{30}$ has a larger surface area than the type-VIII $Eu_8Ga_{16}Ge_{30}$ (Fig. 9.13), the $RC$ value for the type-I clathrate is higher than that of the type-VIII clathrate.

A large reversible MCE and enhanced $RC$ were observed in a composite material composed of the type-I $Eu_8Ga_{16}Ge_{30}$ and EuO. This composite material has a relatively large $RC$ value compared to that of the compositions in Table 9.2. By adjusting the $Eu_8Ga_{16}Ge_{30}$ to EuO ratio, composites with a relatively constant $-\Delta S_M$ over a wide temperature range were produced, desirable for ideal Ericsson-cycle magnetic refrigeration [55]. Figure 9.14 (left) shows the temperature dependence of $-\Delta S_M$ for different applied field changes, up to 6 T, for the

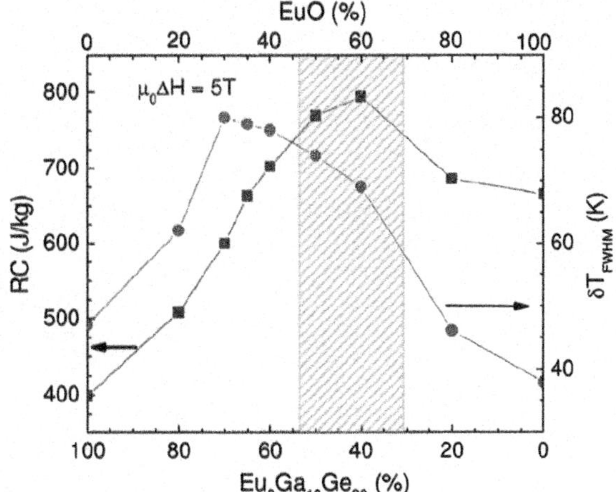


**Fig. 9.15** $RC$ and $\delta T_{FWHM}$ for type-I $Eu_8Ga_{16}Ge_{30}$-EuO composites as a function of the weight percent of the constituents. Reprinted with permission from Chaturvedi et al. [55]. Copyright (2011) American Institute of Physics

40–60 % type-I-EuO composite. Figure 9.14 (right) shows the temperature dependence of $-\Delta S_M$ for the field change of $\mu_0\Delta H = 6$ T for the 40–60 % and 70–30 % composites. From this figure one can see that for a proper ratio between the clathrate and the oxide, the $-\Delta S_M$ curves are not only broadened, but also large values of $[-\Delta S_M]_{max}$ are retained in the composite specimens.

A relatively constant $-\Delta S_M$ over a wide temperature range is advantageous for active magnetic refrigeration. At sufficiently high magnetic fields, this is achieved in the multiphase composite materials composed of type-I $Eu_8Ga_{16}Ge_{30}$ and EuO (Fig. 9.14). By tuning the ratio between the clathrate and the oxide, $RC$ values higher than those of the clathrate and the oxide individually could be achieved. This is a result of the broadening of the $-\Delta S_M$ curves which in turn results in higher $RC$ values. The $RC$ as a function of the weight percent of the clathrate and the oxide is shown in Fig. 9.15. The $RC$ values of the composite specimens are greater than that of $Eu_8Ga_{16}Ge_{30}$. The 40–60 % composite (Fig. 9.15) shows the largest $RC$ value ($\approx$800 J/kg for $\mu_0 H = 5$ T) among the different compositions shown in the figure, and is larger than that of EuO ($\approx$ 665 J/kg for $\mu_0 H = 5$ T). The $RC$ of the 65–35 % composite is almost equal to that of EuO, however the fact that relatively constant $-\Delta S_M$ is relatively high over a wide temperature range for this composite makes it a better choice for Ericsson-cycle based magnetic refrigeration [55].

A detailed comparison of the $RC$ values for the 40–60 % composite with other magnetocaloric candidate materials for active magnetic refrigeration in the temperature range of 10–100 K is given in Fig. 9.16. The composite shows the highest $RC$ value in this group of magnetocaloric materials while possessing nearly zero thermal and field hysteresis losses, making it one of the best candidate materials for active magnetic refrigeration at 70 K [55].

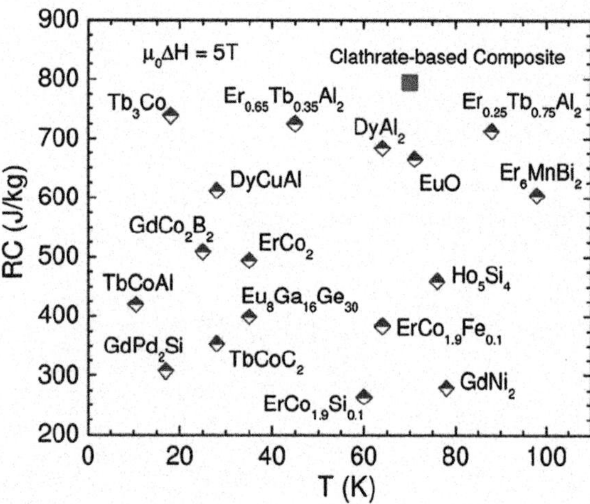


**Fig. 9.16** The *RC* values of the 40–60 % type-I $Eu_8Ga_{16}Ge_{30}$-EuO composite and other magnetocaloric candidate materials in the temperature interval of 10–100 K. Reprinted with permission from Chaturvedi et al. [55]. Copyright (2011) American Institute of Physics

## 9.3 Other Rare-Earth-Containing Clathrates

Rare-earth-containing clathrates other than type-I and type-VIII $Eu_8Ga_{16}Ge_{30}$ have also been studied. In the large majority of these investigations, Eu remains the only 4*f* element which is partially replaced by Sr or Ba. The magnetic properties of type-I $Sr_{8-x}Eu_xGa_{16}Ge_{30}$ were already discussed in Sect. 9.2.6. Also substitutions on the framework of Eu-based type-I clathrates have been investigated. In Sect. 9.3.1, Ge-based clathrates are discussed, in Sect. 9.3.2 a very brief survey of Si-based clathrates is given, a topic that is treated in greater depth in Chap. 8. This section also contains the discussion of a P-based type-I clathrate. The few examples where rare-earth elements other than Eu could be incorporated into a type-I clathrate will be discussed in Sect. 9.3.3. For properties of the chiral clathrate $Ba_{6-x}Eu_xGe_{25}$ (*cP*124) [57–59], the compound $EuGa_2Ge_4$ (*oC*28) [60–64] with a composition very close to $Eu_8Ga_{16}Ge_{30}$, as well as intermetallic $R_3Pd_{20}X_6$ ($RE$ = rare earth, $X$ = Si, Ge; cubic space group $Fm3\,m$) phases [65–68], that are sometimes referred to as "clathrate crystals" in the literature, the reader is referred to the original literature.

### 9.3.1 *Partial Substitution of Eu in Ge-based Type-I Clathrates*

Partial substitution of Sr for Eu has been observed in the pioneering work of Cohn et al. [6], where the type-I clathrate $Sr_4Eu_4Ga_{16}Ge_{30}$ was investigated. Below 40 K, the lattice thermal conductivity of $Sr_4Eu_4Ga_{16}Ge_{30}$ was shown to be distinctly reduced with respect to $Eu_8Ga_{16}Ge_{30}$, being two times smaller at 6 K, the lowest temperature of that experiment. Synchrotron X-ray powder diffraction showed that in spite of the small difference of the covalent radii of Eu (185 pm) and Sr (192 pm), Eu occupies preferentially (with an occupancy of about 76 %) the smaller $E_{20}$ cage [69].

A comparative study of $Ba_{8-x}Eu_xAu_6Ge_{40}$ ($x = 1, 2$) and $Ba_{8-x}Sr_xAu_6Ge_{40}$ ($x = 0, 1, 2, 4$) revealed that the lattice thermal conductivity is greatly reduced by substitution with Eu [70]. $Ba_{8-x}Eu_xAu_6Ge_{40}$ was shown to have the highest $ZT$ between 200 and 900 K of both sample series, with $ZT > 0.5$ above 670 K.

With $K_6Eu_2Ga_{10}Ge_{36}$, $K_6Eu_2Zn_5Ge_{41}$, and $K_6Eu_2Cd_5Ge_{41}$, three new type-I clathrates were found [42]. Single crystal structure refinements revealed that the Eu atoms occupy the smaller $E_{20}$ cage, thus leading to a fully ordered arrangement of the guest atoms. All three compounds order feromagnetically at low temperatures (at 3.8, 4.0, and 9.3 K, respectively). The main purpose of the study was to use mono-valent K to push Eu away from its stable divalent state. However, both magnetization and Eu $L_{III}$ X-ray absorption measurements revealed that this could not be achieved and that Eu stayed in its apparently very stable divalent state [42].

Finally, by high-pressure synthesis, Eu was also incorporated in the binary defect clathrate $Ba_8Ge_{43}\square_3$, resulting in the compounds $Ba_{8-x}Eu_xGe_{43}\square_3$ with $x = 0.3$ and 0.6 [71]. The superstructure revealed in $Ba_8Ge_{43}\square_3$ [72] was absent in these samples [71]. Eu was found to preferentially occupy the smaller $E_{20}$ cages. Also here, the magnetic susceptibility revealed that Eu is divalent. Both samples are metallic and remain paramagnetic down to at least 350 mK [73].

Thus, in mixed Eu–K, Eu–Ba, and Eu-Sr type-I clathrates, Eu always preferentially or even fully occupies the smaller $E_{20}$ cage. Nevertheless, the hybridization with host states is not strong enough to induce any sizable moment reduction that might suggest an appreciable Kondo interaction. In Si-based clathrates, the cages are smaller and a stronger guest-host interaction can be anticipated.

### *9.3.2 Eu-containing Si-based Type-I Clathrates*

Most of the early studies on Si-based clathrates focussed on superconductivity [74]. Today, the quest for inexpensive materials for mass-market applications of thermoelectric converters has moved the focus to the thermoelectric properties of Si-based type-I clathrates (see Chap. 8 for more details).

Concerning Eu-containing representatives, the situation is quite different from the Ge-based case discussed above. In none of the Si-based clathrates studied so far, the incorporation of Eu has improved the thermoelectric properties appreciably [43, 75–78].

The solid solubility of Eu appears to be limited to two Eu per formula unit. It was only reached in the compounds $Ba_6Eu_2Ga_8Si_{36}$, $Ba_6Eu_2Al_8Si_{36}$, $Ba_6Eu_2Cu_4Si_{42}$, and $Ba_6Eu_2Cu_4Ga_4Si_{38}$ [43, 75]. As in the Ge-based clathrates, Eu tends to occupy the smaller $E_{20}$ cage. A fully ordered guest atom arrangement, with Eu in the $E_{20}$ and Ba in the $E_{24}$ cage, was revealed from single crystal X-ray data for all four above compounds [43, 75].

The distance between Eu and the nearest neighbor cage atom, $d_{NN}$, is typically smaller in Si- than in Ge-based clathrates. For instance, $d_{NN} = 3.284$ Å for $Ba_6Eu_2Cu_4Ga_4Si_{38}$ [43], and $d_{NN} = 3.376$ Å for $Ba_{7.4}Eu_{0.6}Ge_{43}\square_3$ [71]. Nevertheless, Eu remains fully divalent also in Si-based clathrates, as shown by Eu $L_{III}$ X-ray absorption measurements on $Ba_6Eu_2Cu_4Si_{42}$ at 10 and 300 K [43]. The application of pressure up to 18 kbar was shown to enhance the magnetic ordering temperature of $Ba_6Eu_2Al_8Si_{36}$ [43]. Thus, also in Si-based clathrates with the rare-earth element Eu there is no sign of Kondo interaction.

Even smaller $d_{NN}$ values are found in $Ba_{8-x}Eu_xCu_{16}P_{30}$ ($x = 0.5, 1, 1.4$), which crystallizes in an orthorhombic superstructure (space group *Pbcn* with a fully ordered framework) of the type-I clathrate structure [79]. Here, Eu occupies only the smaller $E_{20}$ cages (there are two different ones, $Cu_6P_{14}$ and $Cu_8P_{12}$, in the superstructure). For $Ba_{6.6}Eu_{1.4}Cu_{16}P_{30}$, a very small distance $d_{NN} = 3.125$ Å was found. Nevertheless, also in this compound, Eu $L_{III}$ X-ray absorption measurements give no indication for mixed-valent behavior of Eu [79]. The low, presumably ferromagnetic ordering temperatures of 2.9 and 3.1 K for $Ba_7EuCu_{16}P_{30}$ and $Ba_{6.6}Eu_{1.4}Cu_{16}P_{30}$, respectively, are most likely due to the small amount of Eu in the clathrate. Interestingly, in contrast to $Eu_8Ga_{16}Ge_{30}$ [80], the ordering temperatures decrease with increasing magnetic field, suggesting that a field-induced quantum critical point [81, 82] might be accessed at about 1 T. Clearly, measurements at lower temperatures are needed to test this conjecture.

### 9.3.3 Type-I Clathrates Containing Other Rare-Earth Elements than Eu

Many attempts have been undertaken to synthesize clathrates with rare-earth elements other than Eu. At least some of them were motivated by the idea [10, 42, 83–85] to create a "strongly correlated" clathrate that might show a correlation enhanced thermopower similar to strongly correlated $4f$ electron systems. Most of the synthesis attempts were not successful and typically remained unpublished. A noticeable exception is the promising case of $Ba_8Cu_{16}P_{30}$ where the failure to substitute any amount of Ba by Ce, Sm, or Yb was explicitly described [79]. The assumed synthesis of $Ba_6Ce_2Au_4Si_{42}$ [86] was taken up with great interest by the scientific community. Unfortunately, shortly later it was demonstrated [87] and acknowledged by the authors of the original paper [88] that Ce was only present in a foreign phase and not in the main clathrate phase. Another example where a failed synthesis was reported is $Eu_{8-x}Yb_xGa_{16}Ge_{30}$ [85]. The proof that the sizable amount of Yb detected by energy dispersive X-ray spectroscopy (EDX) was not present in the type-I clathrate phase $Eu_8Ga_{16}Ge_{30}$ but in the closely related phase $EuGa_2Ge_4$ was only possible with an advanced procedure: samples for transmission electron microscopy investigations were cut out from the areas where Yb was detected by EDX with a focussed ion beam in a dual beam scanning electron microscope. The symmetry of the Yb containing phase was shown to be tetragonal by selected area electron diffraction in a transmission electron microscope, in agreement with this phase being $Eu_{1-x}Yb_xGa_2Ge_4$ but not $Eu_{8-x}Yb_xGa_{16}Ge_{30}$ [85].

Shortly later, the successful synthesis of $Ba_{8-x}Yb_xGa_{16}Ge_{30}$ ($x = 0.3, 0.5, 0.7, 1$) was reported [89]. The motivation for this study was to further lower the lattice thermal conductivity of type-I clathrates by using a heavier guest atom (Yb instead of Eu, Ba, Sr) and by employing the double-filling technique (mixed filling with Ba and Yb) [89]. Indeed, the authors were right with their idea: the lattice thermal conductivity of $Ba_{7.3}Yb_{0.7}Ga_{16}Ge_{30}$ was more than a factor of 2 smaller than that of $Ba_8Ga_{16}Ge_{30}$. The thermopower, however, decreased with increasing $x$ [89]. This suggests that correlation effects due to Yb 4*f* electrons are absent in this material, which may hint at a two-valent state of Yb.

Also density functional theory was used to support the search for new rare-earth clathrates. Zhu et al. [90] calculated the binding energy and chemical reaction drive energy for type-I $(Ba,RE)_8Ga_xGe_{46-x}$ clathrates and compared them to the respective energies of the competing phases $(Ba,RE)Ge_2$ and Ge (diamond phase). They found that the increase of the total number of valence electrons upon substitution of two-valent Ba by a three-valent *RE* atom destabilizes the clathrate structure and favors the formation of the foreign phase $(Ba,RE)Ge_2$. This trend is weakened with increasing $x$ because Ga reduces the number of valence electrons. The compounds $Sm_8Ga_{16}Ge_{30}$, $Pm_8Ga_{16}Ge_{30}$, and $Nd_8Ga_{16}Ge_{30}$ are predicted to be stable against the formation of the impurity phases. A partial replacement of Ba is predicted to be possible also for other *RE* elements [90].

Other theoretical studies have postulated the existence of $Ba_6La_2Au_6Ge_{40}$ [91], $Yb_8Ga_{16}Ge_{30}$ [92], and $Ba_6Yb_2Au_6Ge_{40}$ [93] and calculated their electronic band structure and thermoelectric properties. $Ba_6La_2Au_6Ge_{40}$ was suggested to have a broader band gap and an enhanced band mass compared to $Ba_8Au_6Ge_{40}$ [91]. An interesting prediction is that *n*-type $Ba_6Yb_2Au_6Ge_{40}$ would have a much larger (negative) thermopower than *n*-type $Ba_8Au_6Ge_{40}$ [93]. For the hypothetical type-I clathrate $Yb_8Ga_{16}Ge_{30}$, a sharp peak, dominated by Yb 4*f* states, was found to be situated in the upper valence band region near the Fermi level [92]. This feature might lead to a strongly enhanced thermopower in *p*-type $Yb_8Ga_{16}Ge_{30}$. The above-described experiments on $Ba_{8-x}Yb_xGa_{16}Ge_{30}$ [89], however, do not seem to support these predictions. Clearly, more experimental and theoretical work is needed to clarify the situation.

Very recently, the successful synthesis of the first Ce- and La-containing clathrates was reported [94]. Single-phase single crystals of the approximate compositions $Ba_7CeAu_6Si_{40}$ and $Ba_7LaAu_6Si_{40}$ were obtained by off-stoichiometric growth using the floating-zone melting technique in a 4-mirror furnace. Both compounds reach much higher thermopower values than reported in the rare-earth-free reference material series $Ba_8Au_xSi_{46-x}$ [95, 96]. For $Ba_7LaAu_6Si_{40}$ this could be attributed to a reduced charge carrier concentration. For $Ba_7CeAu_6Si_{40}$, on the other hand, a 50 % thermopower enhancement over the value expected from the

"Pisarenko" relation $S \sim n^{-2/3}$ for a simple parabolic band of charge carrier concentration $n$ is observed. This was attributed to a rattling-enhanced Kondo interaction [97], a picture supported by additional transport and thermodynamic measurements [94].

## 9.4 Conclusions

Rare-earth clathrates remain a highly active field of research. The potential for both thermoelectric and magnetocaloric applications is well documented, however furtherwork on trivalent rare-earth containing compounds is warranted. Research on both the optimization of the relevant physical properties and aspects related to large scale applications, such as price and long-term stability, is also of interest. We trust that the recent discovery of Ce-containing clathrates with correlation-enhanced thermopower will boost these activities.

**Acknowledgements** SP and MI acknowledge support by the Priority Program "Nanostructured Thermoelectric Materials: Theory, Model Systems and Controlled Synthesis" (SPP 1386) of the Deutsche Forschungsgemeinschaft and from the Austrian Science Fund (projects TRP 176-N22 and I623-N16). SS and GSN acknowledge support from the Army Research Office under Grant No. W911NF-08-1-0276 for research on Eu-clathrates for magnetocaloric applications.

## References

1. J.S. Kasper, P. Hagenmüller, M. Pouchard, C. Cros, Clathrate structure of silicon $Na_8Si_{46}$ and $Na_xSi_{136}$ ($x < 11$). Science **150**, 1713 (1965)
2. M. Von Stackelberg, H.R. Müller, On the structure of gas hydrates. J. Chem. Phys. **19**, 1319 (1951)
3. G.A. Jeffrey, Hydrate inclusion compounds, in *Inclusion Compounds*, 135, ed. by J.L. Atwood, J.E.D. Davies, D.D. MacNicol (Academic Press, New York, 1984)
4. T.C.W. Mak, G.D. Zhou, *Crystallography in Modern Chemistry* (Wiley, New York, 1992)
5. H. Schäfer, On the problem of polar intermetallic compounds: the stimulation of E. Zintl's work for the modern chemistry of intermetallics. Ann. Rev. Mater. Sci. **15**, 1 (1985)
6. J.L. Cohn, G.S. Nolas, V. Fessatidis, T.H. Metcalf, G.A. Slack, Glasslike heat conduction in high-mobility crystalline semiconductors. Phys. Rev. Lett. **82**, 779 (1999)
7. G.S. Nolas, J.L. Cohn, G.A. Slack, S.B. Schujman, Semiconducting Ge clathrates: promising candidates for thermoelectric applications. Appl. Phys. Lett. **73**, 178 (1998)
8. J.S. Tse, M.A. White, Of glassy crystalline behavior in the thermal properties of clathrate hydrates: a thermal conductivity study of tetrahydrofuran hydrate O. J. Phys. Chem. **92**, 5006 (1988)
9. B. Eisenmann, H. Schäfer, R. Zahler, Die Verbindungen $A^{II}_8B^{III}_{16}B^{IV}_{30}$ ($A^{II} \equiv$ Sr, Ba; $B^{III} \equiv$ Al, Ga; $B^{IV} \equiv$ Si, Ge, Sn) und ihre Käfigstrukturen. J. Less-Common Met. **118**, 43 (1986)
10. S. Paschen, W. Carrillo-Cabrera, A. Bentien, V.H. Tran, M. Baenitz, Yu. Grin, F. Steglich, Phys. Rev. B **64**, 214404 (2001)
11. K. Momma, F. Izumi, VESTA 3 for three-dimensional visualization of crystal, volumetric and morphology data. J. Appl. Crystallogr. **44**, 1272 (2011)

12. S. Leoni, W. Carrillo-Cabrera, Y. Grin, Modelling of the $\alpha$ (clathrate VIII) $\rightleftharpoons$ $\beta$ (clathrate I) phase transition in $Eu_8Ga_{16}Ge_{30}$. J. Alloys Compd. **350**, 113 (2003)
13. G.S. Nolas, T.J.R. Weakley, J.L. Cohn, R. Sharma, Structural properties and thermal conductivity of crystalline Ge clathrates. Phys. Rev. B **61**, 3845 (2000)
14. B.C. Chakoumakos, B.C. Sales, D.G. Mandrus, Structural disorder and magnetism of the semiconducting clathrate $Eu_8Ga_{16}Ge_{30}$. J. Alloys Compd. **322**, 127 (2001)
15. B.C. Sales, B.C. Chakoumakos, R. Jin, J.R. Thompson, D. Mandrus, Structural, magnetic, thermal, and transport properties of $X_8Ga_{16}Ge_{30}$ ($X$ = Eu, Sr, Ba) single crystals. Phys. Rev. B **63**, 245113 (2001)
16. G.K.H. Madsen, K. Schwarz, P. Blaha, D.J. Singh, Electronic structure and transport in type-I and type-VIII clathrates containing strontium, barium, and europium. Phys. Rev. B **68**, 125212 (2003)
17. V. Pacheco, A. Bentien, W. Carrillo-Cabrera, S. Paschen, F. Steglich, Yu. Grin, Phys. Rev. B **71**, 165205 (2005)
18. A. Bentien, V. Pacheco, S. Paschen, Y. Grin, F. Steglich, Transport properties of composition tuned $\alpha$- and $\beta$-$Eu_8Ga_{16\text{-}x}Ge_{30+x}$. Phys. Rev. B **71**, 165206 (2005)
19. A. Prokofiev, S. Paschen, H. Sassik, S. Laumann, P. Pongratz, Method for producing clathrate compounds, (2009). patents JP 5248916 (2013), US appl. 12/231,183; Gebrauchsmuster AT 10749 (2009), DE 20 2008 006 946.7
20. S. Laumann, M. Ikeda, H. Sassik, A. Prokofiev, S. Paschen, Melt-spun $Eu_8Ga_{16\text{-}x}Ge_{30+x}$ clathrates. Z. Anorg. Allg. Chem. **638**, 294 (2012)
21. R.C. O'Handley, *Hall Effect Formulae and Units* (Plenum Press, New York, 1980), pp. 417–419
22. E.L. Nagaev, E.B. Sokolova, Anomalous Hall effect in ferromagnetic semiconductors. Sov. Phys. Solid State **19**, 425 (1977)
23. J.M. Ziman, *Electrons and Phonons* (Clarendon press, Oxford, 1960)
24. S. Paschen, B. Wand, G. Sparn, F. Steglich, Y. Echizen, T. Takabatake, J. Magn. Magn. Mater. **226–230**, 57 (2001)
25. A. Bentien, M. Christensen, J.D. Bryan, A. Sanchez, S. Paschen, F. Steglich, G.D. Stucky, B.B. Iversen, Phys. Rev. B **69**, 045107 (2004)
26. J.S. Tse, V.P. Shpakov, V.V. Murashov, V.R. Belosludov, Low frequency vibrations in clathrate hydrates. T. J. Chem. Phys. **107**, 9271 (1997)
27. M. Christensen, A.B. Abrahamsen, N.B. Christensen, F. Juranyi, N.H. Andersen, K. Lefmann, J. Andreasson, C.R.H. Bahl, B.B. Iversen, Nature Mater. **7**, 811–815 (2008)
28. H. Euchner, S. Pailhès, L.T.K. Nguyen, W. Assmus, F. Ritter, A. Haghighirad, Y. Grin, S. Paschen, M. de Boissieu, Phys. Rev. B **86**, 224303 (2012)
29. K. Suekuni, M.A. Avila, K. Umeo, T. Takabatake, Cage-size control of guest vibrations and thermal conductivity in $Sr_8Ga_{16}Si_{30\text{-}x}Ge_x$. Phys. Rev. B **75**, 195210 (2007)
30. K. Suekuni, S. Yamamoto, M.A. Avila, T. Takabatake, Relation between guest free space, U. & lattice thermal conductivity reduction by anharmonic rattling in type I clathrates. J. Phys. Soc. Jpn. **77SA**, 61 (2008)
31. G.S. Nolas, T.J.R. Weakley, J.L. Cohn, Structural, chemical, and transport properties of a new clathrate compound: $Cs_8Zn_4Sn_{42}$. Chem. Mater. **11**, 2470 (1999)
32. Y. Saiga, K. Suekuni, B. Du, T. Takabatake, Thermoelectric properties and structural instability of type-I clathrate $Ba_8Ga_{16}Sn_{30}$ at high temperatures. Solid State Commun. **152**, 1902 (2012)
33. L.T.K. Nguyen, U. Aydemir, M. Baitinger, E. Bauer, H. Borrmann, U. Burkhardt, J. Custers, A. Haghighirad, R. Höfler, K.D. Luther, F. Ritter, W. Assmus, Yu. Grin, S. Paschen, Dalton Trans. **39**, 1071 (2010)
34. H. Zhang, H. Borrmann, N. Oeschler, C. Candolfi, W. Schnelle, M. Schmidt, U. Burkhardt, M. Baitinger, J.-Tai Zhao, Yu. Grin, Inorg. Chem. **50**, 1250 (2011)
35. R.P. Hermann, W. Schweika, O. Leupold, R. Rüffer, G.S. Nolas, F. Grandjean, G.J. Long, Phys. Rev. B **72**, 174301 (2005)

36. R.P. Hermann, V. Keppens, P. Bonville, G.S. Nolas, F. Grandjean, G.J. Long, H.M. Christen, B.C. Chakoumakos, B.C. Sales, D. Mandrus, Phys. Rev. Lett. **97**, 017401 (2006)
37. G.S. Nolas, C.A. Kendziora, Raman scattering study of Ge and Sn compounds with type-I clathrate hydrate crystal structure. Phys. Rev. B **62**, 7157 (2000)
38. Y. Takasu, T. Hasegawa, N. Ogita, M. Udagawa, M.A. Avila, K. Suekuni, I. Ishii, T. Suzuki, T. Takabatake, Phys. Rev. B **74**, 174303 (2003)
39. R. Baumbach, F. Bridges, L. Downward, D. Cao, P. Chesler, B. Sales, Phys. Rev. B **71**, 024202 (2005)
40. M.H. Phan, G.T. Woods, A. Chaturvedi, S. Stefanoski, G.S. Nolas, H. Srikanth, Appl. Phys. Lett. **93**, 252505 (2008)
41. G.T. Woods, J. Martin, M. Beekman, R.P. Hermann, F. Grandjean, V. Keppens, O. Leupold, G.J. Long, G.S. Nolas, Phys. Rev. B **73**, 174403 (2006)
42. S. Paschen, S. Budnyk, U. Köhler, Yu. Prots, K. Hiebl, F. Steglich, Yu. Grin, Phys. B **383**, 89 (2006)
43. Ya. Mudryk, P. Rogl, C. Paul, S. Berger, E. Bauer, G. Hilscher, C. Godart, H. Noël, J. Phys. Condens. Matter **14**, 7991 (2002)
44. M.H. Phan, V. Franco, A. Chaturvedi, S. Stefanoski, H. Kirby, G.S. Nolas, H. Srikanth, J. Appl. Phys. **107**, 09A910 (2010)
45. B.K. Banerjee, On a generalised approach to first and second order magnetic transitions. Phys. Lett. **12**, 16 (1964)
46. J. Mira, J. Rivas, F. Rivadulla, C. Vazquez, M.A. Lopez-Quintela, Change from first- to second-order magnetic phase transition in $La_{2/3}Ca$, $Sr_{1/3}MnO_3$ perovskites. Phys. Rev. B **60**, 2998 (1999)
47. M.H. Phan, V. Franco, A. Chaturvedi, S. Stefanoski, G.S. Nolas, H. Srikanth, Phys. Rev. B **84**, 054436 (2011)
48. K. Umeo, H. Yamane, M.A. Avila, T. Onimaru, T. Takabatake, Pressure effect on the ferromagnetism of an off-center rattling system $Eu_8Ga_{16}Ge_{30}$. J. Phys. Conf. Ser. **391**, 012075 (2012)
49. E. Bruck, Developments in magnetocaloric refrigeration. J. Phys. D. Appl. Phys. **38**, R381 (2005)
50. O. Tegus, E. Brück, L. Zhang, Dagula, K.H. J Buschow, F.R. de Boer, Phys. B **319**, 174 (2002)
51. K.A. Gschneidner, J.V.K. Pecharsky, A.O. Tsokol, Recent developments in magnetocaloric materials. Prog. Phys. **68**, 1479 (2005)
52. A.H. Morrish, *The physical principles of magnetism* (Wiley, New York, 1964)
53. K. Ahn, A.O. Pecharsky, K.A. Gschneidner, V.K. Pecharsky, Preparation, heat capacity, magnetic properties, and the magnetocaloric effect of EuO. J. Appl. Phys. **97**, 063901 (2004)
54. C.E. Reid, J.A. Barclay, J.L. Hall, S. Sarangi, Selection of magnetic materials for an active magnetic regenerative refrigerator. J. Alloys Compd. **366**, 207 (1994)
55. A. Chaturvedi, S. Stefanoski, M.H. Phan, G.S. Nolas, H. Srikanth, Table-like magnetocaloric effect and enhanced refrigerant capacity in $Eu_8Ga_{16}Ge_{30}$-EuO composite materials. Appl. Phys. Lett. **99**, 162513 (2011)
56. S. Gorsse, B. Chevalier, G. Orveillon, Magnetocaloric effect and refrigeration capacity in $Gd_{60}Al_{10}Mn_{30}$ nanocomposite. Appl. Phys. Lett. **92**, 122501 (2008)
57. S. Paschen, V.H. Tran, M. Baenitz, W. Carrillo-Cabrera, Yu. Grin, F. Steglich, The clathrate $Ba_6Ge_{25}$: thermodynamic, magnetic, and transport properties. Phys. Rev. B **65**, 134 (2002)
58. W. Carrillo-Cabrera, R. Cardoso Gil, S. Paschen, Y. Grin, Crystal structure of barium europium germanide, $Ba_{6-x}Eu_xGe_{25}$ ($x = 0.6$), a chiral clathrate. Z. Kristallogr. NCS **218**, 397 (2003)
59. J. Sichelschmidt, W. Carrillo-Cabrera, V.A. Ivanshin, Y. Grin, F. Steglich, spin resonance of $Eu^{2+}$ in the Eu doped clathrate $Ba_6Ge_{25}$. Eur. Phys. J. B **46**, 201 (2005)
60. J.D. Bryan, G.D. Stucky, $Eu_4Ga_8Ge_{16}$: a new four-coordinate clathrate network. Chem. Mater. **13**, 253 (2001)

61. W. Carrillo-Cabrera, S. Paschen, Yu. Grin, $EuGa_{2\pm x}Ge_{4\mp x}$: preparation, crystal chemistry and properties. J. Alloys Compd. **333**, 4 (2002)
62. H. Birkedal, J.D. Bryan, G.D. Stucky, M. Christensen, B.B. Iversen, Magnetic structure and thermal expansion of $Eu_4Ga_8Ge_{16}$, in *Solid-State Chemistry of Inorganic Materials IV*, vol. 755, ed. by M.A. Alario-Franco, M. Greenblatt, G. Rohrer, M.S. Whittingham. Materials Research Society Symposium Proceedings, 363 (2003)
63. J.D. Bryan, H. Trill, H. Birkedal, M. Christensen, V.I. Srdanov, H. Eckert, B.B. Iversen, G.D. Stucky, Phys. Rev. B **68**, 174429 (2003)
64. J. Romhányi, K. Penc, Phys. Rev. B **68**, 174428 (2003)
65. A.V. Gribanov, Y.D. Seropegin, O.I. Bodak, Crystal structure of the compounds $Ce_3Pd_{20}Ge_6$ and $Ce_3Pd_{20}Si_6$. J. Alloys Compd. **204**, L9 (1994)
66. J. Kitagawa, T. Takabatake, E. Matsuoka, F. Takahashi, K. Abe, M. Ishikawa, J. Phys. Soc. Jpn. **71**, 1630 (2002)
67. T. Yanagisawa, N. Tateiwa, T. Mayama, H. Saito, H. Hidaka, H. Amitsuka, Y. Haga, Y.Nemoto, T. Goto, J. Phys. Soc. Jpn. **80**, SA105 (2011)
68. J. Custers, K.-A. Lorenzer, M. Müller, A. Prokofiev, A. Sidorenko, H. Winkler, A.M.Strydom, Y. Shimura, T. Sakakibara, R. Yu, Q. Si, S. Paschen, Nature Mater. **11**, 189 (2012)
69. Y. Zhang, P.L. Lee, G.S. Nolas, A.P. Wilkinson, Gallium distribution in the clathrates $Sr_8Ga_{16}Ge_{30}$ and $Sr_4Eu_4Ga_{16}Ge_{30}$ by resonant diffraction. Appl. Phys. Lett. **80**(16), 2931 (2006)
70. H. Anno, H. Fukushima, K. Koga, K. Okita, K. Matsubara, Effect of guest substitution on thermoelectric properties of clathrate compounds, in *Proceedings of ICT'06: XXV International Conference on Thermoelectrics*, pp. 36–39. IEEE, 25th International Conference on Thermoelectrics (ICT'06), Vienna, 6–10 Aug 2006
71. R. Demchyna, U. Köhler, Yu. Prots, W. Schnelle, M. Baenitz, U. Burkhardt, S. Paschen, U. Schwarz, Z. Anorg. Allg. Chem. **632**, 73 (2006)
72. W. Carrillo-Cabrera, S. Budnyk, Y. Prots, Y. Grin, $Ba_8Ge_{43}$ revisited: a $2a' \times 2a' \times 2a'$ superstructure of the clathrate-I type with full vacancy ordering. Z. Anorg. Allg. Chem. **630**, 2267 (2004)
73. U. Köhler, R. Demchyna, S. Paschen, U. Schwarz, F. Steglich, Schottky anomaly in the low-temperature specific heat of $Ba_{8-x}Eu_xGe_{43}\square_3$. Physica B **378–380**, 263 (2006)
74. H. Kawaji, H. Horie, S. Yamanaka, M. Ishikawa, Superconductivity in the silicon clathrate compound $(Na,Ba)_xSi_{46}$. Phys. Rev. Lett. **74**, 1427–1429 (1995)
75. Ya. Mudryk, P. Rogl, C. Paul, S. Berger, E. Bauer, G. Hilscher, C. Godart, H. Noël, A. Saccone, R. Ferro, Physica B **328**, 44 (2003)
76. K. Koga, K. Suzuki, M. Fukamoto, H. Anno, T. Tanaka, S. Yamamoto, J. Electron. Mater. **38**(7), 1427 (2009)
77. C.L. Condron, S.M. Kauzlarich, G.S. Nolas, Structure and thermoelectric characterization of $A_xBa_{8-x}Al_{14}Si_{31}$ ($A = Sr$, Eu) single crystals. Inorg. Chem. **46**, 2556 (2007)
78. H. Anno, K. Okita, K. Koga, S. Harima, T. Nakabayashi, M. Hokazono, K. Akai, Mater. Trans. **53**, 1220 (2012)
79. K. Kovnir, U. Stockert, S. Budnyk, Yu. Prots, M. Baitinger, S. Paschen, A.V. Shevelkov, Yu. Grin, Inorg. Chem. **50**, 10387 (2011)
80. T. Onimaru, S. Yamamoto, M.A. Avila, K. Suekuni, T. Takabatake, Multiple ferromagnetic structures in an off-center rattling system $Eu_8Ga_{16}Ge_{30}$. J. Phys. Conf. Ser. **200**, 022044 (2010)
81. H. van Löhneysen, A. Rosch, M. Vojta, P. Wölfle, Fermi-liquid instabilities at magnetic quantum critical points. Rev. Mod. Phys. **79**, 1015 (2007)
82. Q. Si, S. Paschen, Quantum phase transitions in heavy fermion metals and Kondo insulators. Phys. Status Solidi B **250**, 425 (2013)
83. S. Paschen, M. Baenitz, V.H Tran, A. Rabis, F. Steglich, W. Carrillo-Cabrera, Yu. Grin, A.M. Strydom, P. de V du Plessis, J. Phys. Chem. Solids **63**, 1183 (2002)

84. S. Paschen, *Thermoelectrics Handbook*, chap. 15 Thermoelectric aspects of strongly correlated electron systems ed. by D.M. Rowe (CRC Press, Boca Raton, 2006)
85. S. Paschen, C. Gspan, W. Grogger, M. Dienstleder, S. Laumann, P. Pongratz, H. Sassik, J. Wernisch, A. Prokofiev, J. Cryst. Growth **310**, 1853 (2008)
86. T. Kawaguchi, K. Tanigaki, M. Yasukawa, Silicon clathrate with an *f*-electron system. Phys. Rev. Lett. **85**, 3189 (2000)
87. V. Pacheco, W. Carrillo-Cabrera, V.H. Tran, S. Paschen, Y. Grin, Comment. Phys. Rev. Lett. **87**, 099601 (2001)
88. T. Kawaguchi, K. Tanigaki, M. Yasukawa, Reply. Phys. Rev. Lett. **87**, 099602 (2001)
89. X. Tang, P. Li, S. Deng, Q. Zhang, High temperature thermoelectric transport properties of double-atom-filled clathrate compound $Yb_xBa_{8-x}Ga_{16}Ge_{30}$. J. Appl. Phys. **104**, 013706 (2008)
90. X. Zhu, N. Chen, L. Liu, Y. Li, Study on rare-earth-doped type-I germanium clathrates. J. Appl. Phys. **111**, 07E305 (2012)
91. K. Akai, G. Zhao, K. Koga, K. Oshiro, M. Matsuura, Electronic structure and thermoelectric properties on transition-element-doped clathrates, in *24th International Conference on Thermoelectrics*, ICT 2005, vol. 230 (2005)
92. D.C. Li, L. Fang, S.K. Deng, H.B. Ruan, M. Saleem, W.H. Wei, C.Y. Kong , J. Electron. Mater. **40**, 1298–1303 (2011)
93. K. Koga, H. Anno, K. Akai, M. Matsuura, K. Matsubara, First-principles study of electronic structure and thermoelectric properties for guest substituted clathrate compounds $Ba_6R_2Au_6Ge_{40}$ (R = Eu or Yb). Mater. Trans. **48**, 2108 (2007)
94. A. Prokofiev, A. Sidorenko, K. Hradil, M. Ikeda, R. Svagera, M. Waas, H. Winkler, K. Neumaier, S. Paschen, Nature Mater. **12**, 1096 (2013)
95. C. Candolfi, U. Aydemir, M. Baitinger, N. Oeschler, F. Steglich, Yu. Grin, J. Appl. Phys. **111**, 043706 (2012)
96. I. Zeiringer, M. Chen, A. Grytsiv, E. Bauer, R. Podloucky, H. Effenberger, P. Rogl, Acta Mater. **60**, 2324 (2012)
97. T. Hotta, Enhanced Kondo effect in an electron system dynamically coupled with local optical phonons. J. Phys. Soc. Jpn. **76**, 084702 (2007)

# Chapter 10
# Mechanical Properties of Intermetallic Clathrates

**M. Falmbigl, S. Puchegger and P. Rogl**

**Abstract** The present work provides a comprehensive compilation and discussion covering hardness, experimentally determined elastic properties, thermal expansion, and Debye and Einstein temperatures for intermetallic clathrates. Comparing hardness values and elastic properties a major influence of the framework atoms is observed. Hardness and elastic moduli change linearly with the Si/Ge-ratio. Also a linear temperature dependence of the elastic properties is observed. In the case that vacancies are present in the clathrate framework, the hardness of different compounds decreases almost linearly with an increasing vacancy-content. As Debye and Einstein temperatures are of significance for the vibrational spectra and performance of thermoelectric materials, corresponding values extracted from various measurement techniques such as thermal expansion, sound velocity, specific heat, electrical resistivity or X-ray absorption and diffraction measurements are evaluated. The Debye temperatures correlate with the melting temperatures of the intermetallic clathrates, and the Einstein temperatures for similar guest atoms show a linear dependence on the cage size. Wherever available, experimental data are compared with those from DFT model simulations. In general the experimentally derived values match those theoretically calculated.

## 10.1 Introduction

Intermetallic clathrates, due to their structural characteristics fulfilling the phonon glass-electron crystal concept (PGEC) [1], are considered as promising thermoelectric materials and thus the thermoelectric properties of these compounds have

M. Falmbigl · P. Rogl (✉)
Institute of Physical Chemistry, University of Vienna, Währingerstrasse 42, 1090 Wien, Austria
e-mail: peter.franz.rogl@univie.ac.at

S. Puchegger
Faculty of Physics, University of Vienna, Boltzmanngasse 5, 1090 Wien, Austria

G. S. Nolas (ed.), *The Physics and Chemistry of Inorganic Clathrates*,
Springer Series in Materials Science 199, DOI: 10.1007/978-94-017-9127-4_10,

been studied intensively in recent years [2]. The potentially positive influence of pressure on thermoelectric parameters has been demonstrated for $Sr_8Ga_{16}Ge_{30}$ [3]. Besides a high thermoelectric figure of merit, mechanical properties such as thermal expansion, hardness, and elastic properties are key quantities for any application in thermoelectric generators. Based on an extensive investigation we recently provided a comprehensive collection of data on the thermal expansion of type-I clathrates [4] and the essential information will be included in this work in form of an updated version. Investigations of the structural behaviour of clathrates under pressures up to 50 GPa have provided a valuable source of phase stabilities under pressure as well as bulk moduli derived from pressure dependent X-ray data for (i) unfilled clathrate phases ($Si_{46}$, $Si_{136}$) [5–8] as well as for (ii) filled clathrates such as $Na_8Si_{46}$ and $I_8Si_{46}$ [7], $K_8Si_{46}$ [9], $Rb_{6.15}Si_{46}$ [10], $Ba_8Si_{46}$ [7, 11, 12], $Ba_8Ge_{43}\square_3$ ($\square$ denotes a vacancy) [13], $Ba_6Ge_{25}$ and $Ba_6Si_{25}$ [14, 15], $Na_2Ba_4Ge_{25}$ [14], $I_8Sb_8Ge_{38}$ [16] and $Ba_8Ga_{16}Ge_{30}$ [17]. In most cases these investigations were backed by infrared experiments and/or by ab-initio calculations.

Despite the importance of elastic parameters for a flawless performance of thermoelectric devices, there exist hitherto only a few detailed experimental investigations of mechanical properties, of which the majority concerns type-I clathrates. For $Sr_8Ga_{16}Ge_{30}$ and $Ba_8Ga_{16}Ge_{30}$ thermal expansion and elastic constants were measured, compression tests were carried out and thermal stresses were evaluated [18, 19]. In addition to hardness values for $Ba_8Al_{16}Si_{30}$ [20], thermal expansion and elastic properties for two slightly different samples of $Ba_8Al_{15}Si_{31}$ were published [21]. Temperature dependent elastic response in the low temperature range from 0.3 to 300 K served in many cases to study the off-center rattling motion of Sr, Ba or Eu in $\{Sr,Ba,Eu\}_8Ga_{16}Ge_{30}$ [22–26] and in $Sr_8Ga_{16}Sn_{30}$ [27, 28].

In contrast to the scarce knowledge on hardness and elastic properties of intermetallic clathrates, there is an enormous amount of data on Einstein and Debye temperatures available in the literature. The exceptional vibrational features of these cage compounds were extensively studied in recent years and will be discussed in the frame of this work. Needless to say that DFT calculations of electron and phonon density of states, of thermoelectric properties and of elastic properties have greatly supported clathrate research. In this context attention should be drawn to a recent work of Karttunen et al. [29], who employed the Perdew-Burke-Ernzerhof hybrid density functional with localized atomic basis sets composed of Gaussian-type functions, to calculate the elastic properties of 14 different types of clathrate frameworks (for elemental structures of C, Si, Ge, Sn) predicting bulk and Young's moduli comparing them with their diamond-like, dense so called $\alpha$-phases.

In order to fill the knowledge gap on the mechanical performance of clathrates, the aim of this work is, to provide experimental data on hardness and elastic properties of intermetallic clathrates, covering a wide range of different compositions, and to compile, evaluate and discuss all data hitherto available in the literature on (i) hardness, (ii) elastic properties, (iii) Debye and Einstein temperatures, and (iv) on thermal expansion.

## 10.2 Experimental Procedures

### *10.2.1 Sample Preparation*

Samples were prepared by various melting techniques, i.e. arc melting or high frequency melting under vacuum or melting in sealed quartz ampoules inside a wire-wound furnace. For the measurement of the elastic properties in all cases the samples were crushed to coarse powders inside a glove box (<10 ppm $H_2O$ and $O_2$), ball milled in Ar-atmosphere to a grain size below 500 nm and finally hot pressed applying a uniaxial pressure of 56 MPa within temperature ranges from 500 to 950 °C.

For hardness measurements, pieces of the samples after the synthesis, or after the hot pressing, in some cases followed by annealing, were imbedded in phenolic resin, followed by grinding and polishing using $Al_2O_3$-powders with different grain sizes down to 0.05 μm. Preparation details for the samples used from the literature are given in the corresponding references or in references therein.

### *10.2.2 Hardness Measurement*

The Vickers hardness (HV) was measured using various Vickers indentation devices:

(a) a microhardness tester mounted on a Carl Zeiss Axioplan optical microscope employing a load of 1 N applying a rate of 0.1 N/s, and a loading time of 10 s; (b) an Anton Paar microhardness tester Microduromat 4000 (MD 4000) mounted on a Reichert Jung optical microscope under a load of 0.2 N (Ge-based samples) or 0.3 N (Si-based samples) applying a rate of 0.05 N/s, and a loading time of 5 s. Prior to these measurements two representative samples ($Ba_8Ge_{43}\square_3$, $Ba_8Pt_3Si_{43}$) were measured at various loads in order to determine an indentation force giving reliable results (Fig. 10.1). At the same time tests demonstrated that loads higher than 1 N suffer from severe fracturing of the samples due to the high brittleness, especially in the case of Si-based clathrates. For these reasons a determination of "Vickers fracture resistance (toughness) Kc" from the Vickers fracture lines proved unsuitable.

For all the resulting indentations the hardness *HV* was calculated according to:

$$HV = 2F\frac{\sin\frac{136^\circ}{2}}{\ell^2} = 1.854\frac{F}{\ell^2} \tag{10.1}$$

with $F$ the indentation load, and $\ell$ the diagonal length of the indent (see Fig. 10.1). The accuracy for both measurement techniques is about 5 %.

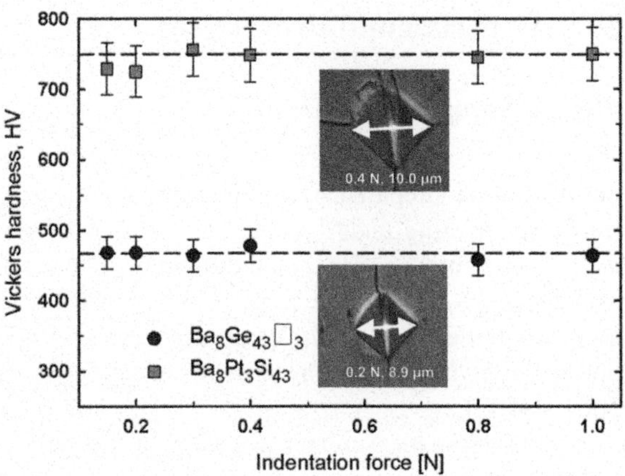


**Fig. 10.1** Hardness of $Ba_8Ge_{43}\square_3$ and $Ba_8Pt_3Si_{43}$ as a function of the indentation force

### *10.2.3 Elastic Properties*

Elastic properties were determined experimentally using various methods:

(i) Resonant ultrasound spectroscopy, RUS, was applied to determine elastic moduli via the eigenfrequencies of the samples. This method, developed by Migliori et al. [30] via establishing the relation between kinetic and elastic energy allows performing least squares fits by minimizing the sum of the squared differences between the measured and calculated eigenfrequencies to derive the elastic properties. The knowledge of the exact sample dimensions and the samples' mass are required. The cylindrical samples were mounted "edge-to-edge" between two piezo-transducers and excited via a HP 8751A network analyzer in the frequency range from 100 to 500 kHz. The spectrum was fitted with Young's modulus, E, and the Poisson's ratio, ν, as fitting variables. Measurements were carried out at room temperature and for selected samples as a function of temperature inside a furnace with protective atmosphere ($N_2$ or Ar). The accuracy of the measurement technique depends on the quality of the samples and is about 5 % for the samples used in this investigation. This method was also applied to a parallelepiped-shaped sample in a temperature range from 2 to 300 K [22]. Zerec et al. used a homemade equipment, which was placed into a PPMS-system (Quantum design) [23].

(ii) The rectangular parallelepiped resonance method [31] was applied to single crystalline specimens from room temperature to 873 K [18]. The elastic moduli for the corresponding polycrystalline material were extracted from the elastic constants applying Hill's method [18].

(iii) The phase comparison type pulse echo method, PE, and an attenuation recorder MATEC 2470B were used for the measurement of (a) $C_{44}$ from 4.2 to 160 K in a frequency range from 20 to 100 MHz [27], (b) $C_{44}$ and $\alpha_{44}$ (ultrasonic attenuation) in a temperature range from 4.2 to 150 K using frequencies from 50 to 150 MHz [25], (c) $(C_{11}\text{–}C_{12})/2$ and α in the same temperature range using frequencies from 20 to 150 MHz [26], and $C_{11}$, $C_{44}$, $(C_{11}\text{–}C_{12})/2$ and α in temperature ranges from 0.4 to 150 K, or 4.2 to 150 K, or 4.2 to 100 K in a frequency range between 24 and 240 MHz [28].

(iv) The longitudinal, $v_L$, and transversal, $v_T$, velocity of sound were measured on specimen discs with a diameter of 10 mm and a thickness of roughly 1.0 mm employing an ultrasonic thickness gauge, UTG, (Olympus 38DL PLUS). The elastic quantities were calculated by Eqs. (10.4)–(10.7) listed below and several others, given in ref. [21].

Bulk moduli were mainly extracted carrying out X-ray powder diffraction (XPD) at room temperature using synchrotron radiation and varying pressures (in most cases using diamond anvil cells for generating the pressure), applying the Murnaghan equation of state [5, 7, 13, 15–17], or a third order Birch-Murnaghan equation of state [10], or after transformation of V(P) data into normalized or reduced stress-strain variables, F (normalized pressure), and f (Eulerian strain parameter), yielding the second order finite strain equation [6, 10]:

$$F = B_0\left[1 - 1.5\left(4 - B_0'\right)f\right]. \tag{10.2}$$

The different elastic moduli, namely Young's modulus, E, bulk modulus, B, and shear modulus, G, of isotropic materials are alternately dependent and linked to each other via the Poisson's ratio

$$\nu = \frac{E}{2G} - 1 = \frac{3B - E}{6B} = \frac{3B - 2G}{6B + 2G} \tag{10.3}$$

and hence knowing two of the quantities allows to calculate all others.

Furthermore for isotropic materials the elastic constants $C_{11}$ and $C_{12}$ can be extracted from

$$C_{11} = 3B - \frac{6B\nu}{1 + \nu} \tag{10.4}$$

and

$$C_{12} = \frac{2G^2 - EG}{E - 3G} \tag{10.5}$$

The elastic constants $C_{11}$ and $C_{44}$ can be derived from the longitudinal sound velocity, $v_L$, and the transversal sound velocity, $v_T$, respectively:

$$C_{11} = dv_L^2 \tag{10.6}$$

$$C_{44} = G = dv_T^2 \tag{10.7}$$

with d being the density of the material. Wherever possible, measured densities of the samples relying on Archimedes principle (distilled water) were used. In all other cases the X-ray density, $d_x$, as defined in Eq. (10.8) was utilized,

$$d_x = \frac{ZM}{N_A V} \tag{10.8}$$

with Z, the number of formula units/unit cell, M the molar weight, $N_A$ the Avogadro number, and V the volume of the unit cell.

Although most investigations focused on the pressure dependence of the clathrate-I structure, data on the bulk moduli are available for different clathrate structures (type-I, type-II, type-VIII, and type-IX), and various filler and framework elements and are collected in Table 10.2. In order to provide also other elastic quantities, in these cases an average Poisson's ratio, $\nu_{avg} = 0.23$, was assumed and Eqs. (10.3)–(10.7) in given or adapted form were used for further calculations.

### 10.2.4 Compression Tests

Compression tests were carried out using an Instron-type testing machine at a strain rate of $1 \times 10^{-4}\ s^{-1}$. The measurements were performed at 1,123 K ($Ba_8Ga_{16}Ge_{30}$) and at 973 K ($Sr_8Ga_{16}Ge_{30}$) under vacuum on samples with roughly $2 \times 2 \times 5\ mm^3$ size. The single crystalline specimens were oriented in a manner that the [110] axis was parallel to the compression axis [18].

### 10.2.5 Debye and Einstein Temperatures

#### 10.2.5.1 Evaluation of the Debye and Einstein Temperatures from Theoretical Calculations

The first theoretical work providing information on the Debye temperature ($\theta_D$) of intermetallic clathrates dates back to the year 1999 [33]. Molecular dynamics calculations for the carbon-framework of type-I and type-II clathrates used a Lennard-Jones potential (later on also for Si-based clathrates [34]). $\theta_D$ for $C_{136}$ [35] and for $Si_{136}$ [34] were estimated from the calculated elastic constant $C_{11}$ applying the empirical relation: $\theta_D = -11.3964 + 0.3475 \times C_{11} - 1.6150 \times 10^{-5} \times C_{11}^2$. Moriguchi et al. [36] used an empirical bond-order potential developed by Tersoff for the calculation of several thermodynamic properties, including the heat capacity, for the type-I and type-II Si networks. From the heat capacity data in the temperature range from 0 to 150 K $\theta_D$ was extracted applying the Debye-model. The heat capacity, $C_V$, was calculated by the density functional theory (DFT),

**Table 10.1** Vickers hardness (HV) of different clathrate-I compounds (this work)

| Compound | HV |
|---|---|
| *Sn/Ge-mixed* | |
| $Ba_8Zn_{7.68}Ge_{36.97}Sn_{1.35}$ | 555 |
| $Ba_8Zn_{7.54}Ge_{37.80}Sn_{0.66}$ | 536 |
| *Ge-based* | |
| $Ba_8Ge_{43}\square_3$ | 468 |
| $Ba_8Cu_5Ge_{41}$ | 525 |
| $Ba_8Ni_{2.5}Zn_{3.7}Ge_{39.8}$ | 596 |
| $Ba_8Ni_{0.8}Zn_{6.1}Ge_{39.1}$ | 526 |
| $Ba_8Ni_{0.5}Zn_{6.5}Ge_{39.0}$[a] | 585 |
| $Ba_8Ni_{0.5}Zn_{6.5}Ge_{39.0}$[a] | 504 |
| $Ba_8Zn_{2.1}Ge_{41.5}\square_{2.4}$ | 516 |
| $Ba_8Zn_{4.6}Ge_{40}\square_{1.4}$ | 520 |
| $Ba_8Zn_{5.7}Ge_{40}\square_{0.3}$ | 566 |
| $Ba_8Zn_{7.68}Ge_{38.32}$ | 563 |
| $Ba_8Zn_{7.7}Ge_{38.3}$[b] | 557 |
| $Ba_8Rh_{0.3}Ge_{43.0}\square_{2.7}$ | 499 |
| $Ba_8Rh_{1.2}Ge_{42.8}\square_{2.0}$ | 543 |
| $Ba_8Pd_2Ge_{42.5}\square_{1.5}$ | 497 |
| $Ba_8Pd_3Ge_{42.5}\square_{0.5}$ | 520 |
| $Ba_8Pd_{3.7}Ge_{42.3}$ | 541 |
| $Ba_8Pd_{3.8}Ge_{42.2}$ | 544 |
| $Ba_8Ag_{2.3}Ge_{41.3}\square_{2.4}$ | 425 |
| $Ba_8Ag_{3.4}Ge_{41.5}\square_{1.1}$ | 456 |
| $Ba_8Ag_{4.1}Ge_{41.4}\square_{0.5}$ | 488 |
| $Ba_8Ag_{5.3}Ge_{40.7}$ | 495 |
| $Ba_8Cd_{4.7}Ge_{40.3}\square_{1.0}$ | 486 |
| $Ba_8Cd_{6.5}Ge_{39.1}\square_{0.4}$ | 510 |
| $Ba_8Cd_{7.6}Ge_{39.1}$ | 515 |
| $Ba_8Ir_{0.4}Ge_{43}\square_{2.6}$ | 562 |
| $Ba_8Pt_2Ge_{42.3}\square_{1.7}$ | 475 |
| $Ba_8Pt_{2.7}Ge_{41.8}\square_{1.5}$ | 495 |
| $Ba_8Pt_{3.3}Ge_{41.6}\square_{1.1}$ | 483 |
| $Ba_8Au_6Ge_{40}$ | 517 |
| *Ge/Si-mixed* | |
| $Ba_8Cu_5Ge_{33}Si_8$ | 596 |
| $Ba_8Cu_5Ge_{23}Si_{18}$ | 694 |
| $Ba_8Zn_8Ge_{19}Si_{19}$[c] | 657 |
| $Ba_8Zn_8Ge_{19}Si_{19}$[c] | 483 |
| *Si-based* | |
| $Ba_8Al_{16}Si_{30}$ | 560[d] |
| $Ba_8Al_{6.7}Ni_1Si_{38.3}$ | 750 |
| $Ba_8Al_{4.2}Ni_{1.3}Si_{42.7}$ | 700 |
| $Ba_8Ni_3Si_{43}$ | 720 |
| $Ba_8Ni_{3.3}Si_{42.7}$ | 730 |
| $Ba_8Ni_{3.8}Si_{42.2}$ | 766 |

(continued)

**Table 10.1** (continued)

| Compound | HV |
|---|---|
| $Ba_8Cu_5Si_{41}$ | 765 |
| $Ba_8Zn_7Si_{39}$ | 706 |
| $Ba_8Rh_{2.9}Si_{43.1}$ | 728 |
| $Ba_8Pd_{2.9}Si_{43.1}$ | 767 |
| $Ba_8Pd_{3.9}Si_{42.1}$ | 716 |
| $Ba_8Ag_5Si_{41}$ | 770 |
| $Ba_8Pt_{3.2}Si_{42.8}$ | 756 |
| $Ba_8Au_{5.1}Si_{40.9}$ | 700 |
| *Inverse clathrates* | |
| $I_8Sn_{19.3}Cu_{4.7}P_{22}$ | 454 |
| $I_8Sn_{19.3}Cu_{3.7}Zn_1P_{21.2}\square_{0.8}$ | 452 |
| $I_8Sn_{19.3}Cu_{2.7}Zn_2P_{20.9}\square_{1.1}$ | 411 |
| $I_8Sn_{19.3}Cu_{1.7}Zn_3P_{19.9}\square_{2.1}$ | 371 |

[a] Different densities (5.86 g/cm$^3$ , 5.84 g/cm$^3$ )
[b] Single crystalline sample
[c] Hand-milled and hot pressed
[d] Ref. [20]

based on the general gradient approximation (GGA) for $Si_{136}$ and $Ge_{136}$ (type-II) and Debye temperatures for both compounds were estimated thereof integrating the Debye-model over all frequencies [37]. Density functional perturbation theory was employed for the calculation of $\theta_D$ for C-, Si-, and Ge-frameworks of type-I as well as type-II clathrates [38]. First principles calculations also served to calculate the Debye temperature as a function of temperature for type-I and type-II silicon clathrates [39], whereas the equilibrium molecular dynamics method was chosen for the type-II Ge-clathrate [40].

DFT calculations yielded the motion frequencies of the Ba- and Sr-guest atoms in $Ba_8Ga_{16}\{Si,Ge\}_{30}$ and $Sr_8Ga_{16}Ge_{30}$, which were extracted from the energy variation of the corresponding atoms, while displacing them from their central position (in case of the tetrakaidecahedral cages perpendicular as well as parallel to the hexagonal faces formed by the framework atoms) [41]. A similar approach was made for several different clathrate compounds [42, 43] in order to gain information on the lattice dynamics and vibrational modes of these compounds. The Einstein temperatures, $\theta_E$, were calculated from the given frequencies. Myles et al. calculated phonon dispersion relations and reported localized low energy vibrational modes for the guest atoms in type-II clathrates [44] employing the local density approximation (LDA). Also Madsen et al. [45] used first principle calculations, but did not treat the anharmonicity perturbatively in their approach. They extracted Einstein frequencies from the calculated mean square displacement amplitude of the guest atoms. Treating the guest atoms as independent Einstein oscillators, allows extracting corresponding Einstein temperatures from calculated atomic displacement parameters (ADP) [46, 47].

#### 10.2.5.2 Experimental Evaluation of the Debye and Einstein Temperatures

Several experimental methods have served to extract and probe Debye and Einstein temperatures. Thermal expansion data measured from 4 to 300 K were analyzed in terms of a semiclassical model of Mukherjee et al. [48], which includes three- and four-phonon interactions, an anharmonic potential, and an electronic contribution. Within this model acoustic phonons are treated in terms of the Debye theory and optical phonons in terms of the Einstein theory. Consequently $\theta_D$ and $\theta_E$ are variables in a fitting procedure of the temperature dependent thermal expansion [4]. Suekuni et al. [49] calculated the Debye temperature from sound velocity values obtained by ultrasound measurements at 4 K. Temperature dependent electrical resistivity data were fitted to the Bloch-Grüneisen model, which considers scattering of charge carriers by thermally excited phonons, accounted for in terms of the Debye-model, and thus yielding $\theta_D$ as one of the variables [50–56]. In several cases deviations from this model, applicable to systems with high charge carrier density and metallic behavior, were considered introducing additional terms like a temperature dependent charge carrier density [52, 54, 57] or Einstein contributions arising from the presence of localized harmonic and/or anharmonic oscillators and an additional anharmonic contribution [58]. The latter model provides information on the Einstein temperatures of the oscillators (guest atoms) too [58]. Another way to estimate the Debye temperature of a compound is to analyze thermal conductivity data, measured in a wide temperature range usually starting close to 0 K up to room temperature [59–65]. All models used for the fitting of the phonon contribution are essentially based on the expression of the thermal conductivity within the kinetic gas theory and Debye temperatures are extracted from modeling the heat capacity term. For $Eu_8Ga_{16}Ge_{30}$ nuclear inelastic scattering (NIS) was carried out at 25 K to probe the phonon density of states (PDOS) [66]. A model for the PDOS including one Einstein oscillator energy for the Eu1-atoms located in the pentagonal dodecahedral cages and three Einstein oscillator energies for the Eu2-atoms located in the tetrakaidecahedral cages was fitted to the measurement results. Later on for this compound also microwave absorption experiments as well as Mößbauer spectroscopy were performed, revealing frequencies correlated to the tunneling of Eu-atoms, from which the Einstein temperature was calculated [67]. Another more frequently used method to probe the PDOS of clathrate compounds is, to apply inelastic neutron scattering (INS). Frequencies corresponding to the vibrations of the guest atoms can be identified directly from the spectra [43, 55, 68] or by comparison with a theoretically calculated phonon density of states [69] or by modeling the PDOS including an Einstein oscillator term [67]. Also Raman spectroscopy was used to extract the frequencies of the "rattling" modes of guest atoms for various types of intermetallic clathrates [16, 42, 70].

Scattering experiments using either neutron or X-ray sources can serve information on the atomic displacement parameters of atoms at specific crystallographic sites. The framework forming atoms can be considered as Debye solid, wherein the isotropic ADPs ($U_{iso}$) are related to $\theta_D$ via,

$$U_{iso} = \frac{3\hbar^2 T}{mk_B\theta_D^2}\left[\frac{T}{\theta_D}\int_0^{\theta_D/T}\frac{x}{e^x-1}dx + \frac{\theta_D}{4T}\right] + d^2 \tag{10.9}$$

where m is the average mass of the atoms, and d a term accounting for static disorder. The guest atoms are considered as independent oscillators, which can be characterized by applying the Einstein model, which correlates the atomic displacement parameters with $\theta_E$ via the relation:

$$U_{xx} = \frac{\hbar^2}{2mk_B\theta_{E,xx}}coth\left(\frac{\theta_{E,xx}}{2T}\right) + d^2 \tag{10.10}$$

Here m is the mass of the "rattling" atom, and d a term for static disorder (further details can be found in [71]). Neutron powder diffraction data collected at different temperatures were used to calculate the Einstein temperatures [72] or both, Einstein as well as Debye temperatures [73, 74], from ADPs. Also single crystals were exposed to neutron diffraction at different temperatures in order to provide $\theta_E$ [75] and $\theta_D$ [73, 76–79]. Another frequently used technique to extract $\theta_D$ and $\theta_E$ is X-ray diffraction. In some cases room temperature data, gained from powder samples measured on conventional diffractometers, were used to calculate the characteristic temperatures [16, 43, 80–85], and also temperature dependent measurements using synchrotron radiation and powder samples were carried out [69, 86–94]. Similarly, X-ray diffraction measurements on single crystals were performed to gain atomic displacement parameters. Whereas in a few cases room temperature data were used to evaluate Einstein and Debye temperatures [72, 95, 96], for the majority of investigations, temperature dependent measurements were carried out [49, 52, 53, 55, 59, 64, 68, 69, 78, 79, 97–109].

The measurement of the heat capacity also supplies a method to analyze the lattice dynamics. Qiu et al. found good agreement to the experimental data applying the Debye-model for the specific heat using $\theta_D$ deduced from density and bulk modulus [110]. Several times linear fits were applied to

$$\frac{C_P}{T} = \gamma + \beta T^2 \quad \text{with } \theta_D = \left(\frac{12\pi^4 Rn}{5\beta}\right)^{\frac{1}{3}} \tag{10.11}$$

with $\gamma$ the Sommerfeld coefficient, R the universal gas constant, and n the number of atoms per unit cell [32, 111–114]. In order to model the heat capacity over a wider temperature range in addition to the electronic contribution, $\gamma T$, and the phonon contribution, accounted for in terms of the Debye-model, additional Einstein terms using one $\theta_E$ [73, 75, 94, 96, 115–118], two $\theta_E$ [49, 52, 55, 56, 65, 68, 69, 84, 86, 87, 93, 105, 119–127], or larger numbers of fitting variables [122, 128] were introduced. Zheng et al. besides $\theta_D$ and two $\theta_E$ also introduced an additional anharmonic contribution and a T-dependent Debye temperature [58]. Also the soft potential model or a mix of several contributions was used to model $C_P(T)$ [63, 122].

### *10.2.6 Thermal Expansion*

The results on the thermal expansion of type-I clathrates are essentially based on an earlier publication of the authors [4]. Supplementary new data on thermal expansion coefficients, $\alpha$, of type-I clathrates are calculated from lattice parameters measured at different temperatures using X-ray single crystal diffraction [52, 53, 104, 107] or X-ray powder diffraction [69, 93, 129]. For $Ba_8Al_{15}Si_{31}$ the thermal expansion was measured from 300 to 700 K (rate of 5 K/min for heating and cooling) using a thermomechanical analyser (Mac Science TMA-4010S) [21]. Theoretical calculations of $C_{46}$ including thermal expansion were carried out via DFT [38] or Monte Carlo simulations [130].

## 10.3 Results and Discussion

### *10.3.1 Hardness*

Hitherto only for one clathrate compound, $Ba_8Al_{16}Si_{30}$, the hardness and its dependence on different annealing temperatures and times has been reported in the literature [20]. The values are varying between HV = 450–560. As no experimental details were provided by the authors [20], a comparison with the data of other measurements is impossible.

Hardness measurements were carried out on type-I clathrates with various compositions and all values are compiled in Fig. 10.2 and Table 10.1. All investigated samples, except the inverse clathrates $I_8Sn_{19.3}Cu_{4.7-x}Zn_xP_{22-y}y$ (with x = 0–3, and y = 0–2.1, respectively), exclusively contain Ba as guest atoms and besides $Ba_8Au_6Ge_{40}$ [56] and $I_8Sn_{19.3}Cu_{4.7-x}Zn_xP_{22-y}y$ [131] all samples exhibit n-type electrical conductivity. In general, the framework forming atoms govern the hardness. Si-based compounds reveal a significantly higher average hardness $HV_{avg}$ = 724, than the Ge-based compounds ($HV_{avg}$ = 516, see Fig. 10.2).

The influence of the Si/Ge-ratio was studied in detail for two different systems, $Ba_8Cu_5Si_{41-x}Ge_x$ [98] and $Ba_8Zn_{7.7}Si_{38.3-x}Ge_x$ [61]. The results are plotted in

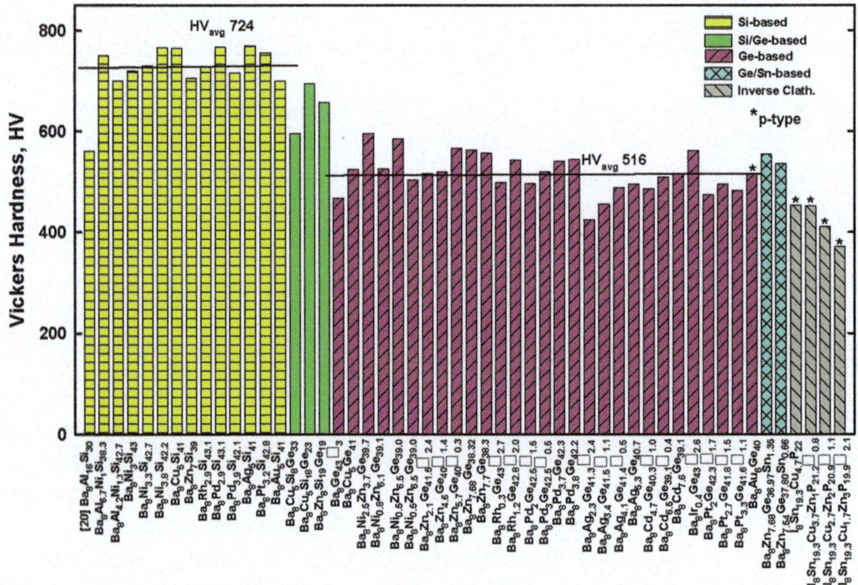


**Fig. 10.2** Vickers hardness of various type-I clathrates

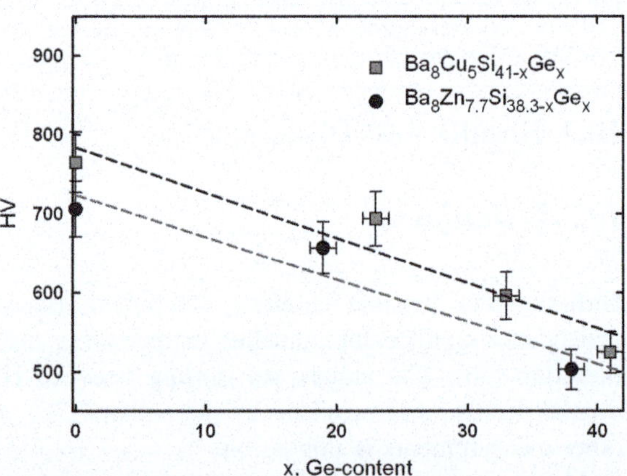


**Fig. 10.3** Vickers hardness as a function of the Si/Ge-ratio for $Ba_8Cu_5Si_{41\text{-}x}Ge_x$ and $Ba_8Zn_{\sim 8}Si_{38\text{-}x}Ge_x$

Fig. 10.3. Indeed HV is continuously decreasing with increasing Ge-content, although in both cases no strictly linear dependence was observed. Nevertheless it is interesting to note, that the slopes of the linear fits (dashed lines in Fig. 10.3) are almost identical (−5.7 for $Ba_8Cu_5Si_{41\text{-}x}Ge_x$, and −5.3 for $Ba_8Zn_{7.7}Si_{38.3\text{-}x}Ge_x$). A similar tendency is observed, when Sn is incorporated into the clathrate framework built up by Zn- and Ge-atoms: HV decreases slightly from 563 ($Ba_8Zn_{7.68}Ge_{38.32}$) to 555 ($Ba_8Zn_{7.68}Ge_{36.97}Sn_{1.35}$) or 536 ($Ba_8Zn_{7.54}Ge_{37.80}Sn_{0.66}$), although from these two samples no continuous dependence of HV on the Sn-content can be concluded.

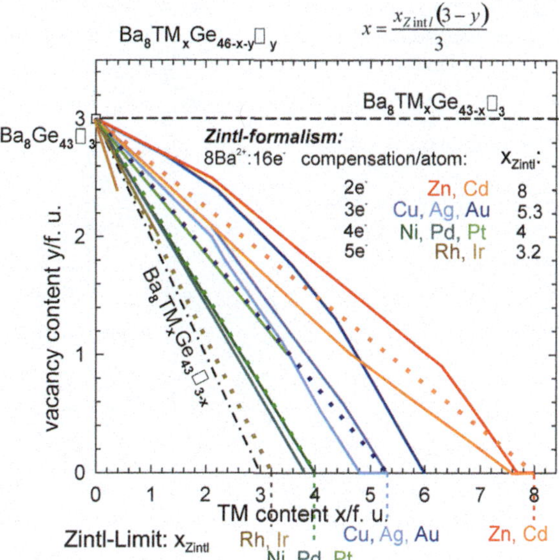


**Fig. 10.4** Mechanism of incorporation of transition metals (TM) into $Ba_8TM_xGe_{46-x-y}\square_y$

For several clathrates $Ba_8TM_x\{Si\text{ or }Ge\}_{46-x-y}\square_y$ with varying transition metal- and vacancy-contents (given in Fig. 10.4), containing Rh [132], Pd [68], Ir [133], Pt [64], Ag [53], Zn [55] and Cd [105] Vickers hardness was measured on the respective bulk single-phase samples. Based on single crystal X-ray diffraction and EPMA data for different transition metals and TM-contents the mechanism of vacancy filling and framework atom substitution was explored in detail for Ba–Ge based type-I clathrates and is shown in Fig. 10.4. In general the maximum solubility for the different elements is well predicted applying the Zintl concept, considering a compensation of the 16 electrons provided by the Ba guest ions with the electron deficiency of the TM-atoms substituting for the tetrahedrally bonded Ge atoms e.g. 2 electrons for Zn (taken as strictly divalent), or 4 for Ni (taken as 0 valent). The formula for a linear relation between x and y starting from binary $Ba_8Ge_{43}\square_3$ to the charge balanced compound (according to the Zintl concept) is provided in Fig. 10.4 and displayed by dashed lines. However, in reality for each element a small deviation from this linear relationship is found suggesting slightly different mechanisms of TM-incorporation.

The results of the hardness measurements as a function of the vacancy content are plotted in Fig. 10.5. In all cases the hardness is increasing with decreasing vacancy-content, which points towards a significantly increased average bonding strength due to the filling of the vacant positions and thus formation of additional bonds between the transition metal and Ge. Another contribution for the increase of HV may arise from solid solution hardening, which is observed, whenever different atomic species are present at one crystallographic site. Interestingly the compounds containing Zn and Pt exhibit a minor decrease of HV close to the maximum solubility (see Table 10.1), which might be explained by a slightly

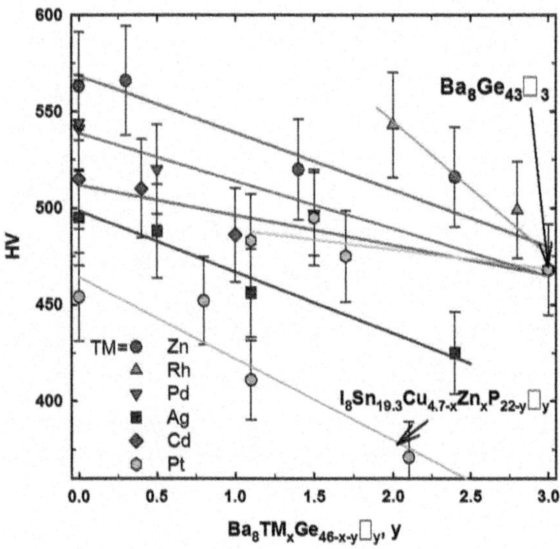


**Fig. 10.5** Dependence of the Vickers hardness on the vacancy-content of $Ba_8TM_xGe_{46\text{-}x\text{-}y}\square_y$ and $I_8Sn_{19.3}Cu_{4.7\text{-}x}Zn_xP_{22\text{-}y}\square_y$. The solid lines show linear fits to the data points. x can be estimated by $x = \frac{x_{Zintl}(3-y)}{3}$, where $x_{Zintl}$ is x for a charge balanced compound (see Fig. 10.4)

weaker bonding strength of TM–Ge compared to Ge–Ge bonds. As the differences are very small, this behavior could also be attributed to slight differences in density or microstructure. In case of Ag the samples with lowest transition metal content exhibit a much lower HV-value compared to HV = 468 measured for the binary $Ba_8Ge_{43}\square_3$, which contains the maximum vacancy content of 3 vacancies per formula unit. This slight deviation may be caused by the measurement using different equipments and indentation forces (see Sect. 2.2). A similar hardening effect by the decrease of the vacancy content is also observed for the inverse type-I clathrate compounds $I_8Sn_{19.3}Cu_{4.7\text{-}x}Zn_xP_{22\text{-}y}\square_y$, where HV also continuously increases with a decreasing amount of vacancies. The generally smaller hardness values of these compounds as compared to the Ba–Ge-based polyanionic clathrates arise from weaker bonding between the framework sites.

Solid solution hardening is observed for $Ba_8Ni_xSi_{46\text{-}x}$, where the Vickers hardness increases with increasing Ni-content, x. This exceptional behavior is also reflected by a decrease of the lattice parameter with increasing transition metal content [32, 52], whereas in case of $Ba_8Pd_xSi_{46\text{-}x}$ [134] the lattice parameter (increase) and HV (decrease) exhibit converse dependences as a function of x.

In a recent review article summarizing the mechanical properties of skutterudites [135] Rogl reported the dependence of HV on the density of the compounds. Although within this work no detailed investigation is provided, a similar trend is observed in two cases: for $Ba_8Ni_{0.5}Zn_{6.5}Ge_{39.0}$ a dramatic decrease of HV from 585 to 504 was observed, whereas the density changes only by 0.4 %, and an even more pronounced reduction of the hardness (HV 657–HV 483) was found for $Ba_8Zn_8Ge_{19}Si_{19}$ for only 0.2 % difference of the densities [61], which at least partially may also arise from significantly different preparation methods (Table 10.1, [61] ).

### 10.3.2 Elastic Properties

#### 10.3.2.1 Elastic Properties Below Room Temperature

RUS measurements on $Eu_8Ga_{16}Ge_{30}$ and $Sr_8Ga_{16}Ge_{30}$ [22, 23] demonstrated that the shear modulus ($C_{44}$) as a function of temperature cannot be satisfactorily modelled calculating the elastic response of a 2-level system employing the Varshni potential function [136] due to an anomaly i.e. the occurrence of a dip in the function $C_{44}(T)$ around 30 K. Zerec et al. introduced a new type of four-well tunneling states, described by two nearly degenerate four level systems [23, 24]. The depression of the elastic constants $C_{11}$, $C_{12}$, and $C_{44}$ observed by the RUS measurements below ~150 K as well as specific heat are reasonably well described applying this model. The results are consistent with the fourfold splitting of the 6c Eu-site observed by neutron diffraction for $Eu_8Ga_{16}Ge_{30}$. Please note, that the Wyckoff sites refer to the setting of the crystal structure standardized with the program Structure Tidy [137]. According to this standardization Wyckoff sites 6c and 6d interchange. For $Sr_8Ga_{16}Ge_{30}$, where the splitting of the 6c-site site is less distinct than in $Eu_8Ga_{16}Ge_{30}$, a modified model was proposed with a quadrupolar interaction of the local Einstein mode with the elastic strain [23].

The influence of rattling motions of the guest atoms in the 6c-site on the lattice for single crystalline $Ba_8Ga_{16}Sn_{30}$ was studied via the elastic moduli of transverse and longitudinal modes [27], and indeed, $C_{44}$ shows an ultrasonic frequency dependence between 30 and 65 K most likely originating from the rattling motions [27]. For $Sr_8Ga_{16}Ge_{30}$ and $Eu_8Ga_{16}Ge_{30}$ a similar effect between 20 and 60 K, or 50 and 100 K was found [25, 26]. The off-center rattling motion and its charge carrier dependence was investigated for a p- and n-type single crystal of $Ba_8Ga_{16}Sn_{30}$, as well as for a single crystal of $K_8Ga_8Sn_{38}$ [28]. Large elastic softening of the transverse modulus $C_{44}$ at low temperatures persisting even below 1 K was observed for the Ba-containing samples, attributed to the off-center rattling motion causing lattice instability. In contrast to this observation, for $K_8Ga_8Sn_{38}$, which exhibits no off-center rattling, a hardening of $C_{44}$ towards low temperatures was reported [28].

#### 10.3.2.2 Elastic Properties at and Above Room Temperature

The elastic properties at room temperature for intermetallic clathrates hitherto measured are presented in Table 10.2 and Fig. 10.6. In order to recognize, which values are reported in the literature and which are calculated/estimated thereof, the reader is referred to Table 10.2. Comparing the results of the Si-based clathrates, significantly smaller values are observed for $Ba_8Au_5Si_{41}$. The measured density of this sample is only 82 % of the X-ray density, calculated applying Eq. (10.8). The dependence of the elastic moduli of skutterudites on the relative density down to ~90 % was investigated revealing a linear decrease of E(d) [$E(d) = 4.16d - 278.5$

**Table 10.2** Elastic properties at room temperature for various intermetallic clathrates

| Composition | Meas. | $\rho_d$ g/cm$^3$ | $v_L$ $\times 10^5$ cm/s | $v_T$ $\times 10^5$ cm/s | $v_m$ $\times 10^5$ cm/s | $C_{11}$ GPa | $C_{12}$ GPa | $C_{44}$ or G GPa | E GPa | B GPa | ν | G/B | $\theta_D$ K | Ref. |
|---|---|---|---|---|---|---|---|---|---|---|---|---|---|---|
| *Si-based* | | | | | | | | | | | | | | |
| $K_8Si_{46}$ ($K_{6.8}Si^{a}_{46}$) | XRD | 2.47[a] | 8.09 | 4.79 | 5.30 | 161.50 | 48.24 | 56.63 | 139.32 | **86(5)** | *0.23* | 0.66 | 577 | [8] |
| $Rb_{6.15}Si_{46}$ | XRD | | | | | 110.80 | 33.1 | 38.85 | 95.58 | **59** | *0.23* | 0.66 | | [10] |
| $Ba_8Si_{46}$ | XRD | 3.62[b] | 6.95 | 4.11 | 4.56 | 174.66 | 52.17 | 61.24 | 150.66 | **93(5)** | *0.23* | 0.66 | 498 | [7] |
| $Ba_8Al_{15}Si_{31}$_A | UTG | **3.196** | **6.04** | **3.50** | **3.88** | **77.33** | **38.11** | **39.21** | **97.75** | **116.54** | **0.246** | 0.40 | 408 | [21] |
| $Ba_8Al_{15}Si_{31}$_B | UTG | **3.296** | **6.18** | **3.58** | **3.97** | **83.78** | **41.54** | **42.23** | **105.40** | **98** | **0.248** | 0.40 | 421 | [21] |
| $Ba_8Ni_3Si_{43}$ | RUS | **3.82** | 6.40 | 3.83 | 4.24 | 156.41 | 44.12 | 56.15 | **137(5)** | 81.55 | **0.22(2)** | 0.69 | 465 | [*] |
| $Ba_8Ni_{3.3}Si_{42.7}$ | RUS | **4.27** | 6.12 | 3.67 | 4.06 | 159.84 | 45.08 | 57.38 | **140(5)** | 83.33 | **0.22(2)** | 0.69 | 462 | [*] |
| $Ba_8Cu_5Si_{41}$ | RUS | **3.83** | 6.19 | 3.73 | 4.13 | 146.69 | 39.94 | 53.38 | **129(4)** | 75.52 | **0.21(1)** | 0.71 | 448 | [*] |
| $Ba_8Ag_5Si_{41}$ | RUS | **3.99** | 5.95 | 3.53 | 3.91 | 135.56 | 42.25 | 49.59 | **122(3)** | 75.31 | **0.23(1)** | 0.66 | 419 | [*] |
| $Ba_8Au_5Si_{41}$ | RUS | **3.94** | 4.67 | 2.76 | 3.06 | 85.79 | 25.62 | 30.08 | **74(8)** | 45.68 | **0.23(1)** | 0.66 | 311 | [*] |
| $Si_{136}$ | XRD | | | | | 169.02 | 50.49 | 59.27 | 145.80 | **90(3)** | *0.23* | 0.66 | | [72] |
| $Si_{136}$ | XRD | | | | | 169.02 | 50.49 | 59.27 | 145.80 | **90(5)** | *0.23* | 0.66 | | [5] |
| $Ba_6Si_{25}$ | XRD | 3.66[c] | 5.78 | 3.42 | 3.79 | 122.07 | 36.46 | 42.80 | 105.30 | **65.0** | *0.23* | 0.66 | 401 | [15] |
| *Si/Ge-based* | | | | | | | | | | | | | | |
| $Ba_8Cu_5Si_{25}Ge_{16}$ | RUS | 4.69 | 5.31 | 3.18 | 3.52 | 132.44 | 37.35 | 47.54 | **116(4)** | 69.05 | **0.22(1)** | 0.69 | 377 | [*] |
| $Ba_8Cu_5Si_{18}Ge_{23}$ | RUS | 5.03 | 5.02 | 3.01 | 3.33 | 126.73 | 35.74 | 45.49 | **111(3)** | 66.07 | **0.22(1)** | 0.69 | 354 | [*] |
| $Ba_8Cu_5Si_6Ge_{35}$ | RUS | 5.65 | 4.56 | 2.73 | 3.02 | 117.59 | 33.17 | 42.21 | **103(6)** | 61.31 | **0.22(1)** | 0.69 | 319 | [*] |
| *Ge-based* | | | | | | | | | | | | | | |
| $Sr_8Ga_{16}Ge_{30}$ | RUS | 5.4[d] | 4.43 | 2.62 | 2.91 | 106.09 | 31.69 | **37.2** | 91.51 | 56.49 | *0.23* | 0.66 | 305 | [22] |
| $Sr_8Ga_{16}Ge_{30}$ | RM | 5.4[d] | 4.30 | 2.51 | 2.78 | **100** | **41** | **36.1** | **91.1** | **64.0** | 0.26 | 0.56 | 292 | [18] |
| $Sr_8Ga_{16}Ge_{30}$ | XRD | 5.4[d] | 6.72 | 3.98 | 4.41 | 244.15 | 72.93 | 85.61 | 210.60 | **130** | *0.23* | 0.66 | 463 | [3] |
| $Ba_8Ge_{43}\square_3$ | XRD | 5.8[e] | 4.13 | 2.44 | 2.71 | **99.16** | 29.62 | 34.77 | 85.54 | **52.8** | *0.23* | 0.66 | 281 | [13] |
| $Ba_8Cu_5Ge_{41}$ | RUS | **5.91** | 4.23 | 2.47 | 2.74 | 105.50 | 49.08 | 36.10 | **93(9)** | 64.00 | **0.28(9)** | 0.56 | 287 | [*] |
| $Ba_8Zn_7Ge_{39}$ | RUS | **5.79** | 4.40 | 2.70 | 2.98 | 112.22 | 34.97 | 42.08 | **101(4)** | 56.11 | **0.23(1)** | 0.75 | 307 | [*] |

(continued)

**Table 10.2** (continued)

| Composition | Meas. | $\rho_d$ g/cm$^3$ | $v_L$ $\times 10^5$ cm/s | $v_T$ $\times 10^5$ cm/s | $v_m$ $\times 10^5$ cm/s | $C_{11}$ GPa | $C_{12}$ GPa | $C_{44}$ or G GPa | E GPa | B GPa | ν | G/B | $\theta_D$ K | Ref. |
|---|---|---|---|---|---|---|---|---|---|---|---|---|---|---|
| $Ba_8Zn_8Ge_{38}$ | RUS | **5.78** | 4.38 | 2.68 | 2.96 | 111.11 | 34.63 | 41.67 | **100(3)** | 55.56 | **0.23(1)** | 0.75 | 306 | [*] |
| $Ba_8Ga_{16}Ge_{30}$ | RM | 5.8[d] | 4.35 | 2.72 | 3.00 | **110** | **39** | **43** | **104** | **62** | 0.22 | 0.69 | 313 | [18] |
| $Ba_8Ga_{16}Ge_{30}$ | XRD | 5.8[d] | 4.66 | 2.76 | 3.06 | 126.20 | **37.7** | 44.25 | 108.86 | **67.2** | *0.23* | 0.66 | 319 | [17] |
| $Ba_8Au_6Ge_{40}$ | RUS | **6.64** | 3.59 | 2.05 | 2.27 | 85.65 | 30.09 | 27.78 | **70(7)** | 48.61 | **0.26(4)** | 0.57 | 233 | [*] |
| $Eu_8Ga_{16}Ge_{30}$ | RUS | 6.1[d] | 4.02 | 2.39 | 2.64 | **99** | **41** | **35** | 86.10 | 53.15 | *0.23* | 0.66 | 277 | [23] |
| $Ba_6Ge_{25}$ | XRD | 5.68[f] | 3.81 | 2.26 | 2.50 | 82.63 | 24.68 | 28.98 | 71.28 | **44(2)** | *0.23* | 0.66 | **255** | [14] |
| *Inverse clathrates* | | | | | | | | | | | | | | |
| $I_8Si_{44.5}I_{1.5}$ | XRD | 3.6[c] | 6.98 | 4.13 | 4.58 | 178.41 | 53.29 | 62.56 | 153.90 | **95(5)** | *0.23* | 0.66 | 495 | [7] |
| $I_8Sb_8Ge_{38}$ | XRD | 6.1 [h] | 4.45 | 2.64 | 2.92 | 121.51 | 36.30 | 42.61 | 104.81 | **64.7** | *0.23* | 0.66 | 302 | [16] |
| $I_8Sn_{19.3}Cu_{4.7}P_{22}$ | RUS | 5.6 | 4.33 | 2.43 | 2.70 | 104.96 | 38.82 | **33.07** | 84(4) | 60.87 | **0.27(3)** | 0.54 | 280 | [*] |

*XRD* X-ray diffraction, *UTG* ultrasonic thickness gage, *RUS* resonant ultrasound spectroscopy, *RM* rectangular parallelepiped resonance method
Densities calculated via Eq. (10.8), using lattice parameters given in [152][a], [153][b], [154][c], [75][d], [13][e], [155][f], [16][g], and [156][h]
[*] this work
**Bold numbers** are published experimental data; others are calculated by equations (see text)
*italic numbers* average value of the Poisson's ratio, $\nu_{avg} = 0.23$

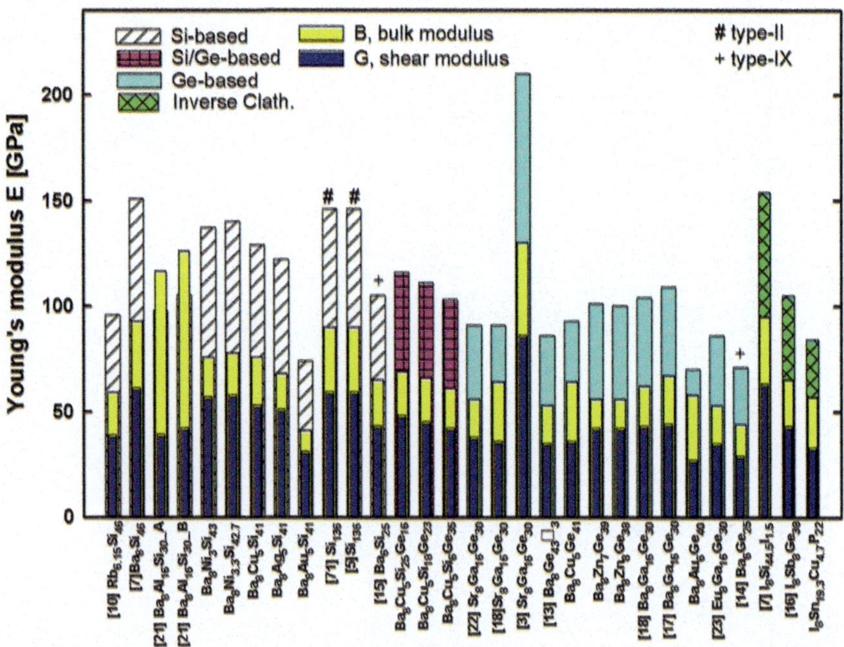


**Fig. 10.6** Elastic moduli of intermetallic clathrates. The sequence follows the order in Table 10.1

for $DD_{0.65}FeCo_3Sb_{12}$ with d in %] [135]. The Young's modulus for a completely dense sample of $Ba_8Au_5Si_{41}$ using this relation extrapolates to a value of E ~ 150 GPa, which is well within the range of the other Ba–Si-based compounds (see Fig. 10.6). In case of the Ge-based samples the bulk modulus for $Sr_8Ga_{16}Ge_{30}$ reported by Meng et al. [3] is much higher than the values found by Okamoto et al. [18] and Zerec et al. [23], based on RM and RUS measurements, respectively. Whereas Okamoto et al. [18] and Zerec et al. [23] devoted their work to the mechanical properties, Meng et al. [3] estimated B using only a few data points from X-ray diffraction at different pressures. Analyzing the data presented in Fig. 10.6 and Table 10.2 results in the general conclusion that Si-based compounds exhibit higher values of the elastic moduli than Ge-based compounds. Comparing the binary compounds of clathrate type-I and type-IX, the bulk modulus of $Ba_8Ge_{43}\square_3$ [26] is only 56 % of the bulk modulus reported for $Ba_8Si_{46}$ [7]. In contrast to this, B of $Ba_6Ge_{25}$ [14] is even 68 % of the value reported for $Ba_6Si_{25}$ [15]. The smaller value for $Ba_8Ge_{43}\square_3$ most likely arises from the presence of three vacancies per formula unit. This suggestion is affirmed, when the Young's moduli of the vacancy free clathrate-I phases $Ba_8Cu_5Si_{46}$ (127(4) GPa) and $Ba_8Cu_5Ge_{46}$ (93(9) GPa) are compared to each other, resulting in a 27 % smaller value for the Ge-based compound. The elastic moduli, E, B, and G are all decreasing linearly in $Ba_8Cu_5Si_{41-x}Ge_x$ with increasing Ge-content (Fig. 10.7), reflecting the weakening of the overall

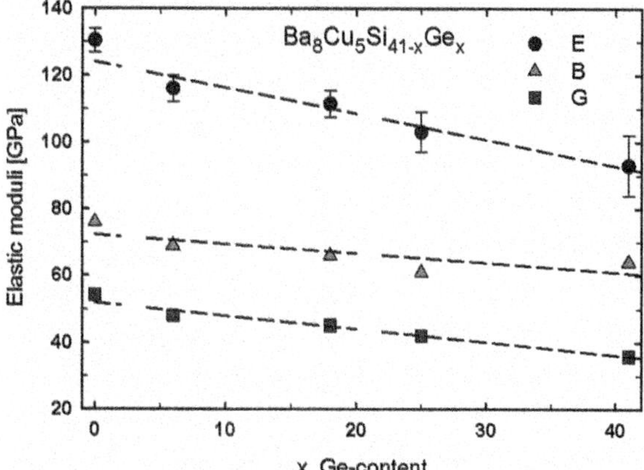


**Fig. 10.7** Elastic moduli as a function of the Ge-content for $Ba_8Cu_5Si_{41-x}Ge_x$. The dashed lines correspond to linear fits to the data points

bonding strength. A similar tendency was already reported for the thermal expansion [4], and the Vickers hardness (see Fig. 10.4). It is interesting to note, that linear fits reveal an almost two times larger change of the Young's modulus as a function of the Ge-content, compared to the influence on B and G, for which almost similar slopes of –0.37 and –0.39 were derived.

Besides the major influence of the framework building atoms also the filler atoms show a significant contribution to the magnitude of the elastic moduli. Okamoto et al., while investigating elastic constants of $Ba_8Ga_{16}Ge_{30}$ ($C_{11}$ at 300 K ~104 GPa) and $Sr_8Ga_{16}Ge_{30}$ ($C_{11}$ at 300 K ~85 GPa), attributed the smaller values of E and G to the smaller radius of Sr-atoms [18]. Taking also the result of Zerec et al. [23] on $Eu_8Ga_{16}Ge_{30}$ ($C_{11}$ at 300 K ~99 GPa) into account this trend is affirmed. A detailed analysis using single crystal neutron diffraction at low temperatures for those three compounds [75] revealed a significant difference between the Sr-, Ba- and Eu-guest atoms in the large tetrakaidecahedral void of the type-I clathrate: whereas the nuclear density of Ba is located right in the center of this cage, Sr- and Eu-atoms tend to an off-center site and tunnelling between the four possible positions [75]. So guest atoms occupying a larger volume of the cages have a stabilizing effect onto the whole clathrate-I structure.

The isotropic elastic moduli, E, B, and G, as a function of temperature above RT are hitherto available only for a few compounds and are displayed in Fig. 10.8. Within the temperature range investigated, E, B, and G are mostly changing linearly with temperature. For $Ba_8Zn_7Ge_{39}$ the data points in the high-T region at 773 and 873 K decline significantly, most pronounced for the bulk modulus (Fig. 10.8b). Interestingly, linear fits to the temperature dependence reveal similar results as observed for the dependence on the Si/Ge-ratio (Fig. 10.7): whereas B(T) and G(T) in all cases exhibit slopes of $\Delta B/\Delta T$ or $\Delta G/\Delta T \sim -0.01$, $\Delta E/\Delta T \sim -0.02$ is much steeper as the Poisson's ratio remains almost constant as a function of temperature. Only $\Delta B/\Delta T = -0.015$ of $Ba_8Cu_5Si_{41}$ is significantly larger than for the other samples also reflected by a slightly smaller Poisson's ratio (see Table 10.2). However, as the temperature dependence of all samples within

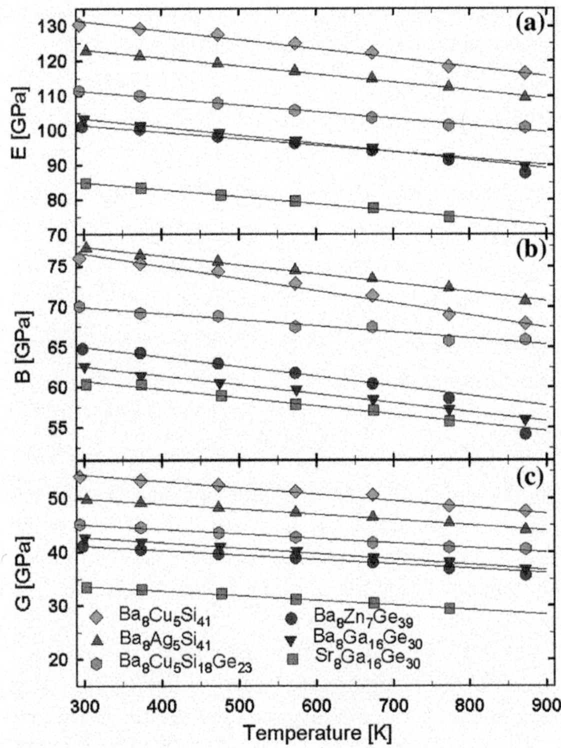


**Fig. 10.8** Elastic moduli as a function of temperature for $Ba_8Cu_5Si_{41}$, $Ba_8Ag_5Si_{41}$, $Ba_8Cu_5Si_{18}Ge_{23}$, $Ba_8Zn_7Ge_{39}$, $Sr_8Ga_{16}Ge_{30}$ [18], and $Ba_8Ga_{16}Ge_{30}$ [18]. The black lines are linear fits to the data points

the T-range investigated is almost similar, on the basis of the few data available it can be concluded, that the T-dependence of the isotropic elastic moduli is completely independent of the nature of the framework as well as the guest atoms.

Vickers hardness (HV) values as a function of the Young's (E) and the Shear moduli (G) are plotted in Fig. 10.9. A general trend for higher HV-values for samples with larger E- and G-moduli can be observed, although no strictly linear dependence for all samples is found. Nevertheless using the simplest linear fit equation ($HV = c_E \times E$ and $HV = c_G \times G$) we find $c_E = 5.8$ (Fig. 10.9a) and $c_G = 14.1$ (Fig. 10.9b). The dotted lines engulf deviations of ±10 %.

### 10.3.3 Compression Tests

Compression tests were performed for two single crystalline specimens, $Ba_8Ga_{16}Ge_{30}$ and $Sr_8Ga_{16}Ge_{30}$, both crystallizing in the type-I structure, for which the most readily operating slip system is $\{001\}\langle 100\rangle$ [18]. According to the authors even at temperatures $T \geq 90$ % of the melting temperature premature

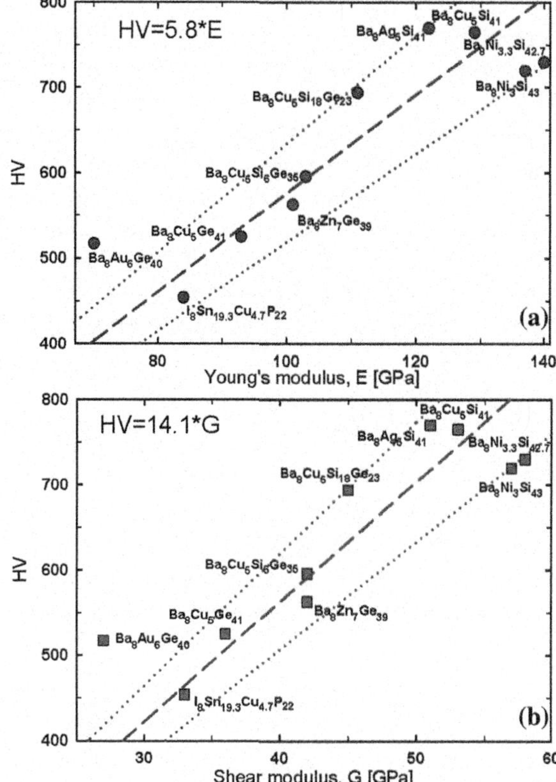


**Fig. 10.9** Vickers hardness as a function of the elastic moduli E (Fig. 10.8a) and G (Fig. 10.8b). The dashed lines correspond to linear fits to the data points (the dotted lines represent deviations of ±10 %)

fracture occurs and no appreciable plastic strain is observed. The samples fracture at applied stresses of 335 and 236 MPa for $Ba_8Ga_{16}Ge_{30}$ and $Sr_8Ga_{16}Ge_{30}$, respectively [18].

### *10.3.4 Comparison to Theoretical Calculations*

Although lots of data on the elastic properties of clathrates are available from theoretical studies (mainly the bulk moduli), within this discussion we confine ourselves to a comparison of the experimentally determined values to the theoretically calculated data wherever available. Based on the local density approximation (LDA) Kitano et al. [138] reported for the bulk modulus B = 85 GPa (88 GPa, assuming $B' = 3.7$, accounting for the strong sensitivity to the pressure derivative B') extracted from a fit of the total energy to the volume using the Murnaghan EOS. This result is in excellent agreement with B = 93(5) GPa found experimentally [7]. The bulk modulus for $I_8Si_{44.5}I_{1.5}$, B = 95(5) GPa reported by the same authors [7], matches nicely with B = 91 GPa calculated via the LDA-approximation by Connetable et al. [139] for $I_8Si_{46}$. For the guest free clathrate-II $Si_{136}$ several theoretically calculated bulk moduli have been reported [6, 37, 140–142], resulting in B ~ 80 GPa, which is slightly smaller than B = 90 GPa found experimentally [5, 6]. This deviation might arise from the synthesis procedure starting from NaSi precursors and a subsequent removal of the Na, which is probably not completely achieved. For $Sr_8Ga_{16}Ge_{30}$ and $Ba_8Ga_{16}Ge_{30}$ calculated bulk moduli are reported in the literature, which are differing from each other. Whereas for $Sr_8Ga_{16}Ge_{30}$ B = 67.7 GPa and for $Ba_8Ga_{16}Ge_{30}$ B = 66.5 GPa [143] are in good agreement with the experimental results of Okamoto et al. [18], the corresponding values of 47.32 and 48.58 GPa [144], respectively, are significantly smaller.

### *10.3.5 Evaluation of Internal Thermal Stresses*

Following the model proposed by Okamoto et al. [18], in order to derive the magnitude of internal thermal stresses ($\tau_{int}$) of a thermoelectric material during operation conditions and assuming a uniform temperature gradient, we use:

$$\tau_{int} = \gamma G = \frac{\alpha \Delta T w}{2\ell} G = \frac{G w \alpha}{2} \times \frac{\Delta T}{\ell} = G\alpha \times \frac{w}{2\ell} \times \Delta T \qquad (10.12)$$

with $\gamma$, the shear stress, G, the shear modulus, $\alpha$, the thermal expansion coefficient, w and $\ell$, the width and the length of the thermoelectric leg, and $\Delta T$, the temperature difference between the hot and the cold site of an thermoelectric generator (TEG). The shear stresses calculated via Eq. (10.12) are generated perpendicular to the applied temperature difference due to the different thermal expansion at the hot and cold site of the TEG, and are dependent on the dimensions of the thermoelectric leg [18]. For our calculations we assume the dimensions of the legs in the TEG to be: w = 2 mm and $\ell$ = 5 mm, and temperature gradients, $\Delta T$, of 100–700 K. The thermal expansion coefficients are taken from the compilation for several clathrate-I compounds [4], also listed in Table 10.4. From the calculated results (plotted in Fig. 10.10) it is interesting to note that although both, thermal expansion coefficients as well as shear moduli, depend on the nature of the framework atoms, the calculated internal stresses do not differ too much for $Ba_8Cu_5Si_{41}$, $Ba_8Cu_5Si_6Ge_{35}$ and $Ba_8Cu_5Ge_{41}$. As can be seen from Eq. (10.12), the product of these two quantities determines $\tau_{int}$ and in case of the Cu-containing clathrates results in rather similar values. Okamoto et al. reported for $Sr_8Ga_{16}Ge_{30}$ and $Ba_8Ga_{16}Ge_{30}$ shear stresses above 100 MPa at which fracture was observed [18]. Using the dimensions of w = 2 mm and $\ell$ = 5 mm for all the compounds investigated up to a temperature gradient $\Delta T$ = 700 K the calculated internal stress remains well below this critical range.

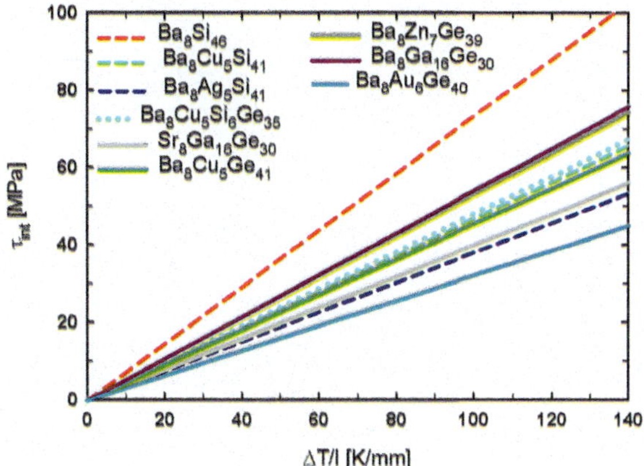


**Fig. 10.10** Internal stresses, $\tau_{int}$, calculated according to Eq. (10.12) for selected clathrates

## 10.3.6 Debye and Einstein Temperatures

Besides all the methods discussed in Sect. 10.2.5 Debye temperatures can be calculated for isotropic bulk samples from elastic properties. According to Anderson [145] $\theta_D$ can be calculated from the sound velocity, $v_M$, by:

$$\theta_D = \frac{h}{k_B}\left(\frac{3nNd}{4M\pi}\right)^{1/3} v_M \tag{10.13}$$

where h is Planck's constant, $k_B$ is the Boltzmann-constant, n is the number of atoms in the asymmetric unit, d is the density, M the molecular weight, and $v_M$ is

$$v_M = \left[\frac{1}{3}\left(\frac{2}{v_T^3} + \frac{1}{v_L^3}\right)\right]^{-1/3} \tag{10.14}$$

with the transversal sound velocity, $v_T$, and the longitudinal sound velocity, $v_L$, defined in Eqs. (10.6) and (10.7). Results of the calculations are presented in Table 10.2. Generally, the Debye temperatures of the Si-based clathrates are significantly higher ($\theta_{D,\ avg} = 433$ K) than the values extracted for the Ge-based clathrates ($\theta_{D,\ avg} = 289$ K). The hitherto reported literature data for Debye and Einstein temperatures are listed together with the measurement method in Table 10.3. The $\theta_D$-values extracted from mechanical properties coincide fairly well with the values derived from other properties. However, in case of the Si-based clathrates the Debye temperatures extracted from fits to thermal conductivity or thermal expansion data are lower than the $\theta_D$-values found within this work from mechanical properties and usually extracted from temperature dependent diffraction experiments (see Tables 10.2 and 10.3). In contrast to this, the Debye temperatures extracted from the ADPs for $Ba_8Ni_3Si_{43}$ [52], $Ba_8Ag_5Si_{41}$ [104], and

**Table 10.3** Debye ($\theta_D$) and Einstein ($\theta_E$) temperatures of intermetallic clathrates (in [K])

| Compound | $\theta_D$ | Meth. | $\theta_E^a$ | | | | Meth. | Ref. |
|---|---|---|---|---|---|---|---|---|
| | | | $\theta_{E,1}$ | $\theta_{E,2}$ | $\theta_{E,3}$ | $\theta_{E,4}$ | | |
| ***C-based*** | | | | | | | | |
| *Type-I* | | | | | | | | |
| $C_{46}$ | 1996 | Calc | | | | | | [38] |
| *Type-II* | | | | | | | | |
| $C_{136}$ | 189 | Calc | | | | | | [33] |
| | 2083 | Calc | | | | | | [38] |
| ***Si-based*** | | | | | | | | |
| *Type-I* | | | | | | | | |
| $Si_{46}$ | 622 | Calc | | | | | | [36] |
| | 551 | Calc | | | | | | [38] |
| $Na_8Si_{46}$ | 386[b] | EXAFS | | 79 | | | EXAFS | [157] |
| | 570 | XPD | 107 | 134 | | | XPD | [80] |
| | 540(25)[c] | | | | | | | |
| | 371 | $C_p$ | | | | | | [110] |
| | | | 209 | 120 | | | INS | [43] |
| | | | 267 | 185 | | | XPD | |
| | | | 218 | 97-150[d] | | | Calc | |
| | 542 | $C_p$ | 162 | 163 | 94 | | $C_p$ | [97] |
| | | | 170 | 147 | 110 | | XSC | |
| $Sr_8Al_{10}Si_{36}$ | 300 | ElRes | | | | | | [51] |
| $Sr_8Al_6Ga_{10}Si_{30}$ | | | | 80 | 54 | 80 | XPD | [70] |
| | | | 62 | 75 | | | RA | |
| $Sr_8Al_{10}Ga_6Si_{30}$ | | | | 80 | | | XPD | [70] |
| | | | 78 | 85 | 108 | | RA | |
| $Sr_8Ga_{16}Si_{30}$ | 370 | $C_p$ | 59 | 120 | | | $C_p$ | [119] |

(continued)

**Table 10.3** (continued)

| Compound | $\theta_D$ | Meth. | $\theta_E^a$ | | | | Meth. | Ref. |
|---|---|---|---|---|---|---|---|---|
| | | | $\theta_{E,1}$ | $\theta_{E,2}$ | $\theta_{E,3}$ | $\theta_{E,4}$ | | |
| $Sr_2Ba_6Ga_{16}Si_{30}$ | | | 104 | 88 | 56 | | XPD | [81] |
| $Ba_8Si_{46}$ | | | 167 | 91 | | | INS | [43] |
| | | | 147 | 93 | | | XPD | |
| | | | 169 | 100-123[d] | | | Calc | |
| $Ba_8Si_{46}$ | 370 | $C_p$ | | | | | | [158] |
| $Ba_8Al_{12}Si_{33}$ | 300 | ElRes | | | | | | [50] |
| $Ba_8Al_{16}Si_{30}$ | 300 | ElRes | | | | | | [50] |
| $Ba_8Ni_{2.6}Si_{43.4}$ | 450 | $C_p$ | | | | | | [114] |
| $Ba_8Ni_{2.7}Si_{43.3}$ | 460 | $C_p$ | | | | | | [114] |
| $Ba_8Ni_{2.8}Si_{43.2}$ | 474 | $C_p$ | | | | | | [114] |
| $Ba_8Ni_{2.9}Si_{43.1}$ | 448 | $C_p$ | | | | | | [114] |
| $Ba_8Ni_3Si_{43}$ | 511 | XSC | | 109 | 91 | | XSC | [52] |
| | 415 | $C_p$ | | | | | | |
| | 357 | $C_p$ | 45.3 | 90.8 | | | $C_p$ | |
| | 440 | Elres | | | | | | |
| $Ba_8Ni_{3.1}Si_{42.9}$ | 424 | $C_p$ | | | | | | [114] |
| $Ba_8Ni_{3.3}Si_{42.7}$ | 560 | Elres | | | | | | [52] |
| $Ba_8Ni_{3.4}Si_{42.6}$ | 430 | $C_p$ | | | | | | [114] |
| $Ba_8Ni_{3.8}Si_{42.2}$ | 431 | $C_p$ | | | | | | [114] |
| $Ba_8Ni_{3.8}Si_{42.2}$ | 450 | Elres | | | | | | [52] |
| $Ba_8Cu_{4.5}Si_{41.5}$ | 287 | FTE | | 59 | | | FTE | [4] |
| $Ba_8Cu_5Si_{41}$ | 420 | XSC | | 103 | 80 | | XSC | [98] |
| $Ba_8Cu_5Si_{41}$ | 456 | XSC | | 102 | 80 | | XSC | [99] |
| $Ba_8Cu_6Si_{40}$ | 300 | ElRes | | | | | | [50] |
| $Ba_8Cu_4Ga_4Si_{38}$ | 300 | ElRes | | | | | | [50] |

(continued)

**Table 10.3** (continued)

| Compound | $\theta_D$ | Meth. | $\theta_E^a$ | | | | Meth. | Ref. |
|---|---|---|---|---|---|---|---|---|
| | | | $\theta_{E,1}$ | $\theta_{E,2}$ | $\theta_{E,3}$ | $\theta_{E,4}$ | | |
| $Ba_8Zn_7Si_{39}$ | 200 | TC | | 99 | 80 | | XSC | [59] |
| $Ba_8Zn_7Si_{39}$ | 205 | FTE | | 80 | | | FTE | [4] |
| $Ba_8Zn_7Si_{39}$ | 455 | XSC | | 98 | 75 | | XSC | [99] |
| $Ba_8Ga_{10}Si_{36}$ | 352.17 | XPD | 104.61 | 61.73 | | | XPD | [82] |
| $Ba_8Ga_{14.64}Si_{31.36}$ | 327 | XPD | 98 | 89 | 57 | | XPD | [83] |
| $Ba_8Ga_{14.86}Si_{31.14}$ | 324 | XPD | 102 | 88 | 57 | | XPD | [83] |
| $Ba_8Ga_{15.10}Si_{30.90}$ | 353 | XPD | 100 | 96 | 55 | | XPD | [83] |
| $Ba_8Ga_{15.88}Si_{30.12}$ | 317 | XPD | 97 | 89 | 53 | | XPD | [83] |
| $Ba_8Ga_{16}Si_{30}$ | | | 146.3[e] | 106.1[e] | 58.7[e] | | Calc | [41] |
| $Ba_8Ga_{16}Si_{30}$ | 336(3) | NSC | 124(2) | 98(7) | 69(1) | | NSC | [76] |
| $Ba_8Ga_8Si_{36}$ | 300 | ElRes | | | | | | [50] |
| $Ba_8Ga_{16}Si_{30}$ | 346.25 | XPD | 114.06 | 76.52 | | | XPD | [82] |
| $Ba_8Ga_{16}Si_{30}$ | 430 | $C_p$ | | | | | | [74] |
| | 420 | NPD | | | | | | |
| $Ba_8Ga_{16}Si_{30}$ | 416 | XPD | 127 | 101 | 77 | | XPD | [86] |
| | | | 146 | 106 | 59 | | Calc | |
| $Ba_8Ga_{16}Si_{30}$ | | | 126 | 88 | 50 | | XPD | [81] |
| $Ba_8Ga_{16}Si_{30}$ | 520(67) | EXAFS | | | | | | [159] |
| $Ba_8Rh_{2.4}Si_{43.6}$ | 508 | XSC | | 115 | 99 | | XSC | [99] |
| $Ba_8Rh_{2.74}Si_{42.61}\square_{0.65}$ | 398 | $C_p$ | 83.1(3) | | | | $C_p$ | [115] |
| $Ba_8Pd_{3.9}Si_{42.1}$ | 242 | FTE | 100 | | | | FTE | [4] |
| $Ba_8Pd_4Si_{42}$ | 452 | XSC | | 114 | 88 | | XSC | [99] |
| $Ba_8Ag_{4.9}Si_{41.1}$ | 427 | XSC | | 108 | 81 | | XSC | [53] |
| | 474 | Elres | | | | | | |
| $Ba_8Ag_5Si_{41}$ | 290 | FTE | 57 | | | | FTE | [4] |

(continued)

**Table 10.3** (continued)

| Compound | $\theta_D$ | Meth. | $\theta_E^a$ | | | | Meth. | Ref. |
|---|---|---|---|---|---|---|---|---|
| | | | $\theta_{E,1}$ | $\theta_{E,2}$ | $\theta_{E,3}$ | $\theta_{E,4}$ | | |
| $Ba_8Pt_{3.6}Si_{42.4}$ | 415 | XSC | | 114 | 92 | | XSC | [99] |
| $Ba_8Pt_{3.8}Si_{42.2}$ | 215 | FTE | | 96 | | | FTE | [4] |
| $Ba_8Au_5Si_{41}$ | 290 | FTE | | 50 | | | FTE | [4] |
| $Ba_8Au_{4.10}Si_{41.90}$ | 412 | $C_p$ | | | | | | [120] |
| $Ba_8Au_{4.85}Si_{41.15}$ | 418 | $C_p$ | | | | | | [120] |
| $Ba_8Au_{5.10}Si_{40.90}$ | 404 | $C_p$ | | | | | | [120] |
| $Ba_8Au_{5.14}Si_{40.84}$ | 407 | $C_p$ | | | | | | [120] |
| $Ba_8Au_{5.43}Si_{40.57}$ | 390 | $C_p$ | | | | | | [120] |
| $Ba_8Au_{5.59}Si_{40.41}$ | 371 | $C_p$ | | | | | | [120] |
| $Ba_8Au_{5.76}Si_{40.24}$ | 362 | $C_p$ | | | | | | [120] |
| $Ba_8Au_{6.10}Si_{39.90}$ | 335 | $C_p$ | | | | | | [120] |
| $Ba_8Au_{4.10}Si_{41.90}$ | 358 | $C_p$ | | 79 | | | $C_p$ | [120] |
| | | | | 101 | | | $C_p$ | |
| $Ba_8Au_{4.85}Si_{41.15}$ | 353 | $C_p$ | | 78 | | | $C_p$ | [120] |
| | | | | 94 | | | $C_p$ | |
| $Ba_8Au_{5.59}Si_{40.41}$ | 343 | $C_p$ | | 76 | | | $C_p$ | [120] |
| | | | | 84 | | | $C_p$ | |
| $Ba_8Au_{6.10}Si_{39.90}$ | 326 | $C_p$ | | 73 | | | $C_p$ | [120] |
| | | | | 95 | | | $C_p$ | |
| $Ba_8Au_{5.1}Si_{40.9}$ | 270 | TC | | | | | | [60] |
| $Ba_6Eu_2Ga_{16}Si_{30}$ | | | 79 | 74 | 55 | | XPD | [81] |
| $Te_{6.78}Se_{1.22}[Si,P]_{46}$ | 491(2) | XSC | 153(1) | 108(1) | | | XSC | [100] |
| *Type-II* | | | | | | | | |
| $Si_{136}$ | 633 | Calc | | | | | | [36] |
| $Si_{136}$ | 21.05 | Calc[c] | | | | | | [34] |

(continued)

**Table 10.3** (continued)

| Compound | $\theta_D$ | Meth. | $\theta_E^a$ | | | | Meth. | Ref. |
|---|---|---|---|---|---|---|---|---|
| | | | $\theta_{E,1}$ | $\theta_{E,2}$ | $\theta_{E,3}$ | $\theta_{E,4}$ | | |
| $Si_{136}$ | 516 | Calc | | | | | | [38] |
| $Si_{136}$ | 469 | $C_p$ | | | | | | [111] |
| $Si_{136}$ | 610 | Calc | | | | | | [37] |
| $Na_{24}Si_{136}$ | | | | 186 | | | XSC | [101] |
| | | | | 84 | | | XSC | |
| $Na_{24}Si_{136}$ | 575 | $C_p$ | | 192.1 | | | $C_p$ | [128] |
| | | | | 54.9 | | | $C_p$ | |
| $Rb_8Na_{16}Si_{136}$ | 400 | XSC | 79.1(Rb) | 187.0(Na) | | | XSC | [102] |
| $Rb_8Na_{16}Si_{136}$ | | | 70.5(Rb) | 175(Na) | | | Calc | [46] |
| $Rb_8Ga_8Si_{128}$ | | | | 60 | | | Calc | [46] |
| $Cs_8Ga_8Si_{128}$ | | | | 73 | | | Calc | [46] |
| $Cs_8Na_{16}Si_{136}$ | 416 | XSC | 76.8(Cs) | 202.9(Na) | | | XSC | [102] |
| $Cs_8Na_{16}Si_{136}$ | | | 82(Cs) | | | | RA | [42] |
| | | | 92(Cs) | | | | Calc | |
| $Cs_8Na_{16}Si_{136}$ | | | 94-96(Cs) | 170-174(Na) | | | Calc | [44] |
| *Type-IX* | | | | | | | | |
| $Ba_{24}Si_{100}$ | 515 | $C_p$ | | | | | | [112] |
| ***Si/Ge-based*** | | | | | | | | |
| *Type-I* | | | | | | | | |
| $Sr_8Ga_{16}Si_{25}Ge_5$ | 355 | $C_p$ | 56 | 115 | | | $C_p$ | [119] |
| $Sr_8Ga_{16}Si_{15}Ge_{15}$ | | | | 63 | | | Calc | [47] |
| $Sr_8Ga_{16}Si_{10}Ge_{20}$ | 315 | $C_p$ | 45.5 | 100 | | | $C_p$ | [119] |
| $Sr_8Ga_{16}Si_5Ge_{25}$ | | | 72 | | | | Calc | [47] |
| $Sr_8Ga_{16}Si_4Ge_{26}$ | 300 | $C_p$ | 40.5 | 94 | | | $C_p$ | [119] |

(continued)

**Table 10.3** (continued)

| Compound | $\theta_D$ | Meth. | $\theta_E^a$ | | | | Meth. | Ref. |
|---|---|---|---|---|---|---|---|---|
| | | | $\theta_{E,1}$ | $\theta_{E,2}$ | $\theta_{E,3}$ | $\theta_{E,4}$ | | |
| $Ba_8Cu_5Si_{18}Ge_{23}$ | 310 | XSC | | 96 | 72 | | XSC | [98] |
| $Ba_8Cu_5Si_6Ge_{35}$ | 270 | XSC | | 92 | 69 | | XSC | [98] |
| $Ba_8Zn_{7.3}Si_{28.1}Ge_{10.6}$ | 200 | TC | | | | | | [61] |
| $Ba_8Ga_{16}Si_{15}Ge_{15}$ | | | | 66 | | | Calc | [47] |
| $Ba_8Ga_{16}Si_5Ge_{25}$ | | | | 67 | | | Calc | [47] |
| $Ba_8In_{15.3}Si_{1.6}Ge_{29.1}$ | 203 | $C_p$ | | 100 | | | $C_p$ | [160] |
| | | | | 39 | | | $C_p$ | |
| ***P-based*** | | | | | | | | |
| *Type-I* | | | | | | | | |
| $Ba_8Cu_{16}P_{30}$ | 375 | XPD | 123[f] | 88[f] | | | XPD | [84] |
| | 371 | $C_p$ | 120 | 90 | | | $C_p$ | |
| $Ba_8Cu_{16}P_{30}$ | 235 | $C_p$ | 449 | 85 | | | $C_p$ | [121] |
| ***Ge-based*** | | | | | | | | |
| $Ge_{46}$ | 320 | Calc | | | | | | [38] |
| $Sr_8Zn_8Ge_{38}$ | 290 | $C_p$ | | | | | | [74] |
| | 270 | NPD | | | | | | |
| $Sr_8Ga_{16}Ge_{30}$ | 271 | NSC | | 85 | | | NSC | [73] |
| | 300 | $C_p$ | | 50 | | | $C_p$ | [73] |
| $Sr_8Ga_{16}Ge_{30}$ | 270 | TC | 103 | 74 | | | XSC | [72] |
| $Sr_8Ga_{16}Ge_{30}$ | | | 101.5[e] | 120.1[e] | 107.0[e] | | Calc | [41] |
| $Sr_8Ga_{16}Ge_{30}$ | 358 | $C_p$ | | 79 | | | $C_p$ | [96] |
| $Sr_8Ga_{16}Ge_{30}$ | 300 | $C_p$ | | 53 | | | $C_p$ | [75] |
| $Sr_8Ga_{16}Ge_{30}$ | 320 | $C_p$ | | | | | | [74] |
| | 290 | NPD | | | | | | |
| $Sr_8Ga_{16}Ge_{30}$ | | | 102(10) [g] | 128(10)[g] | 126(10)[g] | 156(10)[g] | EXAFS | [162] |

(continued)

**Table 10.3** (continued)

| Compound | $\theta_D$ | Meth. | $\theta_E^a$ | | | | Meth. | Ref. |
|---|---|---|---|---|---|---|---|---|
| | | | $\theta_{E,1}$ | $\theta_{E,2}$ | $\theta_{E,3}$ | $\theta_{E,4}$ | | |
| | | | 95(10) [g] | 147(10)[g] | | | EXAFS | |
| $Sr_8Ga_{16}Ge_{30}$ | 313 | XPD | 151 | 163 | 104 | | XPD | [86] |
| | | | 33 | 80 | | | $C_p$ | |
| $Sr_8Ga_{16}Ge_{30}$ | | | | 48.7 | | | INS | [66] |
| $Sr_8Ga_{16}Ge_{30}$ | | | | 49.9 | 42.9 | | Calc | [45] |
| $Sr_8Ga_{16}Ge_{30}$ | 288 | $C_p$ | 35 | 90 | | | $C_p$ | [119] |
| $Sr_8Ga_{16}Ge_{30}$ | | | | 60 | | | Calc | [47] |
| $Sr_8Ga_{16}Ge_{30}$ | 310 | $C_p$ | | 50 | | | $C_p$ | [122] |
| $Ba_8Ge_{43}\square_3$ | 225 | $C_p$ | 57 | 72 | | | $C_p$ | [123] |
| $Ba_8Ge_{43}\square_3$ | 299 | FTE | | 63 | | | FTE | [4] |
| $Ba_8Al_4Ge_{42}$ | 380 | Elres | | | | | | [161] |
| $Ba_8Al_8Ge_{38}$ | 330 | Elres | | | | | | [161] |
| $Ba_8Al_{16}Ge_{30}$ | | | | | | | | [103] |
| Czo | 291(3) | XSC | 106(1) | 85(1) | 61(1) | 84(1) | XSC | |
| Norm | 277(6) | XSC | 101(1) | 81(1) | 64(1) | 81(1) | XSC | |
| Flux | 293(4) | XSC | 101(1) | 85(1) | 69(1) | 83(1) | XSC | |
| $Ba_8Al_{16}Ge_{30}$ | 380 | Elres | | | | | | [161] |
| $Ba_8Ni_3Ge_{43}$ | 275.8(4) | $C_p$ | 57.0(5) | 80.5(3) | | | $C_p$ | [124] |
| $Ba_8Ni_4Ge_{42}$ | 281.5(2) | $C_p$ | 57.0(3) | 82.0(5) | | | $C_p$ | [124] |
| $Ba_8Ni_{5.8}Ge_{40.2}$ | 269 | $C_p$ | 109 | 95 | 79 | | XPD | [87] |
| | | | 108 | 108 | 63.4 | | $C_p$ | |
| $Ba_8Ni_6Ge_{40}$ | 268(8) | XSC | 103(3) | 85(2) | 75.4(1.7) | | XSC | [78] |
| $Ba_8Ni_6Ge_{40}$ | 307 | $C_p$ | | 75 | | | $C_p$ | [116] |
| $Ba_8Ni_6Ge_{40}$ | 253.2(5) | $C_p$ | 50.5(5) | 77.0(5) | | | $C_p$ | [124] |
| $Ba_8Ni_{0.8}Zn_{6.4}Ge_{38.8}$ | 300 | Elres | | | | | | [54] |

(continued)

Table 10.3 (continued)

| Compound | $\theta_D$ | Meth. | $\theta_E^a$ | | | | Meth. | Ref. |
|---|---|---|---|---|---|---|---|---|
| | | | $\theta_{E,1}$ | $\theta_{E,2}$ | $\theta_{E,3}$ | $\theta_{E,4}$ | | |
| $Ba_8NiZn_{6.6}Ge_{38.4}$ | 200 | FTE | | 62 | | | FTE | [4] |
| $Ba_8NiZn_7Ge_{38}$ | 233 | FTE | | 64 | | | FTE | [4] |
| $Ba_8Ni_{2.8}Zn_{1.4}Ge_{38.8}$ | 236 | Elres | | | | | | [54] |
| $Ba_8Cu_5Ge_{41}$ | 294 | FTE | | 80 | | | FTE | [4] |
| $Ba_8Cu_{5.3}Ge_{40.7}$ | 289.12 | $C_p$ | | 50.85 | 80 | | $C_p$ | [125] |
| $Ba_8Cu_6Ge_{40}$ | | | 92(2) | 64.3(4) | | | XPD | [88] |
| $Ba_8Cu_6Ge_{40}$ | 258(11) | XSC | 106(5) | 82(2) | 63.1(8) | | XSC | [78] |
| $Ba_8Cu_6Ge_{40}$ | 306 | $C_p$ | | 70 | | | $C_p$ | [117] |
| $Ba_8Cu_6Ge_{40}$ | 258(11) | XSC | 106(5) | 85(2) | 63(1) | 74(1) | XSC | [69] |
| | 283(2) | XPD | 115(1) | 94(3) | 66(1) | 75(1) | XPD | |
| | | | | 69.4(2) | 57.2(1) | | INS | |
| $Ba_8Zn_{2.1}Ge_{41.5}\square_{2.4}$ | 210 | FTE | | 54 | | | FTE | [4] |
| $Ba_8Zn_{5.7}Ge_{40}\square_{0.3}$ | 220 | FTE | | 51 | | | FTE | [4] |
| $Ba_8Zn_{4.6}Ge_{40.4}\square_{1.4}$ | 235 | Elres | | | | | | [55] |
| $Ba_8Zn_{5.7}Ge_{40.0}\square_{0.3}$ | 210 | Elres | | | | | | [55] |
| $Ba_8Zn_{7.33}Ge_{38.67}$ | 220 | Elres | | | | | | [54] |
| $Ba_8Zn_{7.62}Ge_{38.38}$ | 250 | Elres | | | | | | [54] |
| $Ba_8Zn_{7.7}Ge_{38.3}$ | 225 | $C_p$ | | 85 | 64 | | XSC | [55] |
| | | | | 48.4 | 84.8 | | $C_p$ | |
| | | | | 55 | 90 | | INS | |
| $Ba_8Zn_{7.7}Ge_{38.3}$ | 226 | FTE | | 53 | | | FTE | [4] |
| $Ba_8Zn_{7.7}Ge_{38.3}$ | 267 | XSC | | 84 | 63 | | XSC | [99] |
| $Ba_8Zn_8Ge_{38}$ | | | | | | | | [77] |
| Slow cooled | 279 | NSC | 118 | 98 | 62 | | NSC | |
| Fast cooled | 280 | NSC | 149 | 98 | 67 | | NSC | |

(continued)

**Table 10.3** (continued)

| Compound | $\theta_D$ | Meth. | $\theta_E^a$ | | | | Meth. | Ref. |
|---|---|---|---|---|---|---|---|---|
| | | | $\theta_{E,1}$ | $\theta_{E,2}$ | $\theta_{E,3}$ | $\theta_{E,4}$ | | |
| $Ba_8Zn_8Ge_{38}$ | 251(5) | XSC | 105(3) | 84(1) | 60.0(3) | | XSC | [78] |
| $Ba_8Zn_8Ge_{38}$ | 282.29 | $C_p$ | | 47.3 | 78.5 | | $C_p$ | [125] |
| $Ba_8Ga_{16}Ge_{30}$ | 275 | XSC | | 51 | | | XSC | [95] |
| $Ba_8Ga_{16}Ge_{30}$ | | | 128.9[e] | 78.9[e] | 55.0[e] | | Calc | [41] |
| $Ba_8Ga_{16}Ge_{30}$ | 300 | XSC | 121 | 72 | | | XSC | [96] |
| | 355 | $C_p$ | | 80 | | | $C_p$ | |
| $Ba_8Ga_{16}Ge_{30}$ | 300 | $C_p$ | | 64 | | | NSC | [75] |
| | | | | 60 | | | $C_p$ | |
| $Ba_8Ga_{16}Ge_{30}$ | 324(4) | $C_p$ | 78(2) | 38(1) | | | $C_p$ | [126] |
| $Ba_8Ga_{16}Ge_{30}$ | 312 | XPD | 124 | 101 | 73 | | XPD | [113] |
| | | | | 80 | 42 | | $C_p$ | |
| $Ba_8Ga_{16}Ge_{30}$ | | | | 54.5 | | | INS | [66] |
| $Ba_8Ga_{16}Ge_{30}$ | | | | 20.9 | 38.3 | | Calc | [45] |
| $Ba_8Ga_{16}Ge_{30}$ | 288 | $C_p$ | 74 | 42 | | | $C_p$ | [127] |
| n-$Ba_8Ga_{16}Ge_{30}$ | 288 | TC, $C_p$ | 87 | 49 | | | TC, $C_p$ | [63] |
| p-$Ba_8Ga_{16}Ge_{30}$ | 288 | TC, $C_p$ | 87 | 49 | | | TC, $C_p$ | [63] |
| p-$Ba_8Ga_{16}Ge_{30}$ | 276(2) | NSC | 108(1) | 90(2) | 62(1) | 88(2) | NSC | [79] |
| p-$Ba_8Ga_{16}Ge_{30}$ | 267(1) | XSC | 104(1) | 87(2) | 60(1) | 82(1) | XSC | [79] |
| n-$Ba_8Ga_{16}Ge_{30}$ | 259(2) | NSC | 109(1) | 89(4) | 59(1) | 88(6) | NSC | [79] |
| n-$Ba_8Ga_{16}Ge_{30}$ | 274(3) | XSC | 108(1) | 84(1) | 60(1) | 81(1) | XSC | [79] |
| n-$Ba_8Ga_{16}Ge_{30}$ | 410 | EXAFS | | 80(5) | | | EXAFS | [146] |
| p-$Ba_8Ga_{16}Ge_{30}$ | 416 | EXAFS | | 80(5) | | | EXAFS | [146] |
| $Ba_8Ga_{16}Ge_{30}$ | | | | 63 | | | Calc | [47] |
| $Ba_8Ga_{16}Ge_{30}$ | | | | 50 | | | RA | [16] |
| $Ba_8Ga_{16}Ge_{30}$ | 288 | $C_p$ | | 42 | 74 | | $C_p$ | [122] |

(continued)

**Table 10.3** (continued)

| Compound | $\theta_D$ | Meth. | $\theta_E^a$ | | | | Meth. | Ref. |
|---|---|---|---|---|---|---|---|---|
| | | | $\theta_{E,1}$ | $\theta_{E,2}$ | $\theta_{E,3}$ | $\theta_{E,4}$ | | |
| $Ba_8Ga_{16}Ge_{30}$ | 401 | EXAFS | | | | | | [159] |
| $Ba_8Ga_{16}Ge_{30}$ | 288 | $C_p$ | | 42.8 | 76 | | $C_p$ | [125] |
| $Ba_8Rh_{1.2}Ge_{42.7}\square_{2.1}$ | 245 | XSC | | 88 | 70 | | XSC | [99] |
| $Ba_8Pd_{3.3}Ge_{42.5}\square_{0.2}$ | 260 | $C_p$ | | | | | | [68] |
| $Ba_8Pd_{3.8}Ge_{42.2}$ | 268 | $C_p$ | | 95 | | | XSC | [68] |
| | 226 | $C_p$ | 53.8 | 86.6 | | | $C_p$ | |
| | | | 60 | 90 | | | INS | |
| $Ba_8Pd_{3.8}Ge_{42.2}$ | 224 | FTE | | 100 | | | FTE | [4] |
| $Ba_8Pd_{3.8}Ge_{42.2}$ | 265 | XSC | | 98 | 95 | | XSC | [99] |
| $Ba_8Ag_{2.1}Ge_{41.9}\square_{2.0}$ | 210 | Elres | | | | | | [53] |
| $Ba_8Ag_{2.3}Ge_{41.9}\square_{1.8}$ | 269 | XSC | | 88 | 67 | | XSC | [53] |
| $Ba_8Ag_{3.4}Ge_{41.5}\square_{1.1}$ | 214 | $C_p$ | 49.6 | 81.2 | | | $C_p$ | [53] |
| | 260 | $C_p$ | | | | | | |
| $Ba_8Ag_{4.1}Ge_{41.4}\square_{0.5}$ | 202 | TC | | | | | | [53] |
| $Ba_8Ag_{4.4}Ge_{41.3}\square_{0.3}$ | 261 | XSC | | 88 | 68 | | XSC | [53] |
| $Ba_8Ag_5Ge_{41}$ | 263 | FTE | | 67 | | | FTE | [4] |
| $Ba_8Ag_6Ge_{40}$ | 240(1) | XSC | 107(2) | 78(1) | 60(1) | 67(1) | XSC | [69] |
| | 231(2) | XPD | 110(1) | 71(1) | 64(1) | 57(1) | XPD | |
| | 323(4) | $C_p$ | 56(1) | 82(1) | | | $C_p$ | |
| | | | | 68.8(2) | 53.6(1) | | INS | |
| $Ba_8Cd_{7.6}Ge_{38.4}$ | 230 | $C_p$ | | 78 | 58 | | XSC | [105] |
| | 192 | $C_p$ | 46 | 78.5 | | | $C_p$ | |
| $Ba_8Cd_{7.6}Ge_{38.4}$ | 200 | FTE | | 55 | | | FTE | [4] |
| $Ba_8Cd_{7.6}Ge_{38.4}$ | 251 | XSC | | 77 | 58 | | XSC | [99] |
| $Ba_8In_{15.5}Ge_{30.5}$ | 203 | $C_p$ | 38 | 99 | | | $C_p$ | [160] |

(continued)

**Table 10.3** (continued)

| Compound | $\theta_D$ | Meth. | $\theta_E^a$ | | | | Meth. | Ref. |
|---|---|---|---|---|---|---|---|---|
| | | | $\theta_{E,1}$ | $\theta_{E,2}$ | $\theta_{E,3}$ | $\theta_{E,4}$ | | |
| $Ba_8In_{16}Ge_{30}$ | 300 | ElRes | | | | | | [50] |
| $Ba_8In_{16}Ge_{30}$ | 203 | XPD | 87 | 65 | 65 | | XPD | [86] |
| $Ba_8Pt_{2.7}Ge_{41.7}\square_{1.6}$ | 230 | TC | | 96 | 82 | | XSC | [64] |
| $Ba_8Pt_{2.7}Ge_{41.8}\square_{1.5}$ | 231 | FTE | | 85 | | | FTE | [4] |
| $Ba_8Pt_{2.7}Ge_{41.7}\square_{1.6}$ | 253 | XSC | | 95 | 82 | | XSC | [99] |
| $Ba_8Au_6Ge_{40}$ | 279 | FTE | | 54 | | | FTE | [4] |
| $Ba_8Au_6Ge_{40}$ | 235(1) | XSC | 110(1) | 84(1) | 62(1) | 75(1) | XSC | [69] |
| | 251(2) | XPD | 125(1) | 79(1) | 67(1) | 66(1) | XPD | |
| | 249(2) | $C_p$ | 56(1) | 86(1) | | | $C_p$ | |
| | | | | 72.6(4) | 57.5(2) | | INS | |
| $Eu_8Ga_{16}Ge_{30}$ | 270 | TC | 82 | 53 | | | XSC | [72] |
| $Eu_8Ga_{16}Ge_{30}$ | 245 | XSC | 45 | 75 | | | XSC | [96] |
| | 302 | $C_p$ | | | | | | |
| $Eu_8Ga_{16}Ge_{30}$ | 300 | $C_p$ | | 30 | | | $C_p$ | [75] |
| $Eu_8Ga_{16}Ge_{30}$ | | | 78(10)[g] | 82(10)[g] | 93(10)[g] | 96(10)[g] | EXAFS | [162] |
| $Eu_8Ga_{16}Ge_{30}$ | 159 | $C_p$ | | | | | | [113] |
| $Eu_8Ga_{16}Ge_{30}$ | | | 77.8 | 34.8 | 56.9 | 87 | NIS | [66] |
| $Eu_8Ga_{16}Ge_{30}$ | | | | 31(4) | | | MOE, MA | [67] |
| $I_8Sb_8Ge_{38}$ | 185 | XPD | | 78 | | | XPD | [85] |
| $I_8Sb_8Ge_{38}$ | | | 76 | 83 | | | RA | [16] |
| *Type-II* | | | | | | | | |
| $Ge_{136}$ | 341.5 | Calc | | | | | | [37] |
| $Ge_{136}$ | 301 | Calc | | | | | | [38] |
| $Rb_8Na_{16}Ge_{136}$ | 220 | XSC | 61.7(Rb) | 182.7(Na) | | | XSC | [102] |
| $Cs_8Na_{16}Cu_5Ge_{131}$ | 427(32) | EXAFS | Cu-Ge | | | | | [163] |

(continued)

**Table 10.3** (continued)

| Compound | $\theta_D$ | Meth. | $\theta_E^a$ | | | | Meth. | Ref. |
|---|---|---|---|---|---|---|---|---|
| | | | $\theta_{E,1}$ | $\theta_{E,2}$ | $\theta_{E,3}$ | $\theta_{E,4}$ | | |
| | 428(40) | EXAFS | Ge-Ge | | | | | |
| $Cs_8Na_{16}Ge_{136}$ | | | 30-33(Cs) | 128-135(Na) | | | Calc | [44] |
| $Cs_8Na_{16}Ge_{136}$ | 230 | XSC | 60.1(Cs) | 168.3(Na) | | | XSC | [102] |
| $Cs_8Na_{16}Ge_{136}$ | | | 30(Cs) | | | | RA | [42] |
| | | | 26(Cs) | | | | Calc | |
| *Type-VIII* | | | | | | | | |
| $Eu_8Ga_{16}Ge_{30}$ | 214 | XSC | 45 | | | | XSC | [96] |
| | 364 | $C_p$ | 72 | | | | $C_p$ | |
| $Eu_8Ga_{16}Ge_{30}$ | 251 | $C_p$ | | | | | | [113] |
| *Type-IX* | | | | | | | | |
| $Ba_4Na_2Ge_{25}$ | 250 | $C_p$ | | 32 | | | $C_p$ | [118] |
| $Ba_6Ge_{25}$ | 250 | $C_p$ | | 37.2 | | | $C_p$ | [118, 164] |
| $Ba_6Ge_{25}$ | 240 | $C_p$, TC | 44 | 104 | | | $C_p$, TC | [65] |
| $Ba_{24}Ge_{100}$ | 310 | $C_p$ | | | | | | [112] |
| $Ba_{24}Al_{12}Ge_{88}$ | 339 | XPD | 89 | 117 | 48 | 91 | XPD | [89] |
| $Ba_{24}In_{12}Ge_{88}$ | 232 | XPD | 99 | 119 | 55 | 88 | XPD | [90] |
| *Clathrate-related* | | | | | | | | |
| $BaGe_5$ | 233 | $C_p$ | | | | | | [114] |
| ***Ge/Sn-based*** | | | | | | | | |
| *Type-I* | | | | | | | | |
| $Ba_8Zn_{7.66}Ge_{36.55}Sn_{1.79}$ | 350 | Elres | | | | | | [133] |
| $Ba_8Zn_{7.74}Ge_{36.74}Sn_{1.52}$ | 200 | Elres | | | | | | [133] |
| $Ba_8In_{15.6}Ge_{28.2}Sn_{2.2}$ | 203 | $C_p$ | 35 | 95 | | | $C_p$ | [160] |
| ***Sn-based*** | | | | | | | | |

(continued)

**Table 10.3** (continued)

| Compound | $\theta_D$ | Meth. | $\theta_E^a$ | | | | Meth. | Ref. |
|---|---|---|---|---|---|---|---|---|
| | | | $\theta_{E,1}$ | $\theta_{E,2}$ | $\theta_{E,3}$ | $\theta_{E,4}$ | | |
| *Type-I* | | | | | | | | |
| $K_8Ga_8Sn_{38}$ | 174-175 | XSC | 112 | 97 | 65 | | XSC | [106] |
| $Rb_8Sn_{44\ 2}$ | 152 | XPD | 92(1) | 81(1) | 54.2(3) | | XPD | [91] |
| $Cs_8Zn_4Sn_{42}$ | | | | 61 | | | NPD | [72] |
| $Cs_8Cd_4Sn_{42}$ | 179 | XPD | 91 | 51 | | | XPD | [92] |
| $Ba_8Ga_{14.5}Sn_{31.5}$ | | | | 47 | | | XSC | [107] |
| $Ba_8Ga_{16}Sn_{30}$ | 194-203 | XSC | | 78 | 60 | | XSC | [49] |
| | 218 | RUS | | | | | | |
| n-$Ba_8Ga_{16}Sn_{30}$ | 315(20) | EXAFS | | | | | | [147] |
| p-$Ba_8Ga_{16}Sn_{30}$ | 310(20) | EXAFS | | | | | | [147] |
| $Ba_8Ga_{16}Sn_{30}$ | 230 | Elres | 54 | 66 | | | Elres | [58] |
| | 230 | $C_p$ | 12 | 55 | 70 | | $C_p$ | |
| $Br_8Sn_{24}P_{19.3\ 2.7}$ | 220 | XSC | | | | | | [108] |
| $I_8Sn_{24}P_{19.3\ 2.7}$ | 225 | XPD | 76 | 79 | 63 | | XPD | [93] |
| | 265 | $C_p$ | 60 | 78 | | | $C_p$ | |
| $I_8Sn_{19.3}Cu_{1.7}Zn_3P_{19.9\ 2.1}$ | 280 | FTE | | 50 | | | FTE | [131] |
| $I_8Sb_8Sn_{38}$ | 159 | XPD | | 52 | | | XPD | [85] |
| $I_8Sb_8Sn_{38}$ | | | 42 | 50 | | | RA | [16] |
| *Type-II* | | | | | | | | |
| $K_{9.2}Ba_{14.8}Ga_{38.0}Sn_{95.2\ 2.8}$ | 188-199 | XSC | | 69 | | | XSC | [109] |
| $Cs_{24}Sn_{136}$ | | | 36-43(Cs1) | 7-10(Cs2) | | | Calc | [44] |
| *Type-VIII* | | | | | | | | |
| $Ba_8Ga_{16}Sn_{30}$ | 195 | XSC/ XPD | 64 | | | | XSC/XPD | [165] |
| | 200 | $C_p$ | 50 | | | | $C_p$ | |

(continued)

**Table 10.3** (continued)

| Compound | $\theta_D$ | Meth. | $\theta_E^a$ | | | | Meth. | Ref. |
|---|---|---|---|---|---|---|---|---|
| | | | $\theta_{E,1}$ | $\theta_{E,2}$ | $\theta_{E,3}$ | $\theta_{E,4}$ | | |
| $Ba_8Ga_{16}Sn_{30}$ | 195 | XPD | 64 | | | | XPD | [94] |
| | 200 | $C_p$ | 50 | | | | $C_p$ | |
| n-$Ba_8Ga_{16}Sn_{30}$ | 200 | TC, $C_p$ | 50 | | | | TC, $C_p$ | [63] |
| p-$Ba_8Ga_{16}Sn_{30}$ | 200 | TC, $C_p$ | 50 | | | | TC, $C_p$ | [63] |
| $Ba_8Ga_{16}Sn_{30}$ | 207 | RUS | | | | | | [49] |
| | 210 | $C_p$ | 50 | | | | $C_p$ | |

*XSC* X-ray single crystal diffraction (ADPs), *XPD* X-ray powder diffraction (ADPs), *NSC* neutron single crystal diffraction (ADPs), *NPD* neutron powder diffraction (ADPs), *Cp* heat capacity data, *TC* thermal conductivity data, *INS* inelastic neutron scattering, *NIS* nuclear inelastic scattering, *MOE* Mößbauer spectroscopy, *MA* microwave absorption spectroscopy, *EXAFS* extended X-ray absorption fine structure spectroscopy, *RA* Raman spectroscopy, *FTE* fits to thermal expansion data, *ElRes* fit to electrical resistivity data, *RUS* resonant ultrasound spectroscopy

[a] In order to list the Einstein temperature we decided to give a maximum of 4 different values. In case of diffraction techniques $\theta_{E,1}$ corresponds to the ADPs for the guest atoms in the 2a site of the clathrate-I structure (in the pentagonal dodecahedron), $\theta_{E,2}$ corresponds either to the isotropic ADPs for the guest atom in the 6c-site (in the tetrakaidecahedron) or, whenever provided, to the anisotropic $U_{11}$-values, $\theta_{E,3}$ to the respective $U_{22}$-values, and $\theta_{E,4}$ is calculated from isotropic ADPs in case the guest atom in the larger tetrakaidecahedral void is described by an off-center position. In type-II clathrates the $\theta_{E,1}$ corresponds to the rattling modes of the guest atoms inside the hexakaidecahedron and the $\theta_{E,2}$ to the pentagonal dodecahedron. In the type-VIII clathrate only one kind of cages is present [166]. For type-IX clathrates three different cages are occupied by guest atoms (4a-, 8c-, and 12d-site). $\theta_{E,1}$ corresponds to *the rattling of the guest in the 4a-site,* $\theta_{E,2}$ *in the 8c-site,* and $\theta_{E,3}$ and $\theta_{E,4}$ *to the anisotropic atomic displacement of the guest atoms in the 12-d site. For all* other measurement techniques the Einstein temperatures are listed starting from the highest value, or wherever mentioned $\theta_{E,1}$ gives the value for the atom in the pentagonal dodecahedral void and the following ones correspond to the rattling of the guest atom in the tetrakaidecahedral voids, dependent on the selected model with a varying number of modes

[b] From self-consistent calculations, from standard overlapped potential calculations: $\theta_D = 190$ (20 K), 260 (100 K), 400 (300 K)

[c] Based on theoretically calculated elastic constants {for $Si_{46}$ [167] }

[d] Range considering different displacements of the guest atoms towards centers of pentagons, hexagons or towards atoms of pentagons, hexagons [43]

[e] Calculated from given frequencies

[f] ADP-values taken from ref. [168]

[g] Different models for the displacement of the off-center position or on-center position, for details see [162]

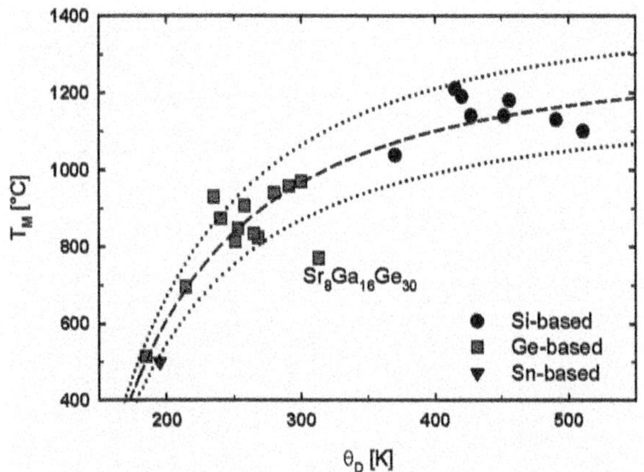


**Fig. 10.11** Melting point, $T_M$, versus Debye temperature, $\theta_D$, for type-I clathrates. The dashed line shows a least squares fit to the data and the dotted lines correspond to a deviation of ±10 %

$Ba_8Cu_5Si_{41}$ [98] are slightly higher than the values gained from the elastic properties of the same material.

The calculated Debye temperatures of $C_{136}$ ($\theta_D = 189$ K, [35] ) and $Si_{136}$ ($\theta_D = 21$ K, [34] ), extracted from calculated elastic properties (which are also out of the range of all other calculated and measured elastic properties, see Table 10.2), are much smaller compared to other calculated and measured Debye temperatures (see Table 10.3). In general the calculated values for guest-free clathrate frameworks are by trend slightly higher than the experimentally observed $\theta_D$-values, which are hitherto only measured for clathrates containing different species of guest atoms (see Table 10.3), but not for guest-free structures.

The theoretically calculated Einstein temperatures are in sound agreement with experimental data, especially with results extracted from temperature dependent ADPs gained from scattering experiments using X-ray or neutron sources (see Table 10.3). Both, $\theta_D$ and $\theta_E$, are independent of the type of electric charge carriers (n- or p-type) as demonstrated for the type-I $Ba_8Ga_{16}Ge_{30}$ compounds by several different methods [63, 79, 146] and the type-I [147] and type-VIII $Ba_8Ga_{16}Sn_{30}$ [63] (see Table 10.3). The melting points of several clathrate-I compounds, summarized recently by Zeiringer et al. [148], versus the corresponding Debye temperatures are displayed in Fig. 10.11. As expected, the melting point increases with $\theta_D$ as both temperatures at least to some extend reflect the strength of interatomic bonding of the compounds, which increases in general from Sn-based via Ge-based to Si-based frameworks. Only for $Sr_8Ga_{16}Ge_{30}$ for some reason a significant deviation from this correlation is observed, which most likely cannot be explained by a different bonding character of the guest atoms to the framework, as $Sr_8Ga_{16}Si_{30}$ and $Eu_8Ga_{16}Ge_{30}$ nicely follow the general trend.

Similar to the elastic properties also Debye and Einstein temperatures decrease in quaternary clathrate-I compounds with varying Si/Ge-ratio displayed in Fig. 10.12 for $BaCu_5Si_{41-x}Ge_x$ [78, 98], where $\theta_D$ and $\theta_E$ are extracted from single crystal X-ray diffraction at different temperatures, and for $Sr_8Ga_{16}Si_{30-x}Ge_x$ [119], where specific heat data, $C_p(T)$, were fitted to the Debye model using two

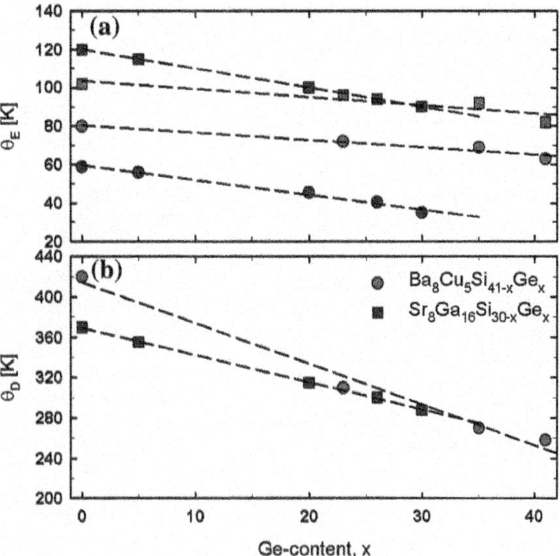


**Fig. 10.12** Einstein and Debye temperatures as a function of the Si/Ge-content for $BaCu_5Si_{41-x}Ge_x$ [78, 98] and $Sr_8Ga_{16}Si_{30-x}Ge_x$ [119]. In Fig. 10.11a the *squares* correspond to $\theta_{E,22}(U_{22})$ for $BaCu_5Si_{41-x}Ge_x$ or $\theta_{EH}$ for $Sr_8Ga_{16}Si_{30-x}Ge_x$, and the *circles* correspond to $\theta_{E,11}(U_{11})$ for $BaCu_5Si_{41-x}Ge_x$ or $\theta_{EL}$ for $Sr_8Ga_{16}Si_{30-x}Ge_x$

additional Einstein modes. Applying linear fits (plotted as dashed lines in Fig. 10.12) reveals a steeper decrease of $\theta_D$ for $BaCu_5Si_{41-x}Ge_x$ (Fig. 10.12b), whereas the Einstein temperatures of $Sr_8Ga_{16}Si_{30-x}Ge_x$ exhibit a more dramatic descent with increasing x (Fig. 10.12a).

In order to evaluate the dependence of the Einstein temperatures on the free space available in the tetrakaidecahedral void of type I clathrates, a simple model is employed: the anisotropy of the atomic displacement parameters of guest atoms in this cage extracted from diffraction experiments reflects the difference of the dimensions of the polyhedron (see inset of Fig. 10.13). Whereas $U_{11}$ is related to the oscillation of the guest atom perpendicular to the two hexagonal planes, $U_{22}$ and $U_{33}$, which by symmetry constraints are equal, result from the motion parallel to these planes. The distance between the two hexagons (d1 in Fig. 10.13) is simply a/2 (a is the lattice parameter). The distance d2 (see Fig. 10.13) is calculated using 2 opposing 24 k-positions of the tetrakaidecahedron. As this position is affected by splitting whenever at the neighboring 6d-site different species with a considerable difference in size are present, we calculated in several cases an average distance taking the different atomic coordinates (general position for 24 k in $Pm\bar{3}n$ is (0,y,z)) weighted by the corresponding occupation factor. In both cases in order to compare $\theta_E$-values of different compounds the ionic diameter of the guest atom, $d_G$, was subtracted and the resulting distance was called "free" distance. Einstein temperatures ($\theta_{E,11}$ from $U_{11}$, and $\theta_{E,22}$ from $U_{22}$) gained from either X-ray or neutron scattering experiments, wherefrom the anisotropic ADP-data were evaluated (see Table 10.3), are plotted as a function of the "free" distance inside the cage in Fig. 10.13. Although this model is a rough simplification neglecting completely the volume of the framework atoms as well as using

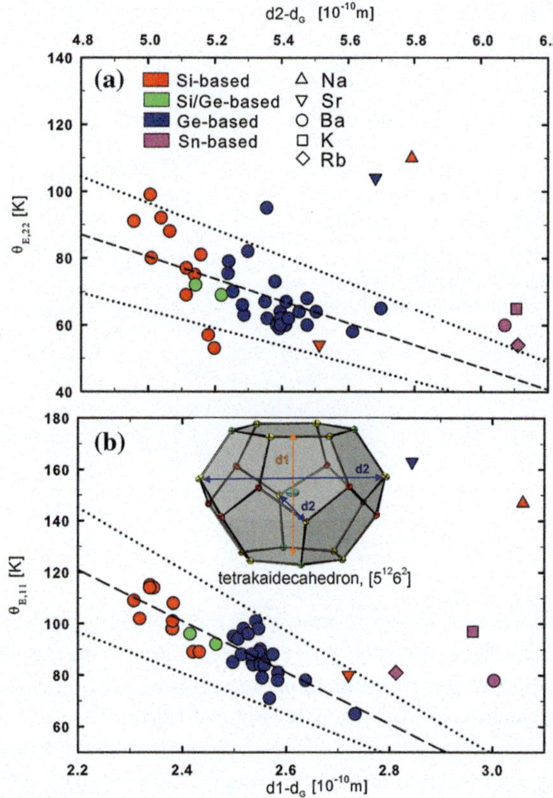


**Fig. 10.13** Einstein temperatures as a function of the "free" distance. Figure 10.12a displays $\theta_{E,22}$ extracted from the slope of $U_{22}(T)$ and Fig. 10.12b displays $\theta_{E,11}$ extracted from the slope of $U_{11}(T)$. d1 and d2 are illustrated in the *inset* of Fig. 10.12b and $d_G$ is the ionic diameter of the guest atoms. The atoms forming the tetrakaidecahedral cage are located in the 6d-(*green*), 16i-(*red*), and 24 k-site (*yellow*). The *dashed lines* correspond to a deviation of ±20 %

d2, which is not the distance in the center of the cage, a linear correlation was found for the Ba-containing type-I clathrates for both, $\theta_{E,11}$ and $\theta_{E,22}$ to the "free" distance. The largest deviations from the linear tendency plotted as a dashed line are found for compounds with other guest atoms than Ba, which might arise from a different bonding to the framework atoms. This difference was extensively studied for $A_8Ga_{16}Ge_{30}$ with A = Sr, Ba, Eu experimentally as well as theoretically showing an off-center position for Sr and Eu in the tetrakaidecahedral cages [75, 149–151]. The biggest shortcoming of this model, namely considering point-like framework atoms, should be more prominent for heavier framework atoms and indeed all data of Sn-based clathrates deviate positively from the general trend indicating a smaller real distance than the "free" distance calculated.

## *10.3.7 Thermal Expansion*

In a recent publication by the authors [4] the thermal expansion behavior of type-I clathrates was discussed including a comprehensive compilation of the literature

**Table 10.4** Thermal expansion coefficients (α) of C-, Si-, Ge- and Sn-based type-I-clathrate compounds and lattice parameters at room temperature [RT]

| Compound | $\alpha \times 10^{-6}$ [$K^{-1}$] | Temperature range [K] for the calculation | Method | Ref. | Latt. Param.[a] at RT (nm) | Ref. |
|---|---|---|---|---|---|---|
| *C-based clathrates* | | | | | | |
| $C_{46}$ | 5.1 | 1500 | CAL | [38] | 0.665 (0 K) | [38] |
| $C_{46}$ | 5.43 | 100-600 | CAL | [130] | 0.66108 (0 K) | [130] |
| *Si-based clathrates* | | | | | | |
| $Na_8Si_{46}$ | 20 | 300 | XLP | [110] | 1.0200 | [110] |
| | 11.96 | 140-300 | XLP | [a] | | |
| $Ba_{6.2}Si_{46}$ | 10 | 300-873 | XLP | [129] | 1.02623 | [129] |
| $Ba_8Si_{46}$ | 12 | 300 | DM | [169] | 1.03177 | [154] |
| $Ba_8Al_{16}Si_{30}$_A | 8 | 320 | TMA | [21] | 1.06400 | [21] |
| $Ba_8Al_{16}Si_{30}$_B | 9 | 320 | TMA | [21] | 1.06291 | [21] |
| $Ba_8Ni_3Si_{43}$ | 8.1 | 100-300 | XSCLP | [52] | 1.02947 | [52] |
| $Ba_8Cu_{4.5}Si_{41.5}$ | 7.07 | 150-300 | DM | [a] | 1.03249 | [a] |
| $Ba_8Cu_5Si_{41}$ | 8.87 | 100-300 | XSCLP | [a] | 1.03168 | [98] |
| $Ba_8Zn_7Si_{39}$ | 8.83 | 150-300 | DM | [a] | 1.04372 | [59] |
| $Ba_8Rh_{2.4}Si_{43.6}$ | 7.3 | 100-300 | XSCLP | [133] | 1.03564 | [133] |
| $Ba_8Pd_{3.9}Si_{42.1}$ | 6.96 | 150-300 | DM | [a] | 1.03629 | [134] |
| $Ba_8Ag_{4.9}Si_{41.1}$ | 9 | 100-300 | XSCLP | [104] | 1.0446 | [104] |
| $Ba_8Ag_5Si_{41}$ | 7.70 | 150-300 | DM | [a] | 1.04608 | [104] |
| $Ba_8Pt_{3.8}Si_{42.2}$ | 6.99 | 150-300 | DM | [a] | 1.03629 | [134] |
| $Ba_8Au_5Si_{41}$ | 7.30 | 150-300 | DM | [a] | 1.04148 | [60] |
| $Ba_8Ga_{16}Si_{30}$ | 16.2 | 300-900 | NSCLP | [a] | 1.04432 | [75/85] |
| | 8.3 | 20-300 | XLP | [86] | 1.05271 | [86] |
| | 10.67 | 300 | NLP | [74] | 1.05532 | [74] |
| *Si/Ge-based clathrates* | | | | | | |
| $Ba_8Cu_5Si_{18}Ge_{23}$ | 9.47 | 100-300 | XSCLP | [a] | 1.05300 | [98] |
| $Ba_8Cu_5Si_6Ge_{35}$ | 11.40 | 100-300 | XSCLP | [a] | 1.06096 | [98] |
| *Ge-based clathrates* | | | | | | |
| $Sr_8Zn_8Ge_{38}$ | 12 | 300 | NLP | 28 | 1.07044 | [74] |
| $Sr_8Ga_{16}Ge_{30}$ | 10.60 | 180-295 | NLP | [a] | 1.07237 | [73] |
| | 13.35 | 155-295 | NSCLP | [a] | 1.07430 | [73] |
| | 9 | 300 | NLP | [74] | 1.07380 | [74] |
| | 11.6 | 20-300 | XLP | [86] | 1.07274 | [86] |
| | 14.1 | 300-1000 | DM | [18] | | |
| $Ba_8Ge_{43}\square_3$ | 12.65 | 150-300 | DM | [a] | 1.06565[c] | [55] |
| $Ba_8Ni_{5.8}Ge_{40.2}$ | 12.25 | 150-300 | XLP | [a] | 1.06776 | [87] |
| $Ba_8NiZn_{6.6}Ge_{38.4}$ | 11.72 | 150-300 | DM | [a] | 1.07502 | [54] |
| $Ba_8NiZn_7Ge_{38}$ | 11.84 | 150-300 | DM | [a] | 1.07540 | [54] |
| $Ba_8Cu_5Ge_{41}$ | 12.61 | 150-300 | DM | [a] | 1.06878 | [170] |
| $Ba_8Cu_{5.9}Ge_{40.1}$ | 12.92 | 150-300 | XLP | [a] | 1.06944 | [88] |
| $Ba_8Cu_6Ge_{40}$ | 12.49 | 100-400 | XLP | [b] | 1.07492 | [69] |
| $Ba_8Zn_{2.1}Ge_{41.5}\square_{2.4}$ | 11.28 | 150-300 | DM | [a] | 1.06835 | [55] |
| $Ba_8Zn_{5.7}Ge_{40}\square_{0.3}$ | 13.01 | 150-300 | DM | [a] | 1.07512 | [55] |
| $Ba_8Zn_{7.7}Ge_{38.3}$ | 12.60 | 150-300 | DM | [a] | 1.07647 | [55] |

(continued)

**Table 10.4** (continued)

| Compound | $\alpha \times 10^{-6}$ [$K^{-1}$] | Temperature range [K] for the calculation | Method | Ref. | Latt. Param.[a] at RT (nm) | Ref. |
|---|---|---|---|---|---|---|
| $Ba_8Zn_8Ge_{38}$ | 10.33[d] | 20-300[d] | NLP | [a] | 1.07500 (FC) | [77] |
| | 8.33[d] | 20-300[d] | NLP | [a] | 1.07410 (SC) | [77] |
| $Ba_8Rh_{1.2}Ge_{42.8}\square_{2.0}$ | 14.0 | 100-300 | XSCLP | [132] | 1.07081 | [132] |
| $Ba_8Pd_{3.8}Ge_{42.2}$ | 10.89 | 150-300 | DM | [a] | 1.07747 | [68] |
| $Ba_8Ag_{2.3}Ge_{41.9}\square_{1.8}$ | 12.65 | 100-300 | XSCLP | [53] | 1.0748 | [53] |
| $Ba_8Ag_{4.4}Ge_{41.3}\square_{0.3}$ | 12.56 | 100-300 | XSCLP | [53] | 1.0829 | [53] |
| $Ba_8Ag_5Ge_{41}$ | 11.56 | 150-300 | DM | [a] | 1.08390 | [53] |
| $Ba_8Ag_6Ge_{40}$ | 13.86 | 100-400 | XLP | b | 1.08472 | [69] |
| $Ba_8Cd_{7.6}Ge_{38.4}$ | 12.02 | 150-300 | DM | [a] | 1.09499 | [105] |
| $Ba_8Ir_{0.2}Ge_{43.2}\square_{2.6}$ | 13.7 | 100-300 | XSCLP | [133] | 1.0739[c] | [133] |
| $Ba_8Pt_{2.7}Ge_{41.8}\square_{1.5}$ | 11.49 | 150-300 | DM | [a] | 1.07259 | [64] |
| $Ba_8Au_6Ge_{40}$ | 11.95 | 150-300 | DM | [a] | 1.07968 | [56] |
| | 9.71 | 100-400 | XLP | b | 1.07969 | [69] |
| $Ba_8Al_{16}Ge_{30}$ | 13.64[d] | 20-300[d] | NXLP | [a] | 1.08513(norm) | [103] |
| | 12.29[d] | 20-300[d] | NXLP | [a] | 1.08402(czo) | [103] |
| $Ba_8In_{16}Ge_{30}$ | 13.2 | 105-300 | XLP | [86] | 1.11455 | [86] |
| $Eu_8Ga_{16}Ge_{30}$ | 8.25 | 160-295 | NSCLP | [a] | 1.06886 | [171] |
| $Ba_8Ga_{16}Ge_{30}$ | 12.61 | 155-295 | NSCLP | [a] | 1.07850 | [171] |
| | 6.4 | 15-300 | XLP | [86] | 1.076388 | [86] |
| | 13.15 | 100-400 | XSCLP | [a] | 1.07825(n-type) | [79] |
| | 9.74 | 100-400 | XSCLP | [a] | 1.08025(p-type) | [79] |
| | 14.2 | 300-1000 | DM | [18] | 1.07661 | [18] |
| *Sn-based clathrates* | | | | | | |
| $Rb_8Sn_{44}\square_2$ | 16.66 | 300-475 | XLP | [a] | 1.20510 | [91] |
| $Cs_8Sn_{44}\square_2$ | 15.59 | 300-510 | XLP | [a] | 1.21060 | [91] |
| | 11.33 | 120-300 | NLP | [a] | 1.21054 | [72] |
| $Cs_8Zn_4Sn_{42}$ | 10.96 | 120-300 | NLP | [a] | 1.21252 | [72] |
| $Cs_8Cd_4Sn_{42}$ | 9.7 | 20-298 | XLP | [92] | 1.22357 | [92] |
| $Ba_8Ga_{14.5}Sn_{31.5}$ | 9.6 | 100-200 | XSCLP | [107] | 1.16800 | [107] |
| *Inverse Sn-based clathrates* | | | | | | |
| $Br_8Sn_{24}P_{19.3}\square_{2.7}$ | 11.0 | 90-293 | XSCLP | [131] | 1.08142 | [108] |
| $I_8Sn_{24}P_{19.3}\square_{2.7}$ | 11.1 | 200-300 | XLP | [93] | 1.09554 | [93] |
| $I_8Sn_{19.3}Cu_{1.7}Zn_3P_{19.9}\square_{2.1}$ | 10.0 | 150-300 | DM | [131] | 1.08915 | [131] |

For $Ba_8Zn_8Ge_{38}$ "FC" and "SC" means fast cooled and slow cooled, for details see [77], for $Ba_8Al_{16}Ge_{30}$ "norm" is attributed to the sample prepared by conventional stoichiometric mixing and "czo" to the sample prepared by Czochralski pulling (see also [103] )

*XLP* calculated from X-ray powder diffraction data, *NLP* calculated from neutron powder diffraction data, *DM* calculated from dilatometer measurement, *XSCLP* calculated from single crystal X-ray diffraction data, *NSCLP* calculated from single crystal neutron diffraction data, *CAL* thermal expansion coefficient calculated by DFT [38] or Monte Carlo Simulations [130], *TMA* Thermomechanical analyser

[a] [4]

[b] This work, calculated from temperature dependent lattice parameters

[c] Corresponds to a/2 of the supercell, for details see [172]

[d] Lattice parameters only available for 2 different temperatures and thus low reliability of α

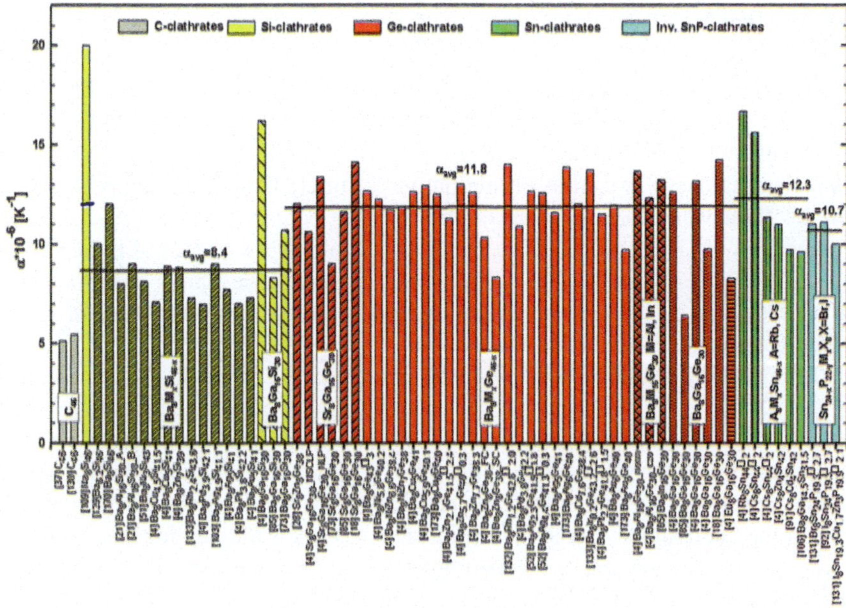


**Fig. 10.14** Thermal expansion coefficients, α, classified with respect to the chemical composition of the compounds. The *blue horizontal bar* indicates the value calculated from the lattice parameters of $Na_8Si_{46}$ (see also Table 10.4). $\alpha_{avg}$ is the average value of the thermal expansion coefficients within one group

data. Recently thermal expansion coefficients have been reported for several Si-based [52, 104, 129] or Ge-based [53, 69] as well as one Sn-based [107] and one inverse clathrate-I compound $I_8Sn_{24}P_{19.3}\square_{2.7}$ [93]. All data are included in Table 10.4. It should be mentioned that LDA calculations predicted for the type II clathrate $Si_{136}$ a region of negative thermal expansion between $T = 10$–$240$ K confirmed by X-ray diffraction [142].

The thermal expansion coefficient of intermetallic clathrates mainly depends on the bonding of the framework, and as already mentioned by Okamoto et al. [18] is rather insensitive to different guest atoms, as the basically ionic bonding to the host structure is weaker than the covalent bonding between the framework atoms. Thus it is not surprising that the average thermal expansion coefficient, $\alpha_{avg}$, of Si-based clathrates ($\alpha_{avg} = 8.4*10^{-6}\ K^{-1}$) is smaller than that of Ge-based clathrates ($\alpha_{avg} = 11.8*10^{-6}\ K^{-1}$). However, although $\alpha_{avg} = 12.3*10^{-6}\ K^{-1}$ for Sn-based clathrates is even slightly higher and nicely follows the expected trend, the scatter between the results from different measurement techniques is much larger (see Fig. 10.14). Thus no conclusion can be gained for Sn-based clathrates from the few thermal expansion coefficients hitherto reported. The same argument holds for the entire class of inverse clathrates.

## 10.4 Conclusions

This work provides a comprehensive compilation of several quantities like hardness and elastic properties for intermetallic clathrate compounds, which are key parameters for practical applications in thermoelectric generators. The results show that hardness as well as elastic properties of clathrates are mainly governed by the framework forming atoms. The stronger bonding between Si-atoms (in comparison to Ge-atoms) leads to larger values of HV as well as of the isotropic elastic moduli E, B, and G. For samples with varying Si/Ge-ratio a linear decrease of hardness and elastic properties with increasing Ge-content was observed. In Ge-based clathrates also reducing the amount of vacancies increases HV.

Temperature dependent RUS and RM measurements revealed a linear decrease of the Young's, the bulk and the shear modulus above room temperature. Interestingly the decrease of E as a function of Ge-content and as a function of temperature is much more pronounced than the decrease of B and G, which exhibit in both cases almost similar slopes. The elastic properties also slightly depend on the size of the guest atoms. Larger atoms inside the cages lead to an increase of E-, B- and G-values. In general the experimentally derived bulk moduli are in sound agreement with the results from DFT-calculations.

Finally the internal stress for intermetallic clathrates arising during the operation of a thermoelectric generator was calculated. Assuming realistic dimensions of the legs of a thermocouple ($w = 2$ mm, $\ell = 5$ mm) and a maximum temperature difference $\Delta T = 700$ K between the cold and the hot side of the leg, we conclude that the internal stress remains below the critical limit, at which fracture will occur.

The bonding strength of the framework is also reflected in the Debye temperatures, which in general decrease from Si-based via Ge-based to Sn-based compounds. Furthermore $\theta_D$ also correlates with the melting temperature, but significantly deviates from a linear dependency. Einstein temperatures decrease with increasing space for the guest atoms inside the larger tetrakaidecahedral cage of type-I clathrates.

Thermal expansion is mainly influenced by the bonding between the framework atoms, whereas the nature of the guest atoms and type of charge carriers play an underpart for the value of the thermal expansion coefficient.

**Acknowledgments** The authors are thankful to N. Nasir, I. Zeiringer, and X. Yan for the generous supply of samples for hardness and RUS measurements.

## References

1. G.A. Slack, in *CRC Handbook of Thermoelectrics*, ed. by D.M. Rowe (CRC Press, Boca Raton, 1995)
2. H. Kleinke, Chem. Mater. **22**, 604–611 (2010)

3. J.F. Meng, N.V. Charda Shekar, J.V. Badding, G.S. Nolas, J. Appl. Phys. **89**(3), 1730–1733 (2001)
4. M. Falmbigl, G. Rogl, P. Rogl, M. Kriegisch, H. Mueller, E. Bauer, M. Reinecker, W. Schranz, J. Appl. Phys. **108**, 043529 (2010)
5. A. San Miguel, P. Keghelian, X. Blase, P. Melinon, A. Perez, J.P. Itie, A. Polian, E. Reny, C. Cros, M. Pouchard, Phys. Rev. Lett. **83**, 5290 (1999)
6. G.K. Ramachandran, P.F. McMillan, S.K. Deb, M. Somayazulu, J. Gryko, J. Dong, O.F. Sankey, J. Phys.: Condens. Matter **12**, 4013–4020 (2000)
7. A. San Miguel, P. Melinon, D. Connetable, X. Blase, F. Tournus, E. Reny, S. Yamanaka, J.P. Itie, Phys. Rev. B: Condens. Matter Mater. Phys. **65**(5), 054109/1–4 (2002)
8. A. San Miguel, P. Toulemonde, High Press. Res. **25**(3), 159–185 (2005)
9. J.S. Tse, Z. Li, K. Uehara, Europhys. Lett. **56**, 261–267 (2001)
10. D. Machon, P. Toulemonde, P.F. McMillan, M. Amboage, A. Munoz, P. Rodriguez-Hernandez, A. San Miguel, Phys. Rev B **79**, 184101 (2009)
11. A. San Miguel, A. Merlen, P. Toulemonde, T. Kume, S. Le Floch, A. Aouizerat, S. Pascarelli, G. Aquilanti, O. Mathon, T. Le Bihan, J.P. Itie, S. Yamanaka, Europhys. Lett. **69**, 556–562 (2005)
12. L. Yang, Y.M. Ma, T. Iitaka, J.S. Tse, K. Stahl, Y. Ohishi, Y. Wang, R.W. Zhang, J.F. Liu, H.-K. Mao, J.Z. Jiang, Phys. Rev. B **74**, 245209 (2006)
13. H. Shimizu, T. Iitaka, T. Fukushima, T. Kume, S. Sasaki, N. Sata, Y. Ohishi, H. Fukuoka, S. Yamanaka, J. Appl. Phys. **101**(6), 63549-1-7 (2007)
14. H.Q. Yuan, F.M. Grosche, W. Carrillo-Cabrera, V. Pacheco, G. Sparn, M. Baenitz, U. Schwarz, Yu. Grin, F. Steglich, Phys. Rev. B: Condens. Matter Mater. Phys. **70**(17), 174512-1-6 (2004)
15. P. Toulemonde, D. Machon, A. San Miguel, M. Amboage, Phys. Rev. B: Condens. Matter Mater. Phys. **83**(13), 134110/1-134110/8 (2011)
16. H. Shimizu, P. Oe, S. Ohno, T. Kume, S. Sasaki, K. Kishimoto, T. Koyanagi, Y. Ohishi, J. Appl. Phys. **105**, 043522 (2009)
17. T. Kume, S. Ohno, S. Sasaki, H. Shimizu, Y. Ohishi, N.L. Okamoto, K. Kishida, K. Tanaka, H. Inui, J. Appl. Phys. **107**, 013517 (2010)
18. N.L. Okamoto, T. Nakano, K. Tanaka, H. Inui, J. Appl. Phys. **104**, 013529 (2008)
19. N.L. Okamoto, T. Nakano, K. Kishida, K. Tanaka, H. Inui, Mater. Res. Soc. Symp. Proc. **1128**, U01–U11 (2009)
20. J. Lee, J. Park, S. You, J. Lee, Y. Kim, I. Kim, K. Jang, S. Ur, Mater. Sci. Forum **539–543**, 3309-1 (2007)
21. H. Anno, M. Hokazono, R. Shirataki, Y. Nagami, J. Mater. Sci. **48**, 2846–2854 (2013)
22. V. Keppens, M.A. McGuire, A. Teklu, C. Laermans, B.C. Sales, D. Mandrus, B.C. Chakoumakos, Phys. B **316–317**, 95–100 (2002)
23. I. Zerec, V. Keppens, M.A. McGuire, D. Mandrus, B.C. Sales, P. Thalmeier, Phys. Rev. Lett. **92**(18), 185502 (2004)
24. I. Zerec, P. Thalmeier, Phys. B **359–361**, 208–210 (2005)
25. I. Ishii, H. Higaki, S. Morita, M.A. Avila, T. Sakata, T. Takabatake, T. Suzuki, Phys. B **383**, 130–131 (2006)
26. I. Ishii, H. Higaki, S. Morita, M.A. Avila, T. Takabatake, T. Suzuki, J. Magn. Magn. Mater. **310**, 957–959 (2007)
27. I. Ishii, H. Higaki, T. Sakata, D. Huo, T. Takabatake, T. Suzuki, Phys. B **359–361**, 1210–1212 (2005)
28. I. Ishii, Y. Suetomi, T.K. Fujita, K. Suekuni, T. Tanaka, T. Takabatake, T. Suzuki, M.A. Avila, Phys. Rev. B: Condens. Matter Mater. Phys. **85**, 085101 (2012)
29. A.J. Karttunen, V.J. Härkönen, M. Linnolahti, T.A. Pakkanen, J. Phys. Chem. C **115**, 19925–19930 (2011)
30. A. Migliori, J.D. Maynard, Rev. Sci. Instrum. **76**, 121301 (2005)
31. K. Tanaka, M. Koiwa, High Temp. Mater. Proc. **18**, 323 (1999)

32. C. Candolfi, U. Aydemir, A. Ormeci, W. Carrillo-Cabrera, U. Burkhardt, M. Baitinger, N. Oeschler, F. Steglich, Yu. Grin, J. Appl. Phys. **110**, 043715 (2011)
33. W. Sekkal, S.A. Abderrahmane, R. Terki, M. Certier, H. Aourag, Mater. Sci. Eng., B **64**, 123–129 (1999)
34. W. Sekkal, A. Zaoui, S.A. Abderrahmane, Materials Sci. Eng B **97**, 176–181 (2003)
35. W. Sekkal, H. Aourag, M. Certier, Comp. Mater. Sci. **9**, 295 (1998)
36. K. Moriguchi, S. Munetoh, A. Shintani, T. Motooka, Phys. Rev. B: Condens. Matter Mater. Phys. **64**, 195409 (2001)
37. K. Biswas, C.W. Myles, M. Sanati, G.S. Nolas, J. Appl. Phys. **104**, 033535 (2008)
38. D. Connetable, Phys. Rev. B: Condens. Matter Mater. Phys. **82**, 075209 (2010)
39. J.C. Li, C.L. Wang, M.X. Wang, H. Peng, R.Z. Zhang, M.L. Zhao, J. Liu, J.L. Zhang, L.M. Mei, J. Appl. Phys. **105**, 043503 (2009)
40. H.-F. Wang, W.-G. Chu, Y.-J. Guo, H. Jin, Chin. Phys. B **19**(7), 076501 (2010)
41. N.P. Blake, D. Bryan, S. Latturner, L. Mollnitz, G.D. Stucky, H. Metiu, J. Chem. Phys. **114**, 10063 (2001)
42. G.S. Nolas, C.A. Kendziora, J. Gryko, J. Dong, C.W. Myles, A. Poddar, O.F. Sankey, J. Appl. Phys. **92**(12), 7225 (2002)
43. E. Reny, S. Yamanaka, C. Cros, M. Pouchard, J. Phys.: Condens. Matter **14**, 11233–11236 (2002)
44. C.W. Myles, J. Dong, O.F. Sankey, Phys. Status Solidi B **239**(1), 26–34 (2003)
45. G.K.H. Madsen, G. Santi, Phys. Rev. B: Condens. Matter Mater. Phys. **72**, 220301 (2005)
46. K. Biswas, C.W. Myles, J. Phys.: Condens. Matter **19**, 466206 (2007)
47. E.N. Nenghabi, C.W. Myles, Phys. Rev. B: Condens. Matter Mater. Phys. **78**, 195202 (2008)
48. G.D. Mukherjee, C. Bansal, A. Chatterjee, Phys. Rev. Lett. **76**, 1876 (1996)
49. K. Suekuni, M.A. Avila, K. Umeo, H. Fukuoka, S. Yamanaka, T. Nakagawa, T. Takabatake, Phys. Rev. B: Condens. Matter Mater. Phys. **77**, 235119 (2008)
50. Ya. Mudryk, P. Rogl, C. Paul, S. Berger, E. Bauer, G. Hilscher, C. Godart, H. Noël, J. Phys.: Condens. Matter **14**, 7991–8004 (2002)
51. Ya. Mudryk, P. Rogl, C. Paul, S. Berger, E. Bauer, G. Hilscher, C. Godart, H. Noël, A. Saccone, R. Ferro, Physica B **328**, 44–48 (2003)
52. M. Falmbigl, M. Chen, A. Grytsiv, P. Rogl, E. Royanian, H. Michor, E. Bauer, R. Podloucky, G. Giester, Dalton Trans. **41**, 8839–8849 (2012)
53. I. Zeiringer, M. Chen, I. Bednar, E. Royanian, E. Bauer, R. Podloucky, A. Grytsiv, P. Rogl, H. Effenberger, Acta Mater. **59**(6), 2368–2384 (2011)
54. M. Falmbigl, N. Nasir, A. Grytsiv, P. Rogl, S. Seichter, A. Zavarsky, E. Royanian, E. Bauer, J. Phys. D Appl. Phys. **45**, 215308 (2012)
55. N. Melnychenko-Koblyuk, A. Grytsiv, L. Fornasari, H. Kaldarar, H. Michor, F. Röhrbacher, M. Koza, E. Royanian, P. Rogl, M. Rotter, E. Bauer, H. Schmid, F. Marabelli, A. Devishvili, M. Doerr, G. Giester, J. Phys.: Condens. Matter **19**, 216223 (2007)
56. I. Zeiringer, N. Melnychenko-Koblyuk, A. Grytsiv, E. Bauer, G. Giester, P. Rogl, J. Phase Equilib. Diff. **32**(2), 115–127 (2011)
57. M. Falmbigl, A. Grytsiv, P. Rogl, X. Yan, E. Royanian, E. Bauer, Dalton Trans. **42**, 2913–2920 (2013)
58. X. Zheng, S.Y. Rodriguez, L. Saribaev, J.H. Ross Jr, Phys. Rev. B: Condens. Matter Mater. Phys. **85**, 214304 (2012)
59. N. Nasir, A. Grytsiv, N. Melnychenko-Koblyuk, P. Rogl, E. Bauer, R. Lackner, E. Royanian, G. Giester, A. Saccone, J. Phys.: Condens. Matter **21**, 385404 (2009)
60. I. Zeiringer, M. Chen, A. Grytsiv, E. Bauer, R. Podloucky, H. Effenberger, P. Rogl, Acta Mater. **60**, 2324–2336 (2012)
61. N. Nasir, A. Grytsiv, N. Melnychenko-Koblyuk, P. Rogl, I. Bednar, E. Bauer, J. Solid State Chem. **183**, 2329–2342 (2010)
62. G.S. Nolas, T.J.R. Weakley, J.L. Cohn, R. Sharma, Phys. Rev. B: Condens. Matter Mater. Phys. **61**, 3845 (2000)

63. M.A. Avila, K. Suekuni, K. Umeo, H. Fukuoka, S. Yamanaka, T. Takabatake, Phys. Rev. B: Condens. Matter Mater. Phys. **74**, 125109 (2006)
64. N. Melnychenko-Koblyuk, A. Grytsiv, P. Rogl, M. Rotter, E. Bauer, R. Lackner, L. Fornasari, F. Marabelli, G. Giester, Phys. Rev. B: Condens. Matter Mater. Phys. **76**, 144118 (2007)
65. S. Paschen, V.H. Tran, M. Baenitz, W. Carrillo-Cabrera, Yu. Grin, F. Steglich, Phys. Rev. B: Condens. Matter Mater. Phys. **65**, 134435 (2002)
66. R.P. Hermann, W. Schweika, O. Leupold, R. Rüffer, G.S. Nolas, F. Grandjean, G.J. Long, Phys. Rev. B: Condens. Matter Mater. Phys. **72**, 174301 (2005)
67. R.P. Hermann, V. Keppens, P. Bonville, G.S. Nolas, F. Grandjean, G.J. Long, H.M. Christen, B.C. Chakoumakos, B.C. Sales, D. Mandrus, Phys. Rev. Lett. **97**, 017401 (2006)
68. N. Melnychenko-Koblyuk, A. Grytsiv, P. Rogl, M. Rotter, E. Bauer, G. Durand, H. Kaldarar, R. Lackner, H. Michor, E. Royanian, M. Koza, G. Giester, Phys. Rev. B: Condens. Matter Mater. Phys. **76**, 195124 (2007)
69. S. Johnsen, M. Christensen, B. Thomsen, G.K.H. Madsen, B.B. Iversen, Phys. Rev. B: Condens. Matter Mater. Phys. **82**, 184303 (2010)
70. H. Shimizu, Y. Takeuchi, T. Kume, S. Sasaki, K. Kishimoto, N. Ikeda, T. Koyanagi, J. Alloys Compd. **487**, 47–51 (2009)
71. B.T.M. Willis, A.W. Pryor (eds.), *Thermal Vibrations in Crystallography* (Cambridge University Press, Great Britain, 1975)
72. G.S. Nolas, B.C. Chakoumakos, B. Mahieu, G.J. Long, T.J.R. Weakley, Chem. Mater. **12**, 1947 (2000)
73. B.C. Chakoumakos, B.C. Sales, D.G. Mandrus, G.S. Nolas, J. Alloys Compd. **296**, 80 (2000)
74. L. Qiu, I.P. Swainson, G.S. Nolas, M.A. White, Phys. Rev. B: Condens. Matter Mater. Phys. **70**, 053208 (2004)
75. B.C. Sales, B.C. Chakoumakos, R. Jin, J.R. Thompson, D. Mandrus, Phys. Rev. B: Condens. Matter Mater. Phys. **63**, 245113 (2001)
76. A. Bentien, B.B. Iversen, J.D. Bryan, G.D. Stucky, A.E.C. Palmqvist, A.J. Schultz, and R.W. Henning, J. Appl. Phys. **91/9**, 5694 (2002)
77. M. Christensen, B.B. Iversen, J. Phys.:Condens. Matter **20**, 104244 (2008)
78. M. Christensen, S. Johnsen, F. Juranyi, B.B. Iversen, J. Appl. Phys. **105**, 073508 (2009)
79. M. Christensen, N. Lock, J. Overgaard, B.B. Iversen, J. Am. Chem. Soc. **128**, 15657 (2006)
80. G.S. Nolas, J.M. Ward, J. Gryko, L. Qiu, M.A. White, Phys. Rev. B: Condens. Matter Mater. Phys. **64**, 153201 (2001)
81. H. Anno, T. Nakabayashi, M. Hokazano, Adv. Sci. Technol. **74**, 26–31 (2010)
82. D. Nataraj, J. Nagao, M. Ferhat, T. Ebinuma, J. Appl. Phys. **93**(5), 2424 (2003)
83. H. Anno, H. Yamada, T. Nakabayashi, M. Hokazono, R. Shirataki, J. Solid State Chemistry **193**, 94–104 (2012)
84. D. Huo, T. Sasakawa, Y. Muro, T. Takabatake, Appl. Phys. Lett. **82**(16), 2640 (2003)
85. K. Kishimoto, S. Arimura, T. Koyanagi, Appl. Phys. Lett. **88**, 222115 (2006)
86. A. Bentien, E. Nishibori, S. Paschen, B.B. Iversen, Phys. Rev. B: Condens. Matter Mater. Phys. **71**, 144107 (2005)
87. S. Johnsen, A. Bentien, G.K.H. Madsen, M. Nygren, B.B. Iversen, Phys. Rev. B: Condens. Matter Mater. Phys. **76**, 245126 (2007)
88. S. Johnsen, A. Bentien, G.K.H. Madsen, B.B. Iversen, Chem. Mater. **18**, 4633 (2006)
89. J.H. Kim, N.L. Okamoto, K. Kishida, K. Tanaka, H. Inui, J. Appl. Phys. **102**, 034510 (2007)
90. J.H. Kim, N.L. Okamoto, K. Kishida, K. Tanaka, H. Inui, J. Appl. Phys. **102**, 094506 (2007)
91. A. Kaltzoglou, T. Fässler, M. Christensen, S. Johnsen, B.B. Iversen, I. Presniakov, A. Sobolev, A. Shevelkov, J. Mater. Chem. **18**, 5630–5637 (2008)
92. A.P. Wilkinson, C. Lind, R.A. Young, S.D. Shastri, P.L. Lee, G.S. Nolas, Chem. Mater. **14**, 1300 (2002)
93. V.V. Novikov, A.V. Matovnikov, D.V. Avdashchenko, N.V. Mitroshenkov, E. Dikarev, S. Takamizawa, M.A. Kirsanova, A.V. Shevelkov, J. Alloys Compd. **520**, 174–179 (2012)

94. D. Huo, T. Sakata, T. Sasakawa, M.A. Avila, M. Tsubota, F. Iga, H. Fukuoka, S. Yamanaka, S. Aoyagi, T. Takabatake, Cond. Mat. 040953 (2004)
95. B.C. Sales, B.C. Chakoumakos, D. Mandrus, J. Solid State Chem. **146**, 528–532 (1999)
96. S. Paschen, W. Carrillo-Cabrera, A. Bentien, V.H. Tran, M. Baenitz, Yu. Grin, F. Steglich, Phys. Rev. B: Condens. Matter Mater. Phys. **64**, 214404 (2001)
97. S. Stefanoski, J. Martin, G.S. Nolas, J. Phys.:Condens. Matter **22**, 485404 (2010)
98. X. Yan, A. Grytsiv, G. Giester, E. Bauer, P. Rogl, S. Paschen, J. Electron. Mater. **40**(5), 589–596 (2011)
99. This work
100. N.S. Abramchuk, W. Carrillo-Cabrera, I. Veremchuk, N. Oeschler, A.V. Olenev, Y. Prots, U. Burkhardt, E.V. Dikarev, Yu. Grin, A.V. Shevelkov, Inorg. Chem. **51**, 11396–11405 (2012)
101. M. Beekman, R.P. Hermann, A. Möchel, F. Juranyi, G.S. Nolas, J Phys.:Condens. Matter **22**, 355401 (2010)
102. G.S. Nolas, D.G. Vanderveer, A.P. Wilkinson, J.L. Cohn, J. Appl. Physics **91**(11), 8970 (2002)
103. M. Christensen, B.B. Iversen, Chem. Mater. **19**, 4896 (2007)
104. I. Zeiringer, E. Bauer, A. Grytsiv, P. Rogl, H. Effenberger, Jap. J. Appl. Phys. **50**(5, Pt. 3), 05FA01/1-05FA01/4 (2011)
105. N. Melnychenko-Koblyuk, A. Grytsiv, St. Berger, H. Kaldarar, H. Michor, F. Röhrbacher, E. Royanian, P. Rogl, E. Bauer, H. Schmid, G. Giester, J. Phys.: Condens. Matter **19**, 046203 (2007)
106. T. Tanaka, T. Onimaru, K. Suekuni, S. Mano, H. Fukuoka, S. Yamanaka, T. Takabatake, Phys. Rev. B: Condens. Matter Mater. Phys. **81**, 165110 (2010)
107. M.C. Schäfer, Y. Yamasaki, V. Fritsch, S. Bobev, Crystals **1**, 145–162 (2011)
108. K.A. Kovnir, J.V. Zaikina, L.N. Reshetova, A.V. Olenev, E.V. Dikarev, A.V. Shevelkov, Inorg. Chem. **43**, 3230 (2004)
109. S. Mano, T. Onimaru, S. Yamanaka, T. Takabatake, Phys. Rev. B: Condens. Matter Mater. Phys. **84**, 214101 (2011)
110. L. Qiu, M.A. White, Z. Li, J.S. Tse, C.I. Ratcliffe, C.A. Tulk, Phys. Rev. B: Condens. Matter Mater. Phys. **64**, 024303 (2001)
111. G.S. Nolas, M. Beekman, J. Gryko, G.A. Lamberton Jr, T.M. Tritt, P.F. McMillan, Appl. Phys. Lett. **82**(6), 910 (2003)
112. J. Tang, J. Xu, S. Heguri, H. Fukuoka, S. Yamanaka, K. Akai, K. Tanigaki, Phys. Rev. Lett. **105**, 176402 (2010)
113. A. Bentien, V. Pacheco, S. Paschen, Yu. Grin, F. Steglich, Phys. Rev. B: Condens. Matter Mater. Phys. **71**, 165206 (2005)
114. C. Candolfi, U. Aydemir, A. Ormeci, M. Baitinger, N. Oeschler, F. Steglich, Yu. Grin, Physical Review B: Condens. Matter Mater. Phys. **83**(20), 205102/1-205102/14 (2011)
115. W. Jung, H. Kessens, A. Ormeci, W. Schnelle, U. Burkhardt, H. Borrmann, H.D. Nguyen, M. Baitinger, Yu. Grin, Dalton Trans. **41**, 13960 (2012)
116. H. Zhang, J.-T. Zhao, M.-B. Tang, Z.-Y. Man, H.-H. Chen, X.-X. Yang, J. Alloys Compd. **476**, 1–4 (2009)
117. H. Zhang, J.-T. Zhao, M.-B. Tang, Z.-Y. Man, H.-H. Chen, X.-X. Yang, J. Phys. Chem. Solids **70**, 312–315 (2009)
118. F.M. Grosche, H.Q. Yuan, W. Carrillo-Cabrera, S. Paschen, C. Langhammer, F. Kromer, G. Sparn, M. Baenitz, Yu. Grin, F. Steglich, Phys. Rev. Lett. **87**, 247003 (2001)
119. K. Suekuni, M.A. Avila, K. Umeo, T. Takabatake, Phys. Rev. B: Condens. Matter Mater. Phys. **75**, 195210 (2007)
120. U. Aydemir, C. Candolfi, A. Ormeci, Y. Oztan, M. Baitinger, N. Oeschler, F. Steglich, Yu. Grin, Phys. Rev. B: Condens. Matter Mater. Phys. **84**, 195137 (2011)
121. K. Kovnir, U. Stockert, S. Budnyk, Y. Prots, M. Baitinger, S. Paschen, A.V. Shevelkov, Yu. Grin, Inorg. Chem. **50**, 10387–10396 (2011)

122. J. Xu, J. Tang, K. Sato, Y. Tanabe, S. Heguri, H. Miyasaka, M. Yamashita, K. Tanigaki, J. Electron. Mater. **40**(5), 879–883 (2011)
123. U. Aydemir, C. Candolfi, H. Borrmann, M. Baitinger, A. Ormeci, W. Carrillo-Cabrera, C. Chubilleau, B. Lenoir, A. Dauscher, N. Oeschler, F. Steglich, Yu. Grin, Dalton Trans. **39**, 1078–1088 (2010)
124. J. Xu, J. Wu, S. Heguri, G. Mu, Y. Tanabe, K. Tanigaki, J. Electron. Mater. **41**(6), 177–1180 (2012)
125. J. Xu, S. Heguri, Y. Tanabe, G. Mu, J. Wu, K. Tanigaki, J. Phys. Chem. Solids **73**, 1521–1523 (2012)
126. A. Bentien, M. Christensen, J.D. Bryan, A. Sanchez, S. Paschen, F. Steglich, G.D. Stucky, B.B. Iversen, Phys. Rev. B: Condens. Matter Mater. Phys. **69**, 045107 (2004)
127. K. Umeo, M.A. Avila, T. Sakata, K. Suekuni, T. Takabatake, J. Phys. Soc. Jpn. **74**(8), 2145–2148 (2005)
128. M. Beekman, W. Schnelle, H. Borrmann, M. Baitinger, Yu. Grin, G.S. Nolas, Phys. Rev. Lett. **104**, 018301 (2010)
129. Y. Liang, B. Böhme, L. Vasylechko, M. Baitinger, Yu. Grin, J. Phys. Chem. Solids **74**, 225–228 (2013)
130. F. Colonna, A. Fasolino, E.J. Meijer, Solid State Commun. **152**, 180–184 (2012)
131. M. Falmbigl, P. Rogl, E. Bauer, M. Kriegisch, H. Müller, S. Paschen, in *Materials Research Society Symposium Proceedings*, vol. 1166-N06-03, pp. 159–170 (2009)
132. M. Falmbigl, F. Kneidinger, M. Chen, A. Grytsiv, H. Michor, E. Royanian, E. Bauer, H. Effenberger, R. Podloucky, P. Rogl, Inorg. Chem. **52**(2), 931–943 (2013)
133. M. Falmbigl, A. Grytsiv, P. Rogl, G. Giester, Intermetallics **36**, 61–72 (2013)
134. N. Melnychenko.Koblyuk, A. Grytsiv, P. Rogl, E. Bauer, R. Lackner, E. Royanian, M. Rotter, G. Giester, J. Phys. Soc. Jpn. **77**, 54–60 (2008)
135. G. Rogl, P. Rogl, Sci. Adv. Mater. **3**, 517–538 (2011)
136. P. Varshni, Phys. Rev. B: Condens. Matter Mater. Phys. **2**, 3952 (1973)
137. E. Parthé, L. Gelato, B. Chabot, M. Penzo, K. Cenzual, R. Gladyshevskii, *TYPIX Standardized Data and Crystal Chemical Characterization of Inorganic Structure Types* (Springer, Berlin-Heidelberg, 1994)
138. A. Kitano, K. Moriguchi, M. Yonemura, S. Munetoh, A. Shintani, H. Fukuoka, S. Yamanaka, E. Nishibori, M. Takata, M. Sakata, Phys. Rev. B: Condens. Matter Mater. Phys. **64**(4), 045206/1-9 (2001)
139. D. Connetable, V. Timoshevskii, E. Artacho, X. Blase, Phys. Rev. Lett. **87**, 206405 (2001)
140. J. Dong, O.F. Sankey, G. Kern, Phys. Rev. B: Condens. Matter Mater. Phys. **60**, 2 (1999)
141. K. Biswas, C.W. Myles, Phys. Rev. B: Condens. Matter Mater. Phys. 74, 115113-1-5 (2006)
142. X. Tang, J. Dong, P. Hutchins, O. Shebanova, J. Gryko, P. Barnes, J.K. Cockcroft, M. Vickers, P.F. McMillan, Phys. Rev. B: Condens. Matter Mater, Phys. **74**(1), 14109-1-10 (2006)
143. L.H. Li, L. Chen, J.Q. Li, L.M. Wu, Chem. Mater. **22**, 4007–4018 (2010)
144. D.C. Li, L. Fang, S.K. Deng, H.B. Ruan, M. Saleem, W.H. Wei, C.Y. Kong, J. Electron. Mater. **40**(5), 1298–1303 (2011)
145. O.L. Anderson, J. Phys. Chem. Solids **24**, 909 (1963)
146. Y. Jiang, F. Bridges, M.A. Avila, T. Takabatake, J. Guzman, G. Kurczveil, Phys. Rev. B: Condens. Matter Mater. Phys. **78**, 014111 (2008)
147. M. Kozina, F. Bridges, Y. Jiang, M.A. Avila, K. Suekuni, T. Takabatake, Phys. Rev. B: Condens. Matter Mater. Phys. **80**, 212101 (2009)
148. I. Zeiringer, A. Grytsiv, P. Brož, P. Rogl, J. Solid State Chem. **196**, 125–131 (2012)
149. C. Gatti, L. Bertini, N.P. Blake, B.B. Iversen, Chem. Eur. J. **9**, 455–4568 (2003)
150. F. Bridges, L. Downward, Phys. Rev. B: Condens. Matter Mater. Phys. **70**, 140201(R) (2004)
151. A. Fujiwara, K. Sugimoto, C.-H. Shih, H. Tanaka, J. Tang, Y. Tanabe, J. Xu, S. Heguri, K. Tanigaki, M. Takata, Phys Rev. B: Condens. Matter Mater. Phys. **85**, 144305 (2012)
152. C. Cros, M. Pouchard, P. Hagenmuller, J.S. Kasper, Bull. Soc. Chim. Fr. 2737–2742 (1968)

153. S. Yamanaka, E. Enishi, H. Fukuoka, M. Yasukawa, Inorg. Chem. **39**, 56–58 (2000)
154. P. Toulemonde, A. San Miguel, A. Merlen, R. Viennois, S. Le Floch, C. Adessi, X. Blasé, J.L. Tholence, J. Phys. Chem. Solids **67**, 117 (2006)
155. W. Carrillo-Cabrera, J. Curda, H.G. von Schnering, S. Paschen, Yu. Grin, Z. Kristallogr, New Cryst. Struct. **215**, 207 (2000)
156. E. Reny, S. Yamanaka, C. Cros, M. Pouchard, in *Nanonetwork Materials* ed. by S. Saito et al., American Institute of Physics Conference Series, vol. 590 (2001), pp. 499–502
157. F. Brunet, P. Melinon, A. San Miguel, P. Keghelian, A. Perez, A.M. Flank, E. Reny, C. Cros, M. Pouchard, Phys Rev. B: Condens. Matter Mater. Phys. **61**, 16550 (2000)
158. K. Tanigaki, T. Shimizu, K.M. Itoh, J. Teraoka, Y. Moritomo, S. Yamanaka, Nat. Mater. **2**, 653 (2003)
159. A.N. Mansour, J. Martin, W. Wong-Ng, G.S. Nolas, J. Phys.: Condens. Matter **24**, 485503 (2012)
160. K. Suekuni, S. Yamamoto, M.A. Avila, T. Takabatake, J. Phys. Soc. Jpn. **77**, 61–66 (2008)
161. S.Y. Rodriguez, L. Saribaev, J.H. Ross Jr, Phys. Rev. B: Condens. Matter Mater. Phys. **82**, 064111 (2010)
162. R. Baumbach, F. Bridges, L. Downward, D. Cao, P. Chesler, B. Sales, Phys. Rev. B: Condens. Matter Mater. Phys. **71**, 024202 (2005)
163. A.N. Mansour, M. Beekman, W. Wong-Ng, G.S. Nolas, J. Solid State Chem. **182**, 107–114 (2009)
164. F. Steglich, A. Bentien, M. Baenitz, W. Carrillo-Cabrera, F.M. Grosche, C. Langhammer, S. Paschen, G. Sparn, V.H. Tran, H.Q. Yuan, Yu. Grin, Phys. B **318**, 97–105 (2002)
165. D. Huo, T. Sakata, T. Sasakawa, M.A. Avila, M. Tsubota, F. Iga, H. Fukuoka, S. Yamanaka, S. Aoyagi, T. Takabatake, Phys. Rev. B: Condens. Matter Mater. Phys. **71**, 075113 (2005)
166. P. Rogl, in *Thermoelectrics Handbook: Macro to Nano*, ed. by D.M. Rowe (CRC Press, Boca Raton, 2006)
167. D. Kahn, J.P. Lu, Phys. Rev. B: Condens. Matter Mater. Phys. **56**, 13898–13901 (1997)
168. J. Duenner, A. Mewis, Z. Anorg, Allg. Chem. **621**, 191–196 (1995)
169. R. Lortz, R. Viennois, A. Petrovic, Y. Wang, P. Toulemonde, C. Meingast, M.M. Marek, H. Mutka, A. Bossak, A. San, Miguel. Phys. Rev. B: Condens. Matter Mater. Phys. **77**, 224507 (2008)
170. N. Melnychenko-Koblyuk, A. Grytsiv, P. Rogl, H. Schmid, G. Giester, J. Solid State Chem. **182**, 1754 (2009)
171. B.C. Chakoumakos, B.C. Sales, D.G. Mandrus, J. Alloys Compd. **322**, 127 (2001)
172. W. Carrillo-Cabrera, S. Budnyk, Y. Prots, Yu. Grin, Z. Anorg. Allg. Chem. **630**, 2267e76 (2004)

# Index




G. S. Nolas (ed.), *The Physics and Chemistry of Inorganic Clathrates*, Springer Series in Materials Science 199, DOI: 10.1007/978-94-017-9127-4,

The manufacturer's authorised representative in the EU is Springer Nature Customer Service Centre GmbH, Europaplatz 3, 69115 Heidelberg, Germany. If you have any concerns regarding our products, please contact ProductSafety@springernature.com

Printed and bound by CPI Group (UK) Ltd, Croydon, CR0 4YY
15/07/2026
02167640-0001